Y0-ALC-970

NATURAL GAS PRODUCTION ENGINEERING

NATURAL GAS PRODUCTION ENGINEERING

Chi U. Ikoku
The Pennsylvania State University

John Wiley & Sons
New York • Chichester • Brisbane • Toronto • Singapore

To Chinelo,
whose patience,
understanding, and encouragement were essential
to the completion of this book.

Library of Congress Cataloging Publication Data:

Ikoku, Chi U.

Natural gas production engineering.

Includes bibliographies and indexes.

1. Gas, Natural. I. Title.

TN880.I333 1984 622′.3385 83-21617

ISBN 0-471-89483-4

Printed in the United States of America

10 9 8 7 6 5 4 3 2

PREFACE

This book presents a comprehensive and rigorous treatment of the technology of producing and transporting natural gas. The design of a development plan for a natural gas field always depends on the reservoir and well characteristics, tubing and flowline performance, and compressor and processing equipment characteristics. This text emphasizes a systems approach to natural gas production, since change in each component will affect the performance of the other components.

Most of us who teach others have heard the adage that the best way to learn a subject in depth is to teach it. As a case in point, *Natural Gas Production Engineering* is a history of many years of successfully using the material for natural gas engineering courses at The University of Tulsa, The Pennsylvania State University, and adult education courses in the United States and overseas. This book is arranged so that it can be used as a text or reference work for students and practicing engineers, geologists, and managers in the crude oil and natural gas production industry.

Chapters 1 to 3 serve as an introduction to the subject. Chapter 1 traces the development of the natural gas industry and tries to instill in the reader an awareness of the increased role natural gas will play as a source of energy. Chapter 2 reviews the properties of natural gases and condensate systems that are of importance in solving gas well performance, gas production, and gas transmission problems. Chapter 3 discusses some concepts of thermodynamics that are used throughout the book.

Chapters 4 through 8 focus on separation and processing, compression, measurement, and flow of gas in pipelines, tubings and annuli. Multiphase flow in pipes is treated and procedures for the design and selection of surface equipment are clearly outlined. Methods for determining static and flowing bottom-hole pressures from wellhead data are presented for both shallow wells and very deep wells producing sour gas. The problem of liquid loading in gas wells is also addressed.

Much of the material on which this book is based was drawn from the publications of the Society of Petroleum Engineers of the American Institute of Mining, Metallurgical and Petroleum Engineers, the American Gas Association, the Division of Production of the American Petroleum Institute, the Gas Processors Suppliers Association, the Petroleum Extension Service of the University of Texas at Austin, and the Gas Conditioning Conference of the University of

Oklahoma. Tribute is due to these organizations and also to a host of schools and authors who sponsor programs and have contributed to petroleum literature in various other publications.

I am indebted to my students whose enthusiasm for the subject has made teaching a pleasure. To my colleagues who have adopted this material in various petroleum and natural gas engineering departments in the U.S. and overseas, I would like to express my gratitude for their constructive criticisms and comments that became textbook inputs. Special thanks is also owed Peggy Conrad for typing the manuscript.

I am grateful to the editorial staff of John Wiley,including Merrill Floyd and Deborah Herbert, for their patience and politeness. My thanks is also owed to Cindy Stein and the members of Wiley's production staff for a fine job.

Chi U. Ikoku
University Park, Pennsylvania

CONTENTS

NOMENCLATURE

QUANTITIES IN ALPHABETICAL ORDER

(*) Dimensions: L = length, m = mass, q = electrical charge, t = time, and T = temperature.
(**) To avoid conflicting designation in some cases, use of reserve symbols and reserve subscripts is permitted.

Quantity	SPE Standard	Reserve SPE Letter Symbols**	Dimensions*
air requirement	a	F_a	
angle	α alpha	β beta	
angle	θ theta	γ gamma	
angle, contact	θ_c theta	γ_c gamma	
angle of dip	α_d alpha	θ_d theta	
area	A	S	L^2
Arrhenius reaction rate velocity constant	w	z	L^3/m
breadth, width, or (primarily in fracturing) thickness	b	w	L
burning-zone advance rate	v_b	V_b, u_b	L/t
capillary pressure	P_c	P_c, p_c	m/Lt^2
charge	Q	q	q
coefficient, convective heat transfer	h	h_h, h_T	m/t^3T
coefficient, heat transfer, interphase convective (use h, or convective coefficient symbol, with pertinent phase subscripts added)			m/t^3T
coefficient, heat transfer, over-all	U	U_T, U_θ	m/t^3T
coefficient, heat transfer, radiation	I	I_T, I_θ	m/t^3T
components, number of	C	n_c	
compressibility	c	k, κ kappa	Lt^2/m
compressibility factor	z	Z	
concentration	C	c, n	various
condensate or natural gas liquids content	C_L	c_L, n_L	various

Courtesy of Society of Petroleum Engineers of
American Institute of Mining,Metallurgical, and Petroleum Engineers, Inc.

Quantity	SPE Standard	Reserve SPE Letter Symbols**	Dimensions*
conductivity	σ sigma	γ gamma	various
conductivity, thermal (always with additional phase or systemsubscripts)	K_h	λ lambda	mL/t^3T
contact angle	θ_c theta	γ_c gamma	
damage ratio ("skin" conditions relative to formation conditions unaffected by well operations)	F_s	F_d	
density	ρ rho	D	m/L^3
depth	D	y, H	L
diameter	d	D	L
diffusion coefficient	D	μ mu, δ delta	L^2/t
dimensionless fluid influx function, linear aquifer	Q_{LtD}	Q_{ltD}	
dispersion coefficient	K	d	L^2/t
displacement	s	L	L
displacement ratio	δ delta	F_d	
distance between adjacent rows of injection and production wells	d	L_d, L_2	L
distance between like wells (injection or production) in a row	a	L_aL_1	L
distance, length, or length of path	L	s, l script l	L
efficiency	E	η eta, e	
electrical resistivity	ρ rho	R	mL^3/tq^2
electromotive force (voltage)	E	V	mL^2/t^2q
elevation referred to datum	Z	D, h	L
encroachment or influx rate	e	i	L^3/t
energy	E	U	mL^2/t^2
enthalpy (always with phase or system subscripts)	H	I	mL^2/t^2
enthalpy (net) of steam or enthalpy above reservoir temperature	H_s	I_s	mL^2/t^2
enthalpy, specific	h	i	L^2/t^2
entropy, specific	s	σ sigma	L^2/t^2T
entropy, total	S	σ_t sigma	mL^2/t^2T
equilibrium ratio	K	k, F_{eq}	
fluid influx function, linear aquifer, dimensionless	Q_{LtD}	Q_{ltD}	
flow rate or flux, per unit area (volumetric velocity)	u	ψ spi	L/t
flow rate or production rate	q	Q	L^3t
fluid (generalized)	F	f	various
flux	u	ψ psi	various
force	F	Q	mL/t^2
formation volume factor	B	F	
fraction gas	f_g	F_g	
fraction liquid	f_L	F_L, f_l	
frequency	f	ν nu	l/t
fuel consumption	m	F_F	various

Quantity	SPE Standard	Reserve SPE Letter Symbols**	Dimensions*
fuel deposition rate	N_g	N_F	m/L^{3}t
gas (any gas, including air)—always with identifying subscripts	G	g	various
gas in place in reservoir, total initial	G	g	L^3
gas-oil ratio, producing (if needed, the reserve symbols could be applied to other gas-oil ratios)	R	F_g, F_{g0}	
general and individual bed thickness	h	d, e	L
gradient	g	γ gamma	various
heat flow rate	Q	q, Φ phi$_{cap}$	mL^2/t^3
heat of vaporization, latent	L_v	λ_v lambda	L^2/t^2
heat or thermal diffusivity	α alpha	α, η_h eta	L^2/t
heat transfer coefficient, convective	h	h_h, h_T	m/t^{3}T
heat transfer coefficient, interphase convection (use h, or convective coefficient symbol with pertinent subscripts added)			m/t^{3}T
heat transfer coefficient, over-all	U	U_T, U_θ	m/t^{3}T
heat transfer coefficient, radiation	I	I_T, I_θ	m/t^{3}T
height (elevation)	Z	D, h	L
height (other than elevation)	h	d, e	L
hydraulic radius	r_H	R_H	L
index of refraction	n	μ mu	
influx (encroachment) rate	e	i	L^{3}t
influx function, fluid, linear aquifer, dimensionless	Q_{LtD}	Q_{ltD}	
initial water saturation	S_{wi}	ρ_{wi} rho, S_{wi}	
injectivity index	I	i	L^{4}t/m
intercept	b	Y	various
interfacial or surface tension	σ sigma	y, γ gamma	m/t^2
interstitial-water saturation in oil band	S_{wo}	S_{wb}	
irreducible water saturation	S_{iw}	ρ_{iw} rho, S_{iw}	
kinematic viscosity	ν nu	N	L^2/t
length	L	s, I script I	L
length, path length, or distance	L	s, I script I	L
mass flow rate	w	m	m/t
mobility ratio	M	F_λ	
mobility ratio, diffuse-front approximation, $[(\lambda_D + \lambda_d)_{swept}/(\lambda_d)_{unswept}]$; D signifies displacing; d signifies displaced; mobilities are evaluated at average saturation conditions behind and ahead of front	$M_{\bar{S}}$	M_{Dd}, M_{su}	
mobility ratio, sharp-front approximation, (λ_D/λ_d)	M	F_λ	
mobility ratio, total, $[(\lambda_t)_{swept}/(\lambda_t)_{unswept}]$; "swept" and "unswept" refer to invaded and uninvaded regions behind and ahead of leading edge of a displacement front	M_t	$F_{\lambda t}$	

Quantity	SPE Standard	Reserve SPE Letter Symbols**	Dimensions*
mobility, total, of all fluids in a particular region of the reservoir; e.g., $(\lambda_o + \lambda_g + \lambda_w)$	λ_t lambda	Λ $\text{lambda}_{\text{cap}}$	L^3t/m
modulus, bulk	K	K_b	m/Lt^2
modulus of elasticity in shear	G	E_s	m/Lt^2
modulus of elasticity (Young's modulus)	E	Y	m/Lt^2
mole fraction gas	f_g	F_g	
mole fraction liquid	f_L	F_L, f_l	
molecular refraction	R	N	L^3
moles, number of	n	N	
moles of liquid phase	L	n_L	
moles of vapor phase	V	n_v	
moles, total	n	n_t, N_t	
number (of moles, or components, or wells, etc.)	n	N	
oil (always with identifying subscripts)	n	n	various
oil in place in reservoir, initial	N	n	L^3
oxygen utilization	e_{O_2}	E_{O_2}	
path length, length, or distance	L	s, I script I	L
permeability	k	K	L^2
Poisson's ratio	μ mu	ν nu, σ sigma	
porosity	ϕ phi	f, ε epsilon	
pressure	p	P	m/Lt^2
production rate or flow rate	q	Q	L^3/t
productivity index	J	j	L^4t/m
quality (usually of steam)	f_s	Q, x	
radial distance	Δr	ΔR	L
radius	r	R	L
radius, hydraulic	r_H	R_H	L
ratio, damage ("skin" conditions relative to formation conditions unaffected by well operations)	F_s	F_d	
ratio initial reservoir free gas volume to initial reservoir oil volume	m	F_{Fo}, F_{go}	
ratio, mobility	M	F_λ	
ratio, mobility, diffuse-front approximation, $[(\lambda_D + \lambda_d)_{\text{swept}}/(\lambda_d)_{\text{unswept}}]$; D signifies displacing; d signifies displaced; mobilities are evaluated at average saturation conditions behind and ahead of front	$M_{\overline{S}}$	M_{Dd}, M_{su}	
ratio, mobility, sharp-front approximation, (λ_D/λ_d)	M	$F\lambda$	
ratio, mobility, total, $[(\lambda_t)_{\text{swept}}/(\lambda_t)_{\text{unswept}}]$; "swept" and "unswept" refer to invaded and uninvaded regions behind and ahead of leading edge of a displacement front	M_t	$F_{\lambda t}$	
reaction rate constant	k	r, j	L/t

Quantity	SPE Standard	Reserve SPE Letter Symbols**	Dimensions*
reciprocal formation volume factor, volume at standard conditions divided by volume at reservoir conditions	b	f, F	
reciprocal permeability	j	ω omega	$1/L^2$
resistance	r	R	mL^2/tq^2
resistance	r	R	various
resistivity, electrical	ρ rho	R	mL^3tq^2
saturation	S	ρ rho, s	
saturation, water, initial	S_{wi}	ρ_{wi} rho, s_{wi}	
saturation, water, irreducible	S_{iw}	ρ_{iw} rho, s_{iw}	
skin effect	s	S, σ sigma	
skin (radius of well damage or stimulation)	r_s	R_s	L
slope	m	A	various
specific gravity	γ gamma	s, F_s	
specific heat (always with phase or system subscripts)	C	c	L^2/t^2T
specific heats ratio	γ gamma	k	
specific injectivity index	I_s	i_s	L^3t/m
specific productivity index	J_s	j_s	L^3t/m
specific volume	v	v_s	L^3/m
specific weight	F_{wv}	γ gamma	m/L^2T^2
stimulation radius of well (skin)	r_s	R_s	L
strain, normal and general	ε epsilon	e, ε_n epsilon	
strain, shear	γ gamma	ε_s epsilon	
strain, volume	θ theta	Θ_v theta	
stress, normal and general	σ sigma	s	m/Lt^2
stress, shear	τ tau	s_s	m/Lt^2
surface tension	σ sigma	y, γ gamma	m/t^2
temperature	T	θ theta	T
thermal coductivity (always with additional phase or system subscripts)	k_h	λ lambda	mL/t^3T
thermal cubic expansion coefficient	β beta	b	$1/T$
thermal or heat diffusivity	α alpha	a, η_b eta	L^2/t
thickness (general and individual bed)	h	d, e	L
time	t	τ tau	t
total mobility of all fluids in a particular region of the reservoir; e.g., $(\lambda_o + \lambda_g + \lambda_w)$	λ_t lambda	Λ lambda_{cap}	L^3t/m
total mobility ratio, $[(\lambda_t)_{swept}/(\lambda_t)_{unswept}]$; "swept" and "unswept" refer to invaded and uninvaded regions behind and ahead of leading edge of a displacement front	M_t	$F_{\lambda t}$	
transfer coefficient, convective heat	h	h_h, h_T	m/t^3T
transfer coefficient, heat, interphase convective (use h, or convective coefficient symbol with pertinent phase subscripts added)			m/t^3T

Quantity	SPE Standard	Reserve SPE Letter Symbols**	Dimensions*
transfer coefficient, heat, over-all	U	U_T, U_θ	m/t^3T
transfer coefficient, heat, radiation	I	I_T, I_θ	m/t^3T
utilization, oxygen	e_{O_2}	E_{O_2}	
velocity	v	V, u	L/t
viscosity	μ mu	η eta	m/Lt
volume	V	v	L^3
volumetric velocity (flow rate or flux, per unit area)	u	ψ psi	L/t
water (always with identifying subscripts)	W	w	various
water in place in reservoir, initial	W	w	L^3
water saturation, initial	S_{wi}	ρ_{wi} rho, s_{wi}	
water saturation, irreducible	S_{iw}	ρ_{iw} rho, s_{iw}	
wave number	σ sigma	$\tilde{v}$	1/L
weight	W	w, G	mL/t^2
wet-gas content	C_{wg}	c_{wg}, n_{wg}	various
width, breadth, or (primarily in fracturing) thickness	b	w	L
work	W	w	mL^2/t^2

Subscripts

Subscript	SPE Standard	Reserve SPE Letter Subscripts**
air	a	A
atmospheric	a	A
average of mean saturation	$\bar{S}$	ρ rho, $\bar{s}$
band or oil band	b	B
base	b	r, β beta
boundary conditions, external	e	o
breakthrough	BT	bt
bubble point or saturation	b	s
burned or burning	b	B
calculated	C	calc
capillary (usually with capillary pressure, P_c)	c	C
casing or casinghead	c	cg
contact (usually with contact angle, θ_c)	c	C
core	c	C
cumulative influx (encroachment)	e	i
damage or damaged (includes "skin" conditions)	s	d
depleted region, depletion	d	δ delta
dispersed	d	D
dispersion	K	d

Subscript	SPE Standard	Reserve SPE Letter Subscripts**
displaced	d	s, D
displacing or displacement	D	s, σ sigma
entry	e	E
equivalent	eq	EV
estimated	E	est
experimental	E	EX
fill-up	F	f
finger or fingering	f	F
flash separation	f	F
fraction or fractional	f	r
fracture, fractured, or fracturing	f	F
free (usually with gas or gas-oil ratio quantities)	F	f
front, front region, or interface	f	F
gas	g	G
gross	t	T
heat or thermal	h	T, θ theta
hole	h	H
horizontal	H	h
hydrocarbon	h	H
imbibition	I	i script i
influx (encroachment), cumulative	e	i
injected, cumulative	i	I
injection, injected, or injecting	i	inj
inner or interior	i	ι iota, i script i
interface, front region, or front	f	F
interference	I	i, i script i
invaded	i	I
invaded zone	i	I
invasion	I	i
irreducible	i	i script i, ι iota
linear, lineal	L	I script I
liquid or liquid phase	L	I script I
lower	I script I	L
mean or average saturation	$\bar{S}$	$\bar{\rho}$ rho, $\bar{s}$
mixture	M	m
mobility	λ lambda	M
nonwetting	nw	NW
normalized (fractional or relative)	n	r, R
oil	o	n
outer or exterior	e	o
permeability	k	K
pore (usually with volume, V_p)	p	P
production period (usually with time, t_p)	p	P
radius, radial, or radial distance	r	R
reference	r	b, ρ rho

Subscript	SPE Standard	Reserve SPE Letter Subscripts**
relative	r	R
reservoir	R	r
residual	r	R
saturation, mean or average	$\overline{S}$	$\bar{\rho}$ rho, $\bar{s}$
saturation or bubble point	b	s
segregation(usually with segregation rate, q_s)	s	S, σ sigma
shear	s	τ tau
skin (stimulation or damage)	s	S
slip or slippage	s	σ sigma
solid(s)	s	σ sigma
stabilization (usually with time)	s	S
steam or steam zone	s	S
stimulation (includes "skin" conditions)	s	S
storage or storage capacity	S	S, σ sigma
strain	ε epsilon	e
surface	s	σ sigma
swept or swept region	s	S, σ sigma
system	s	σ sigma
temperature	T	h, θ theta
thermal (heat)	h	T, θ theta
total, total system	t	T
transmissibility	T	t
treatment or treating	t	τ tau
tubing or tubin head	t	tg
unswept or unswept region	u	U
upper	u	U
vaporization, vapor, or vapor phase	v	V
velocity	v	V
vertical	V	v
volumetric or volume	V	v
water	w	W
weight	W	w
wellhead	wh	th
wetting	w	W

1

INTRODUCTION

1.1 DEVELOPMENT OF NATURAL GAS

Natural gas, which was once an almost embarrassing and unwanted by-product—or more correctly a coproduct—of crude oil production, now provides about one-fifth of all the world's primary energy requirements (Table 1.1). This remarkable development has taken place in only a few years with the increased availability of the gas resources of the countries shown in Table 1.2, and the construction of long-distance, large-diameter steel pipelines which, have brought these ample supplies of gaseous fuel to domestic, commercial, and industrial users many miles away from the fields themselves.

Since its discovery in the United States at Fredonia, New York, in 1821, natural gas has been used as fuel in areas immediately surrounding the gas fields. In the 1920s and 1930s, a few long-distance pipelines from 22 to 24 in. in diameter, operating at 400 to 600 psi, were installed to transport gas to industrial areas remote from the fields. In the early years of the natural gas industry, when gas accompanied crude oil, it had to find a market or be flared; in the absence of effective conservation practices, oilwell gas was often flared in huge quantities. Consequently, gas production at that time was often short-lived, and gas could be purchased for as little as 1 or 2 cents per 1000 cu ft in the field.

The natural gas industry of today did not emerge until after World War II. The consumption of natural gas in all end-use classifications (residential, commercial, industrial, and power generation) has increased rapidly since then. This growth has resulted from several factors, including development of new markets, replacement of coal as a fuel for providing space and industrial process heat, use of natural gas in making petrochemicals and fertilizers, and strong demand for low-sulfur fuels which emerged in the middle 1960s. The resultant expansion of natural gas service has been remarkable.

The rapidly growing energy demands of Western Europe, Japan, and the United States could not be satisfied without importing gas from far afield. Natural

TABLE 1.1
World Energy Consumption and Fuel Shares: History, 1960, 1973, and 1978
(quadrillion Btu)

	1960					1973					1978				
		Fuel Shares (Percent)					Fuel Shares (Percent)					Fuel Shares (Percent)			
Region or Country	Total Energy Consumed	Coal	Oil	Gas	Other	Total Energy Consumed	Coal	Oil	Gas	Other	Total Energy Consumed	Coal	Oil	Gas	Other
United States[a]	44.2	23	45	28	4	75.1	18	47	30	5	78.8	18	49	25	8
Canada	3.9	14	49	10	27	8.1	8	46	20	26	9.0	6	42	22	30
Japan	3.9	48	36	1	15	15.3	14	79	2	5	14.9	13	73	5	9
Western Europe[b]	26.9	55	35	2	8	54.0	19	63	10	8	54.7	19	56	14	11
Finland/Norway/Sweden	1.8	13	49	0	38	3.9	5	58	0	37	4.1	5	50	2	43
United Kingdom/Ireland	8.0	70	29	0	1	10.4	35	51	11	3	10.1	34	44	17	5
Benelux/Denmark[c]	2.7	53	47	0	0	7.3	11	67	22	0	6.4	10	57	30	3
West Germany	6.2	73	25	0	2	11.5	30	58	10	2	12.0	26	54	16	4
France	3.7	51	35	3	11	8.2	15	70	7	8	8.2	15	61	11	13
Austria/Switzerland	0.9	24	34	6	36	2.0	8	58	7	27	2.1	5	53	11	31
Spain/Portugal	1.0	41	41	0	18	3.0	14	70	2	14	3.6	15	67	2	16
Italy	2.2	14	54	11	21	6.2	5	78	10	7	6.2	6	69	17	8
Greece/Turkey	0.4	41	55	0	4	1.4	21	76	0	3	1.9	30	64	0	6
Austria/New Zealand	1.4	54	43	0	3	2.9	35	50	6	9	3.5	40	42	10	8
Total OECD[b]	80.4	35	42	16	7	155.4	18	56	19	7	160.9	18	53	19	10
Total Non-OECD[b]	10.5	31	57	6	6	25.0	21	60	11	8	30.7	20	66	10	4
OPEC	1.7	3	75	21	1	5.0	1	66	30	3	6.2	0	71	24	5
Other	8.8	37	53	4	6	20.0	25	59	7	9	24.5	24	65	7	4
Total free world[b]	90.9	35	43	15	7	180.3	18	56	18	8	191.6	18	55	18	9

Source: From D.O.E.'s 1980 Annual Report to Congress, Volume 3.
Courtesy API.

[a]Includes Puerto Rico, Virgin Islands, and purchases for the Strategic Petroleum Reserve.
[b]Numbers may not add to totals due to rounding.
[c]Benelux countries are Belgium, the Netherlands, and Luxembourg.

TABLE 1.1 (Continued)

World Energy Consumption and Fuel Shares: Base Scenario Midprice Projections* 1985, 1990 and 1995

(quadrillion Btu)

Region or Country	1985					1990					1995				
	Total Energy Consumed	Fuel Shares (Percent)				Total Energy Consumed	Fuel Shares (Percent)				Total Energy Consumed	Fuel Shares (Percent)			
		Coal	Oil	Gas	Other		Coal	Oil	Gas	Other		Coal	Oil	Gas	Other
United States[a]	79.6	26	40	23	11	88.0	32	36	19	13	94.7	37	32	17	14
Canada	10.0	7	36	20	37	10.9	4	34	21	41	11.8	3	33	21	43
Japan	19.3	13	62	14	11	24.1	14	54	20	12	28.2	16	51	19	14
Western Europe[b]	55.3	21	47	16	16	59.2	21	45	16	18	63.7	20	43	17	20
Finland/Norway/Sweden	4.9	5	45	1	49	5.2	5	41	2	52	5.6	4	40	2	54
United Kingdom/Ireland	9.1	35	40	19	6	9.3	33	39	19	9	9.4	33	38	19	10
Benelux/Denmark[c]	6.5	17	47	31	5	7.1	19	47	29	5	7.4	22	44	29	5
West Germany	12.4	31	42	19	8	13.3	33	38	19	10	14.2	34	36	19	11
France	8.8	15	51	11	23	9.5	14	49	11	26	10.3	11	46	13	30
Austria/Switzerland	2.1	7	42	15	36	2.3	7	40	17	36	2.4	7	37	19	37
Spain/Portugal	3.4	16	55	4	25	3.7	11	52	6	31	3.9	10	49	8	33
Italy	6.3	11	59	20	10	7.1	12	57	20	11	8.2	13	55	21	11
Greece/Turkey	1.8	27	58	1	14	2.0	19	56	1	24	2.2	18	54	0	28
Austria/New Zealand	3.7	35	40	16	9	4.1	33	38	19	10	4.6	32	38	20	10
Total OECD[b]	167.9	22	44	20	14	186.3	24	42	18	16	203.0	27	38	18	17
Total Non-OECD[b]	45.3	21	55	13	11	59.3	21	54	13	12	74.3	23	54	13	10
OPEC	10.1	1	67	31	1	13.4	1	66	32	1	16.8	1	68	30	1
Other	35.2	27	52	8	13	45.9	27	51	7	15	57.5	29	50	8	13
Total free world[b]	213.2	22	47	18	13	245.6	23	45	17	15	277.3	26	42	16	16

Source: From D.O.E.'s 1980 Annual Report to Congress, Volume 3.
Courtesy API.

*Projection ranges are based on assumed price paths for imported oil stated in 1979 constant dollars. The low price scenario assumes a delivered world oil price of $32 per barrel, the midrange (used above) $41 per barrel, and the high range $49 per barrel.

[a]Includes Puerto Rico, Virgin Islands, and purchases for the Strategic Petroleum Reserve.

[b]Numbers may not add to totals due to rounding.

[c]Benelux countries are Belgium, the Netherlands, and Luxembourg.

TABLE 1.2 World Natural Gas Production—Twenty Leading Nations
(billion cubic feet)

	1974[r]					1975[r]			
	Marketed Production[a]					Marketed Production[a]			
Nation	Total	Billion cu ft/day	Percent of Free World Production	Percent of Total World Production	Nation	Total	Billion cu ft/day	Percent of Free World Production	Percent of Total World Production
1. United States	21,600.5	59.18	61.6	45.8	1. United States	20,108.7	55.09	59.3	42.6
2. USSR	9,201.3	25.21	26.2	19.5	2. USSR	10,205.9	27.96	30.0	21.6
3. Canada	3,045.5	8.34	8.7	6.5	3. Netherlands	3,208.4	8.79	9.5	6.8
4. Netherlands	2,956.7	8.10	8.4	6.3	4. Canada	3,075.7	8.43	9.1	6.5
5. United Kingdom	1,230.0	3.37	3.5	2.6	5. People's Republic of China	1,400.0	3.84	4.1	3.0
6. People's Republic of China	1,200.0	3.29	3.4	2.5	6. United Kingdom	1,208.2	3.31	3.6	2.6
7. Romania	1,011.5	2.77	2.9	2.1	7. Romania	953.5	2.61	2.8	2.0
8. Iran	787.4	2.16	2.2	1.7	8. Iran	771.1	2.11	2.3	1.6
9. West Germany	713.2	2.00	2.0	1.5	9. West Germany	639.4	1.75	1.9	1.4
10. Mexico	560.9	1.54	1.6	1.2	10. Mexico	583.9	1.60	1.7	1.2
11. Italy	540.4	1.48	1.5	1.1	11. Italy	514.3	1.41	1.5	1.1
12. Venezuela	476.0	1.30	1.4	1.0	12. Venezuela	450.3	1.23	1.3	1.0
13. Libya	345.2	0.95	1.0	0.7	13. Libya	382.6	1.05	1.1	0.8
14. East Germany	273.1	0.75	0.8	0.6	14. East Germany	280.0	0.77	0.8	0.6
15. France	269.4	0.74	0.8	0.6	15. Argentina	271.6	0.74	0.8	0.6
16. Argentina	255.7	0.70	0.7	0.5	16. France	259.8	0.71	0.8	0.6
17. Saudi Arabia	219.0	0.60	0.6	0.5	17. Poland	210.6	0.58	0.6	0.4
18. Poland	202.7	0.56	0.6	0.4	18. Algeria	210.0	0.58	0.6	0.4
19. Algeria	198.5	0.54	0.6	0.4	19. Saudi Arabia	200.0	0.55	0.6	0.4
20. Kuwait	186.9	0.51	0.5	0.4	20. Brunei	186.5	0.51	0.6	0.4
Total free world	35,054.2	96.04	100.0	74.3	Total free world	33,931.1	93.00	100.0	71.9
Total world	47,171.5	129.24	—	100.0	Total world	47,207.3	129.34	—	100.0

Source: U.S. Bureau of Mines World Natural Gas, May 27, 1977.
Courtesy API.

[r] Revised.
[a]Comprises all gas collected and utilized as fuel or as a chemical industry raw material, including gas used in oil and/or gas fields as a fuel by producers even though it is not actually sold.

TABLE 1.2 (Continued)

	1976[r]					1977[p]			
	Marketed Production[a,b]					Marketed Production[a,b]			
Nation	Total	Billion cu ft/day	Percent of Free World Production	Percent of Total World Production	Nation	Total	Billion cu ft/day	Percent of Free World Production	Percent of Total World Production
1. United States	19,952.4	54.64	55.4	39.6	1. United States	20,337.5	55.72	57.5	37.7
2. USSR	11,331.3	30.96	—	22.6	2. USSR	12,395.2	33.96	—	23.0
3. Netherlands	3,542.7	9.68	9.9	7.1	3. Canada	3,301.4	9.04	9.3	6.1
4. Canada	3,157.2	8.63	8.8	6.3	4. Netherlands	2,932.5	8.03	8.3	5.4
5. Iran	1,699.6	4.64	4.7	3.4	5. People's Republic of China	1,787.0	4.90	—	3.3
6. People's Republic of China	1,433.7	3.92	—	2.9	6. Iran	1,756.1	4.81	5.0	3.3
7. United Kingdom	1,202.3	3.28	3.4	2.4	7. United Kingdom	1,499.9	4.11	4.2	2.8
8. Romania	854.6	2.33	—	1.7	8. Romania	1,005.5	2.75	—	1.9
9. Mexico	790.8	2.16	2.2	1.6	9. West Germany	907.7	2.49	2.6	1.7
10. Nigeria	632.6	1.73	1.8	1.3	10. Mexico	745.3	2.04	2.1	1.4
11. Italy	562.1	1.54	1.6	1.1	11. Nigeria	717.4	1.97	2.0	1.3
12. West Germany	544.0	1.49	1.5	1.1	12. Libya	488.6	1.34	1.4	0.9
13. Libya	439.2	1.20	1.2	0.9	13. Italy	448.6	1.23	1.3	0.8
14. Venezuela	396.0	1.08	1.1	0.8	14. Venezuela	392.8	1.08	1.1	0.7
15. Argentina	280.2	0.77	0.8	0.6	15. Indonesia	351.9	0.96	1.0	0.7
16. Algeria	256.9	0.70	0.7	0.5	16. Algeria	284.5	0.78	0.8	0.5
17. France	250.7	0.68	0.7	0.5	17. France	256.8	0.70	0.7	0.5
18. Indonesia	248.6	0.68	0.7	0.5	18. Argentina	250.0	0.68	0.7	0.5
19. Australia	209.3	0.57	0.6	0.4	19. Australia	236.7	0.65	0.7	0.4
20. Chile	175.8	0.48	0.5	0.4	20. Chile	208.9	0.57	0.6	0.4
Total free world	36,006.5	98.50	100.0	71.5	Total free world	35,393.2	96.97	100.0	65.7
Total world	50,367.1	137.74	—	100.0	Total world	53,883.7	147.63	—	100.0

Sources: U.S. Energy Information Administration, United States only; other nations, *Oil and Gas Journal*.
Courtesy API.

[r] Revised.
[p] Preliminary.
[a]Comprises all gas collected and utilized as a fuel or as a chemical industry raw material, including gas used in oil and/or gas fields as a fuel by producers, even though it is not actually sold.
[b]United States is reported marketed production; all others may include some gross production.

TABLE 1.2 (Continued)

Nation (1978)	1978 Marketed Production[a,b] Total	1978 Marketed Production[a,b] Billion cu ft/day	1978 Percent of Free World Production	1978 Percent of Total World Production	Nation (1979)	1979 Marketed Production[a,b] Total	1979 Marketed Production[a,b] Billion cu ft/day	1979 Percent of Free World Production	1979 Percent of Total World Production
1. United States	19,974.0	54.72	56.6	37.1	1. United States	20,471.3	56.09	53.8	35.5
2. USSR	13,131.6	35.98	—	24.4	2. USSR	14,367.1	39.36	—	24.9
3. Canada	3,133.1	8.58	8.9	5.8	3. Canada	3.646.5	9.99	9.6	6.3
4. People's Republic of China	2,325.2	6.37	—	4.3	4. People's Republic of China	2,832.7	7.76	—	4.9
5. Iran	1,746.9	4.79	5.0	3.2	5. Netherlands	2,717.8	7.45	7.1	4.7
6. Netherlands	1,624.5	4.45	4.6	3.0	6. United Kingdom	1,965.7	5.39	5.2	3.4
7. United Kingdom	1,262.4	3.46	3.6	2.3	7. Romania	1,210.5	3.32	—	2.1
8. Romania	1,211.6	3.32	—	2.3	8. Mexico	974.9	2.67	2.6	1.7
9. Mexico	887.5	2.43	2.5	1.7	9. Iran	913.3	2.50	2.4	1.6
10. West Germany	643.2	1.76	1.8	1.2	10. West Germany	826.3	2.26	2.2	1.4
11. Nigeria	586.6	1.61	1.7	1.1	11. Indonesia	816.3	2.24	2.1	1.4
12. Indonesia	565.8	1.55	1.6	1.1	12. Italy	533.6	1.46	1.4	0.9
13. Italy	540.8	1.48	1.5	1.0	13. Norway	414.9	1.14	1.1	0.7
14. Venezuela	401.4	1.10	1.1	0.7	14. Venezuela	406.0	1.11	1.1	0.7
15. Libya	364.0	1.00	1.0	0.7	15. Nigeria	365.6	1.00	1.0	0.6
16. Norway	351.8	0.96	1.0	0.7	16. Chile	306.4	.84	0.8	0.5
17. Algeria	328.7	0.90	0.9	0.6	17. Brunei-Malaysia	302.7	.83	0.8	0.5
18. Brunei-Malaysia	325.7	0.89	0.9	0.6	18. Algeria	301.5	.83	0.8	0.5
19. Chile	320.7	0.88	0.9	0.6	19. Australia	254.1	.70	0.7	0.4
20. Argentina	254.3	0.70	0.7	0.5	20. Argentina	241.7	.66	0.6	0.4
Total free world	35,272.8	96.64	100.0	65.4	Total free world	38,075.1	103.58	100.0	66.0
Total world	53,911.1	147.70	—	100.0	Total world	57,666.5	157.26	—	100.0

Sources: U.S. Energy Information Administration, United States only; other nations, *Oil and Gas Journal*. Courtesy API.

[a]Comprises all gas collected and utilized as a fuel or as a chemical industry raw material, including gas used in oil and/or gas fields as a fuel by producers, even though it is not actually sold.

[b]United States is reported marketed production; all others may include some gross production.

TABLE 1.2 (Continued)

	1980[p]			
	Marketed Production[a,b]			
Nation	Total	Billion cu ft/day	Percent of Free World Production	Percent of Total World Production
1. United States	20,090.0	55.04	53.7	34.4
2. USSR	15,355.5	42.07	—	26.3
3. People's Republic of China	3,469.0	9.50	—	5.9
4. Netherlands	2,799.6	7.67	7.5	4.8
5. Canada	2,668.3	7.31	7.1	4.6
6. United Kingdom	1,500.0	4.11	4.0	2.6
7. Mexico	1,190.5	3.26	3.2	2.0
8. Romania	1,176.4	3.22	—	2.0
9. Indonesia	1,028.4	2.82	2.8	1.8
10. West Germany	738.6	2.02	2.0	1.3
11. Norway	705.2	1.93	1.9	1.2
12. Pakistan	600.0	1.64	1.6	1.0
13. Italy	524.6	1.44	1.4	0.9
14. Venezuela	518.0	1.42	1.4	0.9
15. Algeria	517.0	1.42	1.4	0.9
16. Saudi Arabia	310.2	0.85	0.8	0.5
17. Argentina	296.9	0.81	0.8	0.5
18. Iran	292.0	0.80	0.8	0.5
19. Kuwait	291.3	0.80	0.8	0.5
20. Brunei	282.0	0.77	0.8	0.5
Total free world	37,388.9	102.44	100.0	64.0
Total world	58,458.8	160.16	—	100.0

Source: U.S. Energy Information Administration, United States only; other nations, *Oil and Gas Journal*.

Courtesy API.

[p] Preliminary.

[a]Comprises all gas collected and utilized as a fuel or as a chemical industry raw material, including gas used in oil and/or gas fields as a fuel by producers, even though it is not actually sold.

[b]United States is reported marketed production; all others may include some gross production.

gas, liquefied by a refrigeration cycle, can now be transported efficiently and rapidly across the oceans of the world by insulated tankers. The use of refrigeration to liquefy dry natural gas, and hence reduce its volume to the point where it becomes economically attractive to transport across oceans by tanker, was first attempted on a small scale in Hungary in 1934 and later used in the United States for moving gas in liquid form from the gas fields in Louisiana up the Mississippi River to Chicago in 1951.

The first use of a similar process on a large scale outside the United States was the liquefaction by a refrigerative cycle of some of the gas from the Hassi R'Mel gas field in Algeria and the export from 1964 onward of the resultant liquefied natural gas (LNG) by specially designed insulated tankers to Britain and France. Natural gas is in this way reduced to about one six-hundredth of its original volume and the nonmethane components are largely eliminated.

At the receiving terminals, the LNG is reconverted into a gaseous state by passage through a regasifying plant, whence it can be fed as required into the normal gas distribution grid of the importing country. Alternatively, it can be stored for future use in insulated tanks or subsurface storages. Apart from its obvious applications as a storable and transportable form of natural gas, LNG has many applications in its own right—particularly as a nonpolluting fuel for aircraft and ground vehicles.

Current production from conventional sources is not sufficient to satisfy all demands for natural gas; however, there has been lack of agreement as to the extent of the gas shortage. With the exception of past production, all resource base parameters (Table 1.3) are subject to some uncertainty. Standardized definitions for natural gas supply indicators are not always used, estimation procedures differ, and professional judgment must be exercised in making resource estimates. Estimates of the undiscovered natural gas reserves that may eventually be found also differ greatly. McKelvey's system of nomenclature and definitions provides an excellent guide (Fig. 1.1).

The following definitions will help distinguish between the terms *proved reserves* and *potential resources*:

Proved reserves are those quantities of gas that have been found by the drill. They can be proved by known reservoir characteristics such as production data, pressure relationships, and other data, so that volumes of gas can be determined with reasonable accuracy.

Potential resources constitute those quantities of natural gas that are believed to exist in various rocks of the earth's crust but have not yet been found by the drill. They are future supplies beyond the proved reserves.

Different methodologies have been used in arriving at estimates of the future potential of natural gas. Some estimates were based on growth curves, extrapolations of past production, exploratory footage drilled, and discovery rates. Empirical models of gas discoveries and production have also been developed and converted to mathematical models. Future gas supplies as a ratio of the amount of oil to be discovered is a method that has been used also. Another approach is a volumetric appraisal of the potential of undrilled areas. Different limiting as-

TABLE 1.3

Estimated Proved World Reserves of Natural Gas Annually as of January 1

(billions of cubic feet)

	Western Hemisphere												
Year	United States	Canada	Latin America	Western Hemisphere Total	Middle East	Africa	Asia-Pacific	Western Europe	Total Free World	Communist Nations	Total World	U.S. as a Percent Total World	Year
1967	289,333	43,450	64,550	397,333	215,070	158,155	32,450	88,582	891,590	150,000	1,041,590	27.8	1967
1968	292,908	45,682	67,101	405,691	220,670	167,223	40,050	133,965	967,599	215,500	1,183,099	24.8	1968
1969	287,350	47,666	62,900	397,916	223,775	168,345	52,724	141,176	983,936	343,000	1,326,936	21.7	1969
1970	275,109	51,951	163,150	490,210	235,275	197,143	67,500	150,800	1,140,928	350,000	1,490,928	18.5	1970
1971	290,746[a]	53,376	73,100	417,222	354,262	191,516	56,330	147,731	1,167,061	440,000	1,607,061	18.1	1971
1972	278,806[a]	55,462	72,700	406,968	343,930	193,018	69,800	163,250	1,176,966	558,000	1,734,966	16.1	1972
1973	266,085[a]	52,936	79,218	398,239	344,150	189,015	101,236	178,400	1,211,040	664,400	1,875,440	14.2	1973
1974	249,950[a]	52,457	91,321	393,728	413,325	187,720	114,200	193,797	1,302,770	735,400	2,038,170	12.3	1974
1975	237,132[a]	56,708	100,214	394,054	672,670	314,974	115,880	202,826	1,700,404	846,000	2,546,404	9.3	1975
1976	228,200[a]	56,975	90,487	375,662	538,648	207,152	111,560	180,875	1,413,897	835,000	2,248,897	10.1	1976
1977	216,026[a]	58,282	90,325	364,633	536,460	209,077	120,010	141,905	1,372,085	953,000	2,325,085	9.2	1977
1978	208,878[a]	59,472	108,480	376,830	719,660	207,504	122,725	138,190	1,564,909	955,000	2,519,909	8.3	1978
1979	200,302[a]	59,000	112,950	372,252	730,660	186,290	119,850	143,260	1,552,312	945,000	2,497,312	8.0	1979
1980	194,917[a]	85,500	144,500	424,917	740,330	210,350	128,815	135,376	1,639,158	935,000	2,574,158	7.6	1980
1981	199,021[a]	87,300	159,811	446,132	752,415	208,470	126,290	159,315	1,692,622	953,900	2,646,522	7.5	1981
1982	198,000[a]	89,900	176,323	464,223	762,490	211,667	127,616	150,650	1,716,646	1,194,700	2,911,356	6.8	1982

Source: 1967–1980: United States—American Gas Association, Committee on Natural Gas Reserves.
1981: United States—Department of Energy.
1982: *Oil and Gas Journal*, Worldwide Report Issue.
Rest of the World—*Oil and Gas Journal*, "Worldwide Report" issues.

Courtesy API.

[a]Figures include 26 trillion cu ft in Prudhoe Bay, Alaska (discovered in 1968) for which transportation facilities are not yet available.

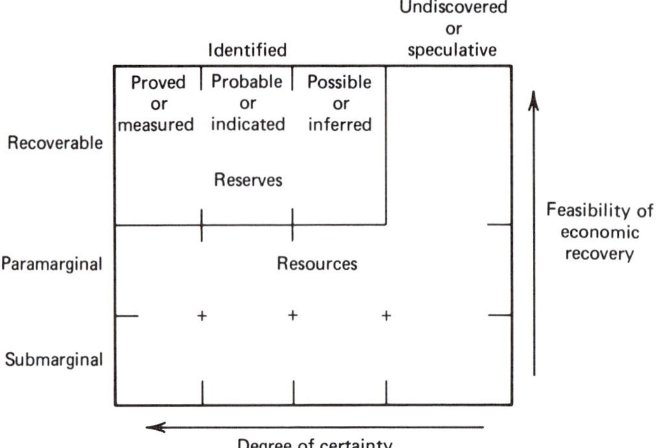

Fig. 1.1 Classification of mineral reserves and resources. (After McKelvey.)

sumptions have been made, such as drilling depths, water depths in offshore areas, economics, and technological factors.

The Potential Gas Committee is working toward standardization of reserve estimates. It compares the geological factors that control known occurrences of gas with geological conditions that are present in areas believed to be prospective for gas occurrence. The figures submitted by the Potential Gas Committee are believed to be the "most likely" or mean values. Limitations on the estimates are to a drilling depth of 30,000 ft and, for offshore areas, out to 1000 m (3281 ft) of water depth.

Three separate categories are used to express the estimates of the Potential Gas Committee. The *probable* potential future supply is that gas expected to be found in close proximity to or associated with known producing fields in known producing formations and with similar geological conditions. The *possible* potential future supply is that gas expected to be found in provinces or basins that are productive but in more remote portions of those areas, further away from the producing fields. The *speculative* future potential supply is that gas expected to be found in frontier areas that have been completely unexplored or inadequately tested.

Potential Gas Committee figures to the end of 1978 for the United States (including Alaska) for supplies of gas expected to be found in conventional reservoirs by conventional exploratory methods and drilling techniques are as follows (after Grow):

Probable	199 trillion cu ft
Possible	399 trillion cu ft
Speculative	421 trillion cu ft
Total	1019 trillion cu ft

(at 14.73 psia and 60°F)

In Table 1.4, these estimates have been broken down according to onshore, both normal and deep, and offshore, with separate figures given for Alaska.

Supplements to natural gas produced from conventional sources may provide portions of the world's future energy. Such supplements may include gas production resulting from stimulation of tight gas reservoirs in the western United States, methane gas occluded in coal, and natural gas contained in geopressured reservoirs. Artificial or substitute natural gas may include gas generated by gasifying coal, oil shale, or hydrocarbon liquids. In addition, gas generated from organic wastes and plant material and hydrogen gas produced from either water or other hydrogen compounds are potential substitute gaseous fuels.

A recent study by the American Gas Association (AGA) indicates that worldwide capability exists for substantially increasing conventional natural gas production in the coming decade and for sustaining production well above today's level until at least the year 2020. This study estimates the level of production that could be achieved, rather than a projected "actual" production volume, since it is not possible to predict the economic and political factors that will influence future production decisions within a region.

The major findings of this study may be summarized as follows:

1. While annual world conventional natural-gas production is now only about 50 trillion cu ft (Tcf), proved reserves are estimated at about 2200 Tcf, and remaining undiscovered resources are estimated at about 7500 Tcf.
2. Cumulative worldwide conventional production of natural gas to 1975 is estimated at about 854 Tcf or about 40% of presently estimated proved reserves and only 11% of remaining undiscovered gas resources.
3. Even at an annual world natural gas production rate of double the present rate, that is, 100 Tcf/year, the estimated world remaining conventional natural gas resource base would be large enough to sustain production at or near this level for at least another 50 years.
4. Under a gas-pricing scenario that would allow a natural gas price of $20/bbl crude oil equivalent (1974 dollars) in the period after 1985, it is estimated that world gas production could rise to about 70 Tcf by 1985 and to about 132 Tcf by the year 2000.
5. At these production rate increases (4.4%/year through 2000), it is estimated that production would peak shortly after the year 2000 and decline to about 115 Tcf by 2020. By that time, about 50% of the presently estimated remaining gas-resource base would have been produced.
6. Key areas of the world where substantial potential exists for greatly increasing production over the next decade include the OPEC groups and the USSR.

These estimates do not assume any production from the numerous sources of conventional and supplemental sources of gas from geopressured resources, tight gas formations, coal beds, shales, and biomass. These represent an additional and substantial gas-resource base, estimated in the range of several thousand trillion cubic feet, which could add significantly to world gas production in the years after 2000.

TABLE 1.4
Potential Gas Committee Estimates of Gas Reserves in United States to 12/31/78
(trillion cubic feet)

	Probable	Possible	Speculative	Total
Onshore—less than 15,000 ft				
48 states	112	188	155	415
Alaska	11	19	29	59
Total	123	207	144	474
Onshore—15,000 to 30,000 ft				
48 states	29	95	75	199
Alaska (no estimates)	—	—	—	—
Total	29	95	75	199
Offshore				
48 states	45	76	95	216
Alaska	2	21	107	130
Total	47	97	202	346
Grand Totals				
48 states	186	359	285	830
Alaska	13	40	136	189
Total	199	399	421	1019

Source: After Grow.

1.2 TYPES OF NATURAL GAS ACCUMULATIONS

1.2.1 Conventional Natural Gas

Dry natural gas is composed primarily of hydrocarbons (compounds containing only hydrogen and carbon). Methane (CH_4), the simplest and most basic compounds of the hydrocarbon series, is the major component. Others, fractionally small but important, include ethane (C_2H_6), propane (C_3H_8), butane (C_4H_{10}), and heavier, more complex hydrocarbons. In processing, most of the butane and heavier hydrocarbons, as well as a portion of the ethane and propane, are frequently removed from the gas in the form of liquids. Most of the water, gaseous sulfur compounds, nitrogen, carbon dioxide, and other impurities found in natural gas are also removed in various processing stages. The composition and the Btu content of unprocessed natural gas produced from different reservoirs vary widely.

In addition to composition and Btu content, gas is commonly designated in terms of the nature of its occurrence underground. It is called nonassociated gas if it is found in a reservoir that contains a minimal quantity of crude oil. It is called either dissolved or associated gas if it is found in a crude oil reservoir. Dissolved gas is that portion of the gas dissolved in the crude oil and associated gas (sometimes called gas-cap gas) is free gas in contact with the crude oil. All crude oil reservoirs contain dissolved gas and may or may not contain associated gas.

Some gases are called gas condensates or simply condensates. Although they occur as gases in underground reservoirs, they have a high content of hydrocarbon liquids. On production, they may yield considerable quantities of hydrocarbon liquids.

The properties of the produced fluids from crude oil and natural gas wells vary widely. Nonassociated gas may be produced at a high pressure and may require minimal treatment before it is transferred into a pipeline. The produced stream from a crude oil reservoir is normally separated in a single stage by passing it through a free-water knockout to remove water and sand and then through a low-pressure separator to split the oil and gas streams. Frequently, the low-pressure gas recovered from a single separator cannot be economically compressed and transmitted to shore or to an existing gas pipeline. As a consequence, this gas is often vented to the atmosphere or flared.

1.2.2 Gas in Tight Sands

Many geologic formations, located for the most part in the Rocky Mountain states of the United States, contain large quantities of submarginal natural gas resources. The prospective reservoirs generally have porosities of 5 to 15%, immobile water saturations of 50 to 70%, and gas permeabilities of 0.001 to 1 millidarcy (md). At higher gas permeabilities, the formations are generally amenable to conventional fracturing and completion methods.

The formations themselves are of two general types with numerous gradations between. One type consists of massive, more or less homogeneous sand bodies of uniform thickness and considerable areal extent. The other consists of shales and clays containing sandy zones or lenticular sandstone members. In either case, the basins containing the tight formations should measure in the thousands of square miles in order to provide suitable targets. In some places aquifers included in the stratigraphic section may strongly influence hydraulic fracturing possibilities.

Within the United States, many areas contain formations that meet the definitions of tight gas sand. The largest portion of the gas resource is found in the Green River Basin of Wyoming, the Piceance Basin of Colorado, and the Uinta Basin of Utah (Fig. 1.2). Table 1.5 shows the estimates of the gas resources of these basins. These resources were determined volumetrically on the basis of total pay interval and net pay thickness. The estimates were also assigned levels of confidence: Category 1, good well control; Category 2, inferred from geological interpretation but having sparse well control; and Category 3, speculative because of lack of testing.

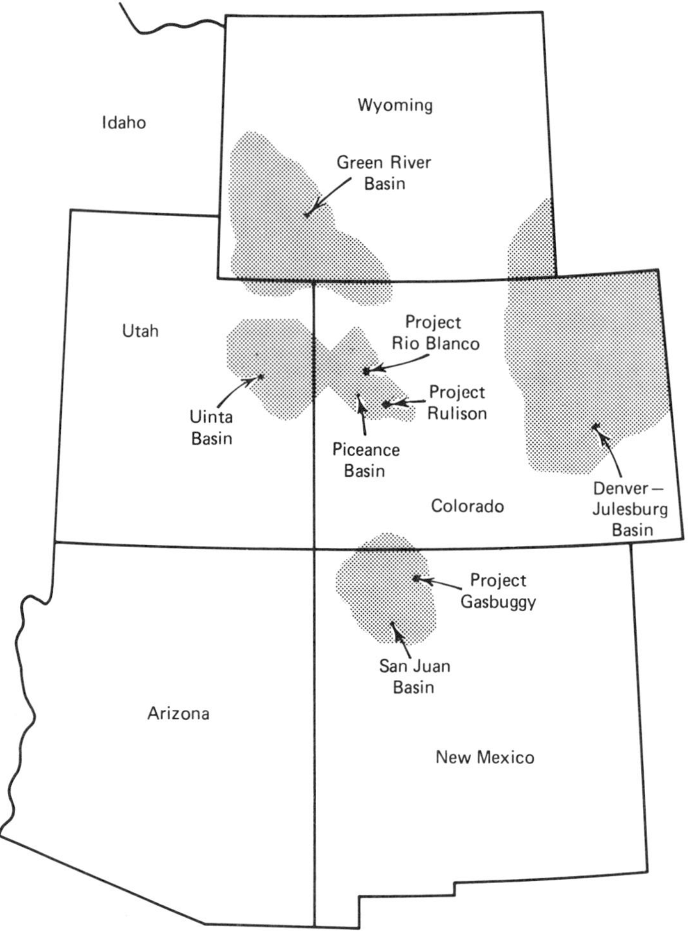

Fig. 1.2 Tight gas-sand basins in the Rocky Mountains of the United States. (After Meyer.)

Artificial Fracturing

Artificially fracturing tight formations increases the area of rock surface in direct communication with the wellbore, thereby creating a pressure sink into which the gas in the low-permeability sand may move. Three techniques accomplish fracturing, or enhancement of fracturing, with varying degrees of success.

The first, nuclear stimulation, is accomplished by detonating a nuclear device underground in a well drilled into the gas sand. The detonation creates a "chimney" of broken rocks and a system of fractures radiating outward from the chimney into the gas-bearing formation. Thus, a borehole of several hundred feet rather than a few inches is created. The locations of three nuclear-stimulation projects, Gasbuggy, Rulison, and Rio Blanco, carried out by the U.S. Atomic Energy Commission during the period 1964–1973, are shown in Fig. 1.2. At

TABLE 1.5
Estimate of Gas Resources in Tight Sandstones

	Depth interval, ft	Area, square miles	Gas in place, 10^{12} cu ft
Piceance basin			
Category 1	5600–8750	550	103.2
Category 2	5600–8750	650	103.9
Total		1200	207.1
Green River basin			
Category 1	8000–12,000	140	37.1
Category 2	9500–15,800	500	94.5
Category 3	9000–12,000	500	94.5
Total		1140	240.0
Uinta basin			
Category 1	8000–11,000	300	101.6
Category 2	8000–11,000	200	47.5
Total		500	149.1
Total		2840	596.2

Source: After Meyer.

present, there are no plans to continue the nuclear-stimulation program. The reasons for this are based on the environmental effects, availability of nuclear devices, and feasibility questions.

Chemical explosive fracturing was tried without much success on tight sands. This technology appears to be most useful in areas with natural fractures, which are lacking in most tight gas sands. Moreover, this technique is quite dangerous to the operator.

Artificial hydraulic fracturing of reservoir rocks has been a useful technique for increasing the productivity of oil and gas wells. Now the technique is being used on a massive scale to release gas bound up in rocks of very low permeability. Most such strata contain few if any natural fractures to provide avenues for migration of the gas to the wellbore. In massive hydraulic fracturing (MHF) a fracturing fluid is pumped into the wellbore under very high pressure for many hours to induce the fracture. The fracturing fluid is followed by a fluid containing a propping agent, such as sand or glass beads. When pumping stops, the fluids are forced back into the wellbore, leaving the proppant behind to hold the fracture apart, thus providing communication over a large area to the wellbore. Most desirable for a tight gas formation is the single propped fracture, vertical or nearly so, extending 1000 to 2000 ft on either side of the borehole and having a height of 100 to 500 ft. The efficacy of MHF has been demonstrated in many areas. However, it is not likely to be widely used in the United States until wellhead gas prices rise to as much as $2/1000 cu ft.

1.2.3 Gas in Tight Shales

The most important black shale–producing areas in the eastern United States are in Kentucky, Ohio, Virginia, and West Virginia. Of these, eastern Kentucky and western West Virginia are considered the most important. Approximately 71,000 wells in these states have been classified as gas producers, of which approximately 9600 or 14% are estimated to be shale producers. Estimated 1975 production from all sources within the four states was 299,465 $\times$ 10^6 cu ft. The Devonian shale of the eastern Kentucky field accounted for 14% of this production (Ray).

The Devonian shales meet the definition of organic-rich shale because they contain 5 to 65% indigenous organic matter. The most prevalent constituent of the shale is quartz; other components are kaolinite, pyrite, and accessory minerals such as feldspar, calcite, gypsum, apatite, zircon, titanite, and muscovite. The shale is generally fissile, finely laminated, and varicolored but predominantly black, brown, or greenish-gray. Core analysis has determined that the shale itself may have up to 12% porosity; however, permeability values are commonly less than 1 md. It is thought, therefore, that the majority of production is controlled by naturally occurring fractures and is further influenced by bedding planes and jointing. The outstanding feature of the eastern Kentucky shales is the long, slow declining productive life of the gas produced from them. Many wells have produced significant amounts of gas for more than 50 years. These production characteristics provide a reliable energy source that can be, and has been, augmented by shallower production with a much faster depletion rate. An added advantage of the eastern Kentucky shale gas is the high Btu value, as high as 1250 Btu/cu ft. The petrochemical potential of this gas also adds to its resource value.

Much research is currently underway in the United States to test the economic effectiveness of various stimulation techniques in shales. Among the more promising are chemical-explosive fracturing and massive hydraulic fracturing. Chemical-explosive fracturing is a process for injecting an explosive into a formation and detonating it chemically. Many variations in types and amounts of explosives have been experimentally tried. Standard practice involved the shooting of the entire section in one shot, the formation being exposed to approximately 10 lb of 80% gelatinated nitroglycerin per foot of section. Massive hydraulic fracturing entails injecting enormous amounts of fluid and sand, of the magnitude of 350,000 gal of fluid and a million pounds of sand, into a formation to artificially fracture it and prop the fractures open. This permits the passage of fluids to the wellbore from the area of the formation fractured.

It is estimated that approximately 20 $\times$ 10^6 sq mi of the earth are underlain by sedimentary rocks and over 5% of the sedimentary rocks in the United States are shales. Projected worldwide, organic shales could have considerable influence on the future energy picture through both conventional drilling and distillation methods.

1.2.4 Methane Gas Occluded in Coal

The total amount of methane gas in minable coal beds with depths less than 3000 ft has been estimated to be 260 Tcf as compared to the total proved gas reserves of the United States, which was 250 Tcf at year-end, 1973. Although the estimated size of the resource base seems significant, the recovery of this type of gas may not exceed 35 to 40 Tcf, owing to practical constraints.

1.2.5 Natural Gas from Geopressured Reservoirs

In a rapidly subsiding basin area, clays often seal underlying formations and trap their contained fluids. After further subsidence, the pressure and temperature of the trapped fluids exceed those normally anticipated at reservoir depth. These reservoirs, commonly called geopressured reservoirs, have been found in many parts of the world during the search for oil and gas. In the United States they are located predominantly both onshore and offshore in a band along the Gulf of Mexico. In length, the band extends from Florida to Texas; in width, it extends from about 100 mi inland to the edge of the continental shelf.

The high-pressure water in the geopressured reservoirs may contain up to 40 scf of gas per barrel of water. The resource base in the Gulf Coast area has been estimated to be 2700 Tcf. However, any prediction of the amount of gas from geopressured reservoirs that may be available to augment the future supply of conventional natural gas would be extremely speculative. The inadequacy of the present description of the resource base precludes analysis of the overall economics or timing of the exploitation of this energy source.

In some applications, hot water from geopressured reservoirs may serve as a source of natural gas, a fresh water supply, and a means of generating electric power. The additional benefits that could be obtained from such multipurpose uses may be a factor in the development of the resource.

REFERENCES

American Petroleum Institute. *Basic Petroleum Data Book: Petroleum Industry Statistics*. Washington: API, 1982.

Grow, G. C. "Future Potential Gas Supply in the United States—Current Estimates and Methods of the Potential Gas Committee." Presented at the American Institute of Chemical Engineers' meeting at Philadelphia, Pennsylvania, June 9, 1980.

McCormick, W. R., Jr., R. B. Kalisch, and T. J. Wander. "AGA Study Assesses World Natural Gas Supply." *Oil and Gas Journal*, pp. 103–106, February 13, 1978.

McKelvey, V. E. "Mineral Resource Estimates and Public Policy." *American Scientist* **60**, pp. 32–40, January–February 1972.

Meyer, R. F. "The Resource Potential of Gas in Tight Formations." *The Future Supply of Nature-made Petroleum and Gas*. Laxenburg, Austria: Pergamon Press, Ch. 36, 1976.

Ray, E. O. "Devonian Shale Production, Eastern Kentucky Field." *The Future Supply of Nature-made Petroleum and Gas*. Laxenburg, Austria: Pergamon Press, Ch. 38, 1976.

2

PROPERTIES OF NATURAL GASES AND CONDENSATE SYSTEMS

2.1 INTRODUCTION

Natural gas is a mixture of hydrocarbon gases and impurities. The hydrocarbon gases normally found in natural gas are methane, ethane, propane, butanes, pentanes, and small amounts of hexanes, heptanes, octanes, and the heavier gases. The impurities found in natural gas include carbon dioxide, hydrogen sulfide, nitrogen, water vapor, and heavier hydrocarbons. Usually, the propane and heavier hydrocarbon fractions are removed for additional processing because of their high market value as gasoline-blending stock and chemical-plant raw feedstock. What usually reaches the transmission line for sale as natural gas is mostly a mixture of methane and ethane with some small percentage of propane.

This chapter reviews those physical properties of natural gases that are important in solving gas well performance, gas production, and gas transmission problems. The properties of a natural gas may be determined either directly from laboratory tests or predictions from known chemical composition of the gas. In the latter case, the calculations are based on the physical properties of individual components of the gas and on physical laws, often referred to as mixing rules, relating the properties of the components to those of the mixture.

2.2 COMPOSITION OF NATURAL GAS

There is no one composition or mixture that can be referred to as the natural gas. Each gas stream produced has its own composition. Two wells from the same reservoir may have different compositions. Also, each gas stream produced from a natural gas reservoir can change composition as the reservoir is depleted. Samples of the well stream should be analyzed periodically, since it may be necessary to change the production equipment to satisfy the new gas composition.

Table 2.1 shows some typical natural gas streams. Well stream 1 is typical of an associated gas, that is, gas produced with crude oil. Well streams 2 and 3 are typical nonassociated low-pressure and high-pressure gases, respectively.

Natural gas is normally regarded as a mixture of straight-chain or paraffin hydrocarbon gases. In straight-chain hydrocarbons, the carbon atoms are attached to form chains. However, cyclic and aromatic hydrocarbon gases are occasionally found in natural gas mixtures. In cyclic hydrocarbons, the carbon atoms are arranged to form rings. Figure 2.1 shows the structures of some straight-chain and cyclic hydrocarbons. The hydrocarbons listed in Table 2.1 are straight-chain or paraffinic compounds.

2.3 PHASE BEHAVIOR

Conventional gas reservoirs have been characterized in many different ways but most commonly on the basis of the surface-producing gas-oil ratio. Using this method, any well (or field) that produces at a gas-oil ratio (GOR) in excess of 100,000 cu ft per barrel of oil [standard cubic feet per stock tank barrel (scf/STB)] is considered a gas well; one producing with a GOR of 5000 to 100,000 scf/STB, a gas-condensate well; and one producing with a GOR of zero to several thousand scf/STB, an oil well. In practice, similar surface gas-oil ratios have been obtained for reservoirs containing a variety of hydrocarbon fluid compositions, existing over a wide range of reservoir pressures and temperatures, and producing with natural or artificial mechanisms. This has resulted in both technical and legal misunderstanding of the nature of conventional gas reservoirs. Therefore, the simplified classification described above is considered inadequate.

Conventional gas reservoirs should be defined on the basis of their initial reservoir pressure and temperature on the usual pressure-temperature (*P-T*) phase diagram (Fig. 2.2). *P-T* phase diagrams show the effects of pressure and temperature on the physical state of a hydrocarbon system. However, the phase diagram in Fig. 2.2 is for a specific composition. Although a different fluid would have a different phase diagram, the general configuration is similar.

In Fig. 2.2, the area enclosed by the bubble point (BP) line *A-S-C* and the dew point (DP) line *C-D-T-B* to the lower left is the region of pressure-tempera-

TABLE 2.1
Typical Natural Gas Analyses

Component	Well No. 1 Mole Percent	Well No. 2 Mole Percent	Well No. 3 Mole Percent
Methane	27.52	71.01	91.25
Ethane	16.34	13.09	3.61
Propane	29.18	7.91	1.37
i-butane	5.37	1.68	0.31
n-butane	17.18	2.09	0.44
i-pentane	2.18	1.17	0.16
n-pentane	1.72	1.22	0.17
Hexane	0.47	1.02	0.27
Heptanes and heavier	0.04	0.81	2.42
Carbon dioxide	0.00	0.00	0.00
Hydrogen sulfide	0.00	0.00	0.00
Nitrogen	0.00	0.00	0.00
Total	100.00	100.00	100.00

Source: Courtesy Petroleum Extension Service.

Note: Production from many wells will contain small quantities of carbon dioxide, hydrogen sulfide, and nitrogen.

ture combinations in which both gas and liquid phases will exist. The curves within the two-phase region show the gas-liquid percentages for any temperature and pressure. The line *A-S-C-T-B* separates the two-phase region from the single-phase regions where all the fluids exist in a single phase. The bubble point line *A-S-C* separates the two phase region from the single-phase liquid region, while the dew point line *C-D-T-B* separates it from the single-phase gas region. Point *C* where the bubble point and dew point lines meet is called the critical point, and the corresponding temperature is called critical temperature T_c.

Consider a reservoir initially at 3000 psia and 125°F represented on the figure by point 1_i. (The subscripts *i* and *a* depict the initial and abandonment conditions.) The pressure and temperature conditions are such that the initial state of the hydrocarbons is a liquid, that is, oil. Thus, point 1_i delineates an oil reservoir. This type is called a bubble point reservoir, for as pressure declines in the reservoir (due to production) isothermally, the bubble point will be reached, in this case at 2550 psia, point *S* on the dashed line. Bubble point refers to the highest pressure at which the first bubble gas comes out of solution from the oil and exists as free gas in the reservoir. It is also known as the saturation pressure.

Below this pressure, a free-gas phase will appear. Eventually the free gas evolved begins to flow to the well bore in ever-increasing quantities. Conversely, the oil flows in ever-decreasing quantities. The actual amount of free gas liberated will depend on the composition of the oil. Oils originally containing large amounts of the lighter hydrocarbons [high (API) gravity] will release large amounts of gas, while those oils originally containing only small amounts of the lighter hydrocarbons (low API gravity) will liberate much smaller amounts of gas.

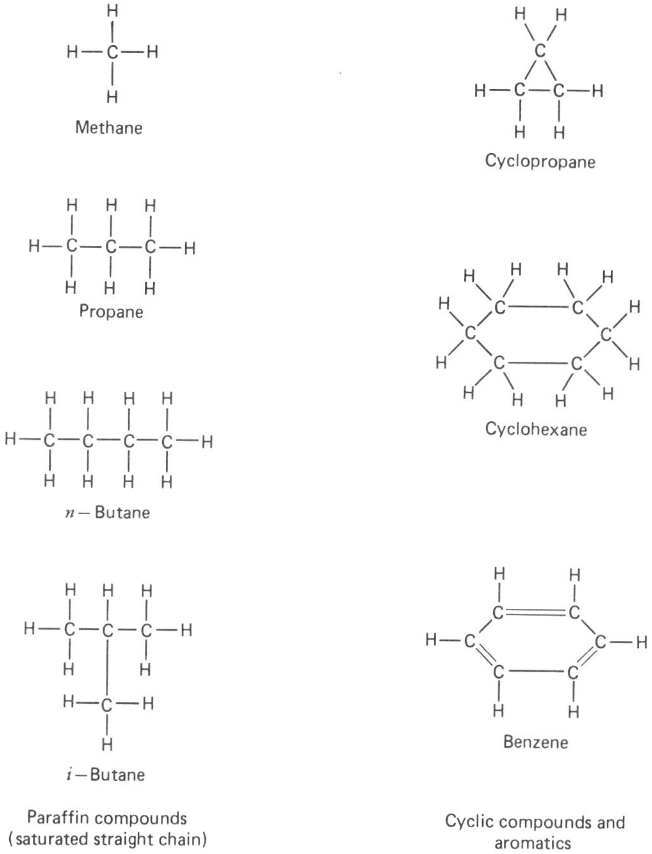

Methane

Cyclopropane

Propane

Cyclohexane

n-Butane

Benzene

i-Butane

Paraffin compounds
(saturated straight chain)

Cyclic compounds and
aromatics

Fig. 2.1 Hydrocarbon gas molecule structures. (Courtesy Petroleum Extension Service.)

Other names for this type of oil reservoir are undersaturated, depletion, dissolved gas, solution gas drive, expansion, and internal gas drive.

If the same hydrocarbon mixture occurred at 2000 psia and 210°F, point 2 in the figure, it would be an oil reservoir with an initial gas cap. Both phases exist in equilibrium. The slightest reduction in pressure causes liberation of gas from oil, making this a saturated oil reservoir. The initial pressure of the reservoir and the saturation pressure of the oil zone will be identical. Hence, if the initial reservoir pressure equals the reservoir saturation pressure, the reservoir has an initial gas cap. The oil zone will be produced as a bubble point reservoir, modified by the presence of the gas cap. The gas cap will be at the dew point and may be either retrograde (discussed later) or nonretrograde (Fig. 2.3).

Next, consider a reservoir at a temperature of 230°F and with an initial pressure of 3300 psia, shown as point 3_i in Fig. 2.2. Since the initial conditions of pressure and temperature are to the right of the critical point and outside the phase envelope, this reservoir exists initially in the gaseous state. As production

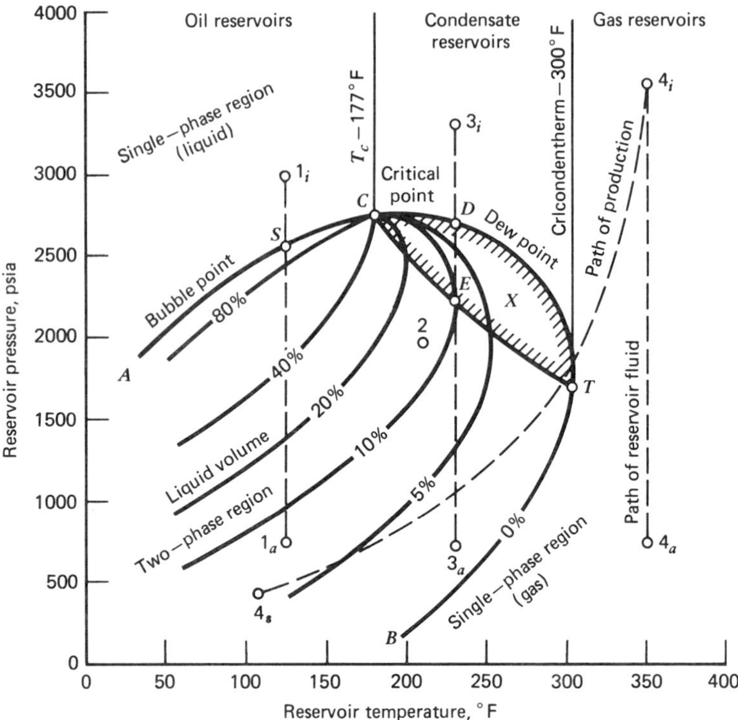

Fig. 2.2 Pressure-temperature phase diagram of a reservoir fluid.

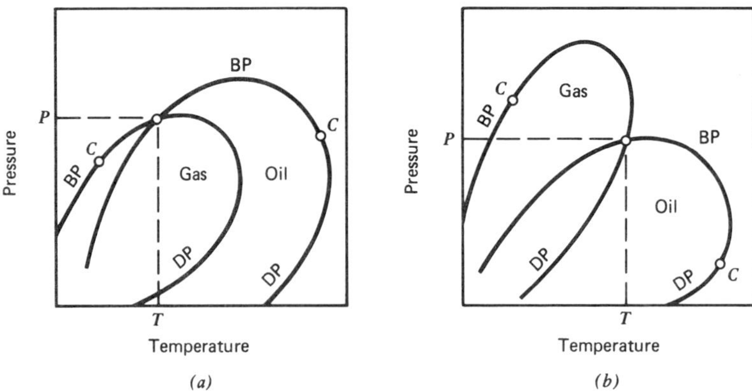

Fig. 2.3 Phase diagrams of a cap gas and oil zone fluid showing (*a*) retrograde cap gas and (b) nonretrograde cap gas. (After Craft and Hawkins.)

begins from the reservoir and pressure declines, no change in the state of the reservoir fluids occurs until the dew point pressure is reached at 2700 psia, point *D*.

Below this pressure a liquid condenses out of the reservoir fluid as a fog or dew. This is not considered to be a normal situation since, for most hydrocarbon fluids, a pressure reduction tends to increase the amount of gas. Therefore, this behavior is usually referred to as retrograde condensation, signifying that vaporization generally occurs during isothermal expansion rather than condensation. The condensation leaves the gas phase with a lower liquid content. As the condensed liquid adheres to the walls of the pore spaces of the rock, it is immobile. Thus, the gas produced at the surface will have a lower liquid content, and the producing GOR will rise. As the liquefiable portions of the reservoir fluids are usually the most valuable components, the loss of part of these fluids could substantially reduce the ultimate income from the property, which must be considered in an economic evaluation.

Examination of Fig. 2.2 will show that for a reservoir fluid to exhibit the phenomenon of retrograde condensation, the initial conditions of pressure and temperature must exist outside the phase envelope to the right of the critical point *C* and to the left of point *T* or within the phase envelope in the region marked *X*. The point *T* is called the cricondentherm and is the maximum temperature at which two phases can exist in equilibrium (300°F for the example). The process of retrograde condensation continues until a point of maximum liquid volume is reached, 10% at 2250 psia (point *E*).

In some cases, a sufficient volume of liquid will be condensed in the reservoir to provide mobility of the liquid phase. In such cases the surface fluid composition depends on the relative mobilities of the vapor and liquid in the reservoir. As production continues from point *E* to the abandonment pressure 3_a, vaporization of the retrograde liquid occurs. This revaporization aids liquid recovery and may be evidenced by decreasing GOR on the surface.

This example assumes that the reservoir fluid composition remains constant. Unfortunately, as retrograde condensation occurs, the reservoir fluid composition changes and the *P-T* envelope shifts, increasing retrograde liquid condensation. Generally, for a particular initial hydrocarbon fluid, retrograde loss increases at lower reservoir temperature, higher abandonment pressure, and for greater shifting of the phase envelope to the right.

As a final illustration, consider a reservoir initially at 350°F and 3600 psia, represented by point 4_i in Fig. 2.2. Since the initial reservoir conditions exist to the right of the critical point *C* and outside the phase envelope, the reservoir fluid will be 100% gas. Furthermore, since the reservoir temperature exceeds the cricondentherm *T*, at no point in the isothermal depletion cycle (along path $4_i - 4_a$) is the phase envelope crossed. Therefore, the fluid *in the reservoir* never changes composition; it is always in the gaseous state.

However, after the reservoir fluid leaves the reservoir and enters the wellbore, the temperature, as well as the pressure, will decline until surface temperature and pressure conditions are reached. The fluid produced through the wellbore and into surface separators at point 4_s, though of the same composition as that in the reservoir, has entered the two-phase region due to the temperature and pressure decline along line $4_i - 4_s$. This accounts for the production of a considerable amount of liquid at the surface from a gas in the reservoir; therefore, this reservoir is referred to as a wet gas reservoir.

If the phase diagram is such that the separator conditions, point 4_s, lies outside the two-phase envelope in the single-phase (gas) region, then only gas will exist on the surface. No liquid will be formed in the reservoir or at the surface and the gas is called dry natural gas. The word *dry* indicates that the liquid does not contain enough of the heavier hydrocarbons to form a liquid at surface conditions. Nevertheless, it may contain liquid fractions, which can be removed by low-temperature separation or by natural gasoline plants.

2.4 THE IDEAL GAS

As a starting point in the study of the properties of real gases, we consider a hypothetical fluid known as an ideal gas. An ideal gas is a fluid in which the volume occupied by the molecules is insignificant with respect to the volume occupied by the total fluid, there are no attractive or repulsive forces between the molecules or between the molecules and the walls of the container, and all collisions of molecules are perfectly elastic, that is, there is no loss in internal energy upon collision.

At low pressures, most gases behave like the ideal gas. In addition, under normal distribution pressures, natural gas follows the ideal gas laws quite closely. Under these conditions it is not normally necessary, therefore, that an accurate determination of any deviation from these laws be made. However, when gas pressures increase, a wide variation between the actual and ideal volumes of the gas may occur. To understand fully what happens when natural gas is subjected to changes in pressure and temperature, the fundamental gas laws must be reviewed. The nomenclature is as follows:

V_1 = volume of gas under original conditions, ft^3

V_2 = volume of gas under changed conditions, ft^3

T_1 = absolute temperature of the gas under original condition, °R (°F + 460)

T_2 = absolute temperature of the gas under changed conditions, °R

p_1 = absolute pressure of the gas under original conditions, psia

p_2 = absolute pressure of the gas under changed conditions, psia

2.4.1 Boyle's Law

Robert Boyle (1627–1691), during the course of experiments with air, observed the following relation between pressure and volume: if the temperature of a given quantity of gas is held constant, the volume of gas varies inversely with the absolute pressure. This relation, written as an equation, is

$$\frac{p_1}{p_2} = \frac{V_2}{V_1} \quad \text{or} \quad p_1V_1 = p_2V_2 \quad \text{or} \quad pV = \text{constant} \tag{2.1}$$

In the application of Boyle's law, volume at a second set of pressure conditions is generally desired. A rearrangement of Eq. 2.1 gives the formula more readily used:

$$V_2 = V_1 \times \frac{p_1}{p_2} \tag{2.2}$$

Example 2.1. A quantity of gas at a pressure of 50 psig has a volume of 1000 cu ft. If the gas is compressed to 100 psig, what volume would it occupy? Assume the barometric pressure is 14.73 psia and the temperature of the gas remains constant.

Solution

$$V_1 = 1000 \text{ cu ft}$$
$$p_1 = (50 + 14.73) = 64.73 \text{ psia}$$
$$p_2 = (100 + 14.73) = 114.73 \text{ psia}$$

Substituting in Eq. 2.2 would give

$$V_2 = 1000 \times \frac{64.73}{114.73} = 564.19 \text{ cu ft}$$

2.4.2 Charles' Law

About 100 years after the discovery of Boyle's law, Jacques A. Charles (1746–1823) and Joseph L. Gay-Lussac (1778–1850) independently discovered the law that is usually called Charles' law. This law is in two parts:

1. If the pressure on a particular quantity of gas is held constant, then, with any change of state, the volume will vary directly as the absolute temperature. Expressed as an equation,

$$\frac{V_1}{V_2} = \frac{T_1}{T_2} \quad \text{or} \quad \frac{T_1}{V_1} = \frac{T_2}{V_2} \quad \text{or} \quad \frac{T}{V} = \text{constant} \tag{2.3}$$

Again, since the volume at a second set of temperature conditions is desired usually more than any other information, a handy arrangement of Eq. 2.3 is given as Eq. 2.4:

$$V_2 = V_1 \times \frac{T_2}{T_1} \tag{2.4}$$

2. If the volume of a particular quantity of gas is held constant, then, with any change of state, the absolute pressure will vary directly as the absolute temperature:

$$\frac{p_1}{p_2} = \frac{T_1}{T_2} \quad \text{or} \quad \frac{T_1}{p_1} = \frac{T_2}{p_2} \quad \text{or} \quad \frac{T}{p} = \text{constant} \tag{2.5}$$

In this instance, the pressure at a second temperature condition would be of more interest. Equation 2.5 could thus be written as

$$p_2 = p_1 \times \frac{T_2}{T_1} \tag{2.6}$$

Example 2.2.

(a) A given mass of gas has a volume of 500 cu ft when the temperature is 50°F and the pressure is 10 psig. If the pressure remains the same, but the temperature is changed to 100°F, what will be the volume of the gas? Atmospheric pressure may be taken as 14.73 psia.

(b) What would be the new pressure of the gas in the above example if the volume remains the same and the temperature is increased from 50 to 100°F as indicated?

Solution

(a) $V_1 = 500$ cu ft
$T_1 = 50 + 460 = 510\ °\text{R}$
$T_2 = 100 + 460 = 560\ °\text{R}$

Using Eq. 2.4,

$$V_2 = 500 \times \frac{560}{510}$$
$$= 549.02 \text{ cu ft}$$

(b) $p_1 = (10 + 14.73) = 24.73$ psia

Using Eq. 2.6,

$$p_2 = 24.73 \times \frac{560}{510}$$
$$= 27.15 \text{ psia or } 12.42 \text{ psig}$$

2.4.3 Boyle's and Charles' Laws

The separate relations of Boyle's and Charles' laws may be combined to give

$$\frac{p_1V_1}{T_1} = \frac{p_2V_2}{T_2} = \text{a constant (nearly)} \tag{2.7}$$

This equation is known as Boyle's–Charles' law and as the simple gas law. It is one of the most widely used relations in gas measurement work, since it approximately represents the behavior of many gases under conditions close to ordinary atmospheric temperatures and pressures. One can substitute known values in the combined formula and solve for any one unknown value.In cases where one of the parameters, such as temperature, is not to be considered, it may be treated as having the same value on both sides of the formula. However, Eq. 2.7 is deficient in one important respect: It does not show the relations connecting volumes and masses of gases.

Example 2.3.

(a) How many cubic feet of an ideal gas, measured at standard conditions of 60°F and 14.73 psia, are required to fill a 100-cu ft tank to a pressure of 40 psia when the temperature of the gas in the tank is 90°F? Atmospheric pressure is 14.4 psia.

(b) What would be the reading on the pressure gauge if the tank in the above example is cooled to 60°F after being filled with the ideal gas?

Solution

(a) $p_1 = 40 + 14.4 = 54.4$ psia
$p_2 = p_{sc} = 14.73$ psia
$T_1 = 90 + 460 = 550$ °R
$T_2 = T_{sc} = 520$ °R
$V_1 = 100$ cu ft
$V_2 = ?$

Using Eq. 2.7,

$$\frac{(54.4)(100)}{550} = \frac{(14.73)(V_{sc})}{520}$$

$$V_{sc} = 349 \text{ scf}$$

(b) $T_2 = 60 + 460 = 520$ °R
$V_2 = 100$ cu ft
$p_2 = ?$

Using Eq. 2.7 again,

$$\frac{(54.4)(100)}{550} = \frac{(p_2)(100)}{520}$$

$$p_2 = 51.4 \text{ psia or } 37.0 \text{ psig}$$

2.4.4 Avogadro's Law

Amadeo Avogadro proposed a law in the nineteenth century which states that, under the same conditions of temperature and pressure, equal volumes of all ideal gases contain the same number of molecules. It has been shown that there are 2.733×10^{26} molecules in 1 pound-mole of any gas.

From Avogadro's law, it may be seen that the weight of a given volume of gas is a function of the weights of the molecules, and that there is some volume at which the gas would weigh, in pounds, the numerical value of its molecular weight. The volume at which the weight of the gas in pounds is equal to the numerical value of its molecular weight is known as the mole-volume. A pound-mole of an ideal gas occupies 378.6 cu ft at 60°F and 14.73 psia. These conditions of temperature and pressure are commonly referred to as standard conditions.

2.4.5 The Ideal Gas Law

The equation of state for an ideal gas may be derived from a combination of Boyle's, Charles'/Gay Lussac's, and Avogadro's laws as

$$pV = nRT \tag{2.8}$$

where

p = absolute pressure, psia
V = volume, cu ft
T = absolute temperature, °R
n = number of pound-moles,where 1 lb-mol is the molecular weight of the gas (lb)
R = the universal gas constant that, for the above units, has the value 10.732 psia cu ft/lb-mol °R

Equation 2.8 is only applicable at pressures close to atmospheric, for which it was experimentally derived, and at which gases behave as ideal.

Since the number of pound-moles of a gas is equal to the mass of the gas divided by the molecular weight of the gas, the ideal gas law can be expressed as

$$pV = \frac{m}{M} RT \tag{2.9}$$

where

m = mass of gas, lb
M = molecular weight of gas lbm/lb-mol

Equation 2.9 may be rearranged to give the mass and density, ρ, of the gas:

$$m = \frac{MpV}{RT} \tag{2.10}$$

and

$$\rho = \frac{m}{V} = \frac{Mp}{RT} \tag{2.11}$$

Example 2.4. Using the fact that 1 pound-mole of an ideal gas occupies 378.6 scf, calculate the value of the universal gas constant, R.

Solution

$p = 14.73$ psia
$V = 378.6$ scf
$n = 1$
$T = 520$ °R

Using Eq. 2.8,

$$(14.73 \text{ psia})(378.6 \text{ cu ft}) = (1 \text{ lb-mol})(R)(520 \text{ °R})$$
$$R = 10.732 \text{ psia cu ft/lb-mol °R}$$

The numerical value of R depends on the units used to express temperature, pressure, and volume. Table 2.2 gives numerical values of R for various systems of units.

Example 2.5. Repeat Example 2.3 using the ideal gas law, Eq. 2.8.

Solution

$$\text{(a)} \quad n = \frac{pV}{pT} = \frac{(54.4 \text{ psia})(100 \text{ cu ft})}{(10.732 \text{ psia cu ft/lb-mol °R})(550 \text{ °R})} = 0.922 \text{ lb-mol}$$

$$V_{sc} = (0.922 \text{ lb-mol})(378.6 \text{ scf/lb-mol}) = 349 \text{ scf}$$

$$\text{(b)} \quad p = \frac{nRT}{V} = \frac{(0.922)(10.732)(520)}{(100)} = 51.4 \text{ psia or } 37.0 \text{ psig}$$

2.5 PROPERTIES OF GASEOUS MIXTURES

Natural gas engineers invariably deal with gas mixtures and rarely with single-component gases. Since natural gas is a mixture of hydrocarbon compounds and because this mixture is varied in types as well as the relative amounts of the compound, the overall physical properties will vary. The overall physical properties of a natural gas determine the behavior of the gas under various processing conditions. If the composition of the gas mixture is known, the overall physical properties can be established from the physical properties of each pure component in the mixture using Kay's mixing rules. Physical properties that are most useful in natural gas processing are molecular weight, boiling point, freezing point, density, critical temperature, critical pressure, heat of vaporization, and specific heat.

Table A.1 is a tabulation of physical constant of a number of hydrocarbon compounds, other chemicals, and some common gases, taken from GPSA *Engi-*

TABLE 2.2

Values of the Gas Constant R in $PV = nRT$

Basis of units listed below is 22.414 0 liters at 0°C and 1 atm for the volume of 1 g mole. All other values calculated from conversion factors listed in tables.

n	Temperature	Pressure	Volume	*R*	*n*	Temperature	Energy	*R*
g mol	K	atm	liter	0.082,057,477	g mol	K	calorie	1.985,9
g mol	K	atm	cm^3	82.057	g mol	K	joule	8.314,5
g mol	K	mm Hg	liter	62.364				
g mol	K	bar	liter	0.083,145	lb mol	°R	Btu	1.985,9
g mol	K	kg/cm^2	liter	0.084,784	lb mol	°R	hp-hr	0.000,780,48
g mol	K	kPa	m^3	0.008,314,5	lb mol	°R	Kw-hr	0.000,582,00
lb-mol	°R	atm	cu ft	0.730,24	lb mol	°R	ft-lb	1545.3
lb-mol	°R	in. Hg	cu ft	21.850				
lb-mol	°R	mm Hg	cu ft	554.98	kmol	K	joule	8314.5
lb-mol	°R	psi	cu ft	10.732				
lb-mol	°R	lb/cu ft	cu ft	1545.3				
lb-mol	K	atm	cu ft	1.314,4				
lb-mol	K	mm Hg	cu ft	998.97				
kmol	K	kPa	m^3	8.314,5				
kmol	K	bar	m^3	0.083,145				

Source: Courtesy NGPSA.

neering Data Book. Table 2.3 gives additional physical constants for the paraffin hydrocarbons methane through *n*-decane, including isobutane and isopentane.

2.5.1 Composition

The composition of a natural gas mixture may be expressed as either the mole fraction, volume fraction, or weight fraction of its components. These may also be expressed as mole percent, volume percent, or weight percent by multiplying the fractional values by 100. Volume fraction is based on gas component volumes measured at standard conditions, so that volume fraction is equivalent to mole fraction.

The mole fraction, y_i, is defined as

$$y_i = \frac{n_i}{\Sigma n_i} \tag{2.12}$$

where

y_i = mole fraction of component i
n_i = number of moles of component i
Σn_i = total number of moles of all components in the mixture

Volume fraction is defined as

$$(\text{volume fraction})_i = \frac{V_i}{\Sigma V_i} = y_i \tag{2.13}$$

where

V_i = volume occupied by component i at standard conditions
ΣV_i = volume of total mixture measured at standard conditions

Weight fraction ω_i, is defined as

$$\omega_i = \frac{W_i}{\Sigma W_i} \tag{2.14}$$

where

ω_i = weight fraction of component i
W_i = weight of component i
ΣW_i = total weight of mixture

It is easy to convert from mole fraction (or volume fraction) to weight fraction and vice versa. These are illustrated in Examples 2.6 and 2.7.

TABLE 2.3
Physical Constants of Paraffin Hydrocarbons and Other Components of Natural Gas

Component	Methane	Ethane	Propane	Isobutane	*N*-Butane	Isopentane	*N*-Pentane	*N*-Hexane	*N*-Heptane	*N*-Octane	*N*-Nonane	*N*-Decane
Molecular weight	16.043	30.070	44.097	58.124	58.124	72.151	72.151	86.178	100.205	114.232	128.259	142.286
Boiling point at 14.696 psia, °F	−258.69	−127.48	− 43.67	10.90	31.10	82.12	96.92	155.72	209.17	258.22	303.47	345.48
Freezing point at 14.696 psia, °F	−296.46	−297.89	−305.84	−255.29	−217.05	−255.83	−201.51	−139.58	−131.05	− 70.18	− 64.28	− 21.36
Vapor pressure at 100°F, psia	(5000)	(800)	190	72.2	51.6	20.44	15.570	4.956	1.620	0.537	0.179	0.0597
Density of liquid at 60°F and 14.696 psia												
Specific gravity at 60°F/60°F	0.3	0.3564	0.5077	0.5631	0.5844	0.6247	0.6310	0.6640	0.6882	0.7068	0.7217	0.7342
°API	340	265.5	147.2	119.8	110.6	95.0	92.7	81.6	74.1	68.7	64.6	61.2
Lb/gal at 60°F, wt in vacuum	2.5	2.971	4.233	4.695	4.872	5.208	5.261	5.536	5.738	5.893	6.017	6.121
Lb/gal at 60°F, wt in air	2.5	2.962	4.223	4.686	4.865	5.199	5.251	5.526	5.728	5.883	6.008	6.112
Density of gas at 60°F and 14.696 psia												
Specific gravity, air = 1.00, ideal gas	0.5539	1.0382	1.5225	2.0068	2.0068	2.4911	2.4911	2.9753	3.4596	3.9439	4.4282	4.9125
Lb/M cu ft, ideal gas	42.28	79.24	116.20	153.16	153.16	190.13	190.13	227.09	264.05	301.01	337.98	374.94
Volume ratio at 60°F and 14.696 psia												
Gal/lb mol	6.4	10.12	10.42	12.38	11.93	13.85	13.71	15.57	17.46	19.39	21.32	23.24
Cu ft gas/gal liquid, ideal gas	59	37.5	36.43	30.65	31.81	27.39	27.67	24.38	21.73	19.58	17.80	16.33
Gas vol/liquid vol, ideal gas	443	280.5	272.51	229.30	237.98	204.93	207.00	182.37	162.56	146.45	133.18	122.13
Critical conditions												
Temperature, °F	−116.63	90.09	206.01	274.98	305.65	369.10	385.7	453.7	512.8	564.22	610.68	652.1
Pressure, psia	667.8	707.8	616.3	529.1	550.7	490.4	488.6	436.9	396.8	360.6	332	304
Gross heat of combustion at 60°F												
Btu/lb liquid	—	22,214	21,513	21,091	21,139	20,889	20,928	20,784	20,681	20,604	20,544	20,494
Btu/lb gas	23,885	22,323	21,665	21,237	21,298	21,040	21,089	20,944	20,840	20,762	20.701	20,649
Btu/cu ft, ideal gas	1009.7	1768.8	2517.5	3252.7	3261.1	4000.3	4009.6	4756.2	5502.8	6249.7	6996.5	7742.1
Btu/gal liquid	—	65,998	91,065	99,022	102,989	108,790	110,102	115,060	118,668	121,419	123,613	125,444

TABLE 2.3 (Continued)

Component	Methane	Ethane	Propane	Isobutane	*N*-Butane	Isopentane	*N*-Pentane	*N*-Hexane	*N*-Heptane	*N*-Octane	*N*-Nonane	*N*-Decane
Cu ft air to burn 1 cu ft gas—ideal gas	9.54	16.70	23.86	31.02	31.02	38.18	38.18	45.34	52.50	59.65	66.81	73.97
Flammability limits at 100°F and 14.696 psia												
Lower, vol % in air	5.0	2.9	2.1	1.8	1.8	1.4	1.4	1.2	1.0	0.96	0.87	0.78
Upper, vol % in air	15.0	13.0	9.5	8.4	8.4	(8.3)	8.3	7.7	7.0	—	2.9	2.6
Heat of vaporization at 14.696 psia												
Btu/lb at boiling point	219.22	210.41	183.05	157.53	165.65	147.13	153.59	143.95	136.01	129.53	123.76	118.68
Specific heat at 60°F and 14.696 psia												
C_p gas—Btu/lb, °F, ideal gas	0.5266	0.4097	0.3881	0.3872	0.3867	0.3827	0.3883	0.3864	0.3875	(0.3876)	0.3840	0.3835
C_v gas—Btu/lb, °F, ideal gas	0.4027	0.3436	0.3430	0.3530	0.3525	0.3552	0.3608	0.3633	0.3677	0.3702	0.3685	0.3695
$N = C_p/C_v$	1.308	1.192	1.131	1.097	1.097	1.078	1.076	1.063	1.054	1.047	1.042	1.038
C_p liquid—Btu/lb, °F	—	0.9256	0.5920	0.5695	0.5636	0.5353	0.5441	0.5332	0.5283	0.5239	0.5228	0.5208
Octane number												
Motor clear	—	+ .05	97.1	97.6	89.6	90.3	62.6	26.0	0.0	—	—	—
Research clear	—	+1.6	+1.8	+0.10	93.8	92.3	61.7	24.8	0.0	—	—	—
Refractive Index n_p at 68°F	—	—	—	—	1.3326	1.35373	1.35748	1.37486	1.38764	1.39743	1.40542	1.4118

Source: GPA Publication 2134-75.

EXAMPLE 2.6.
Conversion from Mole Fraction (or Volume Fraction) to Weight Fraction

Component	Mole Fraction y_i	Molecular Weight M_i	y_iM_i	Weight Fraction $\omega_i = y_iM_i/\Sigma(y_iM_i)$
Methane	0.946	16.043	15.177	0.903
Ethane	0.046	30.070	1.383	0.082
Nitrogen	0.006	28.013	0.168	0.010
Carbon dioxide	0.002	44.010	0.088	0.005
	1.000		16.816	1.000

EXAMPLE 2.7.
Conversion from Weight Fraction to Mole Fraction (or Volume Fraction)

Component	Weight Fraction ω_i	Molecular Weight M_i	ω_i/M_i	Mole Fraction $y_i = \frac{\omega_i/M_i}{\Sigma(\omega_i/M_i)}$
Methane	0.880	16.043	0.0549	0.937
Ethane	0.040	30.070	0.0013	0.022
Propane	0.038	44.097	0.0009	0.015
Nitrogen	0.042	28.013	0.0015	0.026
	1.000		0.0586	1.000

2.5.2 Apparent Molecular Weight

Although, in a strict sense, a gas mixture does not have a unique molecular weight, it behaves as though it does. Thus, the concept of apparent or average molecular weight is quite useful in characterizing a gas mixture. The apparent molecular weight of a gas mixture is a pseudo property of the mixture and is defined as

$$M_a = \Sigma y_i M_i \tag{2.15}$$

where

M_a = apparent molecular weight of mixture
y_i = mole fraction of component i
M_i = molecular weight of component i

The gas laws can be applied to gas mixtures by simply using apparent molecular weight instead of the single-component molecular weight in the formulas.

EXAMPLE 2.8.
Apparent Molecular Weight of a Gas Mixture

Component	Mole Fraction y_i	Molecular Weight M_i	y_iM_i
Methane	0.937	16.043	15.032
Ethane	0.022	30.070	0.662
Propane	0.015	44.097	0.661
Nitrogen	0.026	28.013	0.728
	1.000		17.083

Therefore, the apparent molecular weight of the mixture is 17.08 lbm/lb-mol. Similarly, the apparent molecular weight of the gas mixture in Example 2.6 is 16.82 lbm/lb-mol.

2.6 BEHAVIOR OF REAL GASES

The ideal gas describes the behavior of most gases at pressure and temperature conditions close to atmospheric. Most natural gas engineers and operating personnel, at one time or another, become involved with the erratic behavior of natural gas under pressure. At moderate pressures, the gas tends to compress more than the ideal gas law indicates, particularly for temperatures close to the critical temperature. At high pressures the gas tends to compress less than the ideal gas law predicts. In most engineering problems the pressures of interest fall within the moderate range and the real gases are described as supercompressible.

To correct for the deviation between the measured or observed volume and that calculated using the ideal gas law, an empirical factor z, called the gas deviation factor or the z-factor, is used. In the literature, this factor is sometimes referred to as the compressibility factor, which can result in confusion with another gas property. In order to avoid ambiguity, this factor will be referred to as the gas deviation factor or z-factor throughout this text.

The gas deviation factor is defined as

$$z = \frac{\text{Actual volume of } n \text{ moles of gas at certain } p \text{ and } T}{\text{Ideal (calculated) volume of } n \text{ moles of gas at same } p \text{ and } T} \tag{2.16}$$

2.6.1 Real Gas Equation of State

All gases deviate from ideal gas laws under most conditions. Numerous attempts have been made to account for these deviations of a real gas from the ideal gas equation of state. One of the more celebrated of these is the equation of van der Waals. More recent and more successful equations of state have been derived, for instance, the Beattie–Bridgeman and Benedict–Webb–Rubin equations. But the real gas equation most commonly used in practice by the industry is

$$pV = znRT \tag{2.17}$$

The units are the same as listed for Eq. 2.8; z, which is dimensionless, is the gas deviation factor. The z-factor can be interpreted as a term by which the pressure must be corrected to account for the departure from the ideal gas equation by expressing Eq. 2.17 as

$$\left(\frac{p}{z}\right) V = nRT \tag{2.18}$$

Also, using the ideal gas law (Eq. 2.8) and the definition of z (Eq. 2.16),

$$z = \frac{V}{nRT/p}$$

or

$$pV = znRT \tag{2.17}$$

Equation 2.17 may be written, for a certain quantity of gas, as

$$\frac{p_1 V_1}{z_1 T_1} = \frac{p_2 V_2}{z_2 T_2} \tag{2.19}$$

where

z_1 = gas deviation factor under conditions 1, dimensionless
z_2 = gas deviation factor under conditions 2, dimensionless

The z-factor is a function of both absolute pressure and absolute temperature; but for this book's purposes, the main interest lies in determining z as a function of pressure at constant reservoir or transmission temperature. The $z(p)$ relationship obtained is then appropriate for the description of isothermal reservoir depletion or isothermal transmission problems.

Equation 2.17 may be written in terms of specific volume v or density ρ and gas gravity γ_g

$$pV = \frac{zmRT}{M} \tag{2.20}$$

$$pv = \frac{zRT}{M} \tag{2.21}$$

or

$$\rho = \frac{1}{v} = \frac{pM}{zRT} = \frac{2.7p\gamma_g}{zT} \tag{2.22}$$

where

v = specific volume, cu ft/lbm
m = mass of gas, lbm
M = molecular weight of gas, lbm/lb-mol
ρ = density of the gas, lbm/cu ft
γ_g = specific gravity of the gas (air = 1)

Comparing the density of a gas, at any pressure and temperature, to the density of air at the same conditions gives

$$\frac{\rho_{gas}}{\rho_{air}} = \frac{(M/z)_{gas}}{(M/z)_{air}}$$

And, in particular, at standard conditions:

$$\frac{\rho_{gas}}{\rho_{air}} = \gamma_g = \frac{M_{gas}}{M_{air}} = \frac{M}{29} \tag{2.23}$$

2.6.2 The Theorem of Corresponding States

Beginning with his thesis in 1873, J. D. van der Waals proposed his theorem of corresponding states. Before stating this theorem and discussing some of its uses, the following terms will be defined.

Critical pressure is that pressure which a gas exerts when in equilibrium with the liquid phase and at the critical temperature. It may also be defined as the saturation pressure corresponding to the critical temperature.

Critical temperature is that temperature (of a gas) above which a gas cannot be liquefied by the application of pressure alone, regardless of the amount of pressure.

Critical volume is the volume of 1 pound-mass of gas at the critical temperature and pressure; that is, the specific volume of the gas at critical temperature and critical pressure.

Reduced temperature, reduced pressure, and reduced volume are the ratios of the actual temperature, pressure, and specific volume to the critical temperature, critical pressure, and critical volume, respectively:

$$T_r = \frac{T}{T_c} \tag{2.24}$$

$$p_r = \frac{p}{p_c} \tag{2.25}$$

$$v_r = \frac{v}{v_c} \quad \text{or} \quad \rho_r = \frac{\rho}{\rho_c} \tag{2.26}$$

For any substance, the absolute magnitude of pressure or temperature is not really what counts in locating a state point, that is, in the liquid region or two-phase region or superheat region, etc. The value of the pressure and temperature relative to a corresponding critical value really counts. Thus, the reduced coordinates are most important. The physical characteristics of a substance are controlled by the relative nearness of any state point to the critical point. If the pressure relative to the critical pressure and the temperature relative to the critical temperature are the same for two different substances, then the substances are in corresponding states and any other property, like the density relative to the critical density, will be the same for both substances.

This is the theorem or principle of corresponding states. Stated in other words, the deviation of a real gas from the ideal gas law is the same for different gases at the same corresponding conditions of reduced temperature and reduced pressure. The theorem of corresponding states is accurate within several percent for widely dissimilar types of substances and much more accurate than this for restricted substances having physiochemically similar characteristics.

The theorem of corresponding states has many practical applications. The most common use is in evaluating the deviation of real gases from the equation of state for an ideal gas. This permits evaluation of other thermodynamic property deviations from ideal gas relationships. Equation 2.21 can be expressed in the form

$$p/p_c = (z/z_c)(v_c/v)(T/T_c) = (z/z_c)(\rho/\rho_c)(T/T_c)$$

or

$$p_r = \rho_r T_r (z/z_c) \tag{2.27}$$

By the theorem of corresponding states, reduced pressure and temperature define the reduced density, or

$$\rho_r = f(p_r, T_r) \tag{2.28}$$

Simultaneous solution of Eqs. 2.27 and 2.28 yields the general principle

$$z/z_c = f(p_r, T_r) \tag{2.29}$$

If the form of the algebraic function f in Eq. 2.28 is known, then z_c can be defined. Using Eq. 2.29, the z-factor for any gas mixture is defined solely by reduced temperature and reduced pressure:

$$z = f(p_r, T_r) \tag{2.30}$$

Reduced vapor pressure, reduced enthalpy, reduced entropy, etc. are examples of this theorem's uses for the purposes of generalizing results. The generalization of vapor pressure data permits description of such data for many gases on one chart. Also, the pressure-density-temperature relationship may be generalized for many different liquids by use of reduced coordinates.

2.6.3 Determination of z-factor

Experimental Determination

Some quantity of gas (n moles) is charged into a cylinder, the volume of which can be altered by the movement of a piston. The container is maintained at the desired temperature T throughout the experiment. If V_o is the gas volume at an atmospheric pressure of 14.7 psia then, applying the real gas law Eq. 2.17, 14.7 $V_o = nRT$ since $z \simeq 1$ at atmospheric pressure. At any higher pressure p for which the corresponding volume of the gas is V, $pV = ZnRT$. Dividing these equations gives $z = pV/14.7\ V_o$.

By varying p and measuring V, the isothermal $Z(p)$ function can be readily obtained. This is the most satisfactory method of determining the function; but in the majority of cases, the time and expense involved are not warranted since reliable methods of direct calculation are available.

The *z*-factor Correlation of Standing and Katz

In 1941, Standing and Katz presented a z-factor chart (Fig. 2.4) based on binary mixtures and saturated hydrocarbon vapor data. Figure 2.4 is a correlation of the z-factor as a function of reduced temperature and pressure. This chart is generally reliable for sweet and natural gases and correctable for those containing hydrogen sulfide and carbon dioxide. It has become one of the most widely accepted correlations in the petroleum industry. This correlation requires a knowledge of the composition of the gas or, at least, the gas gravity.

In order to use the Standing–Katz correlation, it is first necessary—from a knowledge of the gas composition—to determine the pseudo critical pressure and temperature, or the apparent molecular weight of the mixture. These pseudo-properties are given by Kay's mixing rules as

$$\text{Pseudocritical pressure, } p_{pc} = \sum_i y_i p_{ci} \tag{2.31}$$

$$\text{Pseudocritical temperature, } T_{pc} = \sum_i y_i T_{ci} \tag{2.32}$$

$$\text{Apparent molecular weight, } M_a = \sum_i y_i M_i \tag{2.15}$$

where

y_i = mole fraction of component i in gaseous state
p_{ci} = pseudocritical pressure of component i
T_{ci} = pseudocritical temperature of component i

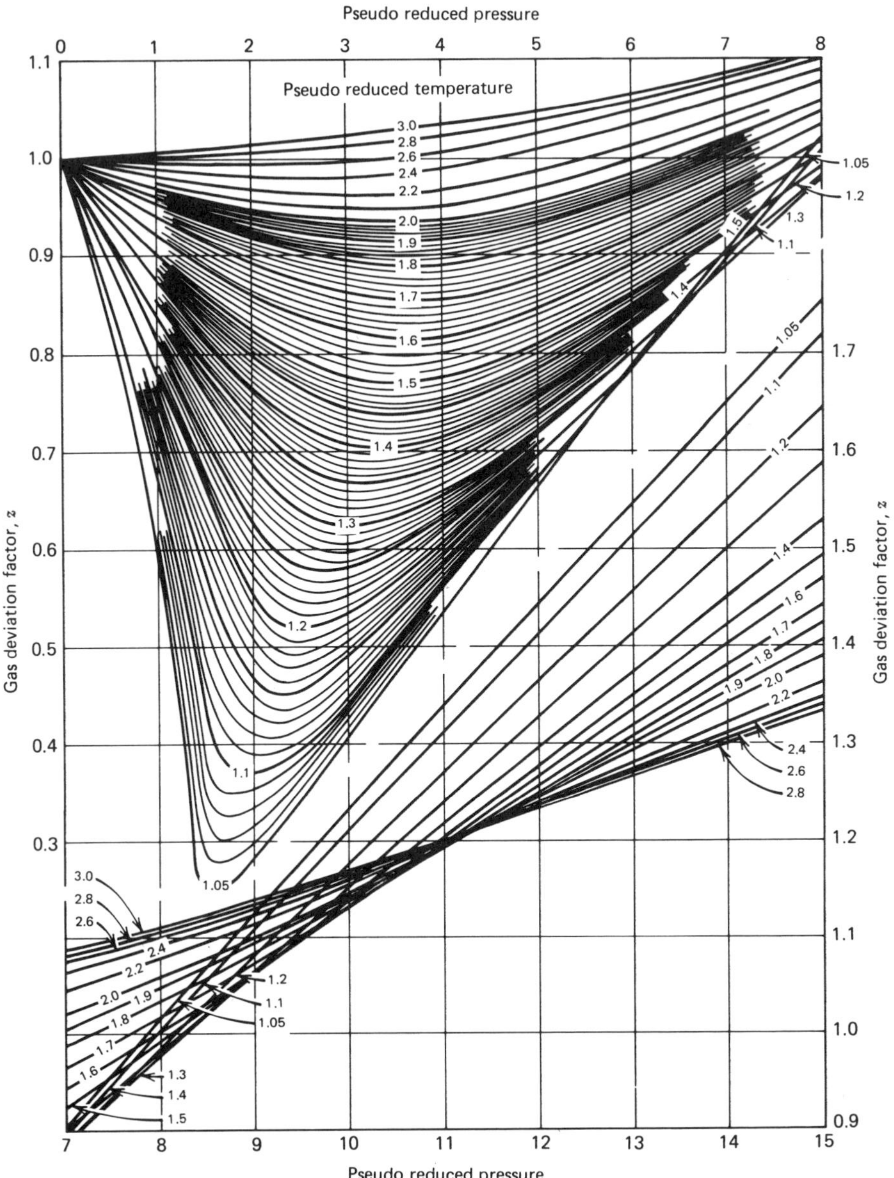

Fig. 2.4 Gas deviation factor for natural gases. (After Standing and Katz.)

In cases where the composition of a natural gas is not available, the pseudocritical pressure and pseudocritical temperature may be approximated from Fig. 2.5 and a knowledge of the gas gravity (Brown et al., Carr et al.). Useful correlations derived from Fig. 2.5 and more recent data from other sources by Thomas, Hankinson, and Phillips are:

$$p_{pc} = 709.604 - 58.718\ \gamma_g \tag{2.33}$$

$$T_{pc} = 170.491 + 307.344\ \gamma_g \tag{2.34}$$

The allowable concentrations of sour gases and other nonhydrocarbons for the above equations are 3% H_2S, 5% N_2, or a total impurity content of 7%.

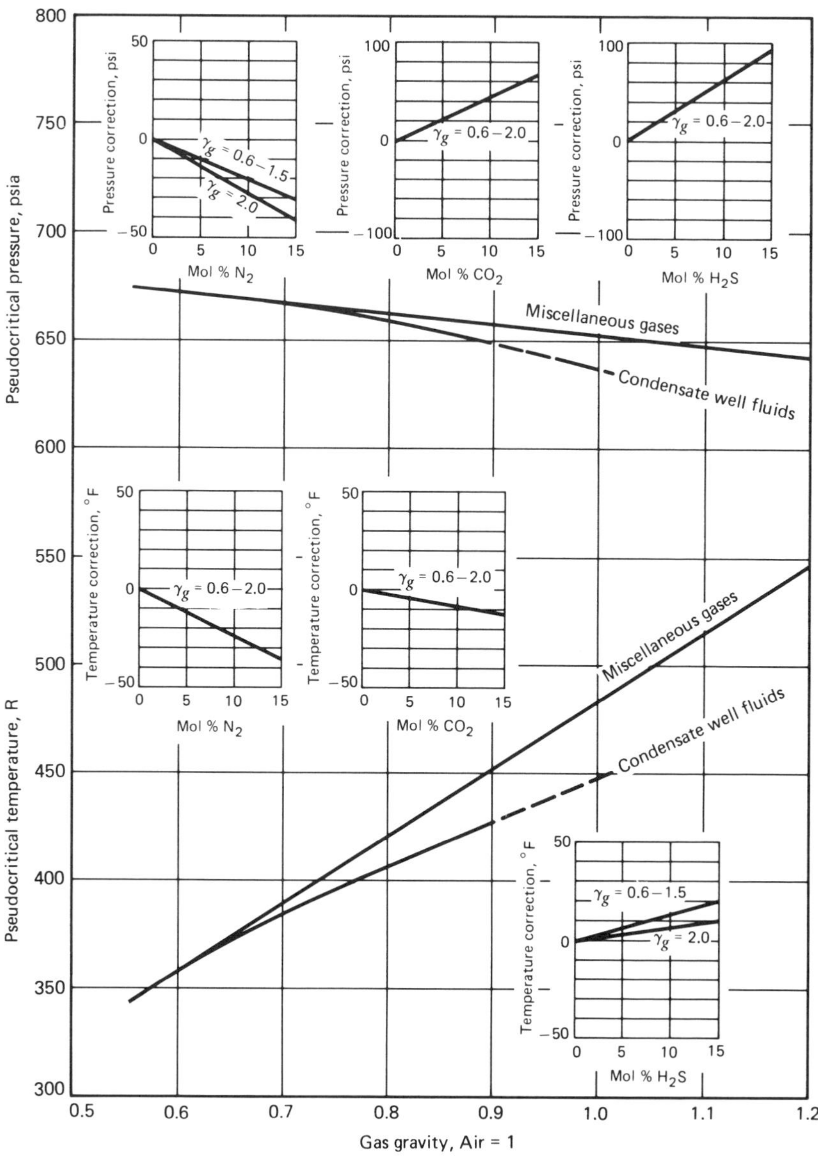

Fig. 2.5 Pseudocritical properties of miscellaneous natural gases. (After Brown et al.)

The next step is to calculate the pseudoreduced pressure and temperature:

$$p_{pr} = p/p_{pc} \tag{2.35}$$

$$T_{pr} = T/T_{pc} \tag{2.36}$$

where p and T are the absolute pressure and absolute temperature at which z-factor is required. With these two reduced parameters the Standing–Katz correlation, which consists of a set of isotherms giving z as a function of the pseudoreduced pressure, can be used to determine the z-factor.

EXAMPLE 2.9 (sweet natural gas)

Component	y_i	M_i	y_iM_i	p_{ci} psia	y_ip_{ci}	T_{ci} R	y_iT_{ci}
N_2	0.0345	28.01	0.9665	493	17.009	227	7.832
CO_2	0.0130	44.01	0.572	1071	13.923	548	7.124
H_2S	0.0000	34.08	0.000	1306	0.000	672	0.000
CH_4	0.8470	16.04	13.586	668	565.796	343	290.521
C_2H_6	0.0586	30.07	1.762	708	41.489	550	32.230
C_3H_8	0.0220	44.10	0.970	616	13.552	666	14.652
i-C_4H_{10}	0.0035	58.12	0.203	529	1.852	735	2.573
n-C_4H_{10}	0.0058	58.12	0.337	551	3.196	765	4.437
i-C_5H_{12}	0.0027	72.15	0.195	490	1.323	829	2.238
n-C_5H_{12}	0.0025	72.15	0.180	489	1.223	845	2.113
n-C_6H_{14}	0.0028	86.18	0.241	437	1.224	913	2.556
n-C_7H_{16}	0.0028	100.20	0.281	397	1.112	972	2.722
n-C_8H_{18}	0.0015	114.23	0.171	361	0.542	1024	1.536
n-C_9H_{20}	0.0018	128.26	0.231	332	0.598	1070	1.926
n-$C_{10}H_{22}$	0.0015	142.29	0.213	304	0.456	1112	1.668
	1.000		19.91		663.29		374.13

$$M_a = 19.91$$

$$\gamma_g = \frac{M_a}{29} = 0.686 \text{ (air = 1)}$$

$$p_{pc} = 663.3 \text{ psia}$$

$$T_{pc} = 374.1\ °\text{R}$$

At a pressure of 2000 psia and temperature of 150°F,

$$p_{pr} = \frac{2000}{663.3} = 3.02$$

$$T_{pr} = \frac{460 + 150}{374.1} = \frac{610}{374.1} = 1.63$$

Using the z-factor chart, Fig. 2.4,

$$z = 0.835$$

For comparison, from Fig. 2.5 for $\gamma_g = 0.686$, $p_{pc} = 667$ psia, $T_{pc} = 376$ °R; and using Eqs. 2.33 and 2.34, $p_{pc} = 667$ psia, $T_{pc} = 381$ °R.

The Standing–Katz z-factor chart is generally reliable for sweet natural gases with small amounts of nonhydrocarbon components, say, less than 5% by volume. For sour natural gases, the Standing–Katz z-factor chart may be used with appropriate adjustment of the pseudocritical temperature and pseudocritical pressure. The pseudocritical temperature adjustment factor ε_3 is given by

$$\varepsilon_3 = 120(A^{0.9} - A^{1.6}) + 15(B^{0.5} - B^{4.0}) \tag{2.37}$$

where

A = sum of the mole fractions of H_2S and CO_2
B = mole fraction of H_2S

Wichert and Aziz have developed a chart (Fig. 2.6) that gives values of ε_3 to be used in adjusting the pseudocritical temperatures and pressure as follows:

$$T'_{pc} = T_{pc} - \varepsilon_3 \tag{2.38}$$

$$p'_{pc} = \frac{p_{pc}T'_{pc}}{T_{pc} + B(1 - B)\varepsilon_3} \tag{2.39}$$

where

T'_{pc} = adjusted pseudocritical temperature, °R
p'_{pc} = adjusted pseudocritical pressure, psia

Using the adjusted pseudocritical temperature and pressure, the reduced temperature and reduced pressure are calculated for use in Fig. 2.4 to predict sour gas z-factors.

EXAMPLE 2.10 (sour natural gas)

Component	y_i	M_i	y_iM_i	p_{ci} psia	y_ip_{ci}	T_{ci} R	y_iT_{ci}
N_2	0.0130	28.01	0.364	493	6.409	227	2.951
CO_2	0.0164	44.01	0.722	1071	17.564	548	8.987
H_2S	0.1841	34.08	6.274	1306	240.435	672	123.715
CH_2	0.7800	16.04	12.511	668	521.040	343	267.540
C_2H_6	0.0043	30.07	0.129	708	3.044	550	2.365
C_3H_8	0.0007	44.10	0.031	616	0.431	666	0.466
i-C_4H_{10}	0.0005	58.12	0.029	529	0.265	735	0.368
n-C_4H_{10}	0.0003	58.12	0.017	551	0.165	765	0.230
i-C_5H_{12}	0.0001	72.15	0.007	490	0.049	829	0.083
n-C_5H_{12}	0.0001	72.15	0.007	489	0.049	845	0.085
C_6H_{14}	0.0001	86.18	0.009	437	0.044	913	0.091
C_7H_{16}+	0.0004	114.23	0.046	361	0.144	1024	0.410
	1.0000		20.15		789.64		407.29

$$M_a = 20.15$$
$$\gamma_g = \frac{M_a}{29} = 0.695$$
$$T_{pc} = 407.3 \text{ °R}$$
$$p_{pc} = 789.6 \text{ psia}$$

From Fig. 2.6 for 18.41% H_2S and 1.64% CO_2,

$$\varepsilon_3 = 25.5 \text{ °R}$$

From Eq. 2.38,

$$T'_{pc} = 407.3 - 25.5 = 381.8 \text{ °R}$$

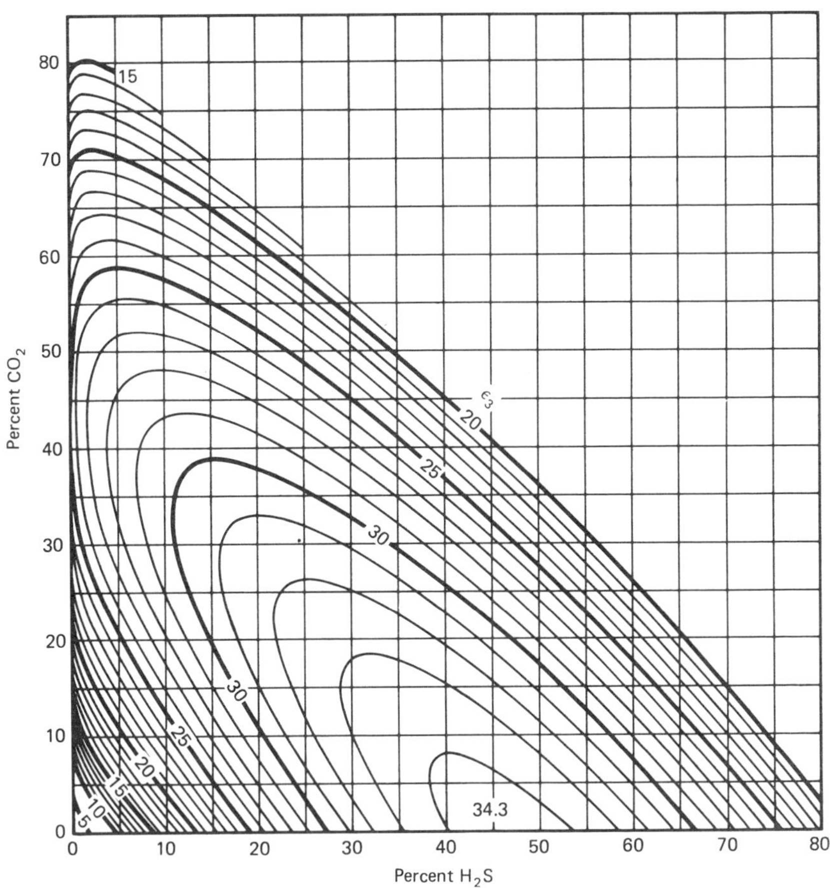

Fig. 2.6 Pseudocritical temperature adjustment factor, ε_3, °R. (After Wichert and Aziz.)

Using Eq. 2.39,

$$p'_{pc} = \frac{(789.6)(381.8)}{(407.3) + (0.1841)(1 - 0.1841)(25.5)} = 733.3 \text{ psia}$$

The reduced properties at 180°F and 2500 psia become

$$T_{pr} = \frac{180 + 460}{381.8} = 1.68$$

$$p_{pr} = \frac{2500}{733.5} = 3.41$$

Using these values in Fig. 2.4 gives a gas deviation factor $z = 0.850$.

In using the mole fraction analysis outlined above, the C_7+ fraction is generally considered as a single component. In Example 2.10, the physical properties of octane (C_8H_{18}) are used for the C_7+ fraction. This is a common practice. If the molecular weight and specific gravity of the C_7+ fraction are known, the pseudocritical temperature and pseudocritical pressure of this fraction may be obtained by Fig. 2.7.

Figures A.1 to A.9 in the Appendix give plots of z-factor versus pressure for methane and gases of various gravities. These may be used instead of the Standing and Katz correlation (Fig. 2.4).

Direct Calculation of *z*-factors

The Standing–Katz z-factor correlation is very reliable and has been used with confidence by the industry for more than 35 years. With the advent of computers, however, the need arose to find some convenient technique for calculating z-factors for use in natural gas engineering programs, rather than feeding in the entire chart from which z-factors could be retrieved by table lookup. Takacs (1976) has compared eight different methods for calculating z-factors, which have been developed over the years. These fall into two main categories: those that attempt to curve-fit the Standing–Katz isotherms analytically and those that compute z-factors using an equation of state. Some of these are given below.

1. The Hall–Yarborough Method (1973) The Hall–Yarborough equations, developed using the Starling–Carnahan equation of state, are

$$z = \frac{0.06125 p_{pr} t e^{-1.2(1-t)^2}}{y} \tag{2.40}$$

where

P_{pr} = the pseudoreduced pressure
t = the reciprocal, pseudoreduced temperature (T_{pc}/T)
y = the reduced density, which can be obtained as the solution of the equation

$$F = -0.06125p_{pr}te^{-1.2(1-t)^2} + \frac{y + y^2 + y^3 - y^4}{(1-y)^3} - (14.76t - 9.76t^2 + 4.58t^3)y^2 + (90.7t - 242.2t^2 + 42.4t^3)y^{(2.18+2.82t)} = 0 \quad (2.41)$$

The nonlinear equation (Eq. 2.41) can be solved for y using Newton–Raphson iterative technique. Substitution of the correct value of y in Eq. 2.40 will give the z-factor.

2. *Dranchuk, Purvis and Robinson Method (1974)* This method fits the Standing–Katz z-factor correlation by means of an eight-coefficient Benedict–Webb–Rubin type equation of state. The z-factor equation is

$$z = 1 + (A_1 + A_2/T_r + A_3/T_r^3)\rho_r + (A_4 + A_5/T_r)\rho_r^2 + A_5A_6\rho_r^5/T_r + A_7\rho_r^2/T_r^3(1 + A_8\rho_r^2)\exp(-A_8\rho_r^2) \quad (2.42)$$

where:

$\rho_r = 0.27p_r/(zt_r)$

$A_1 = 0.31506237 \qquad A_2 = -1.046,709,90 \qquad A_3 = -0.578,327,29$

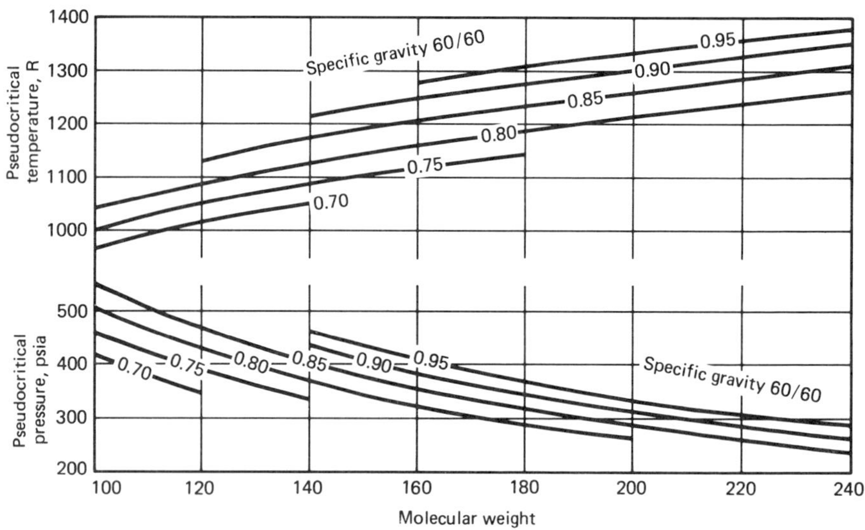

Fig. 2.7 Correlation charts for estimation of the pseudocritical temperature and pressure of heptanes plus fractions from molecular weight and specific gravity. (Courtesy Mathews, Roland, and Katz.)

$$A_4 = 0.53530771 \quad A_5 = -0.612,320,32 \quad A_6 = -0.104,888,13$$
$$A_7 = 0.68157001 \quad A_8 = 0.684,465,49 \tag{2.43}$$

3. Gopal Method (1977) This method fits straightline equations to different portions of the z-factor chart. It uses a general equation of the form

$$z = p_r(AT_r + B) + CT_r + D \tag{2.44}$$

The values of constants A, B, C, and D for various combinations of p_r and T_r are shown in Table 2.4. Note that above p_r of 5.4, an equation of a different form is used.

TABLE 2.4

Equations for z-factors

Reduced Pressure P_r Range Between	Reduced Temperature, T_r Range Between	Equations	Equation Number
0.2 and 1.2	1.05 and 1.2	$P_r(1.6643T_r - 2.2114) - 0.3647T_r + 1.4385$	1
	1.2+ and 1.4	$P_r(0.5222T_r - 0.8511) - 0.0364T_r + 1.0490$	2
	1.4+ and 2.0	$P_r(0.1391T_r - 0.2988) + 0.0007T_r^{a} + 0.9969$	3[b]
	2.0+ and 3.0	$P_r(0.0295T_r - 0.0825) + 0.0009T_r^{a} + 0.9967$	4[b]
1.2+ and 2.8	1.05 and 1.2	$P_r(-1.3570T_r + 1.4942) + 4.6315T_r - 4.7009$	5[c]
	1.2+ and 1.4	$P_r(0.1717T_r - 0.3232) + 0.5869T_r + 0.1229$	6
	1.4+ and 2.0	$P_r(0.0984T_r - 0.2053) + 0.0621T_r + 0.8580$	7
	2.0+ and 3.0	$P_r(0.0211T_r - 0.0527) + 0.0127T_r + 0.9549$	8
2.8+ and 5.4	1.05 and 1.2	$P_r(-0.3278T_r + 0.4752) + 1.8223T_r - 1.9036$	9[b]
	1.2+ and 1.4	$P_r(-0.2521T_r + 0.3871) + 1.6087T_r - 1.6635$	10[b]
	1.4+ and 2.0	$P_r(-0.0284T_r + 0.0625) + 0.4714T_r - 0.0011^{a}$	11
	2.0+ and 3.0	$P_r(0.0041T_r + 0.0039) + 0.0607T_r + 0.7927$	12
5.4+ and 15.0	1.05 and 3.0	$P_r(0.711 + 3.66T_r)^{-1.4667} - 1.637/(0.319\,T_r + 0.522) + 2.071$	13

[a]These terms may be ignored.
[b]For a very slight loss in accuracy, Eqs. 3 and 4 and 9 and 10 can, respectively, be replaced by the following two equations:
$z = P_r(0.0657T_r - 0.1751) + 0.0009\,T_r^{*} + 0.9968$
$z = P_r(-0.2384\,T_r + 0.3695) + 1.4517\,T_r - 1.4580$
[c]Preferably use this equation for P_r up to 2.6 only. For $P_r = 2.6+$, Eq. 9 will give slightly better results. Also, preferably,use Eq. 1 for $1.08 \leqq T_r \leqq 1.19$ and $P_r \leqq 1.4$.

2.6.4 Supercompressibility

Sometimes, another term is used by engineers to compensate for the deviation from the ideal gas law. This term is the supercompressibility factor F_{pv} and is used primarily for high-pressure calculations. A mathematical relationship exists between supercompressibility and gas deviation factor, as follows:

$$z = \frac{1}{(F_{pv})^2} \tag{2.45}$$

The equation of state for real gases can now be written in terms of the supercompressibility factor as

$$pV = \frac{nRT}{(F_{pv})^2} \tag{2.46}$$

Tables of supercompressibilities for natural gases are available from a number of sources (see Chapter 6) and can be readily used with the above equation to predict PVT relationships.

2.6.5 Other Equations of State

An equation of state gives the relationship between the pressure p, molar volume v, and absolute temperature T of a fluid. The functional form may be written as

$$f(p,v,T) = 0 \tag{2.47}$$

Alternatively, Eq. 2.47 may be written explicitly in terms of p:

$$p = \phi(v,T) \tag{2.48}$$

or explicitly in terms of v or T. The ideal gas, and one form of real gas, equations of state have been discussed earlier, Eqs. 2.8 and 2.17, respectively.

The gas deviation factor in the real gas equation of state presented earlier, Eq. 2.17, is not a constant, and needs to be obtained by graphical or numerical techniques. This is a major limitation, which makes direct mathematical manipulations difficult. Many other equations of state have been developed with a constant correction factor thus permitting direct mathematical manipulations like differentiation and integration.

Van der Waals' Equation of State

This is an attempt to modify the ideal gas law so that it will be applicable to nonideal gases. For *one mole* of a pure gas, Van der Waals' equation of state is written:

$$p = \frac{RT}{v - b} - \frac{a}{v^2} \tag{2.49}$$

where a and b are constants. Table 2.5 gives values of a and b for common substances. The quantity a/v^2 accounts for the attractive forces between the molecules. Actual external pressure would need to be larger to produce the same volume than if no attraction existed. The constant b represents the volume of the molecules themselves. Actual volume available to the gas is less than overall volume of gas by the amount taken up by the molecules.

If n moles of gas are involved, Eq. 2.49 becomes

$$p = \frac{nRT}{v - nb} - \frac{n^2 a}{v^2} \tag{2.50}$$

When V is large (low pressure and high temperature), van der Waals' equation reduces to the ideal gas law. If the constants a and b are not known, it is possible to estimate their values from values of critical pressure P_c and critical volume V_c using

$$a = 3p_c V_c^2 \tag{2.51}$$

$$b = \frac{V_c}{3} \tag{2.52}$$

Van der Waals' equation has limited application in engineering. It is accurate only at low pressures.

TABLE 2.5
Van der Waals Constants

Substance	*a* Liters-atm mole	*b* Liters/mole
Methane	2.253	0.04278
Ethane	5.489	0.06380
Propane	8.664	0.08445
Isobutane	12.87	0.1142
n-Butane	14.47	0.1226
Isopentane	18.05	0.1417
n-Pentane	19.01	0.1460
n-Hexane	24.39	0.1735
n-Heptane	31.51	0.2654
Nitrogen	1.390	0.03913
Carbon dioxide	3.592	0.04267
Hydrogen sulfide	4.431	0.04287
Helium	0.03412	0.02370
Water	5.464	0.03049
Hydrogen	0.2444	0.02661
Ethylene	4.471	0.05714
Propylene	8.379	0.08272

Benedict–Webb–Rubin (*B-W-R*) Equation of State

A powerful equation of state describing the behavior of pure, light hydrocarbons over single and two-phase regions, both below and above critical pressure, was developed by Benedict, Webb, and Rubin as early as 1940. The *B-W-R* equation of state is a modification of the Beattie–Bridgeman equation of state and gives better *PVT* predictions than the Beattie–Bridgeman equation. Benedict et al., in a series of articles in 1951, showed the accuracy of the *B-W-R* equation when applied to light hydrocarbons and their simple mixtures and demonstrated the equation's utility for making vapor-liquid equilibrium calculations. Since natural gases are primarily mixtures of light hydrocarbons, the *B-W-R* has been used quite often in computing thermodynamic properties and phase equilibria of natural gases.

The *B-W-R* equation expressing pressure as a function of temperature and molar density is

$$\begin{aligned} p = RTP_m &+ (B_oRT - A_o - C_o/T^2)P_m^2 \\ &+ (bRT - a)P_m^3 + a\alpha P_m^6 \\ &+ \frac{cp_m^3}{T^2}\left[(1 + \gamma P_m^2)\exp(-\gamma P_m^2)\right] \end{aligned} \tag{2.53}$$

The parameters A_o, B_o, C_o, a, b, c, α, and γ are constants for pure substances and functions of composition for mixtures. The pure-component values of the eight constants are listed in Table 2.6. Benedict et al. proposed the following mixing rules to enable Eq. 2.53 be used for mixtures:

$$\begin{aligned} A_o &= (\Sigma y_i A_{oi}^{1/2})^2 \\ B_o &= \Sigma y_i B_{oi} \\ C_o &= (\Sigma y_i C_{oi}^{1/2})^2 \\ a &= (\Sigma y_i a_i^{1/3})^3 \\ b &= (\Sigma y_i b_i^{1/3})^3 \\ c &= (\Sigma y_i c_i^{1/3})^3 \\ \alpha &= (\Sigma y_i \alpha_i^{1/3})^3 \\ \gamma &= (\Sigma y_i \gamma_i^{1/2})^2 \end{aligned} \tag{2.54}$$

P_m is the molar density in lb-mole/cu ft

Redlich–Kwong (*R-K*) Equation of State

The original Redlich–Kwong equation of state is

$$p = \frac{RT}{v - b} - \frac{a}{v(v + b)T^{0.5}} \tag{2.55}$$

where

$$a = \frac{C_a R^2 T_c^{2.5}}{p_c} \tag{2.56}$$

TABLE 2.6
Benedict–Webb–Rubin Constants

Substance	A_o	B_o	$C_o \times 10^{-6}$	a	b	$c \times 10^{-6}$	$\alpha \times 10^3$	$\gamma \times 10^2$
Methane	6,995.25	0.682,401	275.763	2,984.12	0.867,325	498.106	511.172	153.961
Ethane	15,670.7	1.005,54	2,194.27	20,850.2	2.853,93	6,413.14	1,000.44	302.790
Propane	25,915.4	1.558,84	6,209.93	57,248.0	5.773,55	25,247.8	2,495.77	564.524
Isobutane	38,587.4	2.203,29	10,384.7	117,047	10.889,0	55,977.7	4,414.96	872.447
n-Butane	38,029.6	1.992,11	12,130.5	113,705	10.263,6	61,925.6	4,526.93	872.447
Isopentane	4,825.36	2.563,86	21,336.7	226,902	17.144,1	136,025	6,987.77	1,188.07
n-Pentane	45,928.8	2.510,96	25,917.2	246,148	17.144,1	161,306	7,439.92	1,218.86
n-Hexane	5,443.4	2.848,35	40,556.2	429,901	28,003,2	296,077	11,553.9	1,711.15
n-Heptane	66,070.6	3.187,82	57,984.0	626,106	38,991,7	483,427	16,905.6	2,309.42
Nitrogen								
Carbon dioxide								
Hydrogen sulfide								
Helium								
Water								
Hydrogen								
Ethylene	12,593.6	0.891,980	1,602.28	15,645.5	2.206,78	4,133.60	731.661	236.844
Propylene	23,049.2	1.362,63	5,365.97	46,758.6	4.799,97	20,083.0	1,873.12	469.325

$$b = \frac{C_b R T_c}{p_c} \tag{2.57}$$

For temperature in °R and pressure in psia,

$$C_a = 0.427{,}47$$

$$C_b = 0.086{,}64$$

The *R-K* equation involves only two empirical constants, as opposed to the eight required in the *B-W-R* equation.

However, to simplify calculations with the *R-K* equation, especially for application to mixtures, other constants have been defined:

$$A' = \left(\frac{a}{R^2 T^{2.5}}\right)^{0.5} = \left(\frac{C_a T_c^{2.5}}{p_c T^{2.5}}\right)^{0.5} \tag{2.58}$$

$$B' = \frac{b}{RT} = \frac{C_b T_c}{p_c T} \tag{2.59}$$

The original *R-K* equation was modified by Soave in 1972 by replacing the term $1/T^{1/2}$ with a more general temperature dependent term α:

$$p = \frac{RT}{v - b} - \frac{a\alpha}{v(v + b)} \tag{2.60}$$

A gas deviation factor z can be calculated by writing Eq. 2.60 as

$$z^3 - z^2 + z(A - B - B^2) - AB = 0 \tag{2.61}$$

where, for pure substances,

$$A = \frac{C_a \alpha_i p T_{ci}^2}{p_{ci} T^2} \tag{2.62}$$

$$B = \frac{C_b p T_{ci}}{p_{ci} T} \tag{2.63}$$

For mixtures, the constants A and B are calculated using

$$A = C_a \frac{p}{T^2}\left(\Sigma y_i \frac{T_{ci}\alpha_i^{0.5}}{p_{ci}^{0.5}}\right)^2 \tag{2.64}$$

$$B = C_b \frac{p}{T} \Sigma y_i \frac{T_{ci}}{p_{ci}} \tag{2.65}$$

Evaluation of α_i involves calculation of the acentric factor, ω_i:

$$\omega_i = C_a T_{Bi} \left(\frac{\log p_{ci} - 1.167}{T_{ci} - T_{Bi}} \right) - 1.0 \tag{2.66}$$

where T_{Bi} is the boiling point in °R of component i at 14.7 psia. Values of ω_i may be obtained from Table A.1. Then,

$$m_i = 0.480 + 1.57\omega_i - 0.176\omega_i^2 \tag{2.67}$$

and

$$\alpha_i = [1 + m_i(1 - T_R^{0.5})]^2 \tag{2.68}$$

2.7 COMPRESSIBILITY OF NATURAL GASES

The coefficient of isothermal compressibility of a gas is given by

$$c_g = \frac{1}{V}\left(\frac{\partial V}{\partial p}\right)_T \tag{2.69}$$

For an ideal gas,

$$V = \frac{nRT}{p}$$

and

$$\left(\frac{\partial V}{\partial p}\right)_T = -\frac{nRT}{p^2}$$

Thus,

$$c_g = \left(-\frac{p}{nRT}\right)\left(-\frac{nRT}{p^2}\right) = \frac{1}{p} \tag{2.70}$$

For a real gas,

$$V = \frac{nRTz}{p}$$

and

$$\left(\frac{\partial V}{\partial p}\right)_T = nRT\left(\frac{1}{p}\frac{\partial z}{\partial p} - \frac{z}{p^2}\right)$$

Thus,

$$c_g = \left(-\frac{p}{nRTz}\right)\left[nRT\left(\frac{1}{p}\frac{\partial z}{\partial p} - \frac{z}{p^2}\right)\right]$$

or

$$c_g = \frac{1}{p} - \frac{1}{z}\frac{\partial z}{\partial p} \tag{2.71}$$

In terms of reduced variables,

$$\frac{\partial z}{\partial p} = \left(\frac{\partial p_{pr}}{\partial p}\right)\left(\frac{\partial z}{\partial p_{pr}}\right) = \frac{1}{p_{pc}}\left(\frac{\partial z}{\partial p_{pr}}\right)$$

Thus,

$$c_g = \frac{1}{p_{pc}p_{pr}} - \frac{1}{zp_{pc}}\left(\frac{\partial z}{\partial p_{pr}}\right)_{T_{pr}} \tag{2.72}$$

or

$$c_{pr} = c_g p_{pc} = \frac{1}{p_{pr}} - \frac{1}{z}\left(\frac{\partial z}{\partial p_{pr}}\right)_{T_{pr}} \tag{2.73}$$

where $c_{pr} = c_g p_{pc}$ = pseudoreduced compressibility. Using Eq. 2.44,

$$\left(\frac{\partial z}{\partial p_{pr}}\right)_{T_{pr}} = AT_{pr} + B \tag{2.74}$$

and

$$c_{pr} = \frac{1}{p_{pr}} - \frac{1}{z}(AT_{pr} + B) \tag{2.75}$$

Using Eqs. 2.42 and 2.43 along with Eq. 2.71, Mattar, Brar, and Aziz have obtained an expression for c_{pr} given by

$$c_{pr} = \frac{1}{p_{pr}} - \frac{0.27}{z^2 T_{pr}}\left[\frac{(\partial z/\partial \rho_r)_{T_{pr}}}{1 + \rho_r/z(\partial z/\partial \rho_r)_{T_{pr}}}\right] \tag{2.76}$$

where

$$\left(\frac{\partial z}{\partial \rho_r}\right)_{T_{pr}} = (A_1 + A_2/T_{pr} + A_3/T_{pr}^3) + 2(A_4 + A_5/T_{pr})\,\rho_r + 5A_5\,A_6\,\rho_r^4/T_{pr} + \frac{2A_7\rho_r}{T_{pr}^3}(1 + A_8\,\rho_r^2 - A_8^2\,\rho_r^4)\exp(-A_8\,\rho_r^2) \tag{2.77}$$

Eqs. 2.76 and 2.77 can be used directly in computer calculations. These equations were used to develop Figs. 2.8 and 2.9 for manual calculations of the coefficient of isothermal compressibility.

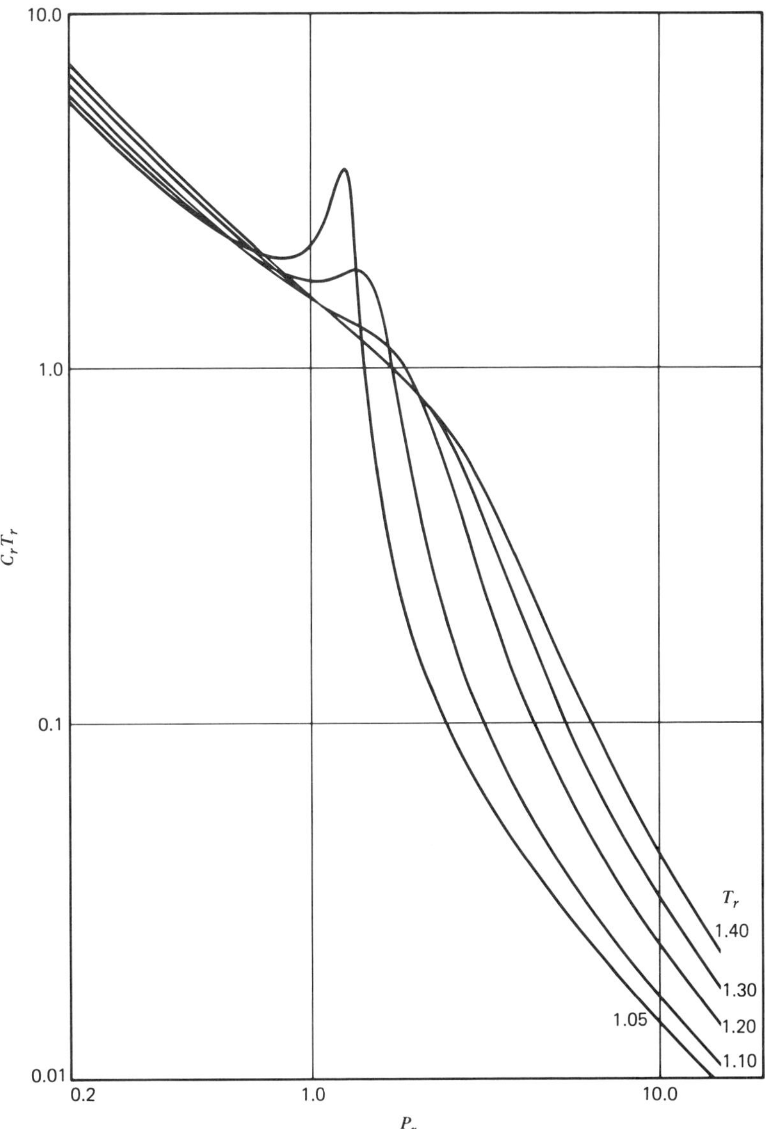

Fig. 2.8 Variation of c_rT_r with reduced temperature and pressure ($1.05 \leq T_r \leq 1.4$; $0.2 \leq p_r \leq 15.0$). (After Mattar, Brar, and Aziz.)

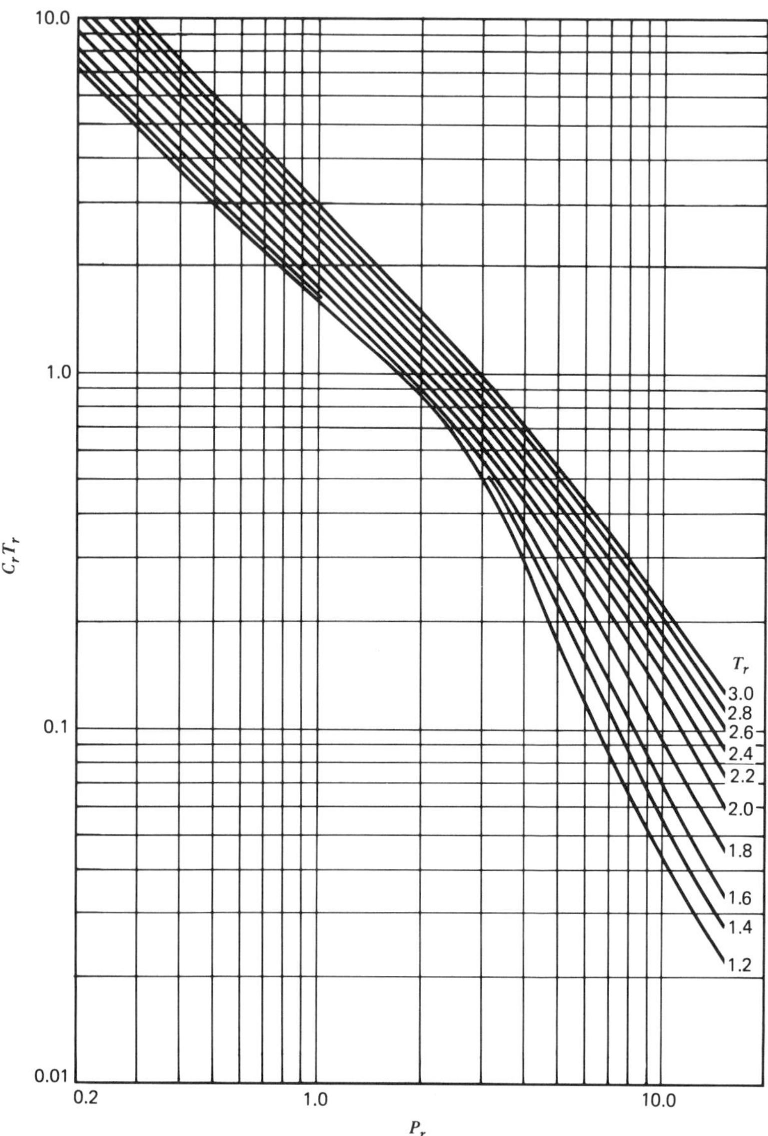

Fig. 2.9 Variation of c_rT_r with reduced temperature and pressure ($1.4 \leqslant T_r \leqslant 3.0$; $0.2 \leqslant p_r \leqslant 15.0$). (After Mattar, Brar, and Aziz.)

Example 2.11.

(a) For the sweet natural gas example (Example 2.9), at 2000 psia and 150°F,

$$T_{pc} = 374.1\ °\text{R}, \qquad T_{pr} = 1.63$$
$$p_{pc} = 663.3\ \text{psia}, \qquad p_{pr} = 3.02$$

From Fig. 2.9, at $T_{pr} = 1.63$ and $p_{pr} = 3.02$,

$$c_{pr}T_{pr} = 0.57$$
$$c_{pr} = 0.57/1.63 = 0.35$$
$$c_g = c_{pr}/p_{pc} = \frac{0.35}{663.3} = 0.000527 \text{ psia}^{-1}$$

(b) For the sour gas example (Example 2.10), using pseudocritical variables that have been adjusted by the method of Wichert and Aziz,

$$T'_{pc} = 381.8\ ^\circ\text{R}$$
$$p'_{pc} = 733.5 \text{ psia}$$

At 60°F and 4000 psia,

$$T_{pr} = 1.36 \text{ and } p_{pr} = 5.45$$

From Fig. 2.8, $c_{pr}T_{pr} = 0.13$:

$$c_{pr} = \frac{0.13}{1.36} = 0.096$$

$$c_g = \frac{c_{pr}}{p_{pc}} = \frac{0.096}{733.5} = 0.000{,}131 \text{ psia}^{-1}$$

2.8 VISCOSITY OF NATURAL GASES

The coefficient of viscosity is a measure of the resistance to flow exerted by a fluid. The dynamic or absolute viscosity μ of a Newtonian fluid is defined as the ratio of the shear force per unit area to the local velocity gradient. Dynamic viscosity is usually given in units of centipoise equivalent to 1 g mass/100 sec cm. Note that 1 cp = 6.72×10^{-4} lbm/ft sec. The kinematic viscosity, not normally used, is defined as

$$\text{Kinematic viscosity, } \nu = \frac{\text{dynamic viscosity, } \mu}{\text{density, } \rho_g} \tag{2.78}$$

Kinematic viscosity is usually given in centistokes equivalent to cm^2/100 sec.

The only accurate way to obtain the viscosity of a gas is to determine it experimentally. However, experimental determination is difficult and slow. Usually, the petroleum engineer must rely on viscosity correlations.

The viscosity of a pure gas depends on the temperature and pressure, but for gas mixtures it is also a function of the composition of the mixture. The following equation may be used to calculate the viscosity of a mixture of gases when the analysis of the gas mixture is known and the viscosities of the components are known at the pressure and temperature of interest:

$$\mu_g = \frac{\Sigma(\mu_{gi} y_i \sqrt{M_i})}{\sum_i (y_i \sqrt{M_i})} \tag{2.79}$$

The use of this equation is somewhat tedious.

For natural gases, the widely used correlations of Carr, Kobayashi, and Burrows take the forms:

$$\mu_1 = f(M, T) \tag{2.80}$$

$$\frac{\mu}{\mu_1} = f(p_r, T_r) \tag{2.81}$$

where

μ_1 = low pressure or dilute-gas viscosity
μ = gas viscosity at high pressure

Figures 2.10, 2.11, and 2.12 give the Carr et al. correlations. Figure 2.10 provides a rapid and reliable method for obtaining the viscosities of mixtures of hydrocarbon gases at 1.0 atm of pressure, given a knowledge of the gravity and temperature of the gas. The inserts are corrections for the presence of nitrogen, carbon dioxide, or hydrogen sulfide. The effect of each of the nonhydrocarbon gases is to increase the viscosity of the gas mixture. In most instances, the viscosities must approach pressures far removed from 1.0 atm. The theorem of corresponding states has been used to develop the correlations given in Figs. 2.11 and 2.12. These figures give the viscosity ratio as a function of pseudoreduced temperature and pseudoreduced pressure.

Example 2.12. Suppose the viscosity of the sour gas (Example 2.10) is required at 180°F and 2500 psia:

$$M_a = 20.15$$
$$\gamma_g = 0.70$$
$$T'_{pc} = 381.8\ °\text{R}$$
$$p'_{pc} = 733.3 \text{ psia}$$

From Fig. 2.10, for $M_a = 20.15$ and $T = 180°\text{F}$,

$$\mu_1 = 0.0121 \text{ cp}$$

From the inserts on Fig. 2.10,

Correction for 1.30 mol % N_2 = 0.000,10 cp
Correction for 1.64 mol % CO_2 = 0.000,10 cp
Correction for 18.41 mole % H_2S = 0.000,40 cp
(extrapolated value)

$\therefore$ Corrected $\mu_1 = 0.0121 + (0.000{,}10 + 0.000{,}10 + 0.000{,}40) = 0.0127$ cp

$$T_{pr} = (180 + 460)/381.8 = 1.68$$
$$p_{pr} = 2500/733.3 = 3.41$$

From Fig. 2.11 or Fig. 2.12, for $T_{pr} = 1.68$ and $p_{pr} = 3.41$,

$$\mu/\mu_1 = 1.40$$
$$\therefore \mu = (1.40)(0.0127) = 0.0178 \text{ cp}$$

An analytical expression for the viscosity of natural gases was presented by Lee, Gonzalez, and Eakin. Their method does not include corrections for impurities, and values obtained would be correct for pure hydrocarbon gases. Their empirical equation for viscosity is

$$\mu_g = K \times 10^{-4} \exp (X\rho_g^y) \tag{2.82}$$

where

$$K = \frac{(9.4 + 0.02M)T^{1.5}}{209 + 19M + T} \tag{2.83}$$

$$X = 3.5 + \frac{986}{T} + 0.01M \tag{2.84}$$

$$y = 2.4 - 0.2X \tag{2.85}$$

In Eqs. 2.82 to 2.85, μ_g = cp, ρ_g = g/cm^3, M = molecular weight of gas, and T = °R.

2.9 GAS FORMATION VOLUME FACTOR AND EXPANSION FACTOR

In gas reservoir engineering, the main use of the real gas equation of state is to relate surface volumes to reservoir volumes of hydrocarbons. This is accomplished by the use of the gas formation volume factor B_g or the gas expansion factor E. The gas formation volume factor is the volume occupied in the reservoir by one standard cubic foot of gas or the ratio of the volume of gas in the reservoir to its volume at standard conditions. Although B_g is usually expressed in units of reservoir cubic feet per standard cubic foot, it is sometimes useful to express it in barrels per standard cubic foot. The gas expansion factor is simply the reciprocal of the gas formation volume factor.

From the definitions and Eq. 2.17,

$$B_g = \frac{V}{V_{sc}} = \frac{p_{sc}}{p} \times \frac{T}{T_{sc}} \times \frac{z}{z_{sc}} \tag{2.86}$$

and

$$E = \frac{1}{B_g} = \frac{V_{sc}}{V} = \frac{p}{p_{sc}} \times \frac{T_{sc}}{T} \times \frac{z_{sc}}{z} \tag{2.87}$$

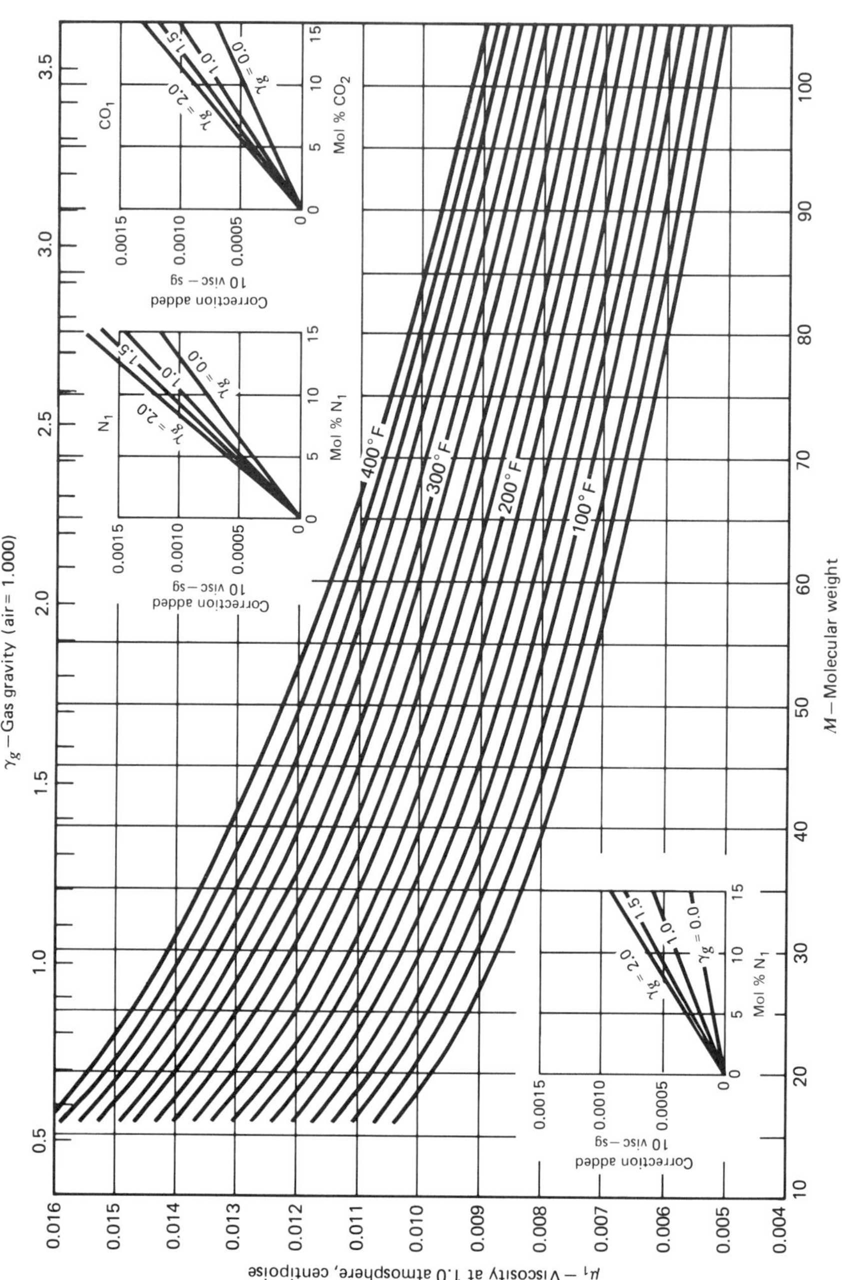

Fig. 2.10 Viscosity of paraffin hydrocarbon gases at 1.0 atm. (After Carr et al.)

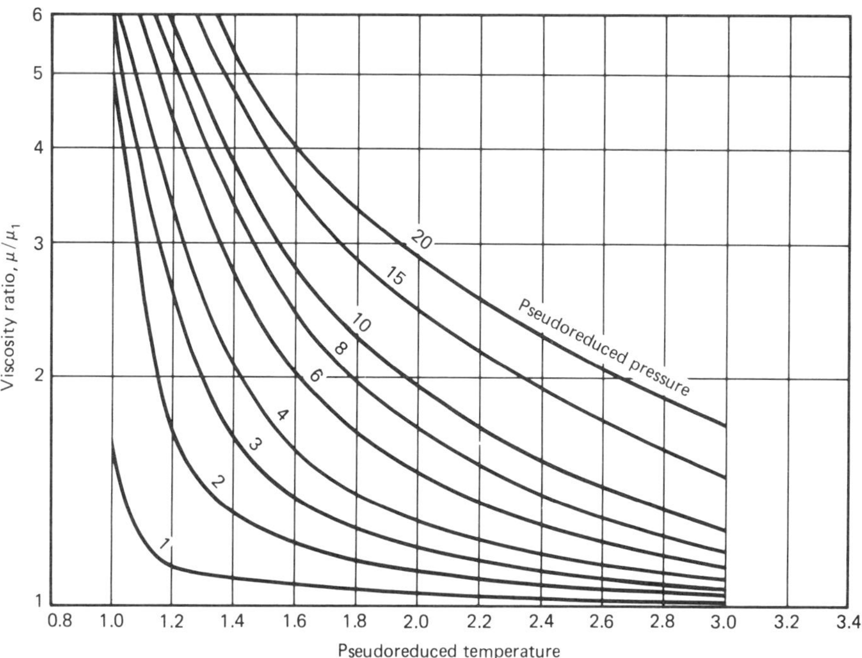

Fig. 2.11 Viscosity ratio versus pseudoreduced temperature. (After Carr et al.)

Thus, at standard conditions of 14.73 psia and 60°F, and assuming $z_{sc} \simeq 1$, Eqs. 2.86 and 2.87 become

$$B_g = 0.0283 \frac{zT}{p} \text{ cu ft/scf} \tag{2.88}$$

and

$$E = 35.30 \frac{p}{zT} \text{ scf/cu ft} \tag{2.89}$$

Dividing reservoir cubic feet by 5.615 to convert to reservoir barrels obtains

$$B_g = 0.005{,}04 \frac{zT}{p} \text{ bbl/scf} \tag{2.90}$$

and

$$E = 198.22 \frac{p}{zT} \text{ scf/bbl} \tag{2.91}$$

Example 2.13. At a pressure of 2500 psia and reservoir temperature of 180°F, the gas deviation factor for the sour natural gas (Example 2.10) is 0.850.

(a) Calculate the formation volume factor and gas expansion factor.

(b) How many standard cubic feet of this gas are contained in a reservoir with a gas pore volume of 1.0×10^9 cu ft?

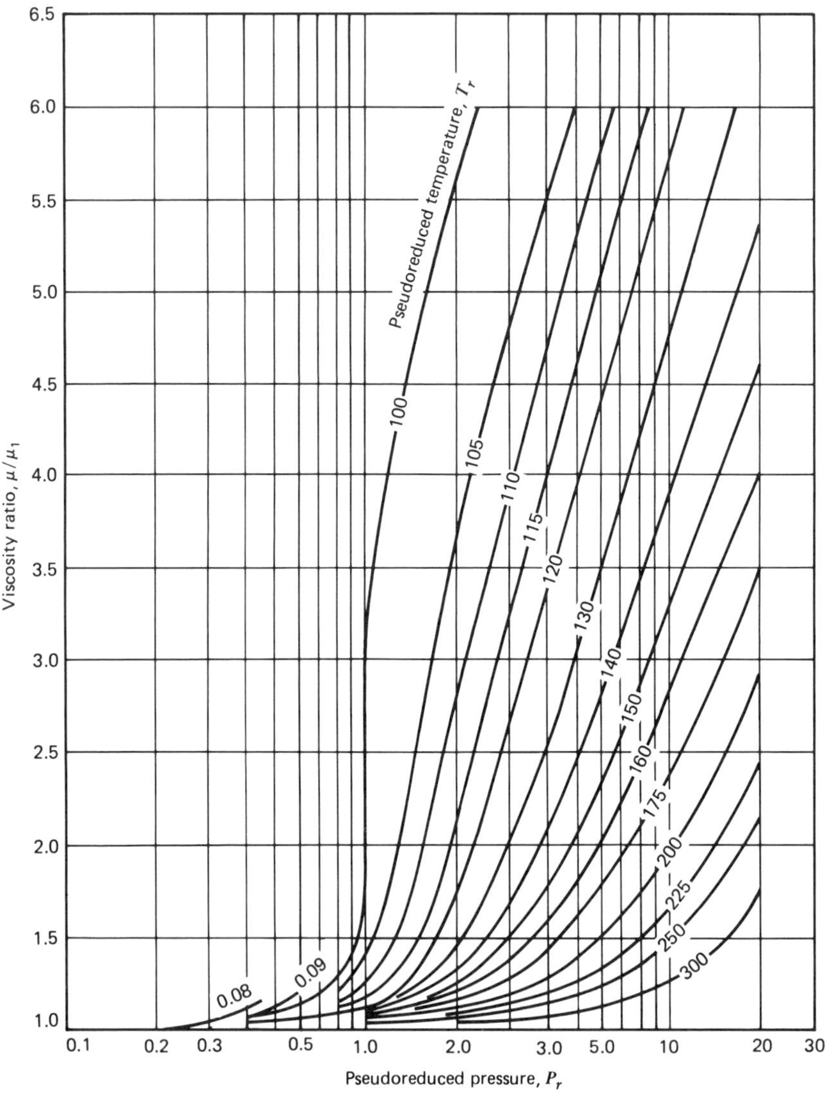

Fig. 2.12 Viscosity ratio versus pseudoreduced pressure. (After Carr et al.)

Solutions

(a) Using Eqs. 2.88 and 2.89,

$$B_g = \frac{(0.0283)(0.850)(640)}{(2500)} = 0.006,16 \text{ cu ft/scf}$$

$$E = \frac{1}{B_g} = 162.39 \text{ scf/cu ft}$$

(b)
$$\text{Gas in place} = \frac{1.0 \times 10^9 \text{ cu ft}}{B_g} = 1.0 \times 10^9 \text{ cu ft} \times \text{E}$$
$$= (1.0 \times 10^9 \text{ cu ft})(162.39 \text{ scf/cu ft})$$
$$= 162.4 \times 10^9 \text{ scf}$$

2.10 WATER VAPOR CONTENT OF NATURAL GAS

Water may be carried along with the gas in the vapor phase or entrained in the gas in droplet form. There exists at any given temperature and pressure a maximum amount of water vapor that a gas is able to hold. A gas is completely saturated when it contains the maximum amount of water vapor for the given temperature and pressure conditions. This saturation temperature at the specified pressure is the dew point of the gas. The water vapor content of natural gas is shown in Fig. 2.13.

It is easy to see from Fig. 2.13 that, keeping the volume and pressure constant on a water vapor-saturated gas, water will condense out at lower temperatures since the capacity of the gas to hold water is less. The same is true if the volume and temperature are kept constant but the pressure is allowed to increase. Keeping the volume and pressure constant on water-saturated gas but lowering the dew point temperature is known as dew point depression; this term indicates to what extent the moisture content of a gas may be lowered.

Example 2.14. A water-saturated gas at 80°F and 400 psia had a 70° dew point depression after passing through a dehydration plant. How many gallons of water were removed per million standard cubic feet (MMscf) of gas measured at 60°F and 14.7 psia?

Solution

From Fig. 2.13,

Water content at 80°F and 400 psia = 68 lbm/MMscf
New temperature = (80 − 70) = 10°F
Water content at 10°F and 400 psia = 5.8 lbm/MMscf

$$\text{Gallons of water removed per MMscf} = \frac{(68 - 5.8) \text{ lbm}}{8.34 \text{ lbm/gal}} = 7.46 \text{ gal}$$

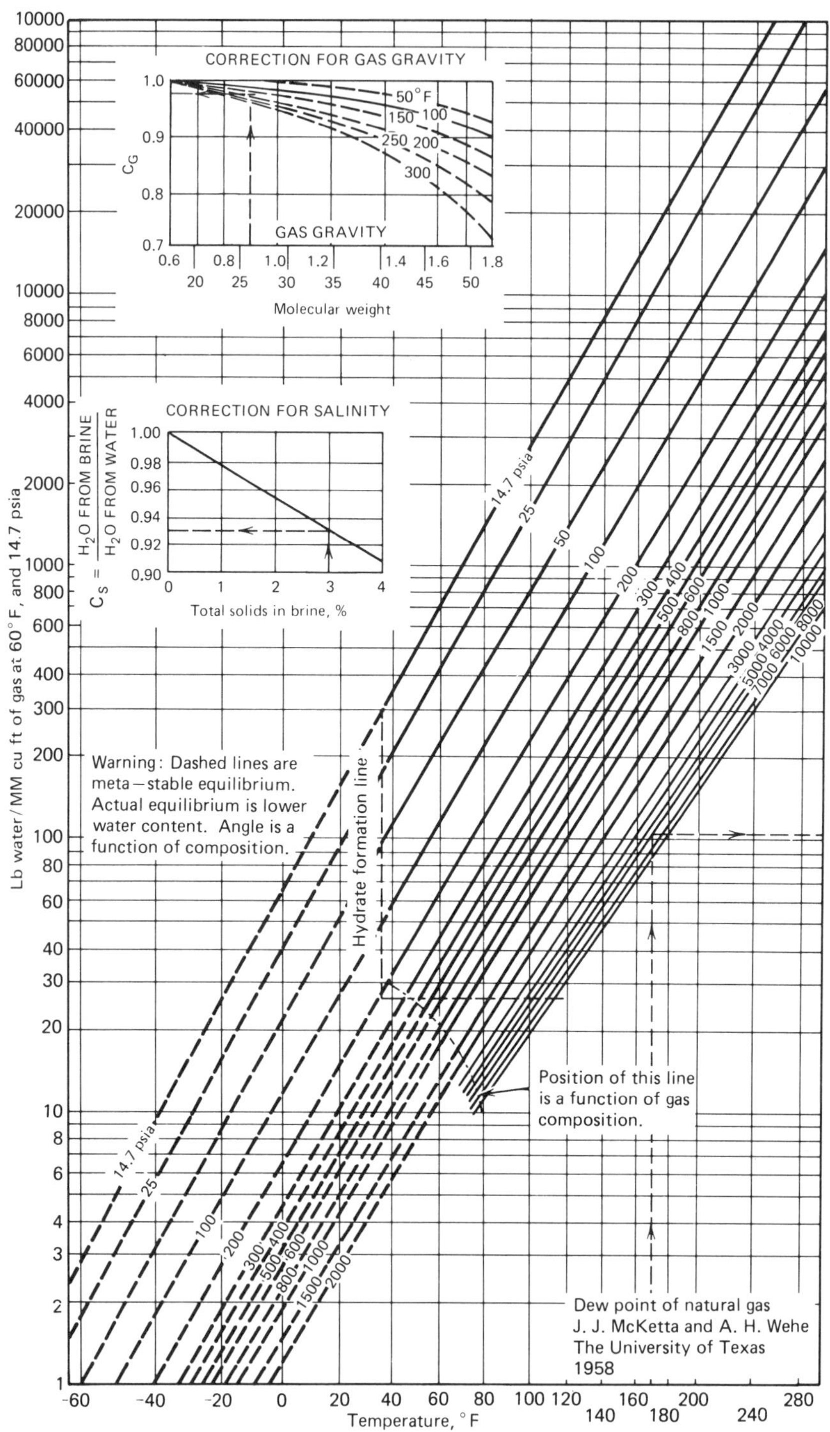

Fig. 2.13 Water contents of natural gases with corrections for salinity and gravity. (Courtesy Gas Processors Suppliers Assoc.)

Figure 2.13 may be used to determine whether a gas well producing a small amount of water is producing liquid-phase water in the reservoir or whether all the produced water is water vapor in the gas at reservoir conditions. An example will illustrate this.

Example 2.15. Consider the follow conditions:

Reservoir pressure = 3000 psia

Reservoir temperature = 200°F

Water vapor content in reservoir = 280 lb water/MMscf of gas

High-pressure separator pressure = 800 psia

High-pressure separator temperature = 100°F

Water content of separator gas = 72 lb water/MMscf of gas

Water produced from well = 1.0 STB water/MMscf of gas

From the above information, water condensed from the separator = 280 − 72 = 208 lb water/MMscf of gas or 0.59 STB water/MMscf of gas. Therefore, there is liquid-phase water production amounting to 1.0 − 0.59 = 0.41 STB water/MMscf of gas.

2.11 TWO-PHASE SYSTEMS

In the case of retrograde and wet gas reservoirs, the surface oil and gas produced exist as an all gaseous phase in the reservoir. Accurate material-balance calculations should be made on the basis of total reservoir gas produced, which is equal to the surface gas produced plus the gas equivalent of the oil produced. Reservoir calculations may be made from generally available field data by recombining the produced gas and oil in the correct ratio to find the average specific gravity (air = 1) of the total well fluid.

2.11.1 API Gravity

The API gravity is another gravity term that is used with hydrocarbon liquids. A special gravity scale was adopted by the American Petroleum Institute for expressing petroleum products:

$$°API = \frac{141.5}{\gamma_o} - 131.5 \tag{2.92}$$

γ_o is the liquid's specific gravity at 60°F, referenced to that of water at 60°F. Thus, a liquid that has the same density as water at 60°F, that is, specific gravity of 1.0, will have an API gravity of 10° API. The gravity of a liquid in °API is determined by its density at 60°F and is independent of temperature. Readings taken at temperatures other than 60°F must be corrected for temperature to give the value at 60°F.

The liquid specific gravity may be obtained by rearranging Eq. 2.92:

$$\gamma_o = \frac{141.5}{°\text{API} + 131.5} \tag{2.93}$$

Table 2.7 is useful for converting API gravities to liquid gravities.

TABLE 2.7
Conversion of API Gravity to Liquid Specific Gravity at 60°F

°API	Specific Gravity	°API	Specific Gravity	°API	Specific Gravity
0	1.076	34	0.8550	68	0.7093
1	1.068	35	0.8498	69	0.7057
2	1.060	36	0.8448	70	0.7022
3	1.052	37	0.8398	71	0.6988
4	1.044	38	0.8348	72	0.6953
5	1.037	39	0.8299	73	0.6919
6	1.029	40	0.8251	74	0.6886
7	1.022	41	0.8203	75	0.6852
8	1.014	42	0.8155	76	0.6819
9	1.007	43	0.8109	77	0.6787
10	1.000	44	0.8063	78	0.6754
11	0.9930	45	0.8017	79	0.6722
12	0.9861	46	0.7972	80	0.6690
13	0.9792	47	0.7927	81	0.6659
14	0.9725	48	0.7883	82	0.6628
15	0.9659	49	0.7839	83	0.6597
16	0.9593	50	0.7796	84	0.6566
17	0.9529	51	0.7753	85	0.6536
18	0.9465	52	0.7711	86	0.6506
19	0.9402	53	0.7669	87	0.6476
20	0.9340	54	0.7628	88	0.6446
21	0.9279	55	0.7587	89	0.6417
22	0.9218	56	0.7547	90	0.6388
23	0.9159	57	0.7507	91	0.6360
24	0.9100	58	0.7467	92	0.6331
25	0.9042	59	0.7428	93	0.6303
26	0.8984	60	0.7389	94	0.6275
27	0.8927	61	0.7351	95	0.6247
28	0.8871	62	0.7313	96	0.6220
29	0.8816	63	0.7275	97	0.6193
30	0.8762	64	0.7238	98	0.6166
31	0.8708	65	0.7201	99	0.6139
32	0.8654	66	0.7165	100	0.6112
33	0.8602	67	0.7128		

$$\text{Specific gravity} = \frac{141.5}{131.5 + °\text{API}}$$

2.11.2 Gas Gravity of Total Well Stream

The total well stream gas specific gravity will differ greatly from the surface gas specific gravity where the gas-oil ratio is low. Many correlations use the specific gravity as an index to various fluid properties. This should be the well stream gas specific gravity. In order to calculate the well stream gas gravity, let:

R_g = total surface gas-oil ratio of the production, scf of dry or residue gas per STB of oil (condensate)

γ_o = specific gravity of the tank oil (water = 1)

M_o = average molecular weight of the tank oil (condensate)

γ_g = average specific gravity of the gas produced from surface separators (air = 1)

Then, the well stream gas specific gravity (air = 1) is given by (Craft and Hawkins)

$$\gamma_w = \frac{R_g\gamma_g + 4584\ \gamma_o}{R_g + 132{,}800\ \gamma_o/M_o} \tag{2.94}$$

When the molecular weight of the tank oil is not known, it may be estimated using the formula:

$$M_o = \frac{44.29\ \gamma_o}{1.03 - \gamma_o} = \frac{6084}{°\mathrm{API} - 5.9} \tag{2.95}$$

γ_w is equal to the average molecular weight of all the hydrocarbons flowing in the well stream (regardless of their phase state) divided by the molecular weight of air.

2.11.3 Two-phase z-factor

If a gas mixture has two phases existing at the pressure and temperature of interest, the gas deviation factor of the two-phase mixture can be estimated from Fig. 2.14. As indicated on this figure, if the z-factor from Fig. 2.14 is essentially the same as the z-factor from Fig. 2.4, this indicates the mixture exists as one single gaseous phase.

The z-factor for a gas-condensate reservoir may also be calculated using the gas law as

$$z\ (\text{two-phase}) = \frac{378.6pV}{(GIIP - G_p)RT} \tag{2.96}$$

where

$GIIP$ = gas initially in place, scf

G_p = cumulative gas produced, scf

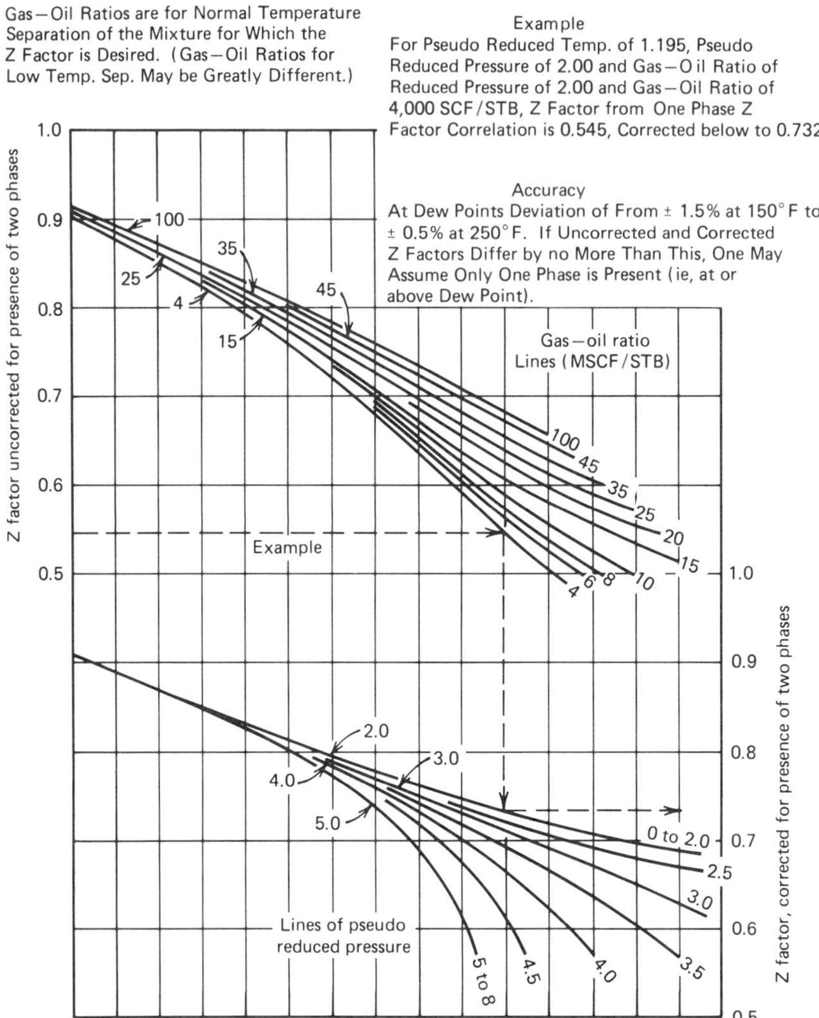

Fig. 2.14 Gas deviation factor corrections for two-phase natural gas systems. (After Elfrik, Sandberg, and Pollard.)

2.12 SOME GAS CONVERSION EQUATIONS

From the basic laws, the following useful conversion equations can be derived. At standard conditions of 14.7 psia and 60°F:

Molecular weight of gas = 28.97 (sp gr)

Density of gas, lbm/cu ft = 0.0764 (sp gr) = mol wt/379 = 28.97 (sp gr)/379

Specific volume of gas, cu ft/lbm = 13.08/sp gr = 379/mol wt

Gas flow, moles/day = Gas flow, cfd/379

Mass flow rate, lbm/hr = 3185 (MMscfd)(sp gr)

At conditions other than 14.7 psia and 60°F:

$$\text{Density of ideal gas, lbm/cu ft} = \frac{2.703\ (\text{sp gr})(\text{pressure, psia})}{(\text{temp, °F} + 460)}$$

$$\text{Density of actual gas, lbm/cu ft} = \frac{2.703\ (\text{sp gr})(\text{pressure, psia})}{(\text{temp, °F} + 460)(z)}$$

$$\text{Ideal gas flow, cfd} = \frac{(\text{gas flow, scfd})(14.7)(\text{temp, °F} + 460)}{(\text{pressure, psia})(520)}$$

$$\text{Actual gas flow, cfd} = \frac{(\text{gas flow, scfd})(14.7)(\text{temp, °F} + 460)(z)}{(\text{pressure, psia})(520)}$$

$$\text{Actual gas flow, cfd} = \frac{0.327\ (\text{MMscfd})(\text{temp, °F} + 460)(z)}{(\text{pressure, psia})}$$

$$\text{Volume of moles, cu ft/mol} = \frac{379\ (\text{temp, °F} + 460)(14.7)}{(520)(\text{pressure, psia})}$$

where

z = gas deviation factor

REFERENCES

Amyx, J. W., D. M. Bass, Jr., and R. L. Whiting. *Petroleum Reservoir Engineering—Physical Properties*. New York: McGraw-Hill, 1960.

Brown, G. G., et al. "Natural Gasoline and the Volatile Hydrocarbons." Tulsa: NGAA, 1948.

Benedict, M., G. B. Webb, and L. C. Rubin. "An Empirical Equation for Thermodynamic Properties of Light Hydrocarbons and Their Mixtures." *Chem. Eng. Prog.* **47**, No. 8, p. 419, August 1951.

Benedict, M., G. B. Webb, and L. C. Rubin. "An Empirical Equation for Thermodynamic Properties of Light Hydrocarbons and Their Mixtures." *Jour. Chem. Phys.* **8**, p. 334, April, 1940.

Benedict, M., G. B. Webb, and L. C.Rubin. "Reduction of Equation to Charts for Prediction of Liquid-Vapor Equilibria." *Chem. Eng. Prog.* **47**, No. 11, p. 571, November 1951.

Burcik, E. J. *Properties of Petroleum Reservoir Fluids* IHRDC, Boston, Mass., 1977.

Buxton, T. S., and J. M. Campbell. "Compressibility of Lean Natural Gas-Carbon Dioxide Mixtures at High Pressure." *Society of Petroleum Engineer Journal*, p. 80, March 1967.

Carr, N. L., R. Kobayashi, and D. B. Burrows. "Viscosity of Hydrocarbon Gases Under Pressure." *Trans. AIME* **201**, pp. 264–272, 1954.

Craft, B. C., and M. F. Hawkins. *Applied Petroleum Reservoir Engineering*. Englewood Cliffs, NJ: Prentice-Hall, 1959.

DeWitt, K. J., and G. Thodos. "Viscosities of Binary Mixtures in the Dense Gaseous State: The Methane-Carbon Dioxide System." SPE Reprint Series No. 13, I, *Gas Technology*, p. 236, 1977.

Dranchuk, P. M., R. A. Purvis, and D. B. Robinson. "Computer Calculation of Natural Gas Compressibility Factors Using the Standing and Katz Correlation." *Institute of Petroleum* IP 74–008, 1974.

Elfrik, E. B., C. R. Sandberg, and T. A. Pollard. "A New Compressibility Correlation for Natural Gases and Its Application to Estimates of Gas-in-place." *Trans AIME* **186**, pp. 219–223, 1949.

Gas Processors Suppliers Assoc. *Engineering Data Book*. Tulsa: GPSA, 1976.

Gatlin, C. *Petroleum Engineering—Drilling and Well Completions*. Englewood Cliffs, NJ: Prentice-Hall, 1960.

Gopal, V. N. "Gas Z-factor Equations Developed for Computer." *Oil and Gas Journal*, pp. 58–60, August 8, 1977.

Hall, K. R., and L. Yarborough. "A New Equation of State for Z-factor Calculations." *Oil and Gas Journal*, pp. 82–92, June 18, 1973.

Kay, W. B. "Density of Hydrocarbon Gases and Vapor." *Industrial and Engineering Chemistry* **28**, pp. 1014–1019, 1936.

Lee, A. L., M. H. Gonzalez, and B. E. Eakin. "The Viscosity of Natural Gases." *Journal of Petroleum Technology*, p. 997, August 1966.

Lohrenz, J., B. G. Bray, and C. R. Clark. "Calculating Viscosities of Reservoir Fluids from their Compositions." *Journal of Petroleum Technology*, p. 1171, December 1964.

Mathews, T. A., C. H. Roland, and D. L. Katz. "High Pressure Gas Measurement." Proceedings 21st Annual Convention, NGAA, p. 41, 1942.

Mattar, L., G. S. Brar, and K. Aziz. "Compressibility of Natural Gases." *Journal of Canadian Petroleum Technology*, pp. 77–80, October–December 1975.

McCain, W. D., Jr. *The Properties of Petroleum Fluids*. Tulsa: Petroleum Publishing, 1973.

Petroleum Extension Service. *Field Handling of Natural Gas*. Austin: Univ. of Texas Press, 1972.

Redlich, O., and Kwong, J. N. S. "On the Thermodynamics of Solutions. An Equation of State. Fugacities of Gaseous Solutions," *Chem. Rev.* **44**, p. 233, 1949.

Robinson, D. B., C. A. Macrygeorgos, and G. W. Grovier. "The Volumetric Behavior of Natural Gases Containing Hydrogen Sulfide and Carbon Dioxide." SPE Reprint Series No. 13, I. *Gas Technology*, p. 211, 1977.

Soave, G. "Equilibrium Constants from a Modified Redlich-Kwong Equation of State." *Chem. Eng. Sc.* **27**, pp. 1197–1203, 1972.

Standing, M. B., and D. L. Katz. "Density of Natural Gases." *Trans AIME* **146**, pp. 140–9, 1942.

Takacs, G. "Comparisons Made For Computer Z-factor Calculations." *Oil and Gas Journal*, pp. 64–66, December 20, 1976.

Thomas, L. K., R. W. Hankinson, and K. A. Phillips. "Determination of Acoustic Velocities for Natural Gas." *Journal of Petroleum Technology* **22**, pp. 889–895, 1970.

Wichert, E., and K. Aziz. "Calculation of Z's for Sour Gases." *Hydrocarbon Processing* **51**(5), pp. 119–122, 1972.

Wolfe, J. F. "Predicting Phase and Thermodynamic Properties of Natural Gases with the Benedict–Webb–Rubin Equation of State." SPE Reprint Series No. 13, Vol 1, pp. 218–226, 1977.

Yarborough, L., and K. R. Hall. "How to Solve Equation of State for Z-factors." *Oil and Gas Journal*, pp. 86–88, February 18, 1974.

PROBLEMS

2.1 A 2 cu ft tank contains gas at 1000 pounds per square inch gauge (psig) and at 60°F. This tank is connected to another tank of unknown volume containing the same gas at 14.7 psia and 60°F. The two tanks are allowed to come to equilibrium, and when this is accomplished and pressure is 650 psi gauge and the temperature is 60°F. What is the volume of the tank?

Atmospheric pressure is at 14.7 psia.

Assume gas behaves as a perfect gas.

2.2 (a) Two cylinders, one containing propane at 100 psig and the other containing propane at 0 psig, are connected and the pressures allowed to equalize. If the volume of each of the cylinders is 5 cu ft and the temperature is held constant at 90°F what would be the equalization pressure, the density of the propane under this pressure, and the weight of gas in the system. Assume gas behaves as a perfect gas.

(b) If the temperature of the system were reduced to 60°F, what would the pressure be?

2.3 A 2 cu ft tank contains propane at 1209 psig and 332°F; taking into account the deviation from the gas laws, how many standard cubic feet of gas have been withdrawn from the cylinder when the pressure has dropped to 602 psig and 332°F? Atmospheric pressure is at 14.7 psia.

2.4 For a gas of the following composition:

	Mol %
Carbon dioxide	0.40
Methane	94.32
Ethane	3.90
Propane	1.17
i-Butane	0.08
n-Butane	0.13
	100.00

Calculate:

(a) The specific gravity, the density, and the specific volume of the gas at 60°F and 14.65 psia.

(b) The composition on weight basis.

2.5 A sample of natural gas from the Bell Field has a specific gravity of 0.665 (air = 1.00). The carbon dioxide and nitrogen content are 0.10 and 2.07 mol %, respectively. Calculate the gas deviation factor, z, at reservoir temperature of 213°F and reservoir pressure of 3250 psia.

2.6 For a gas of the following composition:

	Mol %
Methane	92.67
Ethane	5.29
Propane	1.38
i-butane	0.18
n-butane	0.34
n-pentane	0.14

Calculate:

(a) The pseudocritical temperature and pressure

(b) The gas deviation factor for (1) 400 psig and 80°F and (2) 2500 psig and 200°F.

(c) The density and specific volume of the gas for each of the conditions in 5 (b).

(d) The number of standard cubic feet of gas per acre foot of sand of 25% porosity and 10% connate water for each of the conditions given in 5(b). Atomspheric pressure is at 14.7 psia.

2.7 A natural gas has the following composition:

Component	*Mol %*
C_1	87.09
C_2	4.42
C_3	1.60
i-C_4	0.40
n-C_4	0.52
C_5	0.46
C_6	0.29
C_{7+}	0.06
N_2	4.76
CO_2	0.40
	100.00

Assume a molecular weight of C_{7+}, the same as for C_7.)

Calculate:

(a) Weight percent of each component in the gas.

(b) Apparent molecular weight of the gas.
(c) Specific gravity of the gas.

2.8 A natural gas has the following composition:

Component	*Mol %*
C_1	86.02
C_2	7.70
C_3	4.26
i-C_4	0.57
n-C_4	0.87
i-C_5	0.11
n-C_5	0.14
C_6	0.33
	100.00

For pressure = 750 psia and temperature = 150°F, calculate:
(a) Pseudocritical temperature.
(b) Pseudocritical pressure.
(c) z-factor.

2.9 A natural gas has the following composition: 90% CH_4; 5.0% C_2H_6; 5.0% N_2.
(a) Calculate its pseudocritical temperature and pseudocritical pressure.
(b) Calculate the pseudoreduced temperature and pseudoreduced pressure, and the gas deviation factor for the natural gas at 100°F and 600 psia.
(c) 325,000 cu ft of this natural gas is to be compressed from 14.7 psia and 60°F to 600 psia and 100°F. Calculate the volume the natural gas will occupy under the compressed conditions.

2.10 Calculate the gas reserve in a gas field of 2000 acres, with 40-ft sand thickness, 25% porosity, 15% water saturation, a bottom-hole pressure of 3000 psi. gauge, a temperature of 200°F, and barometric pressure at 14.7 psia. Composition of gas: methane, 94.63%; ethane, 2.54%; propane, 1.46%; isobutane, 0.46%, N-butane, 0.38%; pentanes, 0.36%; and hexanes plus, 0.17%.

2.11 In a recycling plant 250,000 cu ft of gas at 2500 psi. gauge, temperature 100°F, is compressed to 4000 psi. gauge and 150°F and put back into the sand at these conditions. What is the volume of the gas as it is injected into the sand if the gas composition is 75% methane, 15% ethane, and 10% propane?

2.12 A gas of the following composition leaves a high-pressure absorber at 500 psia and 80°F. The gas is then compressed and injected back into the reservoir at 3600 psia and 200°F. Calculate the volume 10 million scf of this gas would occupy in the reservoir.

	Mol %
Methane	93.3
Ethane	3.84
Propane	2.21
n-Butane	0.65
	100.00

2.13 A casing annulus is filled with gas under the following conditions:

$$V = 1500 \text{ cu ft} \qquad p_{avg} = 1000 \text{ psia}$$
$$\text{Gas gravity} = 0.65 \qquad T_{avg} = 200°\text{F}$$

How much gas, in standard cubic feet, must be removed to drop the pressure to 800 psia?

2.14 The original bottom hole pressure in a gas field was 2500 psia and the bottom hole temperature was 90°F. For a gas of the following composition, what would be the bottom hole pressure when one-half of the gas has been withdrawn from the reservoir? Assume constant reservoir volume.

	Mol %
Methane	91.32
Ethane	4.43
Propane	2.12
Butanes	1.36
Pentanes	0.42
Hexanes	0.15
Heptanes and plus	0.20
	100.00

2.15 A natural gas has a specific gravity of 0.85. At a pressure of 3000 psia and a temperature of 230°F, calculate the viscosity of the gas by two different methods.

2.16 Calculate the viscosity of the following gas at reservoir conditions of 2500 psia and 180°F.

Component	*Mol %*
C_1	0.80
C_2	0.10
C_3	0.04
H_2S	0.06
	1.00

2.17 Calculate the viscosity of the following gas at reservoir conditions:

Component	*Mol Fraction*
C_1	0.75
C_2	0.15
C_3	0.04
H_2S	0.06
	1.00

Reservoir pressure = 2000 psia
Reservoir temperature = 200°F

2.18 Calculate the following fluid properties at the given *reservoir* conditions:
(a) Density—lbm/cu ft
(b) Viscosity—cp
(c) Formation volume factor—res. cu ft/surf. cu ft

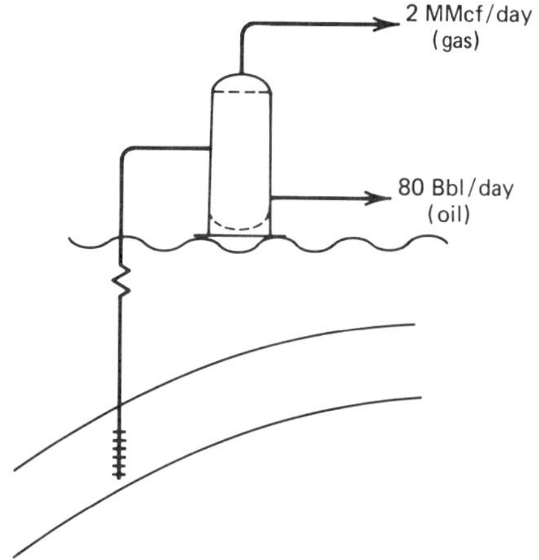

Separator Gas Composition

Component	Mole Fraction
N_2	0.0101
CO_2	0.0204
H_2S	0.2139
C_1	0.6017
C_2	0.0750
C_3	0.0433
$i\text{-}C_4$	0.0061
$n\text{-}C_4$	0.0137
$i\text{-}C_5$	0.0033
$n\text{-}C_5$	0.0052
C_6	0.0053
C_7^+	0.0020
	1.0000

Separator Oil Properties
49° API, assume C_7^+ material.

Reservoir Conditions
Pressure at midperforations = 4537 psig
Temperature at midperforations = 252°F
Assume molecular weight of C_7^+ to be 131 lbm/lb-mol.

2.19 Natural gas is in contact with brine in a reservoir. The brine contains 150,000 ppm dissolved solids. Calculate the water content in lb/MMscf at reservoir conditions of 2000 psia and 200°F.

2.20 The production from a gas well is 50 million cubic feet per day (MMcfd) measured at 14.7 psia and 60°F. The gas is saturated with water vapor under reservoir conditions of 1500 psia and 120°F. Calculate the water collected in the lease separator if the gas is expanded at 800 psia and 60°F through the separator. Give your results in pound-mass per/day and barrels per/day.

3

CONCEPTS OF THERMODYNAMICS

3.1 SYSTEM

A system is a part of the universe set aside for study. It must have enclosing boundaries, outside of which are the surroundings (Fig. 3.1). Examples of systems are reservoir gas flowing toward a wellbore, gas undergoing compression by a mechanical compressor, and gas flowing vertically in a well tubing or horizontally in a pipeline.

First-law analysis is essentially an accounting procedure for measuring energy transfers to and from a system and changes of energy inside the system. The two principal accounting procedures are control-mass analysis and control-volume analysis.

Control mass (Fig. 3.2) is any selected piece of matter. Any assembly is a control mass. An accounting procedure must be carried out over a set accounting period, and an essential step is specifying the time base. This might be a given period of time (the time required for something to happen) or it might be instantaneous.

Control volume (Fig. 3.3) is any defined region in space. This region may be moving, and its shape and volume may be changing. In this case, though, most control volumes have a constant shape and size and are fixed in the reference frame. Control volume is posed in terms of property fields, that is, the distribution of the properties through space and time; the analysis provides information in terms of these fields.

To contrast the control-mass and control-volume perspectives simply, the control-mass method calculates the properties of given pieces of matter as functions of time, while the control-volume method gives the properties of whatever piece of matter happens to be in a given region of space at any given instant.

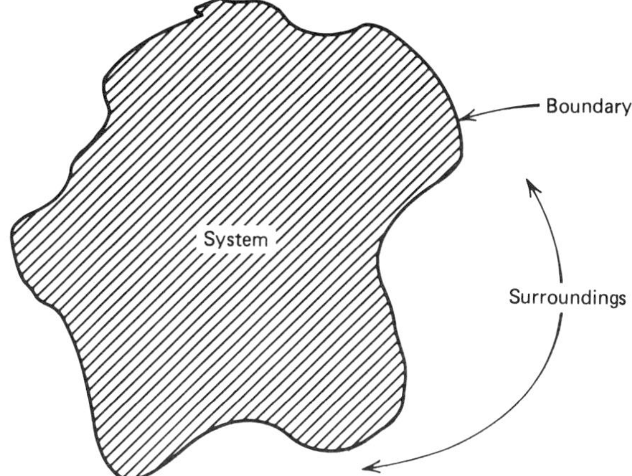

Fig. 3.1 A system and its surroundings.

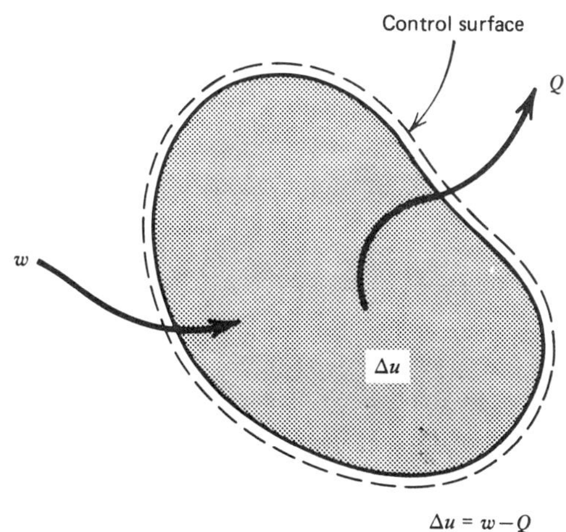

Fig. 3.2 A control mass.

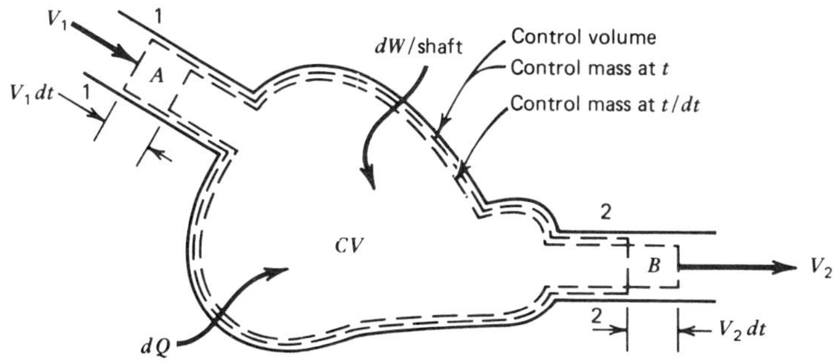

Fig. 3.3 A control volume.

3.2 ENERGY AND ENERGY BALANCES

The energy balance is an important consideration in making engineering calculations. In making a balance of energy, all energy factors must be expressed in the same units if the calculations are to be correct. These units may be either foot-pound force or British thermal units, for example. The relationship of 1 Btu is equivalent to 778.2 ft-lbf, which is useful.

The general methodology of energy-balance analysis is:

1. Define the control mass or control volume, indicating its boundaries on a sketch.
2. Indicate what flows of energy across the boundary will be considered and set up their sign convention on the drawing.
3. Indicate the time basis for the energy balance.
4. Write the conservation of energy in general terms.
5. Make appropriate idealizations and bring in equations of state or other information necessary to allow solution.

This study centers on energy balances of dynamic systems. Energy is carried with flowing fluid and may be transferred from the fluid to the surroundings, and vice versa. Therefore, this chapter deals with the ways that energy may be interchanged between different forms of energy as the system moves through a piece of equipment, a processing plant, or a length of pipe.

Energy carried with the fluid includes the internal energy, U, and all energy that is the peculiar property of the fluid, regardless of its relative location or motion, and the energy carried by the fluid because of its condition of flow or position. This energy carried by the fluid encompasses three other types of energy: energy of motion (kinetic energy, $mu^2/2g_c$), which is the energy associated with movement with respect to a fixed point; energy of position (potential energy, mgZ/g_c), which results from the system's location in the earth's gravitational field; and pressure energy, pV, carried by the system because of its introduction into or exit from flow under pressure.

Energy transferred between a fluid or system in flow and its surroundings is of two kinds. First is heat, Q, absorbed by the flowing material or system as a result of temperature difference between the system and surroundings. Heat gained by the system is positive in sign, while heat lost by the system is negative in sign. Second is work, w, done by the system on the surroundings. This is often called shaft work, w_s, and does not include lost work, lw, caused by friction. Work is positive in sign when the system performs work on the surroundings. Heat and work are the only means of transferring energy between the system and the surroundings.

Consider Fig. 3.4. An energy balance around such a flow system, between points 1 and 2 and the surroundings, assuming no accumulation of material and energy at any point in the system, is given by the equation

$$U_2 + \frac{mu_2^2}{2g_c} + \frac{mgZ_2}{g_c} + p_2V_2 = U_1 + \frac{mu_1^2}{2g_c} + \frac{mgz_1}{g_c} + p_1V_1 + Q - w \tag{3.1}$$

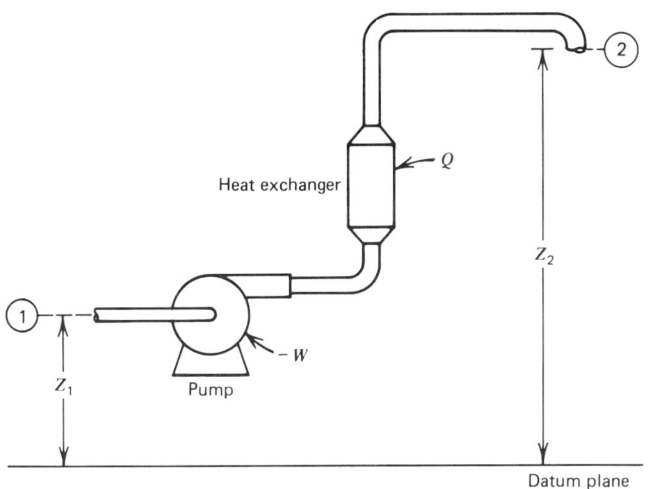

Fig. 3.4 Energy balance.

By definition,

$$\Delta U = U_2 - U_1$$

and

$$\Delta(pV) = p_2V_2 - p_1V_1$$

Therefore,

$$\Delta U + \Delta\left(\frac{1}{2}\frac{mu^2}{g_c}\right) + \Delta\left(\frac{mgZ}{g_c}\right) + \Delta\,(pV) = Q - w \tag{3.2}$$

3.3 ENTHALPY CHANGE, ΔH

The thermodynamic property enthalpy H is defined as

$$H = U + pV \tag{3.3}$$

Thus, the change of enthalpy is given by

$$\Delta H = \Delta U + \Delta(pV) \tag{3.4}$$

Equation 3.2 becomes

$$\Delta H + \Delta\left(\frac{mu^2}{2g_c}\right) + \Delta\left(\frac{mgZ}{g_c}\right) = Q - w \tag{3.5}$$

where

ΔH = total heat content, Btu

In calculating ΔH between two thermodynamic states (pressure and temperature), use specific enthalpy of the fluid, h', in Btu/lbm or the molal specific enthalpy, h, in Btu/lb-mol. Thus,

$$\Delta H = H_2 - H_1 = m(h_2' - h_1') \tag{3.6}$$

or

$$\Delta H = H_2 - H_1 = n(h_2 - h_1) \tag{3.7}$$

where

m = pound-mass of fluid
n = pound-moles of fluid

The units Btu/lb-mol will be used in most of the work in this text.

For natural gas, the molal specific enthalpy is a function of temperature, pressure, and gas composition:

$$h = (T,\ p,\ \text{composition}) \tag{3.8}$$

For a given composition (gas gravity),

$$h = h(T,\ p)$$

Then, taking the differential,

$$dh = \left(\frac{\partial h}{\partial T}\right)_p dT + \left(\frac{\partial h}{\partial p}\right)_T dp \tag{3.9}$$

Integrating Eq. 3.9,

$$\int_1^2 dh = h_2 - h_1 = \Delta h = \int_1^2 \left(\frac{\partial h}{\partial T}\right)_p dT + \int_1^2 \left(\frac{\partial h}{\partial p}\right)_T dp \tag{3.10}$$

3.4 SPECIFIC HEATS

Consider the functional relation $u = u(T, v)$. The difference in energy between any two states separated by infinitesimal temperature and specific-volume differences dT and dv is

$$du = \left(\frac{\partial u}{\partial T}\right)_v dT + \left(\frac{\partial u}{\partial v}\right)_T dv \tag{3.11}$$

The derivative $(\partial u/\partial T)_v$ is called the specific heat at constant volume:

$$C_v \equiv \left(\frac{\partial u}{\partial T}\right)_v \tag{3.12}$$

In Eq. 3.9, the derivative $(\partial h/\partial T)_p$ is called the specific heat at constant pressure:

$$C_p \equiv \left(\frac{\partial h}{\partial T}\right)_p \tag{3.13}$$

The derivatives C_p and C_v constitute two of the most important thermodynamic derivative functions, and values have been experimentally determined as functions of the thermodynamic state for many simple compressible substances.

The specific heats of gases and liquids are determined experimentally in a calorimeter, usually at a pressure of 1 atm. For natural gases, the specific heat at 1 atm pressure is a function of temperature and of gas gravity or molecular weight. Figure 3.5 is a graph of specific heat at constant pressure for natural gases at 1 atm pressure.

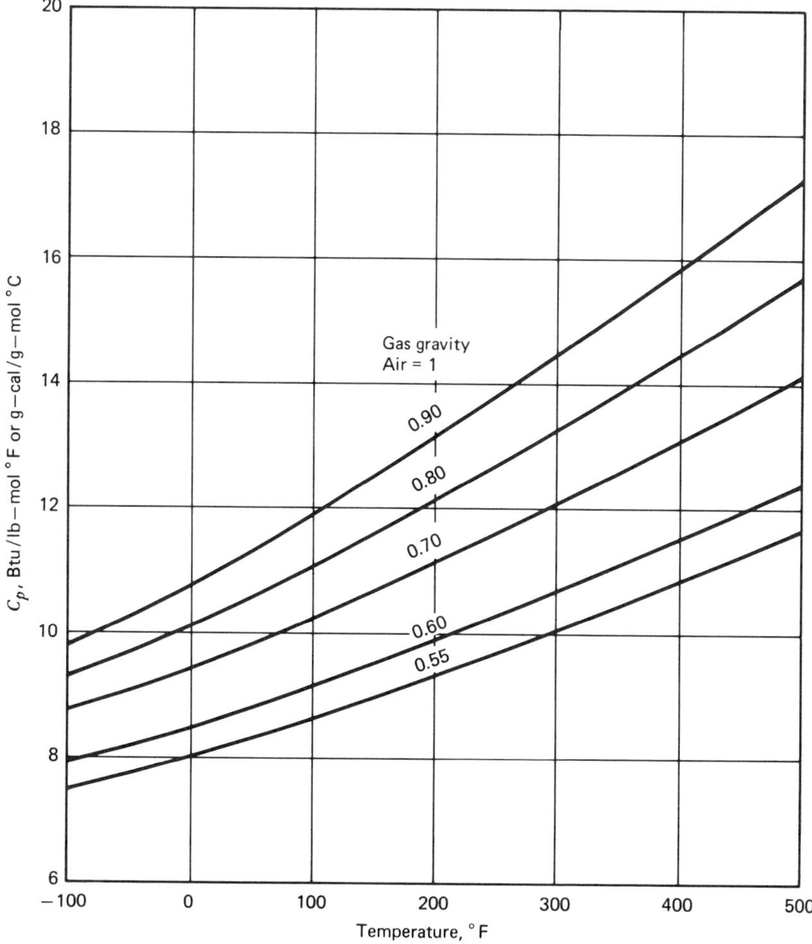

Fig. 3.5 Specific heat of hydrocarbon gases at 1.0 atm pressure. (After Brown.)

The specific heat at constant volume C_v is related to the specific heat at constant pressure C_p for ideal gases as follows:

$$C_p = C_v + R = C_v + 1.99 \text{ Btu/lb-mol °F} \tag{3.14}$$

The ratio $k = C_p/C_v$ of the specific heats is useful in computing the adiabatic compression of gases, for which ideal gases follow the relationship

$$pV^k = \text{constant} \tag{3.15}$$

3.4.1 Effect of Pressure on Specific Heat and Enthalpy

The law of corresponding states can be used to correct specific heat values measured at 1 atm pressure for pressures other than 1 atm. The specific heat at constant pressure at pressures greater than atmospheric is given by

$$C_p(p, T) = C_p(14.7, T) + \Delta C_p(p_r, T_r) \tag{3.16}$$

The correction term $\Delta C_p(p_r, T_r)$ can be calculated from thermodynamics. From thermodynamic relations, the effect of pressure alone is given by

$$\left(\frac{\partial h}{\partial p}\right)_T = v - T\left(\frac{\partial v}{\partial T}\right)_p \tag{3.17}$$

where

v = molal specific volume, cu ft/lb-mol

For real gases,

$$pv = zRT$$

Differentiating,

$$\left(\frac{\partial v}{\partial T}\right)_p = \frac{zR}{p} + \frac{RT}{p}\left(\frac{\partial z}{\partial T}\right)_p$$

Then,

$$\left(\frac{\partial h}{\partial p}\right)_T = \frac{zRT}{p} - \frac{zRT}{p} - \frac{RT^2}{p}\left(\frac{\partial z}{\partial T}\right)_p$$

or,

$$\left(\frac{\partial h}{\partial p}\right)_T = -\frac{RT^2}{p}\left(\frac{\partial z}{\partial T}\right)_p \tag{3.18}$$

Using reduced temperature and reduced pressure,

$$\left(\frac{\partial h}{\partial p_r}\right)_{T_r} = -\frac{RT_cT_r^2}{p_r}\left(\frac{\partial z}{\partial T_r}\right)_{p_r} \tag{3.19}$$

If R = 10.732 psia cu ft/lb-mol °R, p in psia,

$$\left(\frac{\partial h}{\partial p_r}\right)_{T_r} = -\frac{(10.732)(144)}{778.2}\ \frac{T_c T_r^2}{p_r}\left(\frac{\partial z}{\partial T_r}\right)_{p_r} \tag{3.20}$$

or,

$$\left(\frac{\partial h}{\partial p_r}\right)_{T_r} = -\ 1.986\ \frac{T_c T_r^2}{p_r}\left(\frac{\partial z}{\partial T_r}\right)_{p_r} \tag{3.21}$$

where

$$\frac{\partial h}{\partial p_r} = \frac{\text{Btu}}{\text{lb-mol-}p_r}$$

$$p_r = \frac{\text{psia}}{p_c}$$

$$778.2 = \frac{\text{ft-lbf}}{\text{Btu}}$$

Figure 3.6 is a graph of ΔC_p, prepared on a reduced temperature-reduced pressure basis, to avoid preparing many charts for gases of different gravity.

Figure 3.7 is a generalized isothermal pressure correction to enthalpy of gases. It is a graph of $-\Delta h/T$, Btu/lb-mol °R, versus reduced pressure, with lines of constant reduced temperature. This graph uses pseudocritical temperature instead of the actual temperature when dividing $-\Delta h$ to find the $-\Delta h/T_c$ function. Figure 3.7 gives Δh above a reference state of 0 psia and 0 °R.

3.5 ENTROPY, S

The entropy, S, is a property of the system defined as

$$S \equiv \left(\frac{\delta Q}{T}\right)_{\text{internally reversible}} \tag{3.22}$$

The change in this property is analogous to the change in energy. Hence,

$$\Delta S = S_2 - S_1 \propto \int\left(\frac{\delta Q}{T}\right)_{\text{internally reversible}}$$

where

$$\delta Q = \text{change in heat}$$

In making a number of calculations on fluid behavior, consider the changes in entropy that the system may undergo.

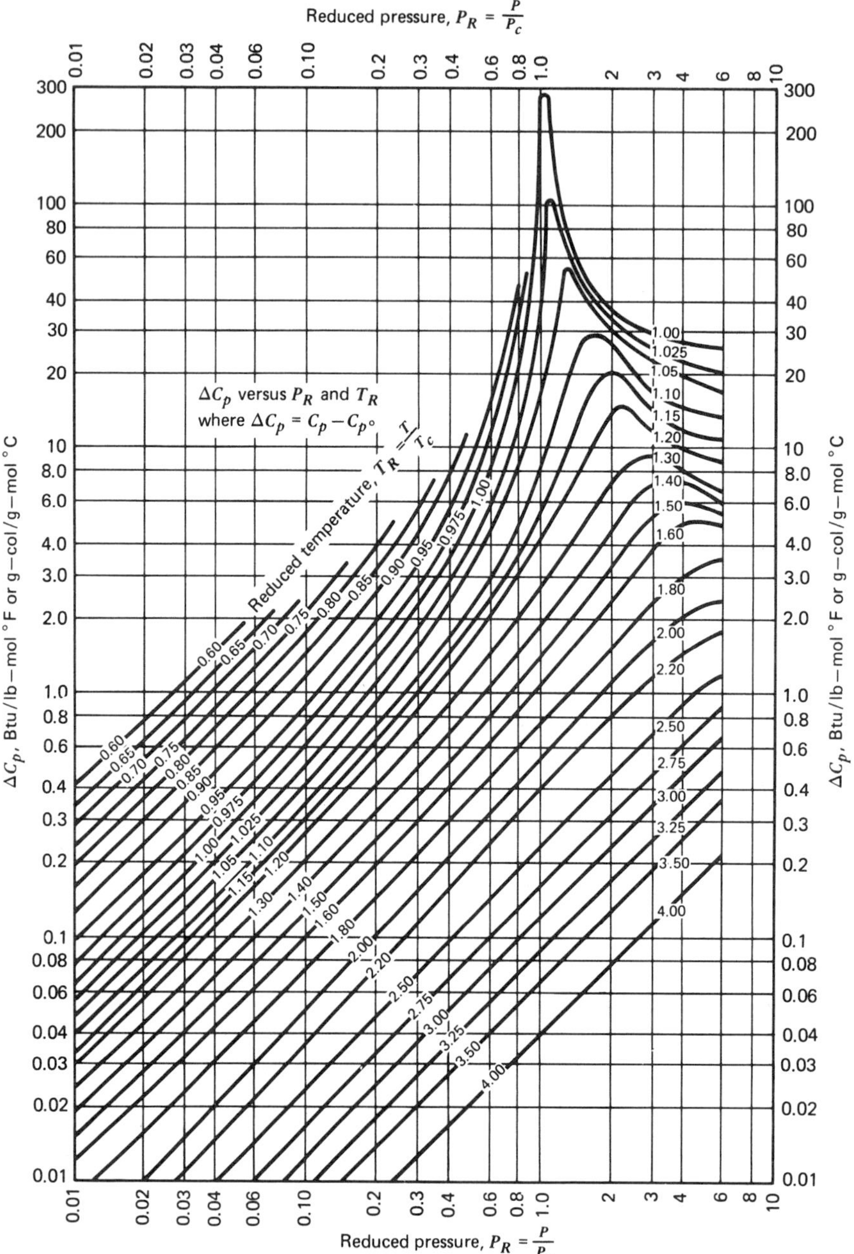

Fig. 3.6 Isothermal pressure correction to heat capacity of vapors. (After Edminster.)

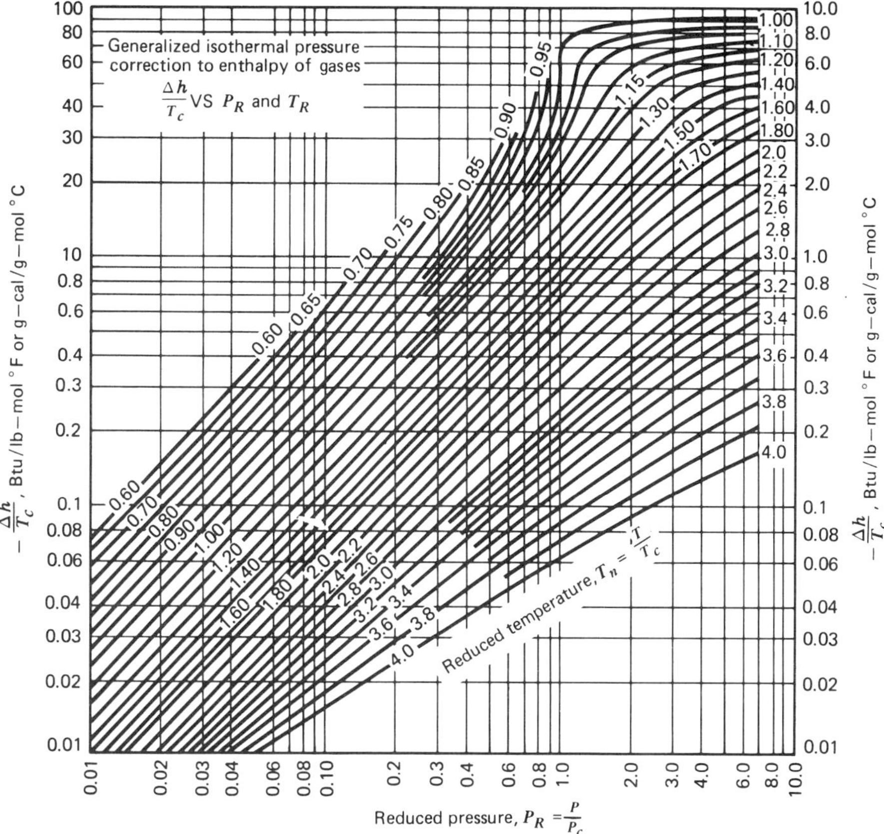

Fig. 3.7 Effect of pressure on enthalpy for natural gases. (After Edminster.)

In general, energy can be defined as the product of an intensive property of a material and the change in the extensive property. Intensive property is independent of the amount of material present, for example, pressure, temperature, density, surface tension, and chemical potential. Extensive property depends on size or amount of material present, for example, mass, area, inertia, and volume.

Example 3.1. Compression of a Quantity of Gas in a Piston. Incremental work done by piston on gas:

$$dw' = pA(-dx') = p(-dV) \tag{3.23}$$

where

$$dV = \text{increase in gas volume}$$

Thus compression work:

$$w_{\text{comp}} = \int_1^2 p(-dV) = -\int_1^2 p\,dV \tag{3.24}$$

Example 3.2. Change in Area of a Gas Bubble. Surface work involved is given by

$$w_{\text{surface}} = \int_1^2 \sigma\,dA \tag{3.25}$$

where

σ = surface (interfacial) tension of the bubble
A = area of the bubble

Example 3.3. Change in Heat Energy

Temperature = intensive property

The extensive property required to go with temperature, in order to express heat energy as a product of an intensive property and change in an extensive property is called entropy, S. In the form of Eqs. 3.24 and 3.25,

$$Q = \int_1^2 T\,dS$$

Change in internal energy, ΔU, is the sum of changes in all forms of energy taking place in the material in flow, including heat effects, compression effects, surface effects, chemical effects, and so on:

$$\Delta U = \int_1^2 T\,dS + \int_1^2 p(-dV) + \int_1^2 \sigma\,dA + \int_1^2 \mu_1\,dm_1 + \int_1^2 \mu_2\,dm_2 + \cdots \tag{3.27}$$

where

T = absolute temperature of the material
S = absolute entropy of the material

$\int_1^2 T\,dS$ = change in internal energy due to heat effects between states 1 and 2 or points 1 and 2 in the flow system

$\int_1^2 p(-dV)$ = change in internal energy due to compression effects between states or points 1 and 2

$\int_1^2 \sigma\,dA$ = change in internal energy due to surface effects between states 1 and 2

$\int_1^2 \mu_1 \, dm_1$ = change in internal energy due to chemical effects or changes in component or substance 1, between states 1 and 2

The energy term $\Delta(pV)$ is a complete differential:

$$\Delta(pV) = \int_1^2 p \, dV + \int_1^2 V \, dp \tag{3.28}$$

Combining Eqs. 3.1, 3.27, and 3.28,

$$\int_1^2 T \, dS + \Delta\left(\frac{mu^2}{2g_c}\right) + \Delta\left(\frac{mgZ}{g_c}\right) + \int_1^2 V \, dp + \int_1^2 \sigma \, dA + \int_1^2 \mu_1 \, dm_1 + \cdots = Q - w \tag{3.29}$$

In any process, the increase in internal energy due to heat effects $\int_1^2 T \, dS$ is equal to the sum of the heat absorbed from the surroundings and all other energy dissipated into heat effects within the system due to irreversibilities such as overcoming friction occurring in the process. Thus,

$$\int_1^2 T \, dS = Q + lw \tag{3.30}$$

where

lw = "lost work," energy that could have done work but was dissipated in irreversibilities within the flowing material

If Eqs. 3.29 and 3.30 are combined and rearranged,

$$\int_1^2 V \, dp + \Delta\left(\frac{mu^2}{2g_c}\right) + \Delta\left(\frac{mgZ}{g_c}\right) + \int_1^2 \sigma \, dA + \int_1^2 \mu_1 \, dm_1 + \cdots = -w - lw \tag{3.31}$$

These equations contain no limiting assumptions other than no accumulation of material in the unit and are unrestricted in application to material flowing or transferred from state 1 to state 2. The equation used is mainly a matter of convenience. In natural gas engineering problems, surface and chemical effects are negligible and these energy terms are usually dropped from the energy equation. For example, a fluid flowing through a pipe will usually be free of chemical changes, surface effects, and so on, and Eq. 3.31 may be written as

$$\int_1^2 V \, dp + \Delta\left(\frac{mu^2}{2g_c}\right) + \Delta\left(\frac{mgZ}{g_c}\right) = -w - lw \tag{3.32}$$

Writing Eq. 3.32 for a unit mass of material,

$$\int_1^2 v\,dp + \frac{\Delta u^2}{2g_c} + \frac{g}{g_c}\,\Delta Z = -\overline{w} - \overline{lw} \tag{3.33}$$

where

$\overline{w}$ = work done per unit mass
$\overline{lw}$ = "lost work" per unit mass

If the flow is also approximately isothermal and the fluid is almost incompressible, the volume of a unit mass may be assumed to be constant and Eq. 3.33 may be further simplified to

$$\frac{\Delta p}{\rho} + \frac{\Delta u^2}{2g_c} + \frac{g}{g_c}\,\Delta Z = -\overline{w} - \overline{lw} \tag{3.34}$$

Equation 3.34 is limited to a material of approximately constant density and is frequently referred to as Bernoulli's equation when $\overline{w}$ and $\overline{lw}$ are zero.

Recall the following equation:

$$\Delta H + \Delta\left(\frac{mu^2}{2g_c}\right) + \Delta\left(\frac{mgZ}{g_c}\right) = Q - w \tag{3.5}$$

Equation 3.29 (without surface and chemical effects) can be written as

$$\int_1^2 T\,ds + \Delta\left(\frac{mu^2}{2g_c}\right) + \Delta\left(\frac{mgZ}{g_c}\right) + \int_1^2 V\,dp = Q - w \tag{3.29a}$$

Comparing Eqs. 3.5 and 3.29a,

$$\Delta H = \int_1^2 T\,dS + \int_1^2 V\,dp \tag{3.35}$$

3.5.1 Calculation of Molal Specific Entropy (s)

As in the case of enthalpy, one can write

$$\int_1^2 T\,dS = Q + lw = n\int_1^2 T\,ds \tag{3.36}$$

where

n = pound moles of fluid considered
s = molal specific entropy, Btu/lb-mol °R

Rewriting Eq. 3.35 in terms of molal values,

$$\Delta h = \int_1^2 T\,ds + \int_1^2 v\,dp \tag{3.37}$$

where

v = molal specific volume, cu ft/lb-mol

Recall that

$$dh = \left(\frac{\partial h}{\partial T}\right)_p dT + \left(\frac{\partial h}{\partial p}\right)_T dp \tag{3.9}$$

$$\left(\frac{\partial h}{\partial T}\right)_p = C_p \tag{3.13}$$

$$\left(\frac{\partial h}{\partial p}\right)_T = v - T\left(\frac{\partial v}{\partial T}\right)_p \tag{3.17}$$

Thus Eq. 3.9 can be integrated to give

$$\Delta h = \int_1^2 C_p\, dT + \int_1^2 \left[v - T\left(\frac{\partial v}{\partial T}\right)_p\right] dp \tag{3.38}$$

Solving Eqs. 3.37 and 3.38 for ds yields

$$ds = \frac{C_p\, dT}{T} - \left(\frac{\partial v}{\partial T}\right)_p dp \tag{3.39}$$

Integration of Eq. 3.39 between two pressure-temperature points will yield the difference in entropy. In Brown's enthalpy-entropy charts, the reference state at which $s° = 0$ is selected as 32°F and 14.7 psia. For illustrative purposes, calculate s using this reference state, using an average C_p and taking it outside the integral sign.

Thus,

$$s(p,\ T,\ \text{composition}) = \bar{C}_p \int_{492}^{T} \frac{dT}{T} - \frac{144}{778.2}\int_{14.7}^{p} \left(\frac{\partial v}{\partial T}\right)_p dp \tag{3.40}$$

(The constant 144 allows pressure to be in psia and 778.2 converts ft-lbf/lb-mol.) For a given composition gas,

$$v\left(\frac{\text{ft}^3}{\text{lb-mol}}\right) = \frac{10.732\,zT(°\text{R})}{p(\text{psia})}$$

Therefore,

$$\left(\frac{\partial v}{\partial T}\right)_p = \frac{10.732z}{p} + \frac{10.732T}{p}\left(\frac{\partial z}{\partial T}\right)_p$$

Substituting this in Eq. 3.40,

$$s = \bar{C}_p \int_{492}^{T} \frac{dT}{T} - \frac{(144)(10.732)}{778.2}\left[\int_{14.7}^{p} \left(\frac{z}{p}\right) dp + \int_{14.7}^{p} \frac{T}{p}\left(\frac{\partial z}{\partial T}\right)_p dp\right] \tag{3.41}$$

In terms of average z and average $\left(\frac{\partial z}{\partial T}\right)$:

$$s = \bar{C}_p \ln \frac{T}{492} - 1.986 \left[\bar{z} \ln \frac{p}{14.7} + T \left(\overline{\frac{\partial z}{\partial T}}\right)_p \ln \frac{p}{14.7} \right] \tag{3.42}$$

When working with reduced parameter z chart, Eq. 3.42 is easier to write in terms of, reduced variables. Change $\partial z/\partial T$ to $1/T_c\ (\partial z/\partial T_r)_{p_r}$ and Eq. 3.42 becomes

$$s(p, T, \text{composition}) = \bar{C}_p \ln \frac{T}{492} - 1.986 \left[\bar{z} + \frac{T}{T_c} \left(\overline{\frac{\partial z}{\partial T_r}}\right)_{p_r} \right] \ln \frac{p}{14.7} \tag{3.43}$$

Equation 3.43 is written for a reference state of 32°F and 14.7 psia. It can be modified for any reference state. Equation 3.43 contains three averages. These can be evaluated as follows (Fig. 3.8):

$$\bar{C}_p = \frac{C_p \text{ area}}{T - 492\ °\text{R}}$$

$$\bar{z} = \frac{\text{z area}}{p - 14.7} \quad \text{or} \quad \frac{z \text{ area}}{p_r - p_r^\circ}$$

$$\left(\overline{\frac{\partial z}{\partial T_r}}\right)_{p_r} = \frac{\text{deze area}}{p_r - p_r^\circ}$$

Example 3.4. Calculation of Thermodynamic Properties from Pressure-Volume-Temperature and Specific Heat Data. Calculate the molal enthalpy and molal entropy change, Δh and Δs, when a quantity of 0.7 specific gravity natural gas is compressed from 14.7 psia and 100°F to 800 psia and 300°F. Check your answers against Brown's enthalpy-entropy chart for 0.7 gravity natural gas given in Section 3.6.

Solution

Enthalpy Change, Δh

For 0.7 specific gravity natural gas,

$$p_{pc} = 667 \text{ psia and } T_{pc} = 392\ °\text{R}$$

Using Eq. 3.9,

$$\int_1^2 dh = \Delta h = \int_1^2 \left(\frac{\partial h}{\partial T}\right)_p dT + \int_1^2 \left(\frac{\partial h}{\partial p}\right)_T dp$$

The change in enthalpy may be divided into two processes: change at constant pressure and change at constant temperature. Consider change from 100°F to 300°F, at a constant pressure of 14.7 psia. The change in enthalpy is given by

$$\Delta h_p = \int_1^2 \left(\frac{\partial h}{\partial T}\right)_p dT = \int_{560°\text{R}}^{760°\text{R}} C_p\, dT = \bar{C}_p\, \Delta T$$

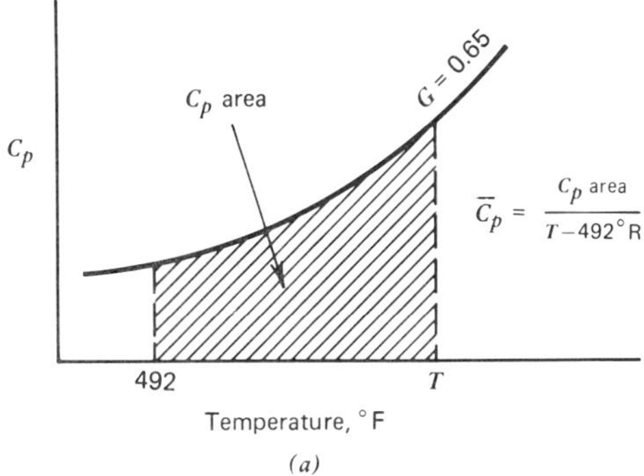

(a)

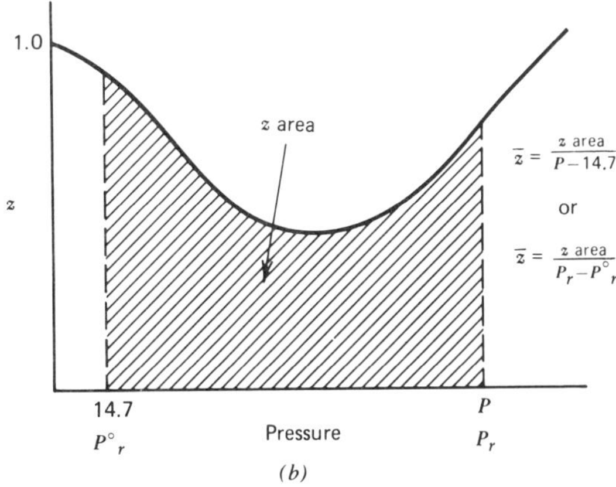

(b)

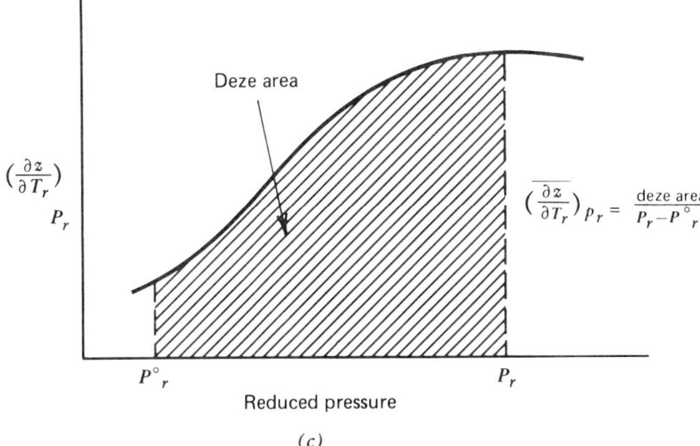

(c)

Fig. 3.8 Averages used in entropy calculation. (After Standing.)

From Fig. 3.5, $\bar{C}_p$ at an average temperature of 200°F is 11.1 Btu/lb-mol °F:

$$\Delta h_p = (11.1)(760 - 560) = 2220 \text{ Btu/lb-mol}$$

Then consider change from 14.7 to 800 psia, at a constant temperature of 300°F. The change in enthalpy is given by

$$\Delta h_T = \int_1^2 \left(\frac{\partial h}{\partial p}\right)_T dp = -\int_1^2 1.986 \frac{T_{pc}T_{pr}^{\ 2}}{p_{pr}} \left(\frac{\partial z}{\partial T_{pr}}\right)_{p_{pr}} dp_{pr}$$

or

$$\Delta h_T = -(1.986)(392)\left(\frac{760}{392}\right)^2 \int_1^2 \frac{1}{p_{pr}} \left(\frac{\partial z}{\partial T_{pr}}\right)_{p_{pr}} dp_{pr}$$

At 800 psia, $$p_{pr} = \frac{800}{667} = 1.20$$

At 300°F, $$T_{pr} = \frac{760}{392} = 1.94$$

From Table 2.4, z-factor equations, for $0.2 \leqslant p_{pr} \leqslant 1.2$ and $1.4 \leqslant T_{pr} \leqslant 3.0$:

$$z \simeq p_{pr}(0.0657T_{pr} - 0.1751) + 0.9968$$

Thus,

$$\left(\frac{\partial z}{\partial T_{pr}}\right)_{p_{pr}} = 0.0657p_{pr}$$

$$\therefore \Delta h_T = -(1.986)(392)\left(\frac{760}{392}\right)^2 \int_1^2 \frac{1}{p_{pr}} (0.0657\, p_{pr})\, dp_{pr}$$

$$= -(1.986)(392)\left(\frac{760}{392}\right)^2 (0.0657)\left(\frac{800}{667} - \frac{14.7}{667}\right) = -226 \text{ Btu/lb-mol}$$

Total enthalpy change is

$$\Delta h = \Delta h_p + \Delta h_T = 2220 - 226 = 1994 \text{ Btu/lb-mol}$$

From Brown's *h-s* chart for 0.7 gravity natural gas (Fig. 3.10), at 14.7 psia and 100°F,

$$h_1 = 680 \text{ Btu/lb-mol}$$

At 800 psia and 300°F,

$$h_2 = 2600 \text{ Btu/lb-mol}$$
$$\Delta h = h_2 - h_1 = 2600 - 680 = 1920 \text{ Btu/lb-mol}$$

Entropy change, Δs

From Eq. 3.43, modified to reflect 100°F rather than 32°F,

$$\Delta s = \bar{C}_p \ln \frac{T}{560} - 1.986 \left[\bar{z} + \frac{T}{T_{pc}} \left(\frac{\overline{\partial z}}{\partial T_{pr}} \right)_{p_{pr}} \right] \ln \frac{p}{14.7}$$

At constant pressure (14.7 psia), $\bar{C}_p$ at an average temperature of 200°F is 11.1 Btu/lb-mol °F. The average z-factor from $p_{pr} = 0.02$ to 1.20 at $T_{pr} = 1.94$, from the z-factor chart (Fig. 2.4), is

$$\bar{z} = \frac{0.998 + 0.960}{2} = 0.979$$

$$\left(\frac{\overline{\partial z}}{\partial T_{pr}} \right)_{p_{pr}} = \frac{\left(\frac{\partial z}{\partial T_{pr}} \right)_{p_{pr} = 0.02} + \left(\frac{\partial z}{\partial T_{pr}} \right)_{p_{pr} = 1.20}}{2}$$

$$\left(\frac{\overline{\partial z}}{\partial T_{pr}} \right)_{p_{pr}} = \frac{0.00 + (0.0657)(1.20)}{2} = 0.0394$$

Thus,

$$\Delta s = (11.1) \ln \frac{760}{560} - 1.986 \left[0.979 + \frac{760}{392}(0.0394) \right] \ln \frac{800}{14.7}$$

$$= -4.99 \text{ Btu/lb-mol °F}$$

From Brown's h-s chart for 0.7 gravity natural gas (Fig. 3.10), at 14.7 psia and 100°F,

$$s_1 = 1.2 \text{ Btu/lb-mol °F}$$

At 800 psia and 300°F,

$$s_2 = -3.7$$
$$\Delta s = s_2 - s_1 = -3.7 - 1.2 = -4.9 \text{ Btu/lb-mol °F}$$

Example 3.5. Calculate the amount of energy involved in taking 1 MMcf (measured at 14.73 psia and 60°F) of the 0.7 specific gravity gas of Example 3.4 from 14.7 psia and 100°F to 800 psia and 300°F, using Edmister's $\Delta h/T_c$ chart and C_p chart.

Solution

Number of moles involved = $(1 \times 10^6)/378.6 = 2641$.

For a constant pressure process (at 14.7 psia), the energy required to go from 100 to 300°F is

$$\Delta h_p = (11.1)(300 - 100) = 2220 \text{ Btu/lb-mol}$$

and

$$\Delta H_p = (2641)(2220) = 5.863 \times 10^6 \text{ Btu}$$

For constant temperature (300°F) compression, using Fig. 3.7, at $T_{pr} = 1.94$ and $p_{pr} = 0.02$,

$$-\frac{\Delta h}{T_c} = 0.015 \qquad \text{or} \qquad \Delta h = -0.015T_c$$

at

$$T_{pr} = 1.94 \qquad \text{and} \qquad p_{pr} = 1.20,$$

$$-\frac{\Delta h}{T_c} = 0.7 \qquad \text{or} \qquad \Delta h = -0.7T_c$$

Thus,

$$\Delta h_T = (-0.7 + 0.015)(392) = -268 \text{ Btu/lb-mol}$$

and

$$\Delta H_T = (2641)(-268) = -7.092 \times 10^5$$

The total energy required is

$$\begin{aligned}\Delta H &= \Delta H_p + \Delta H_T \\ &= 5.863 \times 10^6 - 7.092 \times 10^5 = 5.154 \times 10^6 \text{ Btu}\end{aligned}$$

3.6 THE ENTHALPY-ENTROPY DIAGRAM (MOLLIER DIAGRAM)

In the analysis of steady-flow processes, an enthalpy-entropy diagram is quite useful. The enthalpy is the important thermodynamic property in the steady-flow energy balance, and entropy is the principal property of concern with respect to the second law. Thus, the coordinates of an *h-s* diagram represent the two major properties of interest in the first- and second-law analysis of open systems. The vertical distance between two states on this diagram is a measure of Δh. The enthalpy change, in turn, is related through the adiabatic steady-flow energy balance to the work or kinetic-energy changes for turbines, compressors, nozzles, and so on. The horizontal distance between two states Δs is a measure of the degree of irreversibility for an adiabatic process. The isentropic process conveniently appears as a vertical line, allowing the end state of an idealized adiabatic process to be found easily. As a consequence, the *h-s* diagram is helpful in visualizing process changes for control-volume analyses.

Brown has prepared enthalpy-entropy diagrams for natural gases. The *h-s* diagrams of natural gases of 0.6, 0.7, 0.8, 0.9, and 1.0 gravity are given in Figs. 3.9 to 3.13. The pseudocritical temperatures and pressures used are given on the charts and correspond essentially to hydrocarbon gases. The datum for each chart is 32°F, 1 atm. A chart for a gas of 0.7 gravity but containing 10 mole % nitrogen is given by Fig. 3.14. Campbell presents an *h-s* chart for an average natural gas

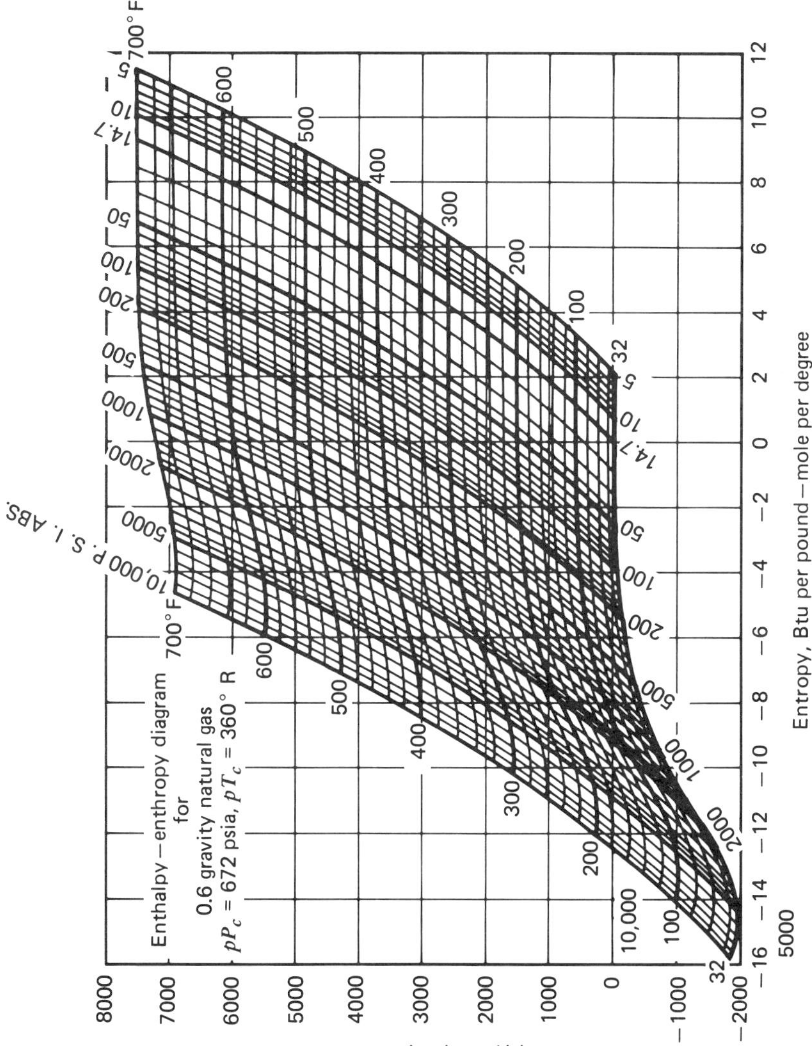

Fig. 3.9 Enthalpy-entropy diagram for 0.6-gravity natural gas. (After Brown.)

(0.65–0.75 gravity) in Fig. 3.15. These charts are useful in finding the temperature change on expanding or compressing gases and for finding the reversible work of compression or expansion.

Enthalpy-entropy charts apply only to the gaseous state; if a gas is cooled below its dew point, condensation occurs and heat removal cannot be determined directly from the charts.

Example 3.6. Cooling or heating a Gas at Constant Pressure.

(a) What is the amount of heat that must be removed in cooling 1 lb-mol of 0.6 specific gravity natural gas at 200 psia from 600 to 100°F?

(b) If this gas is then heated from 100 to 300°F, how much heat is required per pound-mole of gas?

Solution

(a) Using Fig. 3.9 for 0.6 gravity natural gas, read the enthalpy at the intersection of the 600°F line and the 200-psia line, which is $h = 6100$ Btu/lb-mol. Then follow the 200-psia line to its intersection with the 100°F line and read the enthalpy h_2 of 500 Btu/lb-mol.

$$\begin{aligned}\text{Heat removed} &= h_1 - h_2 \\ &= 6100 - 500 \\ &= 5600 \text{ Btu/lb-mol}\end{aligned}$$

(b) At 200 psia and 300 °F the enthalpy $h_3 = 2600$ Btu/lb-mol.

$$\begin{aligned}\text{Heat added} &= h_3 - h_2 \\ &= 2600 - 500 \\ &= 2100 \text{ Btu/lb-mol}\end{aligned}$$

Example 3.7. Adiabatic Reversible Expansion of a Gas. If a 0.7 specific gravity natural gas at 300°F and 500 psia is expanded adiabatically and reversibly to 100 psia, what will be the final temperature of the gas?

Solution

Using the *h-s* diagram for 0.7 gravity natural gas (Fig. 3.10), at the intersection of the 300°F and 500 psia lines, the entropy $s_1 = -2.5$ Btu/lb-mol °F. In an adiabatic process no heat is added or removed. Since the process is reversible,

$$Q = \int T\, ds = 0$$

or

$$ds = 0 \qquad \text{and} \qquad s = \text{constant}$$

Thus a reversible adiabatic process is isentropic. To get the properties at the final state, follow the entropy line for s_1 to its intersection with the 100-psia line and read the temperature T_2 of 105°F.

Example 3.8. Throttling or Joule–Thomson Effect. If a gas of 0.8 specific gravity at 2000 psia and 200°F expands through a small orifice without the addition or

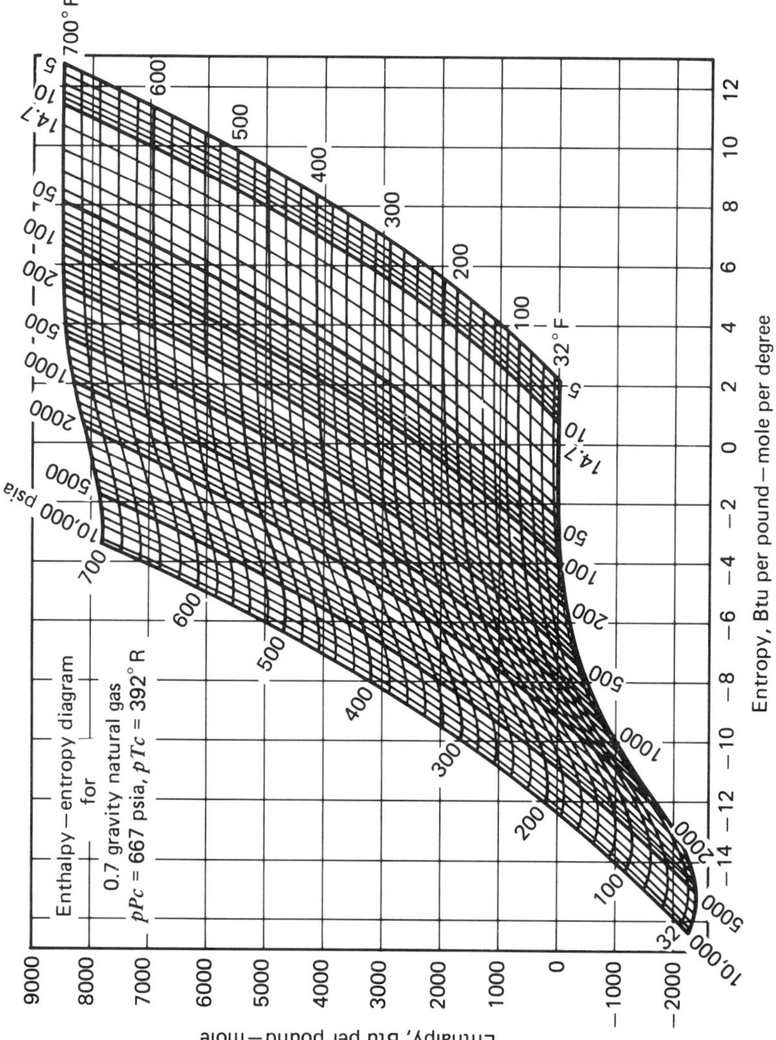

Fig. 3.10 Enthalpy-entropy diagram for 0.7 gravity natural gas. (After Brown.)

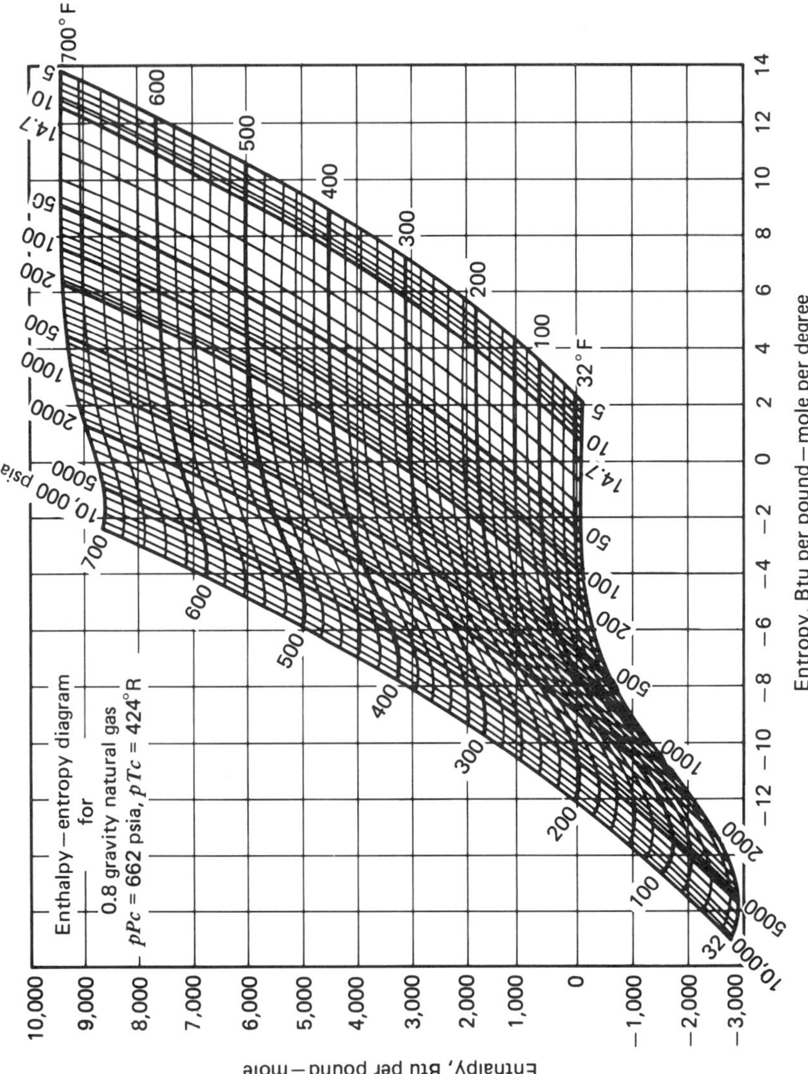

Fig. 3.11 Enthalpy-Entropy diagram for 0.8 gravity natural gas. (After Brown.)

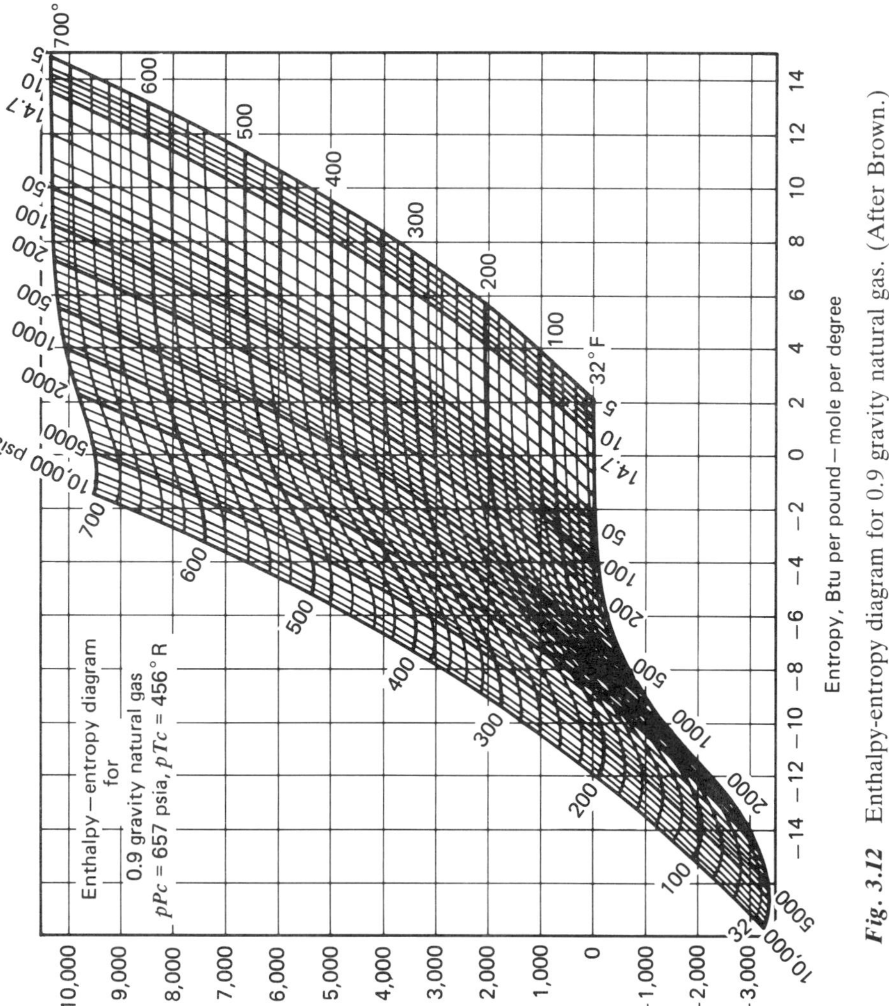

Fig. 3.12 Enthalpy-entropy diagram for 0.9 gravity natural gas. (After Brown.)

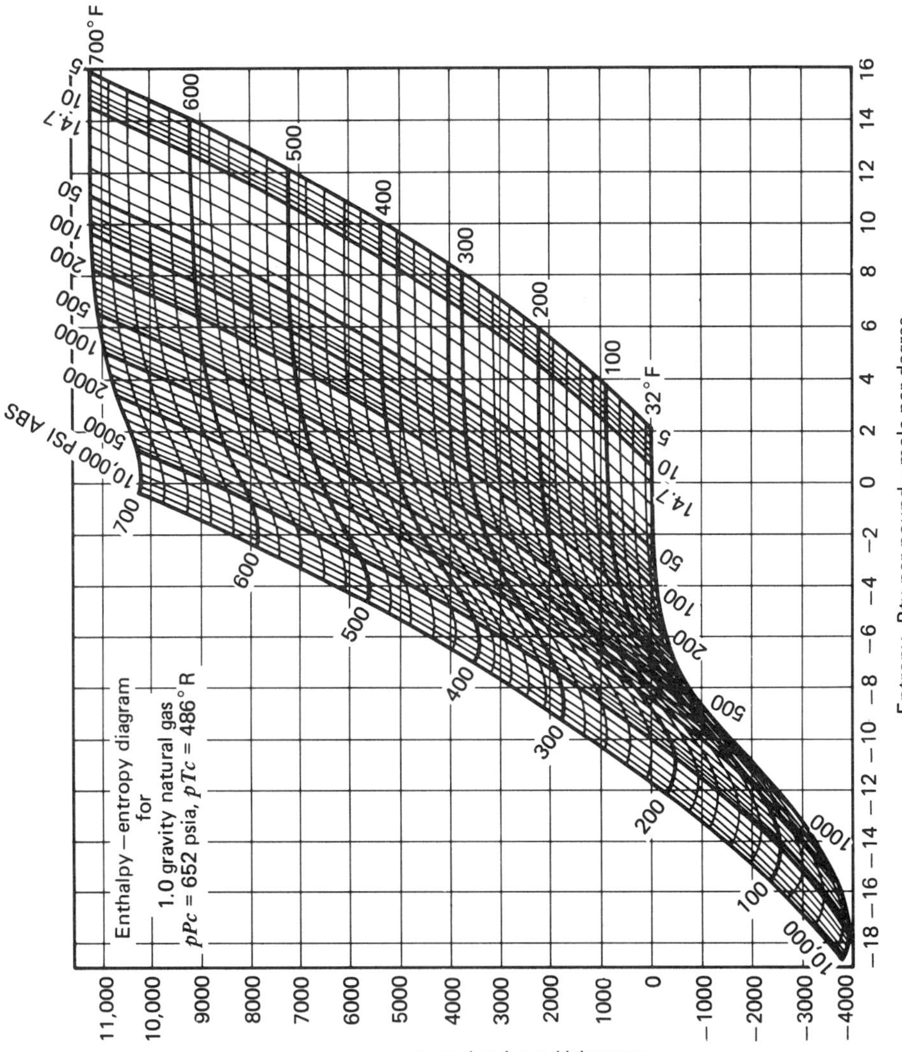

Fig. 3.13 Enthalpy-entropy diagram for 1.0 gravity natural gas. (After Brown.)

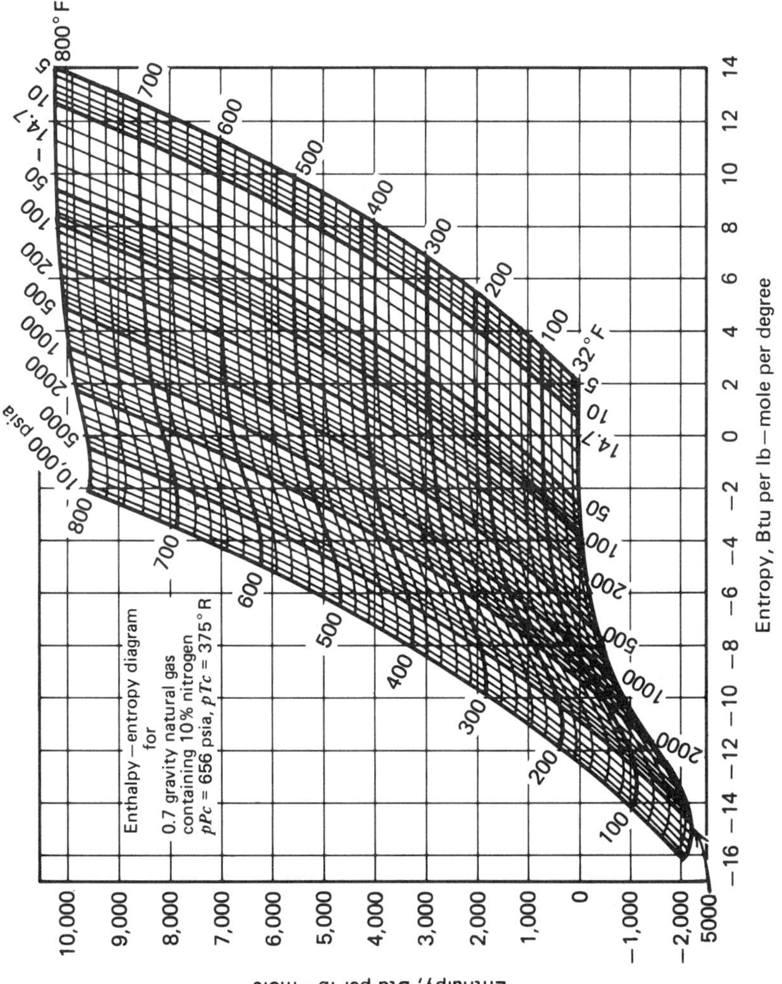

Fig. 3.14 Enthalpy-entropy diagram for 0.7 gravity gas containing 10% nitrogen. (After Brown.)

subtraction of heat and is brought finally to its initial velocity and a pressure of 50 psia, what will be its temperature?

Solution

Refer to Fig. 3.11 for 0.8 gravity gas. At the intersection of the 2000-psia and 200°F lines, $h_1 = 700$ Btu/lb-mol. Since the heat content is constant, follow the 700 Btu/lb-mol to its intersection with the 50-psia line and read the temperature T_2 of 100 °F.

In most cases throttling processes occur so rapidly and in such a small space, that there is neither sufficient time nor a large enough area for much heat transfer. Therefore, such processes may be assumed to be adiabatic. But they are not reversible.

Example 3.9. Adiabatic Reversible Compression. A 0.9 specific gravity natural gas at 50 psia and 80°F is compressed adiabatically and reversibly to 1000 psia. What is the temperature of the compressed gas?

Solution

Refer to Fig. 3.12 for 0.9 gravity gas. At the intersection of the 50-psia line and the 80°F line, the entropy $s_1 = -1.3$ Btu/lb-mol °F. Then follow the $s_1 = -1.3$ line to its intersection with the 1000-psia line and read $T_2 = 410$°F.

Example 3.10. Isothermal Expansion of a Gas

(a) One pound-mole of 1.0 specific gravity natural gas at 2000 psia and 420°F is expanded through a throttling valve or orifice and is brought to its initial velocity and a pressure of 100 psia. How much heat must be added to maintain the temperature of the gas at 420°F?

(b) What would the temperature be if no heat is added?

Solution

(a) *h-s* diagram for 1.0 gravity gas is given in Fig. 3.13. The enthalpy at the intersection of the 420°F and 2000-psia lines is $h_1 = 4700$ Btu/lb-mol. Follow the 420°F line to its intersection with the 100-psia line and read the enthalpy $h_2 = 5650$ Btu/lb-mol. Since no external work has been done, the amount of heat added is

$$h_2 - h_1 = 5650 - 4700 = 950 \text{ Btu/lb-mol}$$

(b) If no heat is added to the gas the final temperature would be obtained by following the $h_1 = 4700$ Btu/lb-mol line to its intersection with the 100-psia line. This gives $T_2 = 362$ °F.

Example 3.11. Forty million cubic feet per day (measured at 60°F and 14.73 psia) of an average natural gas at 240°F are flowing freely through an expansion valve with the pressure dropping from 2000 to 1000 psia.

(a) What is the temperature change across the valve?

(b) What is the maximum work that could be derived from this expanding gas stream without the use of heat?

(c) What would be the temperature change on the gas when expanding through an adiabatic reversible engine?

Solution

Refer to the *h-s* diagram for average natural gas (Fig. 3.15).

(a) At 2000 psia and 240°F,

$$h_1 = 1400 \text{ Btu/lb-mol}$$

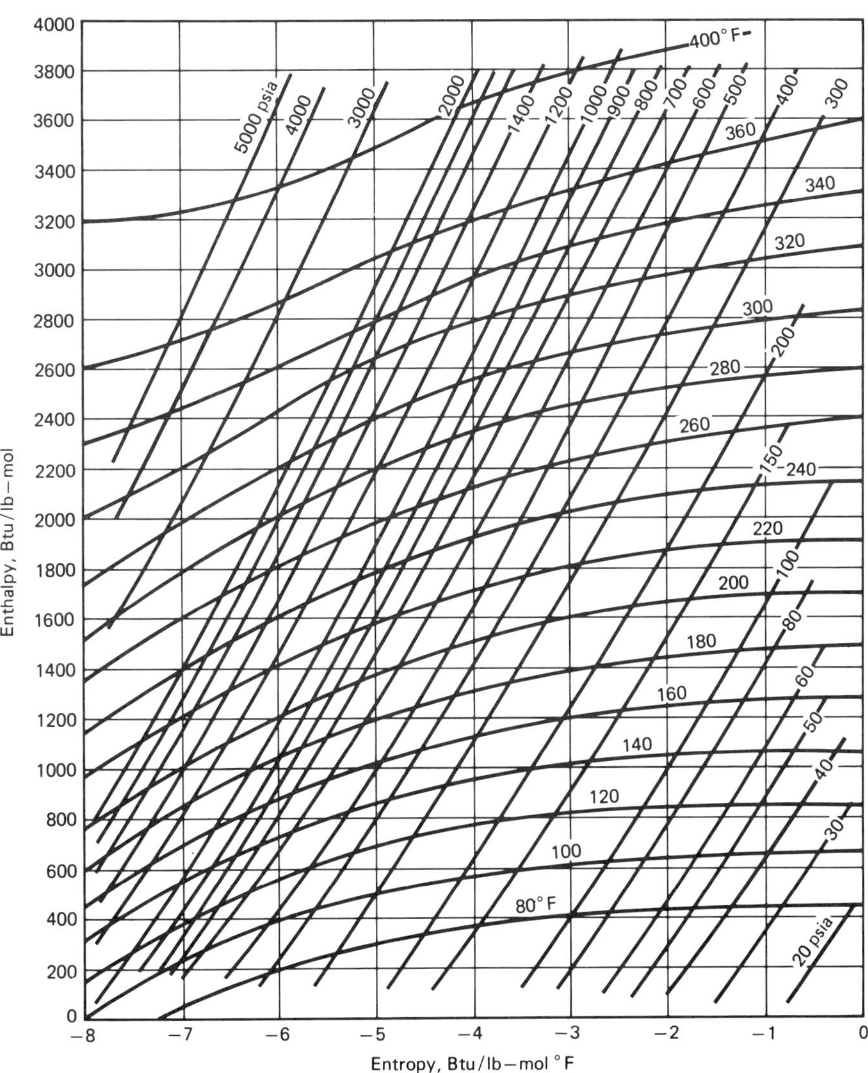

Fig. 3.15 Enthalpy-entropy diagram for a 0.65 to 0.75 specific gravity natural gas. (After Campbell.)

For the throttling process across the valve the enthalpy is constant. At 1000 psia and h = 1400 Btu/lb-mol,

$$T_2 = 212°\text{F}$$

(b), (c) Maximum work would be obtained using an adiabatic and reversible (i.e., isentropic) process. At 2000 psia and 240°F,

$$s_1 = -7.00 \text{ Btu/lb-mol °F}$$

At 1000 psia and $s = -7.00$ Btu/lb-mol °F,

$$T_3 = 140 \text{ °F}$$
$$h_3 = 560 \text{ Btu/lb-mol}$$

Work done by gas $= -\Delta h = h_1 - h_3 = 1400 - 560 = 840$ Btu/lb-mol

$$1 \text{ horsepower (hp)} = \frac{33{,}000 \times 60}{778.2} = 2{,}545 \text{ Btuh}$$

1 lb-mol occupies 378.6 cu ft at 60°F and 14.73 psia

$$\text{Power} = \frac{(40{,}000{,}000)(840)}{(378.6)(24)(2{,}545)} = 1453 \text{ hp}$$

REFERENCES

Brown, G. G. "A Series of Enthalpy-Entropy Charts for Natural Gases." *Trans. AIME* **60**, p. 65, 1945.

Brown, G. G., et al. *Unit Operations*. New York: Wiley, 1950.

Campbell, J. M. *Gas Conditioning and Processing*. Norman, Okla: Campbell Petroleum Series, 1978.

Edminster, W. C. "Application of Thermodynamics to Hydrocarbon Processing." *Petroleum Engineering series* **27–28**; **11**, p. 129; **12**, p. 116; **1**, p. 128; **2**, p. 137. 1948–1949.

Katz, D. L., et al. *Handbook of Natural Gas Engineering*. New York: McGraw-Hill, 1959.

Standing, M. B. "Natural Gas Engineering Notes." Stanford University, 1977.

PROBLEMS

3.1 *Calculation of Thermodynamic Properties from Pressure-Volume-Temperature and Specific Heat Data*

Calculate the molal enthalpy and molal entropy change, Δh and Δs, when a quantity of 0.8 specific gravity natural gas is compressed from 50.0 psia and 100°F to 800 psia and 300°F. Check your answers against Brown's enthalpy-entropy chart for 0.8 gravity natural gas.

3.2 Calculate the amount of energy involved in taking 1 MMcf (measured at 14.73 psia and 60°F) of the 0.8 specific gravity gas of Problem 1 from 14.7 psia and 100°F to 800 psia and 300°F, using Edmister's $\Delta h/T_c$ chart and C_p chart.

3.3 *Cooling or Heating a Gas at Constant Pressure*

(a) What is the amount of heat that must be removed in cooling 1 lb-mol of 0.8 specific gravity natural gas at 200 psia from 600 to 100°F?
(b) If this gas is then heated from 100 to 300°F, how much heat is required per pound-mole of gas?

3.4 *Adiabatic Reversible Expansion of a Gas*

If a 0.8 specific gravity natural gas at 300°F and 500 psia is expanded adiabatically and reversibly to 100 psia, what will be the final temperature of the gas?

3.5 *Throttling or Joule–Thomson Effect*

If a gas of 0.6 specific gravity at 2000 psia and 200°F expands through a small orifice without the addition or subtraction of heat and is brought finally to its initial velocity and a pressure of 50 psia, what will be its temperature?

3.6 *Adiabatic Reversible Compression*

A 0.8 specific gravity natural gas at 50 psia and 80°F is compressed adiabatically and reversibly to 1000 psia. What is the temperature of the compressed gas?

3.7 *Isothermal Expansion of a Gas*

(a) One pound-mole of 0.8 specific gravity natural gas at 2000 psia and 420°F is expanded through a throttling valve or orifice and is brought to its initial velocity and a pressure of 100 psia. How much heat must be added to maintain the temperature of the gas at 420°F?
(b) What would the temperature be if no heat is added?

3.8 Forty million cubic feet per day (measured at 60°F and 14.73 psia) of an average natural gas at 320°F is flowing freely through an expansion valve with the pressure dropping from 4000 to 1000 psia.

(a) What is the temperature change across the valve?
(b) What is the maximum work that could be derived from this expanding gas stream without the use of heat?
(c) What would be the temperature change on the gas when expanding through an adiabatic reversible engine?

4

SEPARATION AND PROCESSING

4.1 INTRODUCTION

Some accumulations furnish natural gas of very high purity (almost methane, CH_4). Natural gases of this composition do not require any processing; they only require a dehydration treatment before being conveyed to the transportation pipeline. Hydrocarbon streams as produced from other reservoirs are complex mixtures of hundreds of different compounds. A typical wellstream is a high-velocity, turbulent, constantly expanding mixture of gases and hydrocarbon liquids, intimately mixed with water vapor, free water, solids, and other contaminants. As it flows from the hot, high-pressure petroleum reservoir, the wellstream undergoes continuous pressure and temperature reduction. Gases evolve from the liquids, water vapor condenses, and some of the wellstream changes in character from liquid to bubbles, mist, and free gas. The high-velocity gas carries liquid droplets, and the liquid carries gas bubbles.

In most cases, the gas, liquid hydrocarbons, and free water should be separated as soon as possible after bringing them to the surface; these phases should be handled and transported separately. This separation of liquids from the gas phase is accomplished by passing the wellstream through an oil-gas or oil-gas-water separator.

Field processing simply removes undesirable components and separates the wellstream into salable gas and hydrocarbon liquids, recovering the maximum amounts of each at the lowest possible overall cost. Field processing of natural gas consists of four basic processes:

1. Separation of the gas from free liquids such as crude oil, hydrocarbon condensate, water, and entrained solids.
2. Processing the gas to remove condensable and recoverable hydrocarbon vapors.

3. Processing the gas to remove condensable water vapor which, under certain conditions, might cause hydrate formation.
4. Processing the gas to remove other undesirable compounds, such as hydrogen sulfide or carbon dioxide.

4.2 GAS AND LIQUID SEPARATION

Separation of wellstream gas from free liquids is by far the most common of all field-processing operations and also one of the most critical of the processes. Composition of the fluid mixture determines the design criteria for sizing and selecting a separator for a hydrocarbon stream. In the case of low-pressure oil wells, the liquid phase will be large in volume compared to the gas phase. In the case of high-pressure gas-distillate wells, the gas volume will be higher compared to the liquid volume. The liquid produced with high-pressure gas is generally a high API gravity hydrocarbon, usually referred to as distillate or condensate. However, both low-pressure oil wells or high-pressure gas-distillate wells may contain free water.

Separators are also used in many locations other than at wellhead production batteries, such as gasoline plants, upstream and downstream of compressors, liquid traps in gas transmission lines, inlet scrubbers to dehydration units, and gas sweetening units. At some of these other locations, separators are referred to as scrubbers, knockouts, and free liquid knockouts. All these vessels serve the same primary purpose: to separate free liquids from the gas stream.

A properly designed wellstream separator must perform the following functions:

1. Cause a primary-phase separation of the mostly liquid hydrocarbons from those that are mostly gas.
2. Refine the primary separation by removing most of the entrained liquid mist from the gas.
3. Further refine the separation by removing the entrained gas from the liquid.
4. Discharge the separated gas and liquid from the vessel and ensure that no reentrainment of one into the other takes place.

Sivalls has presented an excellent discussion of oil and gas separation. This development is taken largely after his paper.

4.2.1 Internal Construction of Separators

The principal items of construction that should be present in a good liquid-gas separator are the same, regardless of the overall shape or configuration of the vessel. Some of these features are as follows:

1. A centrifugal inlet device where the primary separation of the liquid and gas is made.

2. A large settling section of sufficient length or height to allow liquid droplets to settle out of the gas stream with adequate surge room for slugs of liquid.
3. A mist extractor or eliminator near the gas outlet to coalesce small particles of liquid that will not settle out by gravity.
4. Adequate controls consisting of level control, liquid dump valve, gas back pressure valve, safety relief valve, pressure gauge, gauge glass, instrument gas regulator, and piping.

The bulb of the gas-liquid separation occurs in the inlet centrifugal separating section. Here, the incoming stream is spun around the walls of a small cylinder, which would be the walls of the vessel in the case of a vertical or spherical separator. This subjects the fluids to a centrifugal force up to 500 times the force of gravity. This action stops the horizontal motion of the free liquid entrained in the gas stream and forces the liquid droplets together where they will fall to the bottom of the separator into the settling section.

The settling section lets the turbulence of the fluid stream subside and allows liquid droplets to fall to the bottom of the vessel because of the difference in the gravity between the liquid and gas phases. A large open space in the vessel is adequate for this purpose. Introducing special quieting plates or baffles with narrow openings only complicates the internal construction of the separator and provides places for sand, sludge, paraffin, and other materials to collect and eventually plug the vessel and stop the flow. The separation of liquid and gas using the centrifugal inlet feature and a large, open settling section produces a more stable liquid product, which can be contained in atmospheric or low-pressure storage tanks. Minute scrubbing of the gas phase by use of internal baffling or plates may produce more liquid to be discharged from the separator, but the product will not be stable since light ends will be entrained in it, incurring more vapor losses from the storage system.

Sufficient surge room should be allowed in the settling section to handle slugs of liquid without carryover to the gas outlet. This can be accomplished to some extent by placement of the liquid level control in the separator, which in turn determines the liquid level. The amount of surge room required is often difficult, if not impossible, to determine from well test or flowing data. In most cases, the separator size used for a particular application is a compromise between initial cost and possible surging requirements.

Another major item required for good and complete liquid-gas separation is a mist eliminator or extractor near the gas outlet. Small liquid droplets that will not settle out of the gas stream, due to little or no gravity difference between them and the gas phase, will be entrained and pass out of the separator with the gas. This can be almost eliminated by passing the gas through a mist eliminator near the gas outlet, which has a large surface impingement area. The small liquid droplets will hit the surfaces, coalesce, and collect to form larger droplets that will then drain back to the liquid section in the bottom of the vessel. A stainless steel woven-wire mesh mist eliminator is probably the most efficient type since it removes up to 99.9% or more of the entrained liquids from the gas stream. This type offers more surface area for collecting liquid droplets per unit volume than vane types, ceramic packing, or other configurations. The vane mist eliminators apply in areas where there is entrained solid material in the gas phase that may collect and plug a wire mesh mist eliminator.

4.2.2 Types of Separators

There are four major types or basic configurations of separators, generally available from manufacturers: vertical, horizontal single tube, horizontal double tube, and spherical. Each type has specific advantages, and selection is usually based on which one will accomplish the desired results at the lowest cost.

Vertical

A vertical separator is often used on low to intermediate gas-oil ratio well streams and where relatively large slugs of liquid are expected. It will handle greater slugs of liquid without carryover to the gas outlet, and the action of the liquid level control is not as critical (Fig. 4.1). A vertical separator occupies less floor space, an important consideration where space might be expensive as on an offshore platform. Due to the greater vertical distance between the liquid level and the gas outlet, there is less tendency to revaporize the liquid into the gas phase. However, because the natural upward flow of gas in a vertical vessel opposes the falling droplets of liquid, it takes a larger-diameter separator for given gas capacity than a horizontal vessel. Also, vertical vessels are more expensive to fabricate and ship in skid-mounted assemblies.

Horizontal Single Tube

The horizontal separator (Fig. 4.2) may be the best separator for the money. The horizontal separator has a much greater gas-liquid interface area consisting of a large, long, baffled gas-separation section that permits much higher gas velocities. This type of separator is easier to skid-mount and service and requires less piping for field connections and a smaller diameter for a given gas capacity. Several separators can be stacked easily into stage-separation assemblies, minimizing space requirements.

In operation, gas flows horizontally and, at the same time, falls toward the liquid surface. The gas flows in the baffle surfaces and forms a liquid film that is drained away to the liquid section of the separator. The baffles need only be longer than the distance of liquid trajectory travel at the design gas velocity. The liquid level control placement is more critical than in a vertical separator, and surge space is somewhat limited. Horizontal separators are almost always used for high gas-oil ratio well streams, for foaming well streams, or for liquid-from-liquid separation.

Horizontal Double Tube

A horizontal double-tube or double-barrel separator (Fig. 4.3) has all the advantages of a normal horizontal separator plus a much higher liquid capacity. Incom-

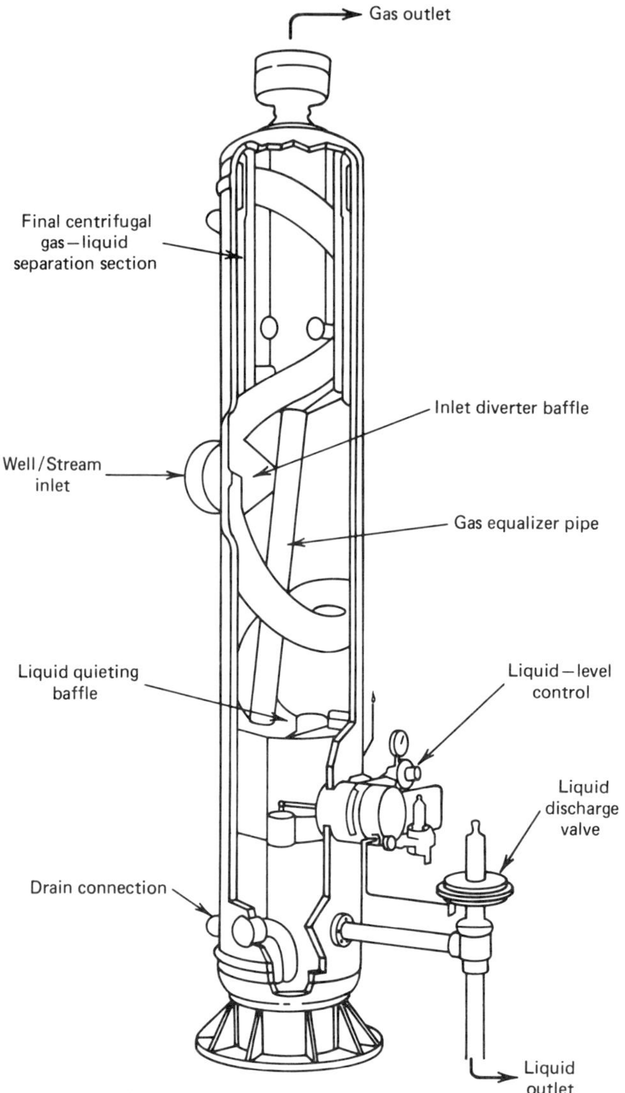

Fig. 4.1 Conventional vertical separator. (Courtesy Petroleum Extension Service.)

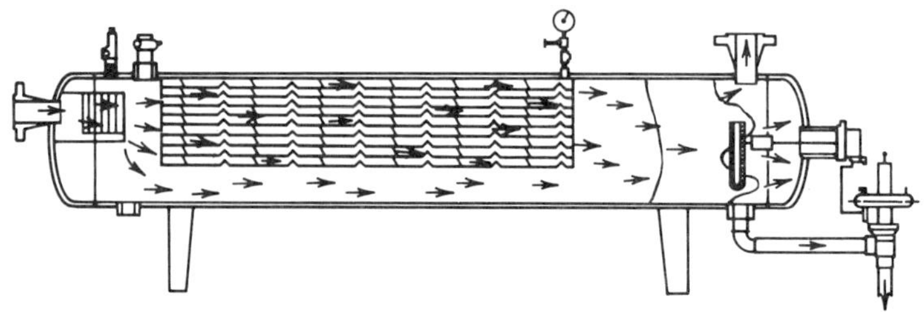

Fig. 4.2 Conventional horizontal separator. (Courtesy Petroleum Extension Service.)

ing free liquid is immediately drained away from the upper section into the lower section. The upper section is filled with baffles, and gas flow is straight through and at higher velocities.

Spherical

Spherical separators offer an inexpensive and compact vessel arrangement (Fig. 4.4). However, these types of vessels have a very limited surge space and liquid settling section. The placement and action of the liquid level control in this type of vessel is very critical.

Three-phase or oil-gas-water separation can be easily accomplished in any type of separator by installing either special internal baffling to construct a water leg or water siphon arrangement, or by using an interface liquid level control. A three-phase feature is difficult to install in a spherical separator because of the limited internal space available. Figure 4.5 is a horizontal three-phase separator.

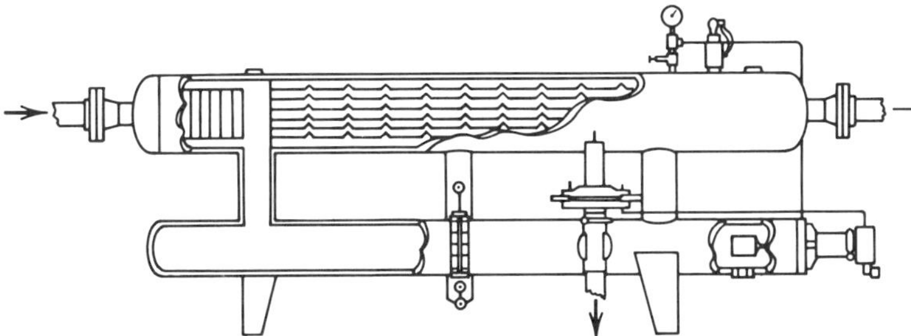

Fig. 4.3 Conventional horizontal double-barrel separator. (Courtesy Petroleum Extension Service.)

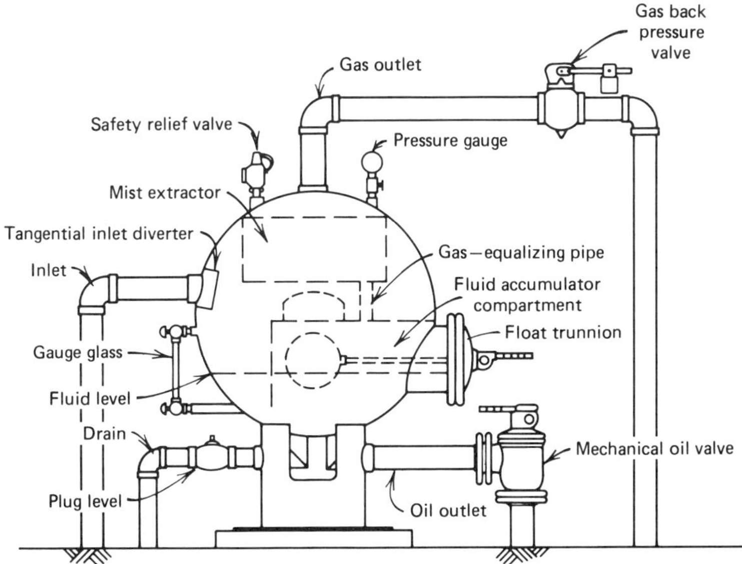

Fig. 4.4 Spherical low-pressure separator. (After Sivalls.)

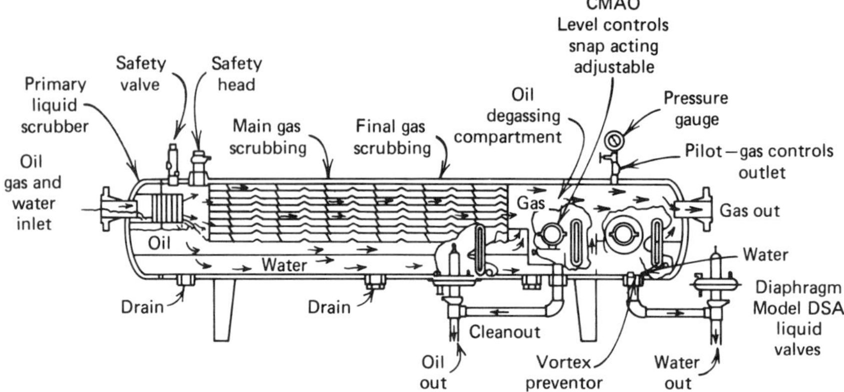

Fig. 4.5 Conventional horizontal three-phase separator. (Courtesy Petroleum Extension Service.)

With three-phase operation, two liquid level controls and two liquid dump valves are required. Three-phase separators are used commonly for well testing and in instances where free water readily separates from the oil or condensate.

From an evaluation of the advantages and disadvantages of the various types of separators, the horizontal single-tube separator has emerged as the one that gives the most efficient operation for initial investment cost for high-pressure gas-distillate wells with high gas-oil ratios. For high liquid loadings, vertical separators should be considered.

4.2.3 Factors Affecting Separation

Separator operating pressure, separator operating temperature, and fluid stream composition affect the operation and separation between the liquid and gas phases in a separator. Changes in any one of these factors on a given fluid wellstream will change the amount of gas and liquid leaving the separator. Generally, an increase in operating pressure or a decrease in operating temperature will increase the liquid covered in a separator. However, there are optimum points in both cases beyond which further changes will not aid in liquid recovery. In fact, storage system vapor losses may become too great before these points are reached.

In the case of wellhead separation equipment, an operator generally wants to determine the optimum conditions for a separator to effect the maximum income. Again, generally speaking, the liquid recovered is worth more than the gas. So, high liquid recovery is a desirable feature, providing it can be held in the available storage system. Also, pipeline requirements for the Btu content of the gas may be another factor in separator operation. Without the addition of expensive mechanical refrigeration equipment, it is often unfeasible to try to lower the operating temperature of a separator. However, on most high-pressure wells, an indirect heater heats the gas prior to pressure reduction to pipeline pressure in a choke. By careful operation of this indirect heater, the operator can prevent overheating the gas stream ahead of the choke, which will adversely affect the temperature of the separator downstream from the indirect heater.

The operator can also control operating pressure to some extent by use of back-pressure valves within the limitation of the flowing characteristics of the well against a set pressure head and the transmission line pressure requirements. As previously mentioned, higher operating pressure will generally result in higher liquid recovery.

An analysis can be made using the wellstream composition to find the optimum temperature and pressure at which a separator should operate to give maximum liquid or gas phase recovery. These calculations, known as flash vaporization calculations, require a trial-and-error solution and are more generally adapted to solution by a programmed computer. An operator can, however, make trial settings within the limitations of the equipment to find the operating conditions that result in the maximum amount of gas or liquids. In the case where separators are used as scrubbers or knockouts ahead of other treating equipment or compressors, it is generally best to remove the maximum amount of liquid from the gas stream to prevent operational damage to the equipment downstream from the scrubber.

4.2.4 Separator Design

The following discussion on oil-gas separator design has been adapted from Sivalls' excellent treatment of the subject. Sivalls' tables, graphs, and procedures are accepted as the standard of the industry.

Gas Capacity

The gas capacity of oil-gas separators has been calculated for many years from the following empirical relationship proposed by Souders-Brown:

$$v = K \left[\frac{\rho_L - \rho_g}{\rho_g} \right]^{0.5} \tag{4.1}$$

and

$$A = \frac{q}{v} \tag{4.2}$$

where

v = superficial gas velocity based on total cross-sectional area of vessel, fps
A = cross-sectional area of separator, sq ft
q = gas flow rate at operating conditions, cfs
ρ_L = density of liquid at operating conditions, lbm/cu ft
ρ_g = density of gas at operating conditions, lbm/cu ft
K = empirical factor

Vertical separators	$K = 0.06$ to 0.35, avg 0.21
Horizontal separators	$K = 0.40$ to 0.50, avg 0.45
Wire mesh mist eliminators	$K = 0.35$
Bubble cap trayed columns	$K = 0.16$ for 24-in. spacing
Valve tray columns	$K = 0.18$ for 24-in. spacing

This relationship is based on a superficial vapor velocity through a vessel. The vapor or gas capacity is then in relationship to the diameter of the vessel. The formula is also used for other designs such as trayed towers in dehydration or gas sweetening units and for the sizing of mist eliminators. Therefore, the K-factor for these is presented along with the factors for vertical and horizontal separators so the relationship that one bears with the other can be seen.

$$q = \frac{2.40(D)^2(K)(p)}{z(T + 460)}\left[\frac{\rho_L - \rho_g}{\rho_g}\right]^{0.5} \tag{4.3}$$

where

q = gas capacity at standard conditions, MMscfd
D = internal diameter, ft
p = operating pressure, psia
T = operating temperature, °F
z = gas deviation factor

Since this equation is empirical, a better determination of separator gas capacity might be made from actual manufacturers' field test data. Figures 4.6 to 4.13 are gas capacity charts for various standard size separators based on operating pressure. These actual manufacturers' gas capacity charts consider height differences in vertical separators and length differences in horizontal separators, which add to the gas capacity of the separators. As can be seen, height and length differences are not taken into account in the Souders–Brown equation. But field experience has proved that additional gas capacity can be obtained by increasing height of vertical separators and length of horizontal separators.

As seen on the sizing charts for horizontal separators, a correction must be made for the amount of liquid in the bottom of the separator. This is for single-tube horizontal vessels. One-half full of liquid is more or less standard for most single-tube horizontal separators. However, the gas capacity can be increased by lowering liquid level to increase the available gas space within the vessel. Gas capacities of horizontal separators with liquid sections one-half full, one-third full, or one-quarter full can be determined from the gas capacity charts.

Liquid Capacity

The liquid capacity of a separator is primarily dependent on the retention time of the liquid within the vessel. Good separation requires sufficient time to obtain an equilibrium condition between the liquid and gas phase at the temperature and pressure of separation. The liquid capacity of a separator or the settling volume required based on retention can be determined from the following equation:

$$W = \frac{1440V}{t} \quad \text{or} \quad t = \frac{1440V}{W} \quad \text{or} \quad V = \frac{Wt}{1440} \tag{4.4}$$

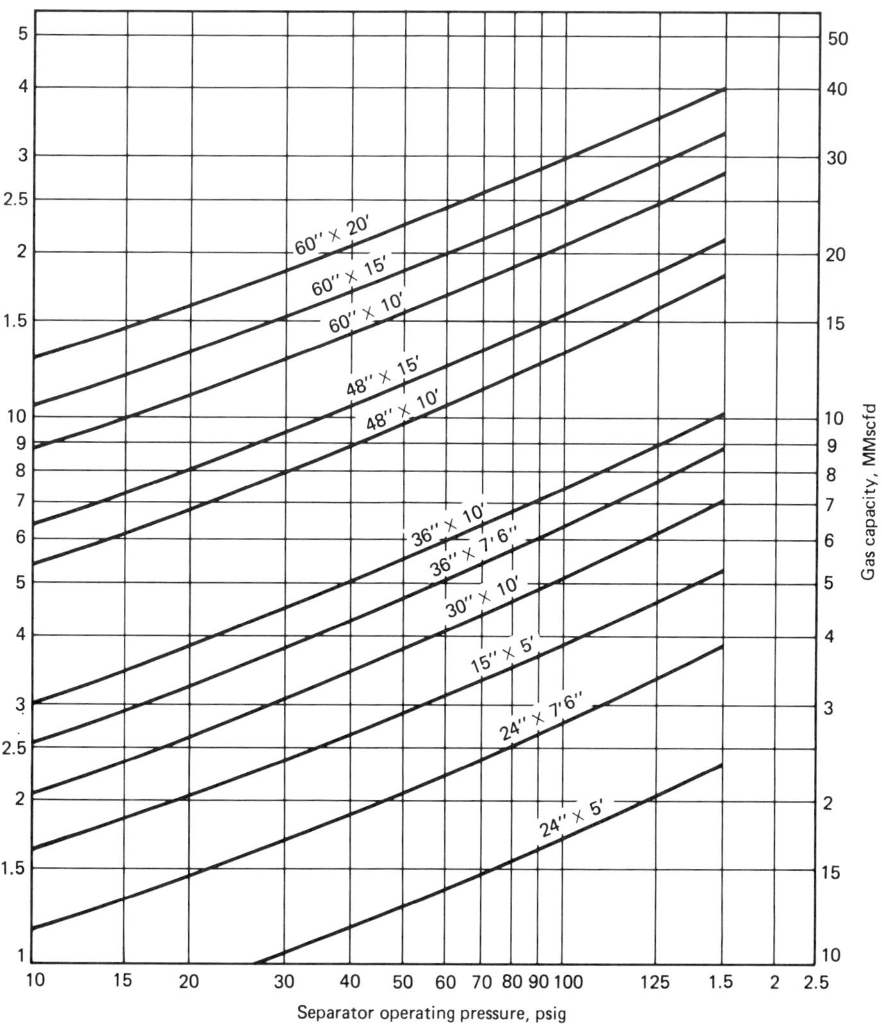

Fig. 4.6 Gas capacity of vertical low-pressure separators. (After Sivalls.)

where

W = liquid capacity, bbl/day
V = liquid settling volume, bbl
t = retention time, min

Basic design criteria for liquid retention times in separators have been determined by numerous field tests:

Oil-gas separation	1 min
High-pressure oil-gas-water separation	2 to 5 min
Low-pressure oil-gas-water separation	5 to 10 min at 100°F and up
	10 to 15 min at 90°F
	15 to 20 min at 80°F
	20 to 25 min at 70°F
	25 to 30 min at 60°F

Figures 4.14 and 4.15 are sizing charts for the liquid capacity of horizontal single-tube high-pressure separators. These are based on the parameters of separator working pressure, size, and the depth of liquid used in the liquid settling section. Tables A.2 to A.12 list the standard specifications of typical oil-gas separators and the liquid-settling volumes with the conventional placement of liquid level controls. The settling volumes may determine the liquid capacity of a particular vessel. For proper sizing of both, the liquid capacity and gas capacity required should be determined.

It may be noted that on most high-pressure gas distillate wells the gas-oil ratio is high, and the gas capacity of a separator is usually the controlling factor. However, the reverse may be true for low-pressure separators used on well-streams with low gas-oil ratios. The liquid discharge or dump valve on the separator should be sized for the pressure drop available and the liquid flow rate.

4.2.5 Stage Separation

Stage separation is a process in which gaseous and liquid hydrocarbons are separated into vapor and liquid phases by two or more equilibrium flashes at consecutively lower pressures. As illustrated in Fig. 4.16, two-stage separation requires two separators and a storage tank; and so on. The tank is always counted as the final stage of vapor-liquid separation because the final equilibrium flash occurs in the tank.

The purpose of stage separation is to reduce the pressure on the reservoir liquids a little at a time, in steps or stages, so that a more stable stock-tank liquid will result. The ideal method of separation, to attain maximum liquid recovery, would be that of differential liberation of gas by means of a steady decrease in pressure from that existing in the reservoir to the stock-tank pressure. However, to carry out this differential process would require an infinite number of separation stages. Differential liberation can be closely approached by using three or more series-connected stages of separation, in each of which flash vaporization takes place.

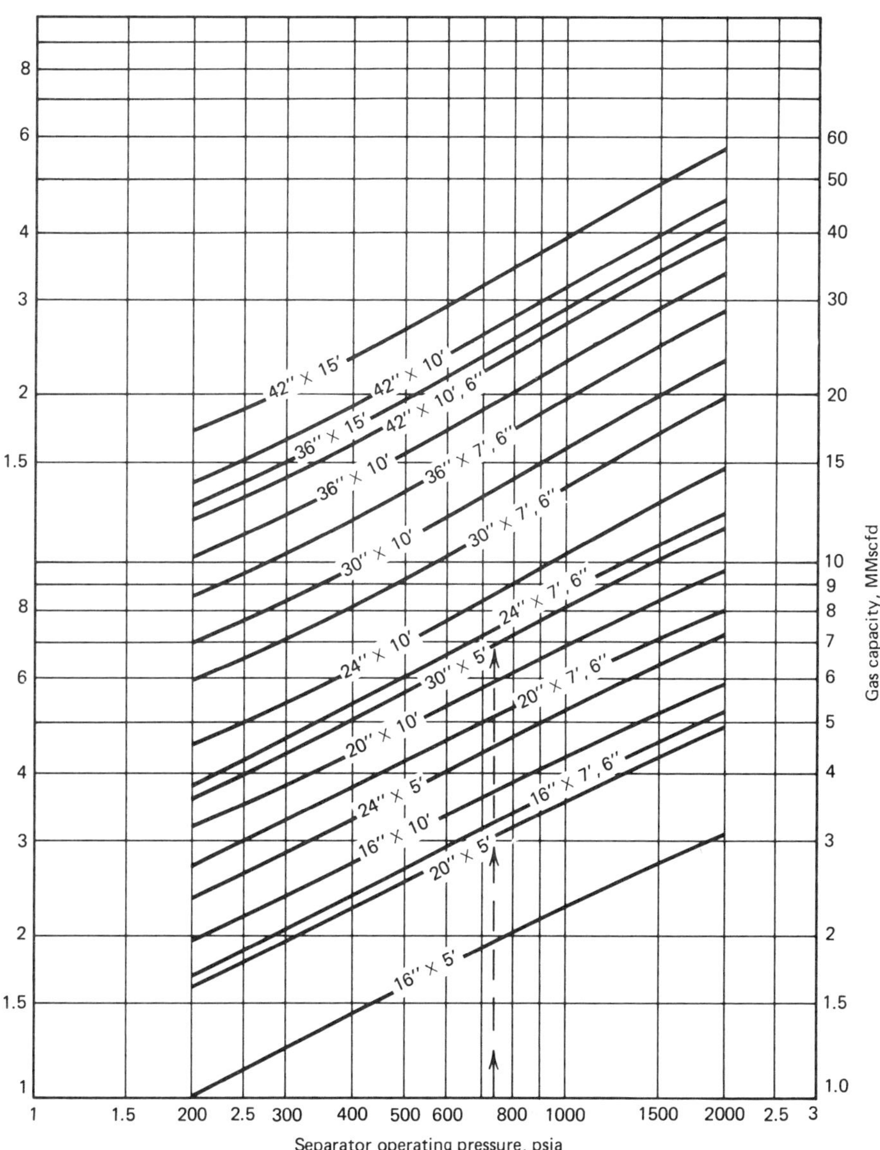

Fig. 4.7 Gas capacity of vertical high-pressure separators. (After Sivalls.)

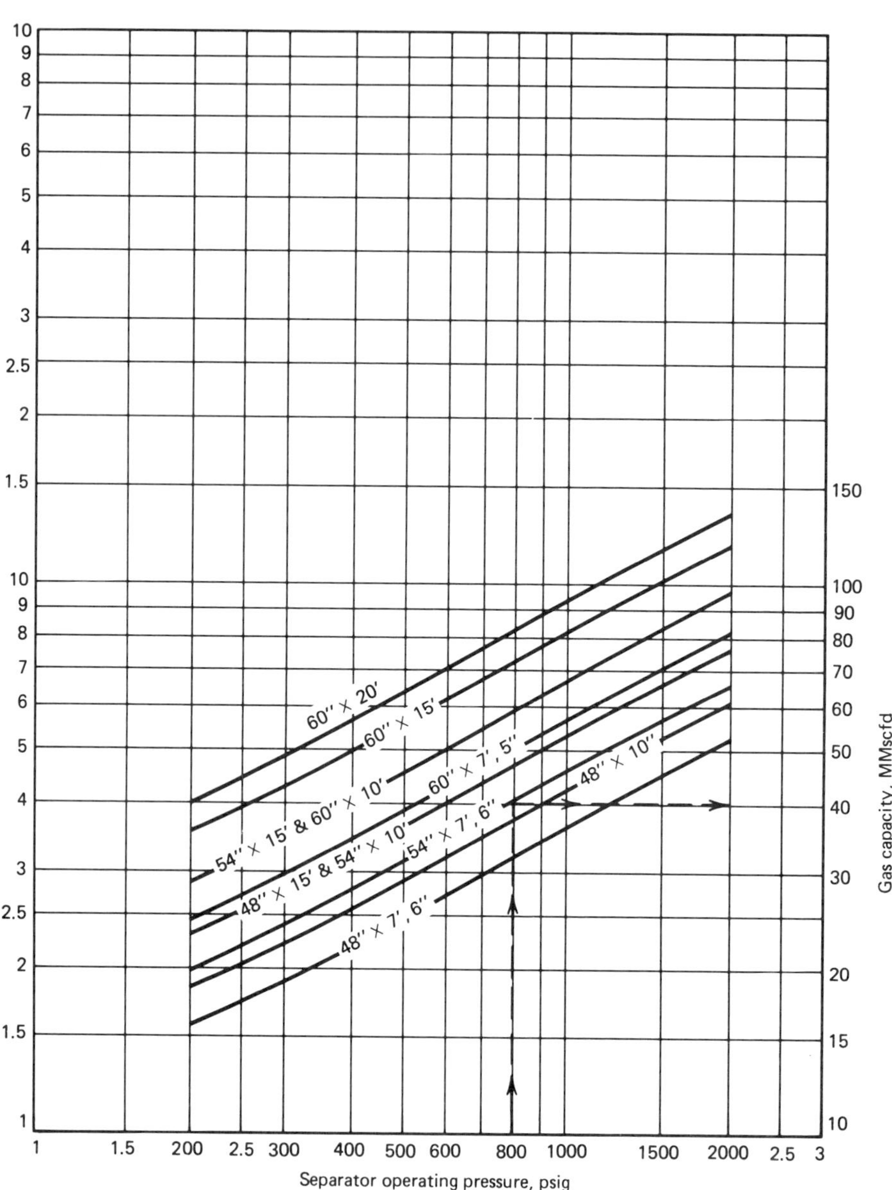

Fig. 4.8 Gas capacity of vertical high-pressure separators. (After Sivalls.)

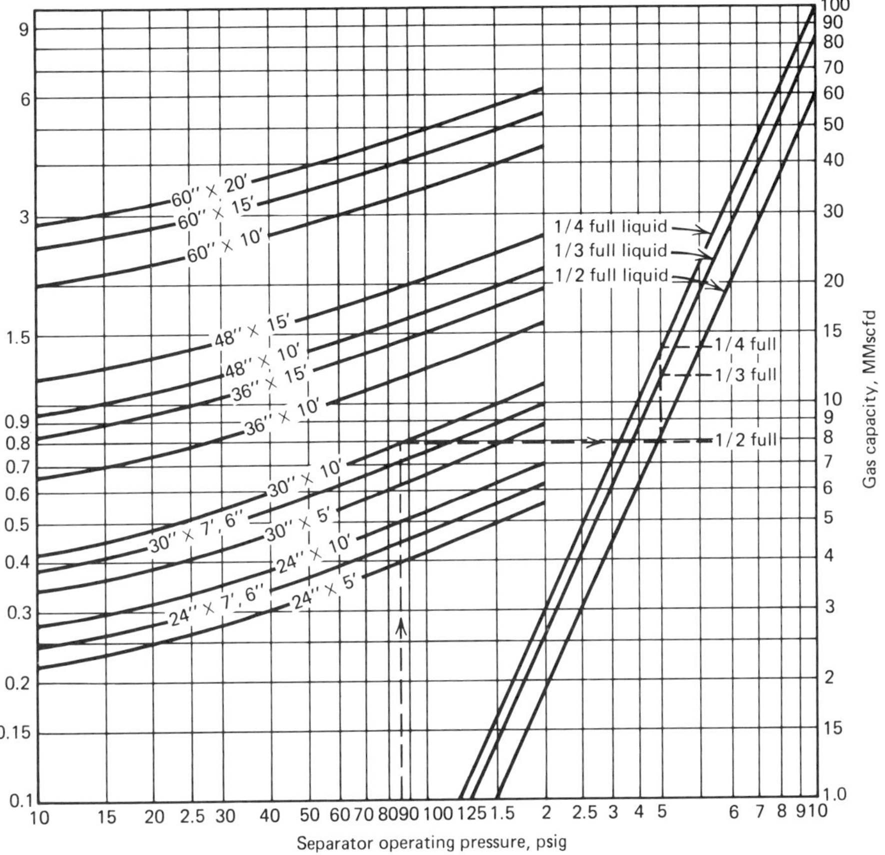

Fig. 4.9 Gas capacity of horizontal low-pressure separators. (After Sivalls.)

In high-pressure gas-condensate separation systems, it is generally accepted that a stepwise reduction of the pressure on the liquid condensate will appreciably increase the recovery of stock tank liquids. The calculation of the actual performance of the various separators in a multistage separation system can be made as described above using the initial wellstream composition and the operating temperatures and pressures of the various stages. However, in the absence of a computer to perform a complete set of flash vaporization calculations, general guidelines can be furnished to estimate the anticipated performance of multistage separation units.

Although three to four stages of separation theoretically would increase the liquid recovery over two stages, the net increase over two-stage separation will rarely pay out the cost of the second or third separator. Therefore, it has been generally accepted that two stages of separation plus the stock tank are considered optimum. Actual increase in liquid recovery for two-stage separation over single-

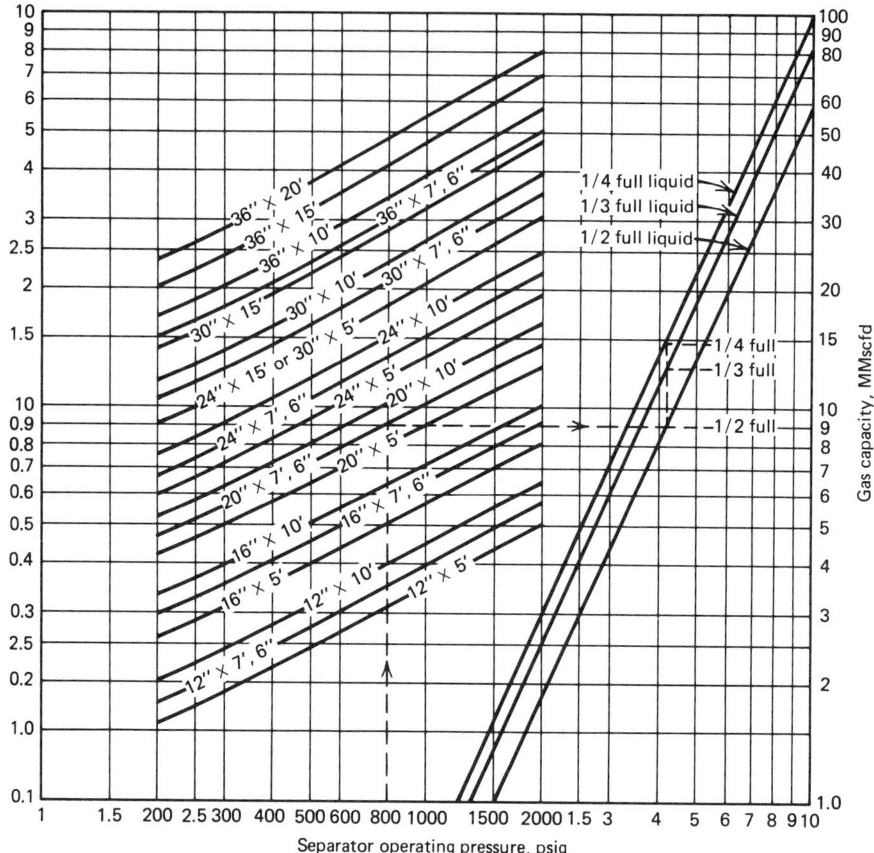

Fig. 4.10 Gas capacity of horizontal high-pressure separators. (After Sivalls.)

stage will vary from 2 to 12%, depending on the wellstream composition, operating pressures, and temperatures. However, 20 to 25% increases in recoveries have been reported.

The optimum high stage or first separator operating pressure is generally governed by the gas transmission line pressure and operating characteristics of the well. These range from 600 to 1200 psi. If the transmission line pressure is at least 600 psi, operators will generally let the first-stage separator ride the line or operate at the transmission line pressure. For each high or first-stage pressure, there is an optimum low-stage separation pressure that will afford the maximum liquid recovery. This operating pressure can be determined from an equation based on equal pressure ratios between the stages (Campbell):

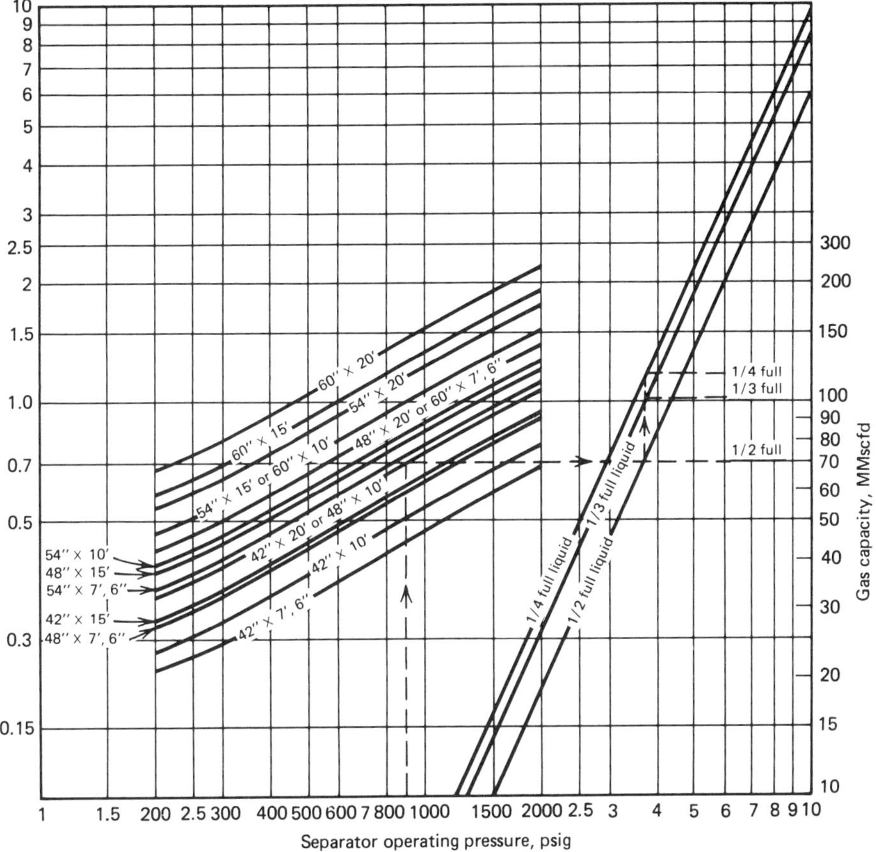

Fig. 4.11 Gas capacity of horizontal high-pressure separators. (After Sivalls.)

$$R = \left(\frac{p_1}{p_s}\right)^{1/n} \tag{4.5}$$

or

$$p_2 = \frac{p_1}{R} = p_s(R)^{n-1} \tag{4.6}$$

where

R = pressure ratio
n = number of stages − 1
p_1 = first-stage or high-pressure separator pressure, psia
p_2 = second-stage or low-pressure separator pressure, psia
p_s = stock tank pressure, psia

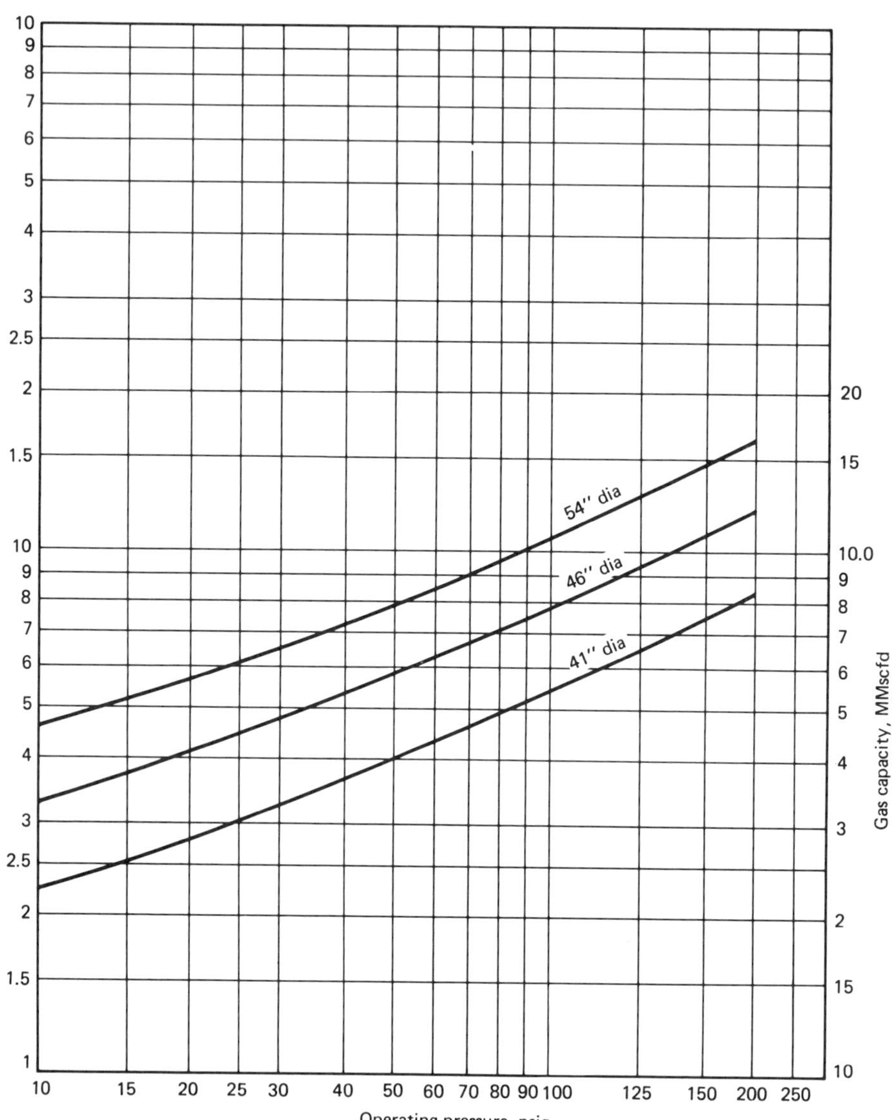

Fig. 4.12 Gas capacities of spherical L.P. separators. (After Sivalls.)

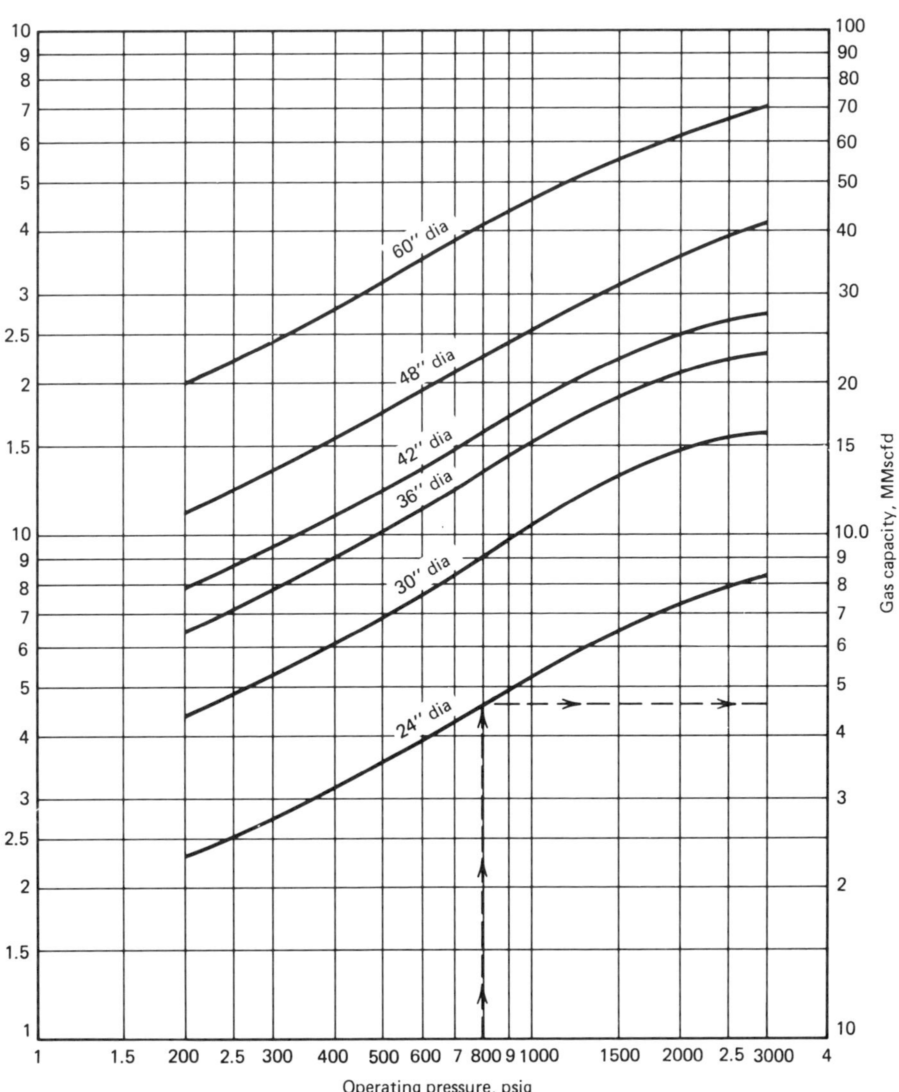

Fig. 4.13 Gas capacities of spherical high-pressure separators. (After Sivalls.)

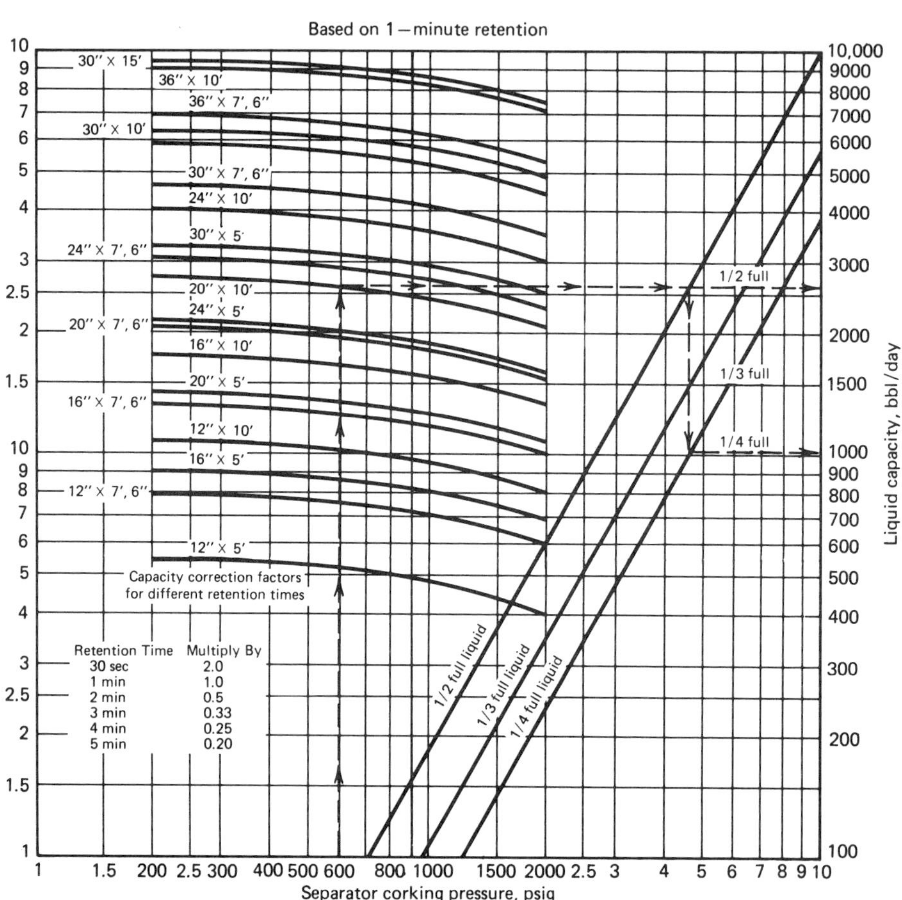

Fig. 4.14 Liquid capacity of horizontal single-tube high-pressure separators. (After Sivalls.)

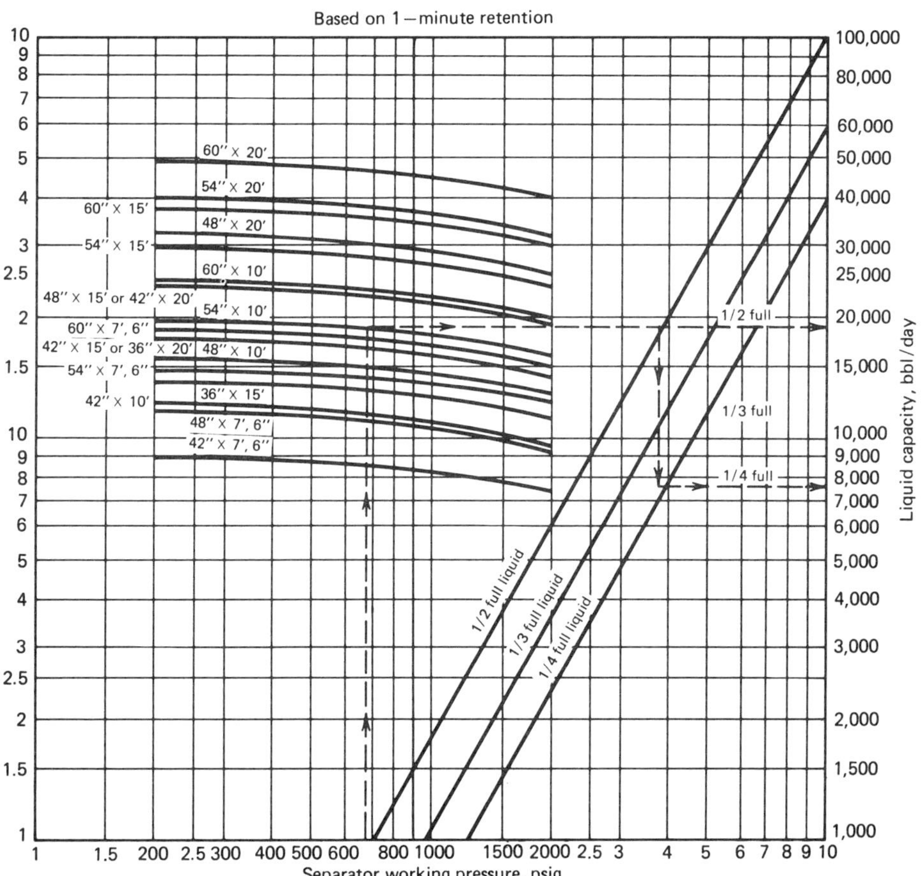

Fig. 4.15 Liquid capacity of horizontal single-tube high-pressure separators. (After Sivalls.)

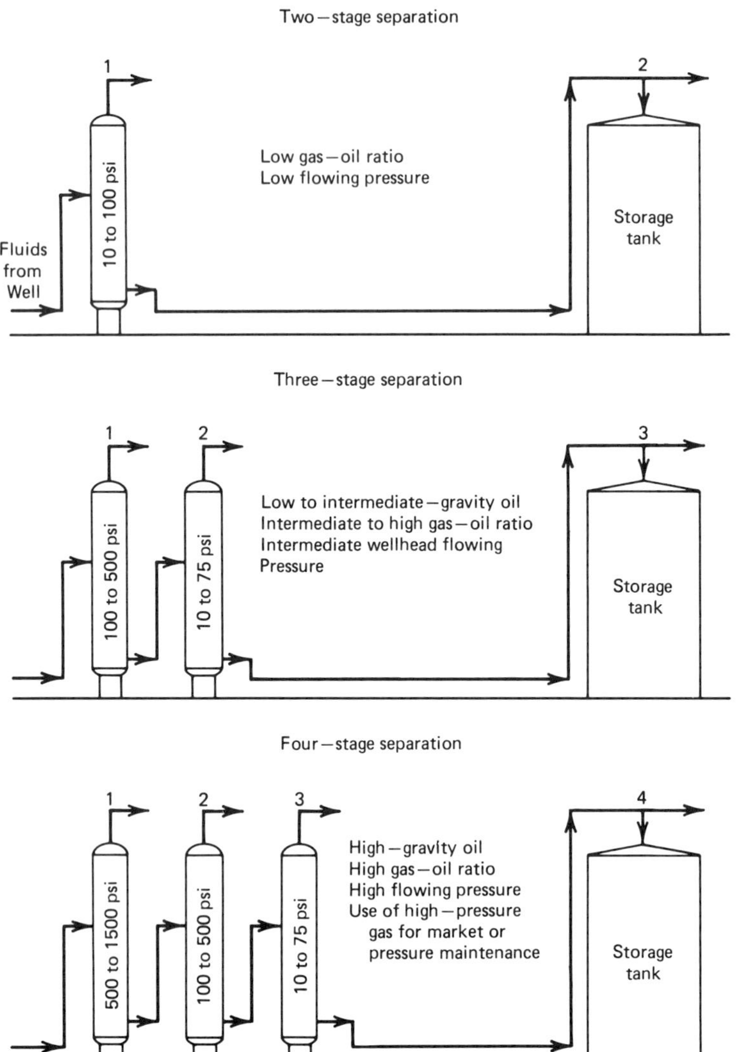

Fig. 4.16 Stage separation flow diagrams. (Courtesy Petroleum Extension Service.)

The magnitude of stock tank liquid recoveries are not considered in the equation.

Figure 4.17 has been prepared to determine the optimum low-stage separator pressure based on the high-stage separator pressure with additional parameters of overall stock tank liquid recovery. This information has been determined from extensive field test data.

Figure 4.18 is a chart illustrating the average percent increase in liquid recovery for two-stage separation over single-stage separation. By using this, an operator can rapidly determine the payout of additional equipment required.

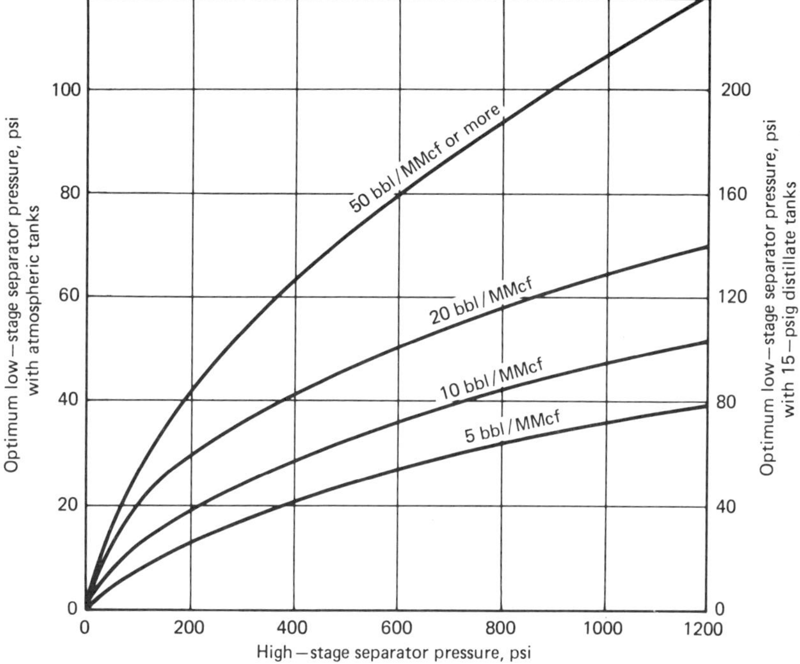

Fig. 4.17 Optimum low-stage separator pressure. (After Sivalls.)

From Fig. 4.18 it also can be seen that additional recovery can be obtained by using 15-psig pressure distillate storage tanks rather than atmospheric storage tanks.

4.2.6 Low-temperature Separation

A low-temperature separation unit is another type of equipment employed for gas-liquid separation that consists primarily of a high-pressure separator, pressure-reducing chokes, and various pieces of heat exchange equipment. As described previously, lowering the operating temperature of a separator increases the liquid recovery. When the pressure is reduced on a high-pressure gas condensate stream by use of a pressure-reducing choke, the fluid temperature also decreases. This is known as the Joule–Thomson or throttling effect, which is an irreversible adiabatic process where the heat content of the gas remains the same across the choke but the pressure and temperature of the gas stream is reduced (Katz et al.).

Low-temperature separation is probably the most efficient means yet devised for handling high-pressure gas and condensate at the wellhead. The process separates water and hydrocarbon liquids from the inlet wellstream, recovers more liquids from the gas than can be recovered with normal-temperature separators, and dehydrates gas, usually to pipeline specifications.

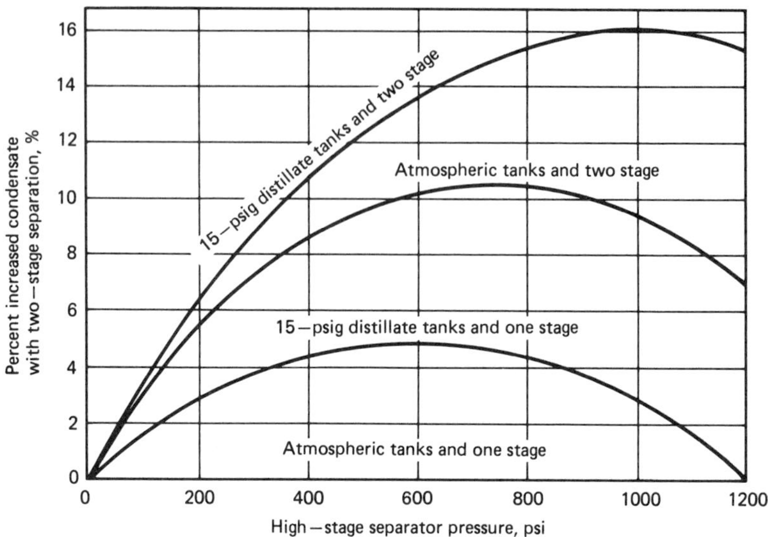

Fig. 4.18 Estimated increase in condensate recovery with two-stage separation. (After Sivalls.)

By special construction of low-temperature separation units, the pressure-reducing choke is mounted directly on the inlet of the high-pressure separator. Hydrates formed downstream of the choke due to the low gas temperature after the pressure reduction are blown into the separator and fall to the bottom settling section where they are heated and melted by liquid heating coils located in the bottom of the vessel. The low-temperature effect is used in low-temperature units to increase the liquid recovery. The lower the operating temperature of the separator, the higher the liquid recovery will be. However, the maximum flowing pressure from the well at a given flow rate and the transmission line pressure will indicate the maximum amount of pressure drop available across the choke.

Enthalpy curves on natural gas can be used to determine the temperature drop expected based on the available pressure drop. Figure 4.19 illustrates natural gas expansion-temperature reduction curves for 0.7 specific gravity gas. This chart shows that a 1500-psi drop from 3000 psi at 120°F will provide a final temperature of only 78°F, while a 1500-psi drop from 2000 psi at 120°F will give a final temperature of 49°F. The actual temperature drop per pounds per square inch of pressure drop will depend on composition of gas stream, gas and liquid flow rates, bath temperature, and ambient temperature.

In general, at least a 2500- and 3000-psi pressure drop should be available from wellhead flowing pressure to pipeline pressure before a low-temperature separation unit will pay out in increased liquid recovery. The lowest operating temperature recommended for low-temperature units is usually around −20°F. Carbon steel embrittlement occurs below this temperature and high-alloy steels

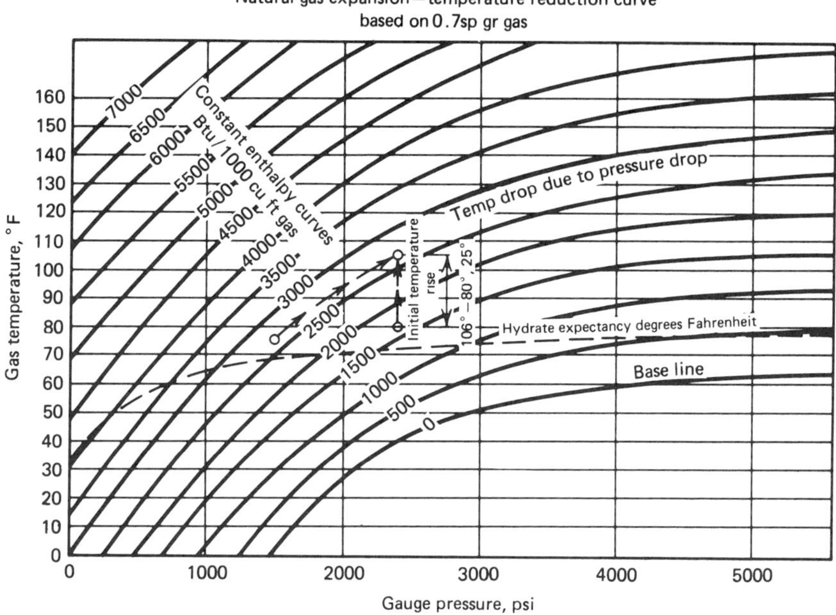

Fig. 4.19 Natural gas expansion-temperature reduction curves. (Courtesy Petroleum Extension Service.)

for lower temperatures are usually not economical for most oil field installations. Generally, low-temperature separation units are operated in the range of 0 to 20°F.

Figure 4.20 is a chart showing the effect of separation temperature on liquid recovery. This chart assists the operator in feasibility studies to determine if low-temperature separation equipment will pay out on a given wellstream. The purpose of this chart is to estimate the liquid recovery at a lower temperature of separation knowing the recovery or gas-oil ratio that can be obtained by conventional separation methods. This chart can also be used to see what effect lowering the operating temperature in conventional separation equipment would have on the liquid recovery.

Example 4.1. Size a standard oil-gas separator both vertically and horizontally for the following conditions:

Gas flow rate	5.0 MMscfd
Operating pressure	800 psig
Condensate flow rate	20 bbl/MMscf

Solution

$$\text{Total liquid capacity} = 20\ (5.0) = 100\ \text{bbl/day}$$

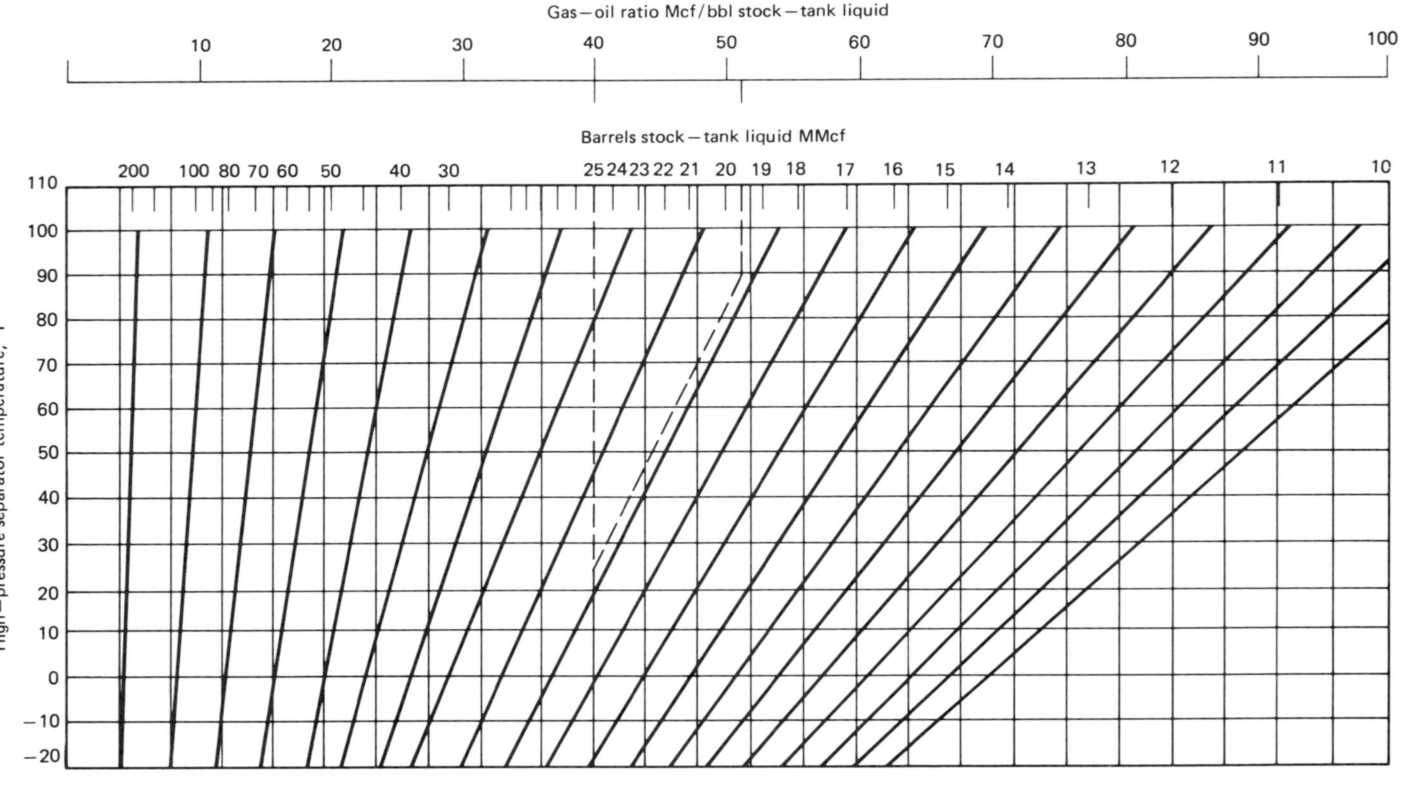

Fig. 4.20 Effect of separator temperature on liquid recovery. (Courtesy Petroleum Extension Service.)

From Fig. 4.7, at 800-psig operating pressure, a 20 in. × 7 ft, 6 in. vertical separator will handle 5.4 MMscfd. From Table A.5, a 20 in. × 7 ft, 6 in. separator will handle the following liquid capacity:

$$W = \frac{1440V}{t} = \frac{1440(0.65)}{1.0} = 936 \text{ bbl/day}$$

From Fig. 4.10, at 800-psig operating pressure and one-half full of liquid, a 16 in. × 5 ft horizontal separator will handle 5.1 MMscfd. From Table A.9, a 16 in. × 5 ft separator will handle

$$W = \frac{1440V}{t} = \frac{1440(0.61)}{1.0} = 878 \text{ bbl/day}$$

Therefore, a smaller horizontal separator would be required and would be more economical. For the operating pressure involved, at least a 1000-psig working pressure separator should be used.

Example 4.2. Size a standard vertical oil-gas separator for the following conditions:

Oil flow rate	2500 bbl/day
Gas-oil ratio	1000
Operating pressure	50 psig

Solution

$$\text{Gas flow rate} = 1000\ (2500) = 2{,}500{,}000 \text{ cfd} = 2.5 \text{ MMscfd}$$

From Fig. 4.6, at 50-psig operating pressure, a 36 in. × 5 ft vertical separator will handle 2.9 MMscfd. From Table A.3, a 36 in. × 5 ft separator will handle

$$W = \frac{1440V}{t} = \frac{1440(1.61)}{1.0} = 2{,}318 \text{ bbl/day}$$

Therefore, a larger separator will be required to handle the liquid load. A 30 in. × 10 ft separator will handle

$$W = \frac{1440(2.06)}{1.0} = 2{,}966 \text{ bbl/day}$$

The gas capacity of a 30 in. × 10 ft separator is 3.75 MMscfd. In fact, the diameter of a separator generally controls the price and a 30 in. × 10 ft separator will probably be cheaper than one that is 36 in. × 5 ft.

Example 4.3. A 20 in. × 10 ft, 100-psi working pressure (WP) horizontal separator is operated at one-half full liquid capacity. Can it be used on a well with the following conditions:

Gas flow rate	9.0 MMscfd
Line pressure	500 psig

Solution

From Fig. 4.10, a 20 in. × 10 ft separator will handle only 8.1 MMscfd at 500 psig. But, if a gas back-pressure valve were put on the separator and held at 800 psig, the separator would handle 10.2 MMscfd. This is accepted procedure, providing the well will flow the desired rate at 800 psig.

Example 4.4. Size a horizontal high-pressure separator for the following conditions:

Gas flow rate	10.0 MMscfd
Operating pressure	800 psig
Condensate load	500 bbl/day
Water load	100 bbl/day

Solution

From Fig. 4.10, at 800-psig operating pressure, a 20 in. × 10 ft horizontal separator will handle 10.2 MMscfd operating one-half full liquid capacity. Where three-phase operation is required in a horizontal separator, the liquid section should be one-half full; otherwise, the level control action becomes too critical.

From Table A.9, the liquid capacity will be

$$W = \frac{1440V}{t} = \frac{1440(1.80)}{5.0} = 518 \text{ bbl/day}$$

Therefore, the 20 in.× 10 ft separator will not handle the combined liquid load of 500 + 100 = 600 bbl/day. Five minute retention time is used as a conservative figure without any additional information.

From Table A.9, a separator with more settling volume is 24 in. × 10 ft. Its liquid capacity is

$$W = \frac{1440(2.63)}{5.0} = 757 \text{ bbl/day}$$

The gas capacity of a 24 in. × 10 ft separator at 800 psig is 15.0 MMscfd.

Example 4.5. A well test was made using a simple high pressure separator and an atmospheric stock tank and the following results were obtained:

Gas flow rate	10.0 MMscfd
Operating pressure	800 psig
Condensate recovery	500 bbl/day

What additional recovery could be expected using two-stage separation and 15-psig pressure storage tanks?

Solution

From Fig. 4.18, using the top curve at 800 psig, a 15.3% increase could be expected. This would result in, 500(0.153) = 76.5 bbl/day additional condensate recovery.

Example 4.6. On the test of a high-pressure gas-condensate well the following data were recorded:

Gas flow rate	10.0 MMscfd
Operating pressure	800 psig
Condensate recovery	200 bbl/day
Separator temperature	85°F

What additional recovery could be expected using a low-temperature separation unit operating at 20°F?

Solution

$$\text{Liquid flow rate} = \frac{200\ \text{bbl/day}}{10.0\ \text{MMscfd}} = 20\ \text{b/MMscf}$$

From Fig. 4.20, following the dotted lines, the recovery would go from 20 bbl/MMcf to 25.7 bbl/MMcf, or an additional 5.7 bbl/MMscf = 57 bbl/day.

4.3 DEHYDRATION OF NATURAL GAS

Natural gas destined for transport by pipeline must meet certain specifications. Items usually included in the United States and Canada are maximum water content (water dew point), maximum condensable hydrocarbon content (hydrocarbon dew point), allowable concentrations of contaminants such as H_2S, CO_2, mercaptans, minimum heating value, and cleanliness (allowable solids content). In addition there will be specifications regarding delivery pressure, rate, and possibly temperature.

The reservoir fluids flowing to the surface during production operations normally contain water regardless of which zone the well has been completed in, as shown in Fig. 4.21. The gas-producing zone of a reservoir tends to contain more water vapor and less liquid or free water than the oil-producing zone. Water vapor is probably the most common undesirable impurity found in untreated natural gas. The principal reasons for the removal of water vapor from natural gas for long-distance transmission include the following:

1. Liquid water and natural gas can form solid, ice-like hydrates that plug equipment.

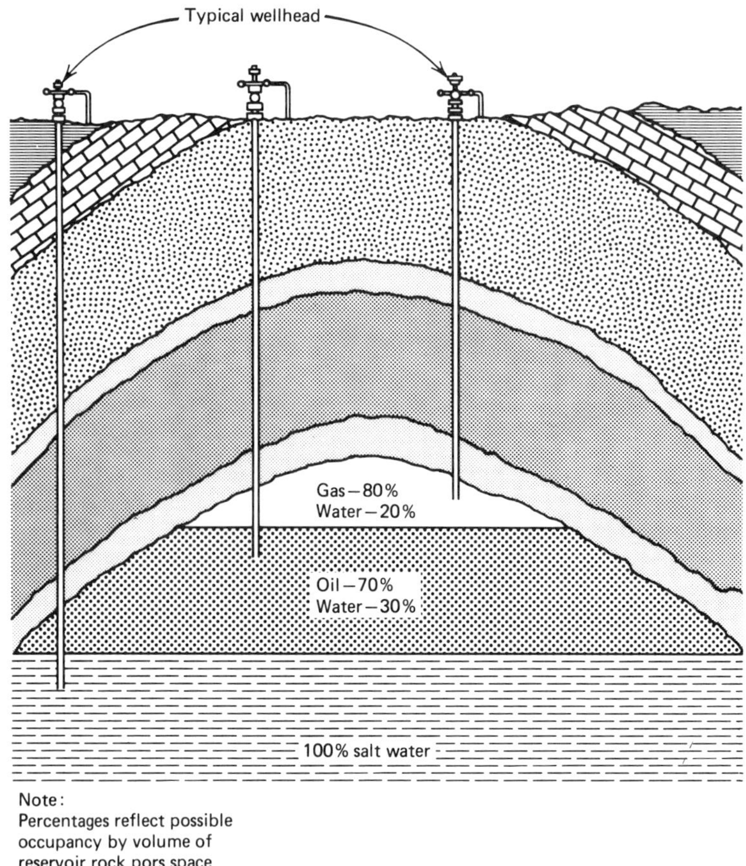

Fig. 4.21 Typical oil and gas reservoir. (After Guenther.)

2. Natural gas containing liquid water is corrosive, particularly if it also contains CO_2 or H_2S.
3. Water vapor in natural gas may condense in pipelines potentially causing slugging flow conditions.
4. Water vapor increases the volume and decreases the heating value of natural gas; this leads to reduced line capacity.

4.3.1 Water Content of Natural Gas Streams (No Deleterious Contaminants)

All natural gases contain water vapor to some degree. Solubility of water increases as temperature increases and decreases as pressure increases. Salts dissolved in the liquid water in equilibrium with natural gas reduce the water content of the gas. Water content is usually expressed as pounds of water per million standard cubic feet of natural gas (lbm/MMscf).Values are best obtained from correlations of experimental data such as the McKetta and Wehe chart given in Fig. 4.22. Typical values are

Reservoir gas (5000 psig/250°F) ≅ 500 lbm/MMscf
Trap gas (500 psig/125°F) ≅ 400 lbm/MMscf
Pipeline gas ≅ 6–8 lbm/MMscf

The term dehydration means removal of water vapor. The water content of natural gas is indirectly indicated by the dew point. The dew point may be defined as the temperature at which the natural gas is saturated with water vapor at a given pressure. At the dew point, natural gas is in equilibrium with liquid water; a decrease in temperature or an increase in pressure will cause the water vapor to begin condensing. The difference between the dew point temperature of a water-saturated gas stream and the same stream after it has been dehydrated is known as the dew point depression.

To illustrate the concept of dew point depression, assume that natural gas at 500 psia and 60°F at the saturation point contains 30 lbm of water per 1 million cu ft. The dew point of this gas is 60°F. Suppose this natural gas is going to be transported in a pipeline at 20°F. The saturation point will then be 7 lbm of water/MMcf. The original 30 lbm of water, if left in the gas, will exist in the form of 7 lbm of water vapor and 23 lbm of free water/MMcf if the pressure remains the same. This free water is a potential source of hydrates to form and plug the line. Suppose the natural gas is processed in a dehydration unit and the dew point is depressed 50°F. This means that no free water will exist in the gas until the temperature goes to 10°F or lower. Gas at 500 psia and 10°F contains about 5 lbm of water vapor/MMcf. The dehydration unit must remove 25 lbm of water from each 1 million cu ft of gas in order to achieve the 50°F dew point depression.

Estimating Water Content

A reliable technique for estimating the saturated water vapor content of natural gas is fundamental to the design and operation of dehydration equipment. Several of these methods that relate water content and dew point are summarized below.

1. The correlation of McCarthy, Boyd, and Reid can be used for most natural gas dehydration applications and will generally result in reasonably conservative designs. Their chart is presented in Fig. 4.23.
2. The correlation of McKetta and Wehe (Fig. 4.22) is similar to McCarthy, Boyd, and Reid and includes correlation factors for produced water salinity and gas specific gravity.
3. At or near atmospheric pressure, Dalton's law of partial pressures is valid and may be used to estimate the water vapor content of gas.
4. The Natural Gas Processors Suppliers Association has described a method for predicting the water content of sour natural gas using the data of Weibe and Gaddy for CO_2 and the data of Selleck, Carmichael, and Sage for H_2S.

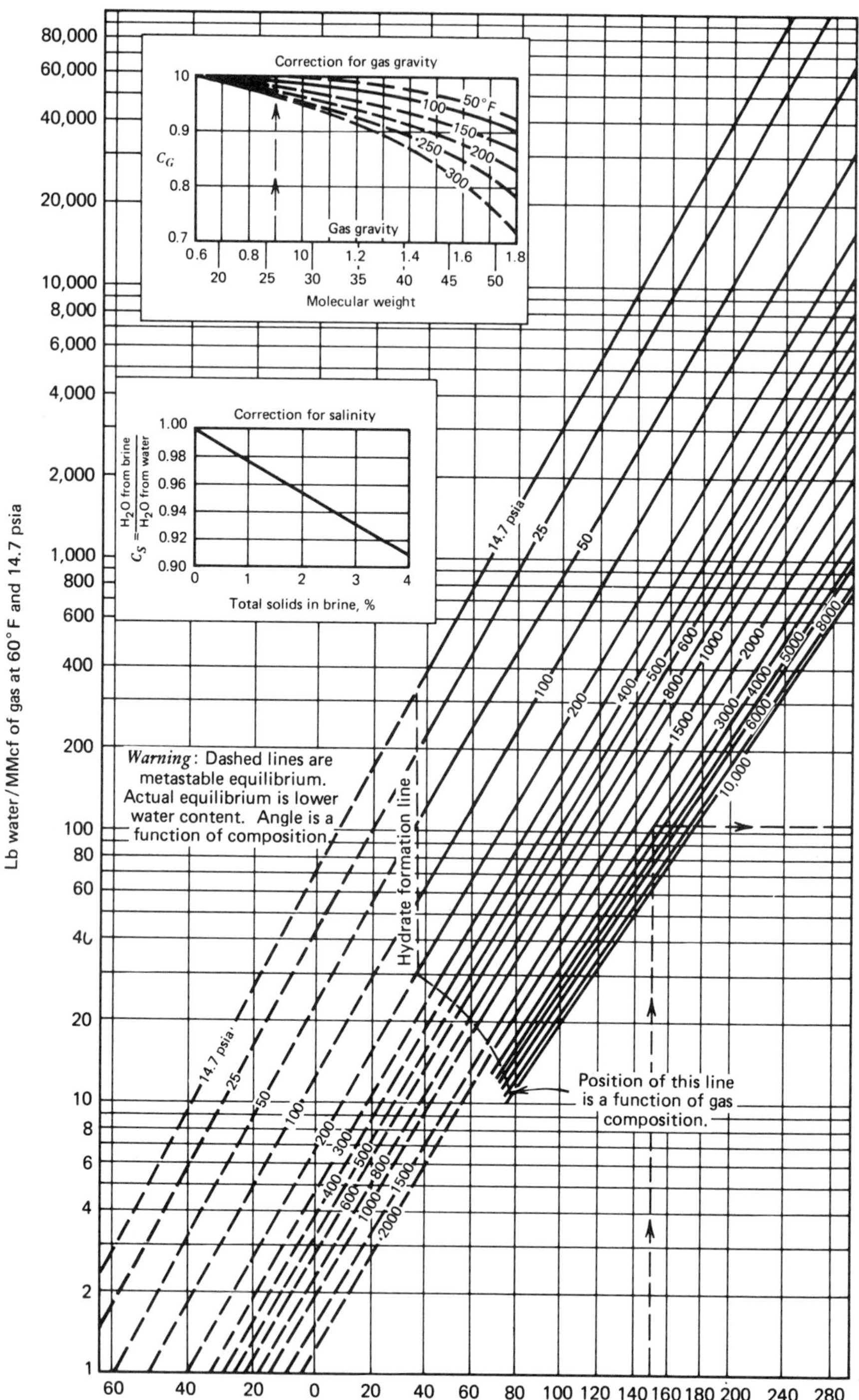

Fig. 4.22 Water contents of natural gases with corrections for salinity and gravity. (After McKetta and Wehe.)

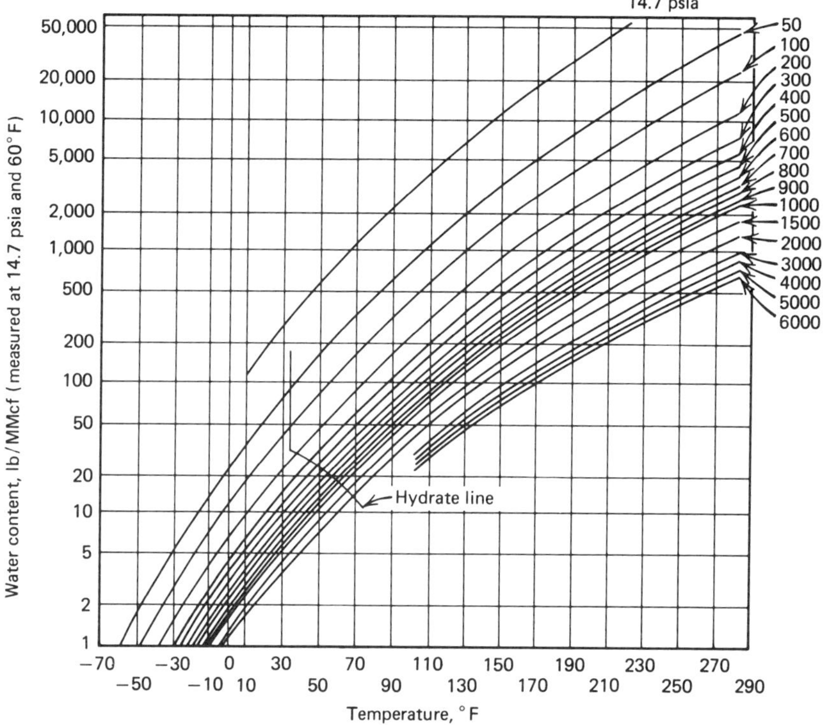

Fig. 4.23 Water vapor content of saturated natural gas. (After Guenther.)

5. The most rigorous approach uses physical chemistry techniques to account for the effect of differences in gas composition on water vapor content. This method was developed by Sharma and Campbell.

In almost all cases, the inlet natural gas to a dehydration system will be saturated with water vapor at process temperature and pressure, but the upstream process configuration should be examined to confirm this. Note, however, that the assumption of saturation is a conservative approach.

4.3.2 Hydrate Control in Gas Production

Gas producers, processors, and pipeline operators have often been plagued by operating difficulties that cause undesirable and costly interruptions in gas production, field processing, and distribution. While many operating difficulties are mechanical in nature, two constantly recurring operating problems, gas hydrate formation and freeze-up in separator water dump lines, are more costly and more serious than others.

These difficulties most often occur in cold whether when seasonal demand for gas is at its highest level, thus making shutdowns or interruptions in production or gas service most expensive. These two operating problems are somewhat related, and their solutions in some cases are quite similar.

Gas Hydrate Formation

Natural gas hydrates are solid crystalline compounds formed by the chemical combination of natural gas and water under pressure at temperatures considerably above the freezing point of water. In the presence of free water, hydrates will form when the temperature is below a certain degree (hydrate temperature). The chemical formulas for natural gas hydrates are:

Methane	$CH_4{\cdot}7H_2O$
Ethane	$C_2H_6{\cdot}8H_2O$
Propane	$C_3H_8{\cdot}18H_2O$
Carbon dioxide	$CO_2{\cdot}7H_2O$

Gas hydrate crystals resemble ice or wet snow in appearance but do not have ice's solid structure, are much less dense, and exhibit properties that are generally associated with chemical compounds. The main framework of their structure is water; the hydrocarbon molecule occupies the void space in a crystalline network held together by chemically weak bonds with the water. The water framework is icelike; unlike ice, however, it has void space and a network structure.

Hydrate formation is often confused with condensation, and the distinction between the two must be clearly understood. Condensation of water from a natural gas under pressure occurs when the temperature is at or below the dew point at that pressure. Free water obtained under such conditions is essential to formation of hydrates, which will occur at or below the hydrate temperature at the same pressure. Hence, the hydrate temperature would be less than or equal to the dew point temperature.

During the flow of natural gas, it becomes necessary to define, and thereby avoid, conditions that promote the formation of hydrates. This is essential since hydrates may choke the flow string, surface lines, and other equipment. Hydrate formation in the flow string results in a lower value for measured wellhead pressures. In a flow rate measuring device, hydrate formation results in lower flow rates. Excessive hydrate formation may also completely block flow lines and surface equipment.

The conditions that tend to promote the formation of natural gas hydrates are:

1. Natural gas at or below its water dew point with liquid water present.
2. Temperatures below the "hydrate formation" temperature for the pressure and gas composition considered.
3. High operating pressures that increase the "hydrate formation" temperature.
4. High velocity or agitation through piping or equipment.
5. Presence of a small "seed" crystal of hydrate.
6. Presence of H_2S or CO_2 is conducive to hydrate formation since these acid gases are more soluble in water than hydrocarbons.

A rigorous technique for predicting conditions for hydrate formation involves the use of vapor-solid equilibrium constants (Katz et al.). The calculations are analogous to a dew point calculation for multicomponent mixtures. This method of hydrate prediction has proved to be rather reliable, as have the methods of McLeod & Campbell and Trekell & Campbell.

Figures 4.22 and 4.23 show hydrate formation lines for 0.6 gravity gas. Hydrates tend to form to the left of these lines.

Figures 4.24 to 4.29 present data that may be used for first approximations of hydrate formation conditions and for estimating permissible expansion of natural gases without the formation of hydrates. It is convenient to divide hydrate formation into two categories: (1) hydrate formation due to a decrease in temperature with no sudden pressure drop, such as in the flow string or surface line, and (2) hydrate formation where a sudden expansion occurs, such as in flow provers, orifices, back-pressure regulators, or chokes.

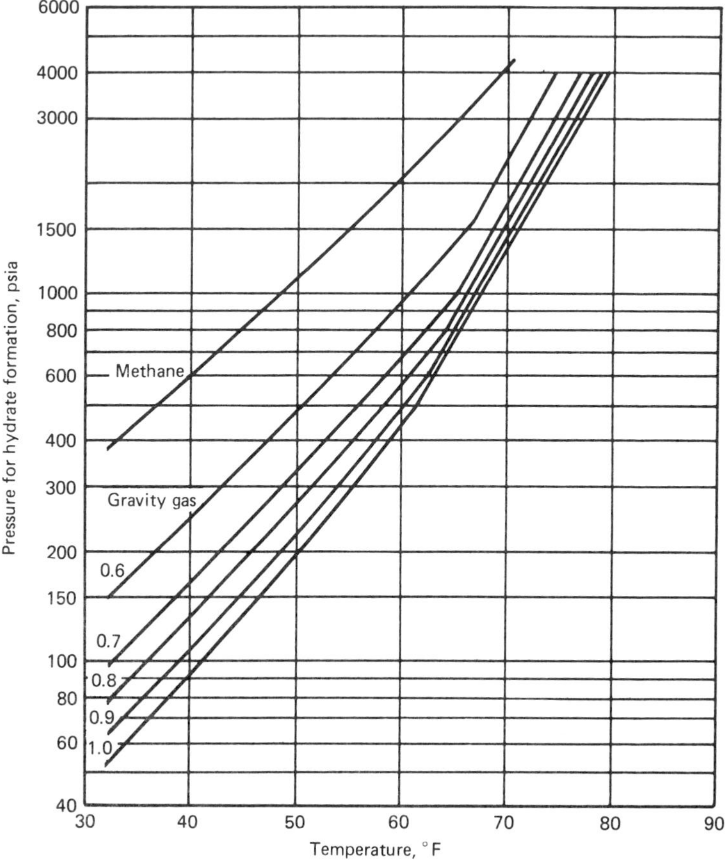

Fig. 4.24 Pressure-temperature curves for predicting hydrate formation. (Courtesy Gas Processors Suppliers Assoc.)

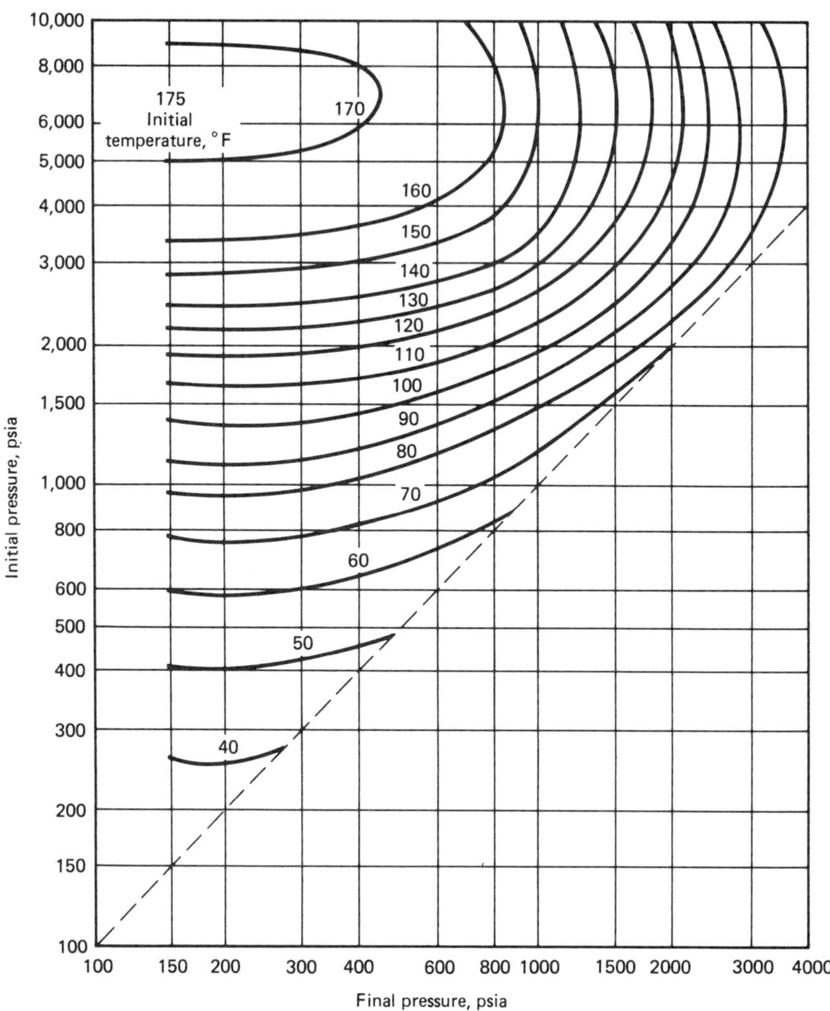

Fig. 4.25 Permissible expansion of a 0.6-gravity natural gas without hydrate formation. (Courtesy Gas Processors Suppliers Assoc.)

1. Hydrate Formation in the Flow String and Surface Lines. The hydrate temperature depends on the pressure and composition, reflected by the gravity of the gas. Figure 4.24 gives approximate values of the hydrate temperature as a function of pressure and specific gravity. Hydrates will form whenever temperature and pressure plot to the left of the hydrate formation line for the gas in question.

Example 4.7.

(a) A certain natural gas has a specific gravity of 0.693. If the gas is at 50°F, what would be the pressure above which hydrates could be expected to form?

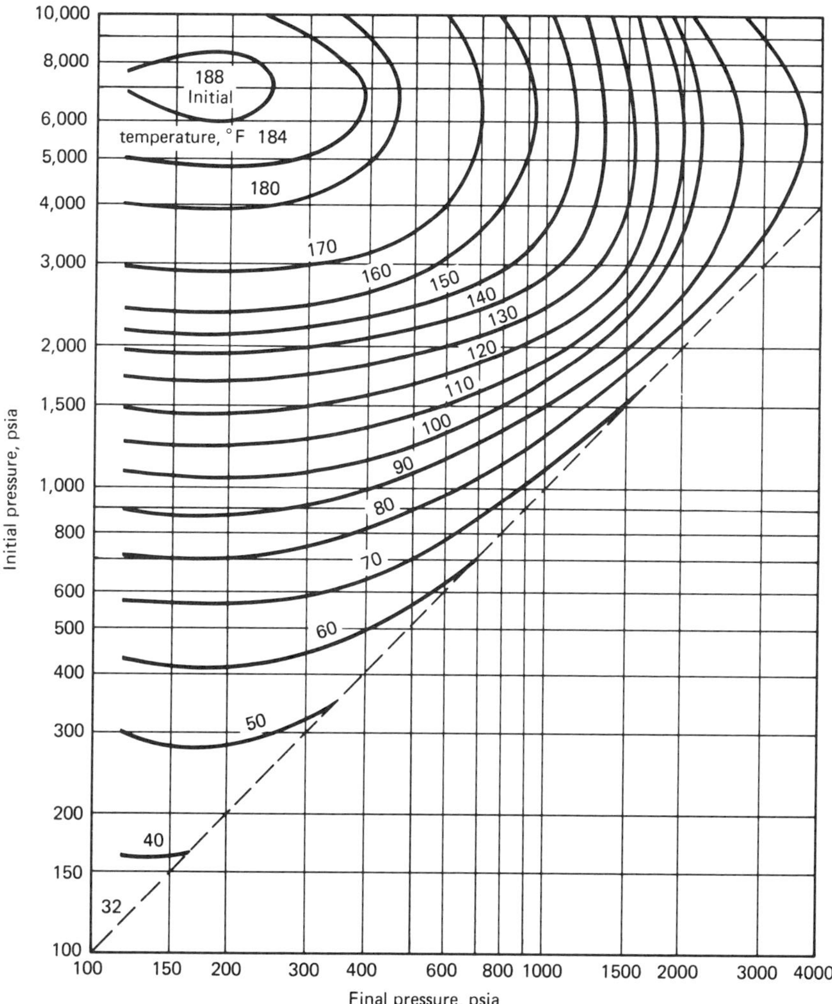

Fig. 4.26 Permissible expansion of a 0.7-gravity natural gas without hydrate formation. (Courtesy Gas Processors Suppliers Assoc.)

(b) A gas of specific gravity 0.8 is at a pressure of 1000 psia. To what extent can the temperature be lowered without hydrate formation, assuming presence of free water?

Solution

(a) From Fig. 4.24 at 50°F,

$$p = 480 \text{ psia for 0.6-gravity gas}$$
$$p = 320 \text{ psia for 0.7-gravity gas}$$

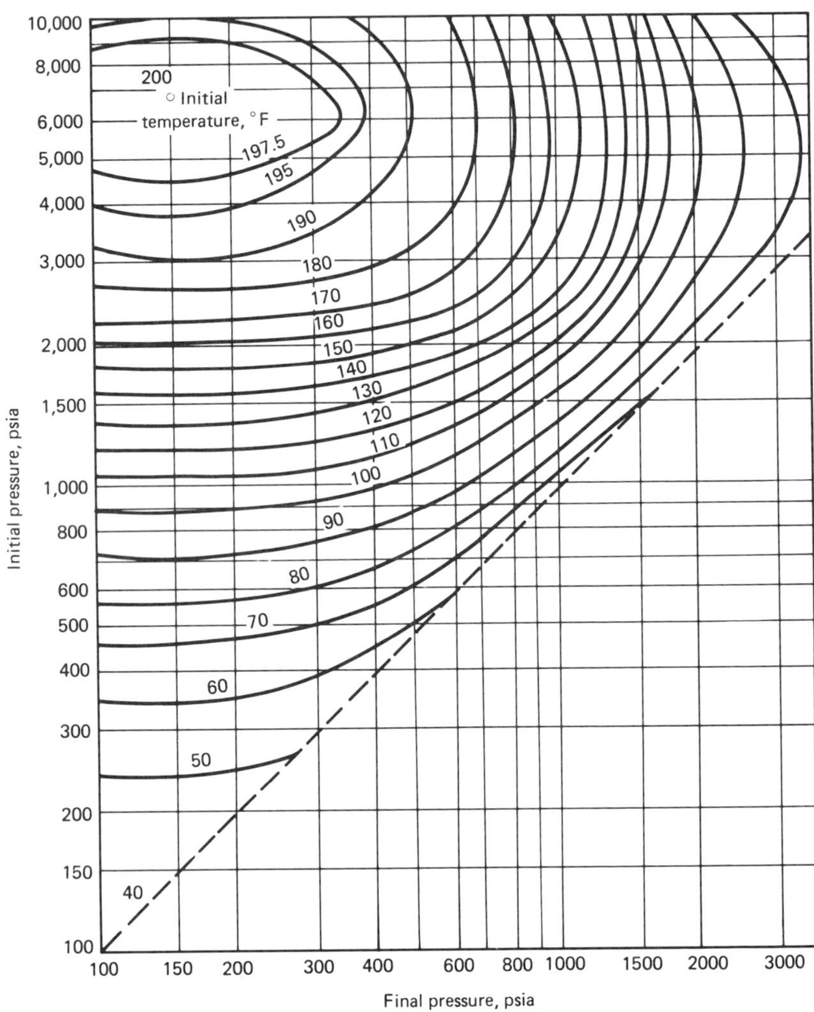

Fig. 4.27 Permissible expansion of a 0.8-gravity natural gas without hydrate formation. (Courtesy Gas Processors Suppliers Assoc.)

Using linear interpolation,

$$p = 480 - \left[(480 - 320) \times \left(\frac{0.693 - 0.6}{0.7 - 0.6}\right)\right] = 480 - 148$$

$$= 332 \text{ psia for 0.693-gravity gas}$$

Hydrates will form above a pressure of 332 psia.

(b) From Fig. 4.24, at a specific gravity of 0.8 and a pressure of 1000 psia, hydrate temperature is 66°F. Thus, hydrates may form at or below 66°F.

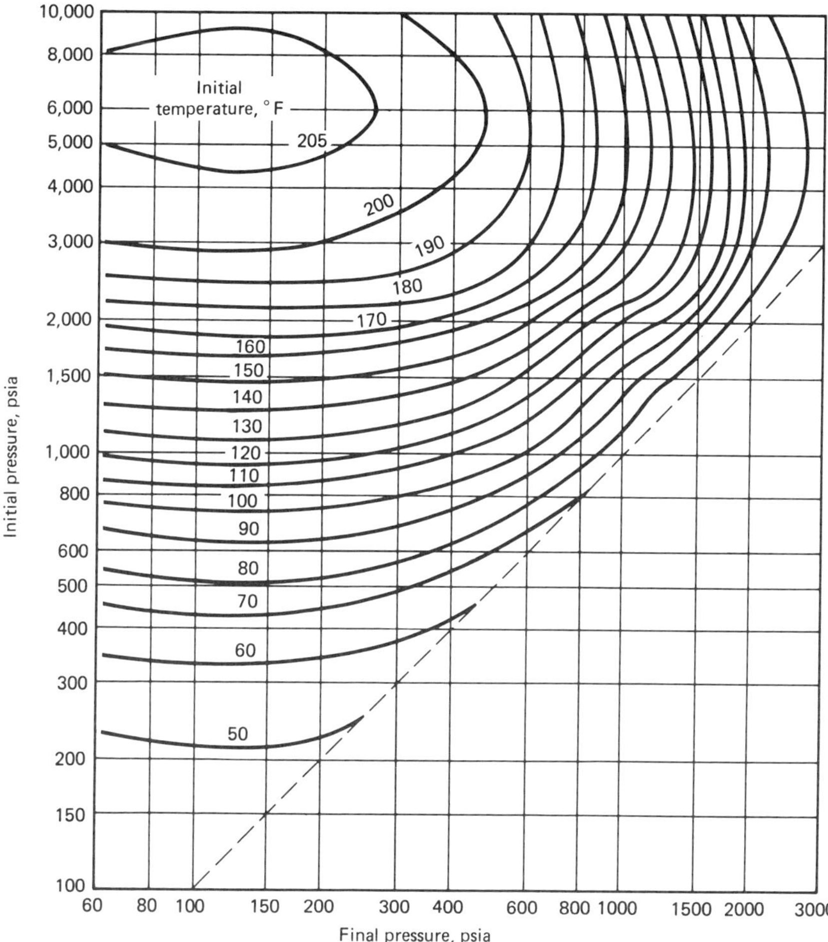

Fig. 4.28 Permissible expansion of a 0.9-gravity natural gas without hydrate formation. (Courtesy Gas Processors Suppliers Assoc.)

2. *Hydrate Formation in Flow Provers, Orifices, and Back-pressure Regulators.* Sudden expansion in one of these devices is accompanied by a temperature drop that may cause hydrate formation. Figures 4.25 to 4.29 may be used to approximate the conditions for hydrate formation.

Example 4.8.

(a) How far can a 0.8-gravity gas at 1000 psia and 100°F be expanded without hydrate formation?

(b) How far can a 0.7-gravity gas at 800 psia and 100°F be expanded without hydrate formation, assuming presence of free water?

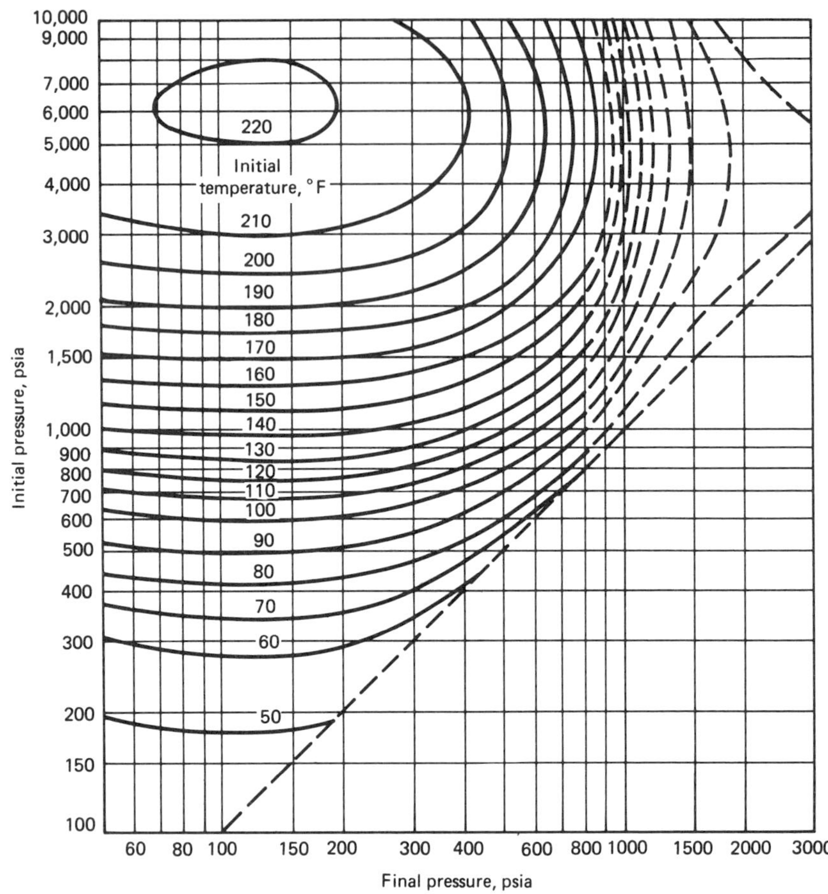

Fig. 4.29 Permissible expansion of a 1.0-gravity natural gas without hydrate formation. (Courtesy Gas Processors Suppliers Assoc.)

(c) A 0.6-gravity gas is to be expanded from 1000 psia to 400 psia. What is the minimum initial temperature that will permit expansion without danger of hydrate formation?

Solution

(a) The intersection of the 1000 psia initial pressure line with the 100°F initial temperature curve gives a final pressure of 440 psia (Fig. 4.27). Hence, this gas may be expanded to a final pressure of 440 psia without a possibility of hydrate formation.

(b) The 100°F initial temperature curve does not intersect the 800-psia initial pressure line (Fig. 4.26). Hence, this gas may be expanded to atmospheric pressure without hydrate formation.

(c) The intersection of the 1000-psia initial pressure line and the 400-psia final pressure line gives an initial temperature of 78°F (Fig. 4.25). Hence, 78°F is the minimum initial temperature to avoid hydrate formation.

Figures 4.24 to 4.29 are strictly applicable to sweet natural gases. For sour gases, they may be used, keeping in mind that the presence of H_2S and CO_2 will increase the hydrate temperature and reduce the pressure above which hydrates will form. In other words, the presence of H_2S or CO_2 increases the possibility of hydrate formation.

Preventing Hydrate Formation

The hydrate formation line in Fig. 4.22 or 4.23 shows that hydrate formation can be prevented by heating the cold, unprocessed wellstream. If the pressure and water content of the gas remain constant during heating, the gas will become undersaturated, thereby eliminating one of the conditions for hydrate formation. If the lines and equipment must be maintained below the hydrate point, the water must be inhibited for trouble-free operation. Methanol, ethylene glycol (EG), and diethylene glycol (DEG) are commonly injected into gas streams to depress the freezing point. All of these inhibitors can be recovered and recycled; however, the recovery of methanol is often uneconomical. Hydrate inhibitor injection does not always provide the ultimate degree of dehydration specified by the purchaser or required by the process conditions.

Methanol injection systems are frequently installed at facilities where low gas volumes prohibit dehydration or other means of hydrate inhibition that require high capital investments, equipment is temporary until a decision regarding a permanent facility is made, or hydrate problems are relatively mild, infrequent, seasonal, or expected during start-up such that hydrate inhibition by methanol supplements a primary dehydration system. EG and DEG are injected primarily at low-temperature processing plants for extracting natural gas liquids. The glycol prevents freezing in these plants during the condensation of water and hydrocarbons. The water phase of the process liquid contains the EG or DEG, which is always recovered and regenerated.

4.3.3 Dehydration Systems

There are essentially four methods in current use for dehydrating gases: direct cooling, compression followed by cooling, absorption, and adsorption. Generally, the first two methods do not result in sufficiently low water dew points to permit injection into a pipeline. Additional dehydration by absorption or adsorption may be required.

Water vapor may be removed from natural gas by bubbling the gas counter currently through certain liquids that have a special attraction or affinity for water. When water vapors are removed by this process, the operation is called absorption. There are also solids that have an affinity for water. When gas flows through a bed of such granular solids, the water is retained on the surface of the particles of the solid material. This process is called adsorption. The vessel in which either absorption or adsorption takes place is called the contactor or sorber. The liquid or solid having affinity for water and used in the contactor in connection with either of the processes is called the desiccant.

There are two major types of dehydration equipment in use at this time: the liquid desiccant dehydrator and the solid desiccant dehydrator. Each has its special advantages and disadvantages and its own field of particular usefulness. Practically all the gas moved through transmission lines is dehydrated by one of these methods.

Dehydration by Cooling

Examination of the water vapor saturation chart, Fig. 4.22 or 4.23, shows that the ability of natural gas to contain water vapor decreases as the temperature is lowered at constant pressure. Cooling for the specific purpose of gas dehydration is sometimes economical if the gas temperature is unusually high. Cooling is often used in conjunction with other dehydration processes. This method is limited in that gases dehydrated by cooling are still at their water dew point unless the temperature is raised again or the pressure is decreased.

Gas processing facilities that cool water may use different methods. In one type, some water is condensed and removed from gas at compressor stations by the compressor discharge coolers. Note that the saturation water content of gases decreases at higher pressure (see Fig. 4.22 or 4.23).

Low-temperature separation systems use the expansion-refrigeration cooling of the Joule–Thomson effect to condense water and hydrocarbons from the gas. In this method, hydrates are intentionally allowed to form by expanding the gas from wellhead pressure to pipeline pressure. The resulting dry gas, hydrates, and gas condensate enter a separator containing coils through which the warm well-stream flows to melt the hydrates. EG or DEG may be injected if the available wellhead pressure is insufficient to produce the required dehydration by expansion alone. Low-temperature separation can be an attractive technique since it can dehydrate gas as well as recover valuable liquid hydrocarbons.

All recent lean oil absorption gas plants use mechanical refrigeration to chill the inlet gas stream. EG is usually injected into the gas chilling section of the plant, which simultaneously dehydrates the gas and recovers liquid hydrocarbons, in a manner similar to the low-temperature separators.

Adsorption of Water Vapor by Solid Desiccants

Adsorption is defined as the ability of a substance to hold gases or liquids on its surface. This property occurs to a greater or lesser extent on all surfaces. Dehydration plants using solid desiccants can remove practically all water from natural gas (as low as 1 ppm). They can be used with temperatures higher than those for which glycol plants are satisfactory. Because of their great drying ability, solid desiccants may be employed where higher efficiencies are required.

In adsorption dehydration, the water vapor from the gas is concentrated and held at the surface of the solid desiccant by forces thought to be caused by residual valency. Solid desiccants have very large surface areas per unit weight to take advantage of these surface forces. The most common solid adsorbents include silica, alumina, and certain silicates known as molecular sieves.

Figure 4.30 shows a typical solid desiccant dehydration plant. The incoming wet gas must be carefully cleaned, preferably by a filter separator, to remove

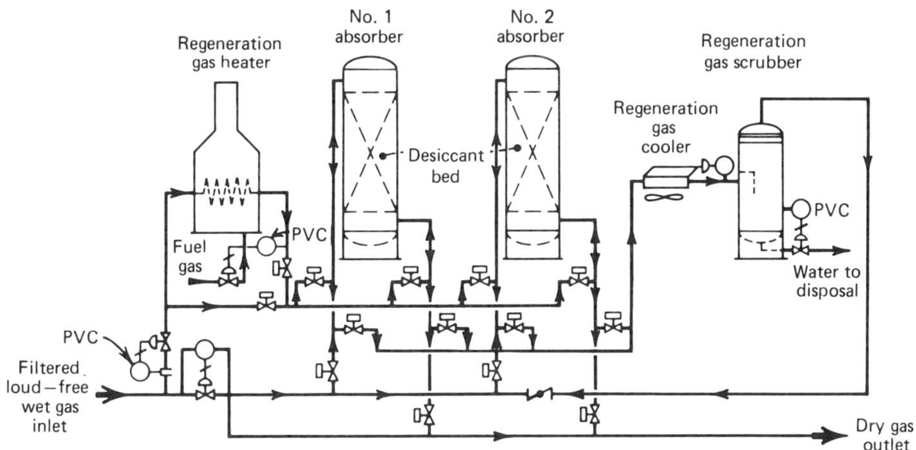

Fig. 4.30 Typical solid desiccant dehydration plant. (After Guenther.)

virtually all solid and liquid contaminants in the gas. The filtered liquid-free gas generally flows downward during dehydration through one adsorber containing a desiccant bed. The downflow arrangement lessens disturbance of the bed due to high gas velocity during the dehydration or adsorption step. While one adsorber is dehydrating, the other adsorber is being regenerated by a hotslip stream of inlet gas from the regeneration gas heater.

Although Fig. 4.30 shows a direct-fired heater, hot oil, steam, or an indirect heater can supply the necessary regeneration heat. The regeneration gas normally flows upward through the bed to ensure thorough regeneration of the bottom of the bed, which is the last area contacted by the gas being dehydrated. The hot regenerated bed is cooled by shutting off or bypassing the heater. The cooling gas flows downward through the bed so that any water adsorbed from the cooling gas will be at the top of the bed and will not be desorbed into the gas during the dehydration step. The still hot regeneration gas and the cooling gas flow through the regeneration gas cooler to condense the desorbed water. Power-operated valves activated by a timing device switch the adsorbers between the dehydration, regeneration, and cooling steps.

Figure 4.31 illustrates the relationship between regeneration-gas temperature and desiccant-bed temperature for a typical 8-hr regeneration cycle. Initially, the hot regeneration gas must heat the tower and desiccant. At about 240°F, water begins boiling or vaporizing and the bed continues to heat up, but more slowly since water is being driven out of the desiccant. After all the water has been removed, heating is continued to drive off any heavier hydrocarbons and contaminants, which would not vaporize at low temperatures. With cycle times of 4 hr or greater, the bed will be properly regenerated when the outlet-gas temperature has reached 350 to 375°F. Following the heating cycle, the bed will be cooled by flowing unheated regeneration gas through the bed. The cooling cycle will normally be terminated when bed temperature has dropped to about 125°F since

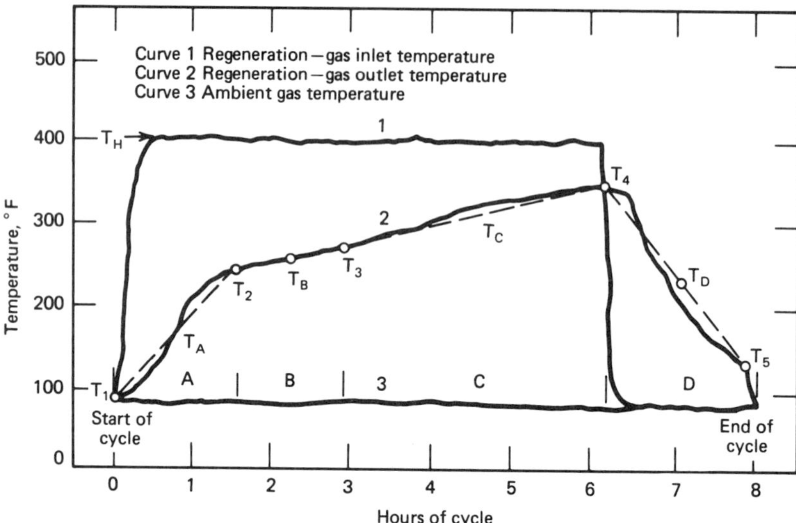

Fig. 4.31 Regeneration-gas temperature versus desiccant-bed temperature in a dry-desiccant dehydrator. (Courtesy Petroleum Extension Service.)

further cooling might cause water to condense from the wet gas stream and presaturate the bed before the next adsorption cycle begins.

The usable life of a desiccant may range from 1 to 4 years in normal service. All desiccants become less effective in normal use through loss of effective surface area. The loss of effective surface area is rapid at first, then becomes more gradual as the desiccant ages. Abnormally fast degradation occurs through blockage of the small pores and capillary openings, which contain most of the effective surface area. Lubricating oils, amines, glycols, corrosion inhibitors, and other contaminants, which cannot be removed during regeneration cycle, will eventually ruin the bed. Hydrogen sulfide may poison the desiccant and reduce its capacity. Activated alumina has good resistance to liquids, but it tends to powder due to mechanical agitation of the flowing gas.

There are several advantages of solid-desiccant dehydration. First, lower dew points are obtainable over a wide range of operating conditions. Second, essentially dry gas (moisture content less than 1.0 lb/MMcf) can be produced. Third, higher contact temperatures can be tolerated with some adsorbents. Fourth, greater adaptability is offered to sudden load changes, especially on starting up. Fifth, the plant may be put in operation more quickly after a shutdown. In some cases adsorption temperatures up to 125°F have been reported for silica gel adsorbents. Finally, adaptability for recovery of certain liquid hydrocarbons is high in addition to dehydration functions.

However, there are also operating problems. Space adsorbents degenerate with use and eventually require replacement. And, the amount of water vapor adsorbed per regeneration decreases with continued use. Loss in capacity may be accelerated by collecting contaminants like compressor cylinder oil deposited in desiccant beds. For a third problem, a tower must be regenerated, cooled, and readied for operation as another tower approaches exhaustion. With more than two towers involved, this operation is relatively complicated. If operation is on a time cycle, maximum allowable time on dehydration gradually shortens because desiccant loses capacity with use.

Unloading towers and recharging them with new desiccant should be completed well ahead of the operating season. In the interest of maintaining continuous operation when most needed, this may require discarding desiccant before the end of its normal operating life. To conserve material, the inlet part of the tower may be recharged and the remainder of the desiccant retained since it may still possess some useful life. Additional service life may be obtained if the direction of gas flow is reversed at a time when the tower would normally be recharged.

Sudden pressure surges should be avoided. They may upset the desiccant bed and channel the gas stream with poor dehydration. If a plant is operated above its rated capacity, pressure loss will increase and some attrition may occur. Attrition causes fines, which may in turn cause excessive pressure loss with resulting loss in capacity. Finally, if water cooling is used during the reactivating cycle, provision must be made against freezing in water coolers. This problem generally occurs only during cooler shutdown.

Absorption of Water Vapor by Liquid Desiccants

In absorption dehydration, the water vapor is removed from the gas by intimate contact with a hygroscopic liquid desiccant. The contacting is usually performed in trayed or packed towers. The glycols have proved to be the most effective liquid desiccants in current use. The four types of glycol that have been successfully used to dehydrate natural gas are ethylene glycol (EG), diethylene glycol (DEG), triethylene glycol (TEG), and tetraethylene glycol (T_4EG).

In almost all cases a single type of pure glycol is used; however, under certain circumstances, a glycol blend may be economically attractive. In over 40 years of use in the gas industry, TEG has gained nearly universal acceptance as the most cost effective of the glycols due to superior dew point depression, operating cost, and operation reliability. Glycol dehydration is usually economically more attractive than solid-desiccant dehydration when both processes are capable of meeting the required dew point. It has been reported that a solid-desiccant plant designed for 10 MMscfd costs approximately 53% more than a TEG plant, and that a solid-desiccant plant designed for 50 MMscfd costs approximately 33% more than a TEG plant (Guenther).

Triethylene glycol has been successfully used to dehydrate sweet and sour natural gases over the following range of operating conditions:

Dew point depression	40–140°F
Gas pressure	25–2500 psig
Gas temperature	40–160°F

The dew point depression obtained depends on the equilibrium dew point temperature for a given TEG concentration and contact temperature. Increased glycol viscosity may be a problem at the lower end of the contact temperature range; consequently, heating of the natural gas may be desirable. Very hot gas streams contain more water, vaporize more TEG, have lower equilibrium dew points, and are therefore often cooled prior to dehydration.

The wet inlet gas must be free of liquid water and hydrocarbons, wax, sand, drilling muds, and so on. The presence of any of these substances can cause severe foaming, flooding, higher glycol losses, poor efficiency, and increased maintenance in the dehydration tower or absorber. These impurites should be removed upstream of the absorber by an efficient scrubber, separator, or even a filter separator for very contaminated gases.

Methanol, injected at the wellhead as a hydrate inhibitor, can cause several problems for glycol dehydration plants. First, methanol increases the heat requirements of the glycol regeneration system, since methanol is coabsorbed with water vapor by glycol. Second, slugs of liquid methanol can cause flooding in the absorber. Finally, methanol vapor vented to the atmosphere with the water vapor from the regeneration system is hazardous and should be recovered or vented so that hazardous concentrations do not occur.

Figure 4.32 illustrates the process and flow through a typical glycol dehydrator. The west inlet gas stream first enters the unit through an inlet gas scrubber where any liquid accumulations are removed. A two-phase or distillate-gas scrubber is illustrated in Fig. 4.32. If any liquid water is in the gas stream, a three-phase scrubber may be used to discharge the distillate and water from the vessel separately. The mist eliminator aids in removing any entrained liquid particles from the wet gas stream leaving the top of the inlet scrubber.

The wet gas then enters the bottom of the glycol-gas contactor and flows upward through the trays as illustrated countercurrent to the glycol flowing downward through the column. The gas contacts the glycol on each tray and the glycol absorbs the water vapor from the gas stream. The dry gas leaves the top of the contactor vessel through another mist eliminator which aids in removing any entrained glycol droplets from the gas steam. The gas then flows down through a vertical glycol cooler, usually fabricated in the form of a concentric pipe heat exchanger, where the outlet dry gas aids in cooling the hot regenerated glycol before it enters the contactor. The dry gas then leaves the unit from the bottom of the glycol cooler.

The dry glycol enters the top of the glycol-gas contactor from the glycol cooler and is injected onto the top tray. The glycol flows across each tray and down through a downcomer pipe onto the next tray. The bottom tray downcomer is fitted with a seal pot to hold a liquid seal on the trays.

The wet glycol, which has now absorbed the water vapor from the gas stream, leaves the bottom of the glycol-gas contactor column, passes through a high-pressure glycol filter, which removes any foreign solid particles that may have been picked up from the gas stream, and enters the power side of the glycol pump.

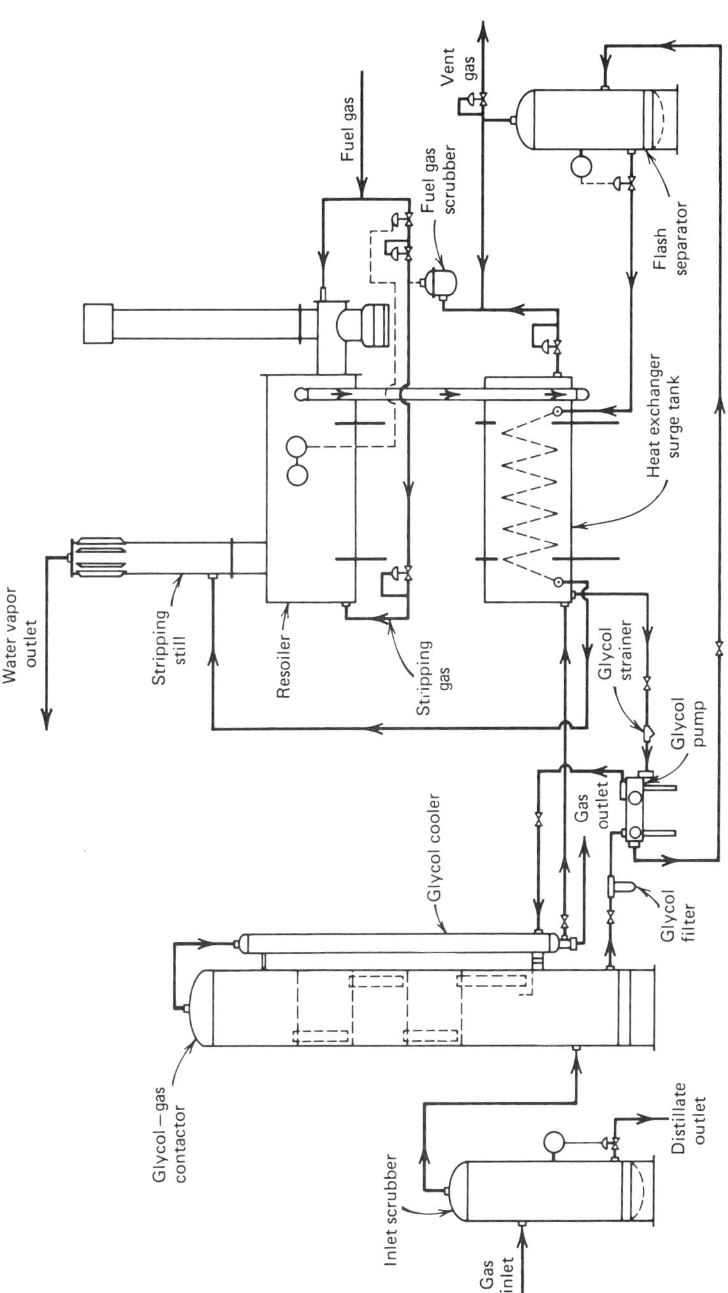

Fig. 4.32 Glycol dehydrator flow diagram. (After Sivalls.)

In the glycol pump the wet high-pressure glycol from the contactor column pumps the dry regenerated glycol into the column. The wet glycol stream flows from the glycol pump to the inlet of the flash separator. The low-pressure flash separator allows for the release of the entrained solution gas, which must be used with the wet glycol to pump the dry glycol into the contactor. The gas separated in the flash separator leaves the top of the flash separator vessel and may be used to supplement the fuel gas required for the reboiler. Any excess vent gas is discharged through a back-pressure valve.

The flash separator is equipped with a liquid level control and diaphragm motor valve that discharges the wet glycol stream through a heat exchange coil in the surge tank to preheat the wet glycol stream. If the wet glycol stream absorbs any liquid hydrocarbons in the contactor, it may be desirable to use a three-phase flash separator to separate the glycol from the liquid hydrocarbons before the stream enters the reboiler. Any liquid hydrocarbons present in the reboiler will cause undue glycol losses from the stripping still.

The wet glycol stream leaves the heat exchange coil in the surge tank and enters the stripping still mounted on top of the reboiler at the feed point in the still. The stripping still is packed with a ceramic intalox saddle-type packing, and the glycol flows downward through the column and enters the reboiler. The wet glycol passing downward through the still is contacted by hot rising glycol and water vapors passing upward through the column. The water vapors released in the reboiler and stripped from the glycol in the stripping still pass upward through the still column through an atmospheric reflux condenser that provides a partial reflux for the column. The water vapor then leaves the top of the stripping still column and is released to the atmosphere.

The glycol flows through the reboiler in essentially a horizontal path from the stripping still column to the opposite end. In the reboiler, the glycol is heated to approximately 350 to 400°F to remove enough water vapor to reconcentrate it to 99.5% or more. In field dehydration units, the reboiler is generally equipped with a direct-fired firebox, using a portion of the natural gas stream for fuel. In plant-type units, the reboiler may be fitted with a hot oil-heated coil or steam coil. A temperature control in the reboiler operates a fuel gas motor valve to maintain the proper temperature in the glycol. The reboiler is also generally equipped with a high-temperature safety overriding temperature controller to shut down the fuel gas system in case the primary temperature control should malfunction.

In order to provide extra-dry glycol, 99% plus, it is usually necessary to add some stripping gas to the reboiler. A valve and small pressure regulator are generally provided to take a small amount of gas from the fuel gas system and inject it into the bottom of the reboiler through a spreader system. This stripping gas will "roll" the glycol in the reboiler to allow any pockets of water vapor to escape that might otherwise remain in the glycol due to its normal high viscosity. This gas will also sweep the water vapor out of the reboiler and stripping still. By lowering the partial pressure of the water vapor in the reboiler and still column, the glycol can be reconcentrated to a higher percentage.

The reconcentrated glycol leaves the reboiler through an overflow pipe and passes into the shell side of the heat exchanger-surge tank. In the surge tank the hot reconcentrated glycol is cooled by exchanging heat with the wet glycol stream passing through the coil. The surge tank also acts as a liquid accumulator for feed for the glycol pump. The reconcentrated glycol flows from the surge tank through a strainer and into the glycol pump. From the pump it passes into the shell side of the glycol cooler mounted on the glycol-gas contactor. It then flows upward through the glycol cooler where it is further cooled and enters the column on the top tray.

Glycol dehydrators have several advantages. Initial equipment costs are lower, and the low-pressure drop across absorption towers saves power. Operation is continuous; this is generally preferred to batch operation. Also, makeup requirements may be added readily; recharging of towers presents no problems. Finally, the plant may be used satisfactorily in the presence of materials that would cause fouling of some solid adsorbents.

There are also several operating problems. First, suspended foreign matter, such as dirt, scale, and iron oxide, may contaminate glycol solutions. Also, overheating of solution may produce both low and high boiling decomposition products. The resultant sludge may collect on heating surfaces, causing some loss in efficiency, or, in severe cases, complete flow stoppage. Placing a bypass mechanical filter ahead of the solution pump usually prevents such troubles. When both oxygen and hydrogen sulfide are present, corrosion may become a problem because of the formation of acid material in glycol solution.

Second, liquids (e.g., water, light hydrocarbons, or lubricating oils) in inlet gas may require installation of an efficient separator ahead of the absorber. Highly mineralized water entering the system with inlet gas may, over long periods, crystallize and fill the reboiler with solid salts. Third, foaming of solution may occur with a resultant carry-over of liquid. The addition of a small quantity of antifoam compound usually remedies this trouble.

Fourth, some leakage around the packing glands of pumps may be permitted since excessive tightening of packing may result in the scouring of rods. This leakage is collected and periodically returned to the system. Fifth, highly concentrated glycol solutions tend to become viscous at low temperatures and, therefore, are hard to pump. Glycol lines may solidify completely at low temperatures when the plant is not operating. In cold weather, continuous circulation of part of the solution through the heater may be advisable. This practice can also prevent freezing in water coolers.

In starting a plant, all absorber trays must be filled with glycol before good contact of gas and liquid can be expected. This may also become a problem at low-circulation rates because weep holes on trays may drain solution as rapidly as it is introduced. Finally, sudden surges should be avoided in starting and shutting down a plant. Otherwise, large carry-over losses of solution may occur.

4.3.4 Glycol Dehydrator Design

Triethylene glycol dehydrators using tray or packed-column contactors may be sized from standard models by using the following procedures and associated graphs and tables (Sivalls). Custom-designed glycol dehydrators for specific applications may also be designed using these procedures. The following information must be available on the gas stream to be dehydrated:

1. Gas flow rate, MMscfd.
2. Specific gravity of gas.
3. Operating pressure, psig.
4. Maximum working pressure of contact, psig.
5. Gas inlet temperature, °F
6. Outlet gas water content required, lbm/MMscf.

Having the above information, it is then necessary to select two points of design critieria:

1. Glycol to water circulation rate based on water removed. A value of 2 to 6 gal TEG/lb H_2O removed is adequate for most glycol dehydration requirements. Use 2.5 to 4 gal TEG/lb H_2O for most field dehydrators.
2. Lean TEG concentration from reconcentrator. 99.0 to 99.9% lean TEG is available from most glycol reconcentrators. A value of 99.5% lean TEG is adequate for most design considerations.

The following procedures may be used to size a glycol dehydrator for a specific set of conditions, evaluate performance, and determine the gas capacity of a given size unit.

Inlet Scrubber

A good inlet scrubber is essential for efficient operation of any glycol dehydrator unit. The required diameter of a vertical inlet scrubber may be selected using Fig. 4.33, based on the operating pressure of the unit and gas capacity required. Two-phase inlet scrubbers are generally constructed with 7½-ft shell heights. Additional data on typical standard vertical inlet scrubbers are contained in Tables A.14 and A.15 in the Appendix.

Glycol-gas Contactor

Select a contactor diameter based on the operating pressure required with the approximate required gas capacity from Fig. 4.34 or 4.35. Figure 4.34 is for glycol contactors using trayed columns, and Fig. 4.35 is for contactors using packed columns. The gas capacities as determined for a given diameter contactor from Fig. 4.34 or 4.35 must be corrected for the operating temperature and gas specific gravity.

Calculate the gas capacity of the gas-glycol contactor selected for the specific operating conditions:

$$q_o = q_s(C_t)(C_g) \tag{4.7}$$

where

q_o = gas capacity of contactor at operating conditions, MMscfd

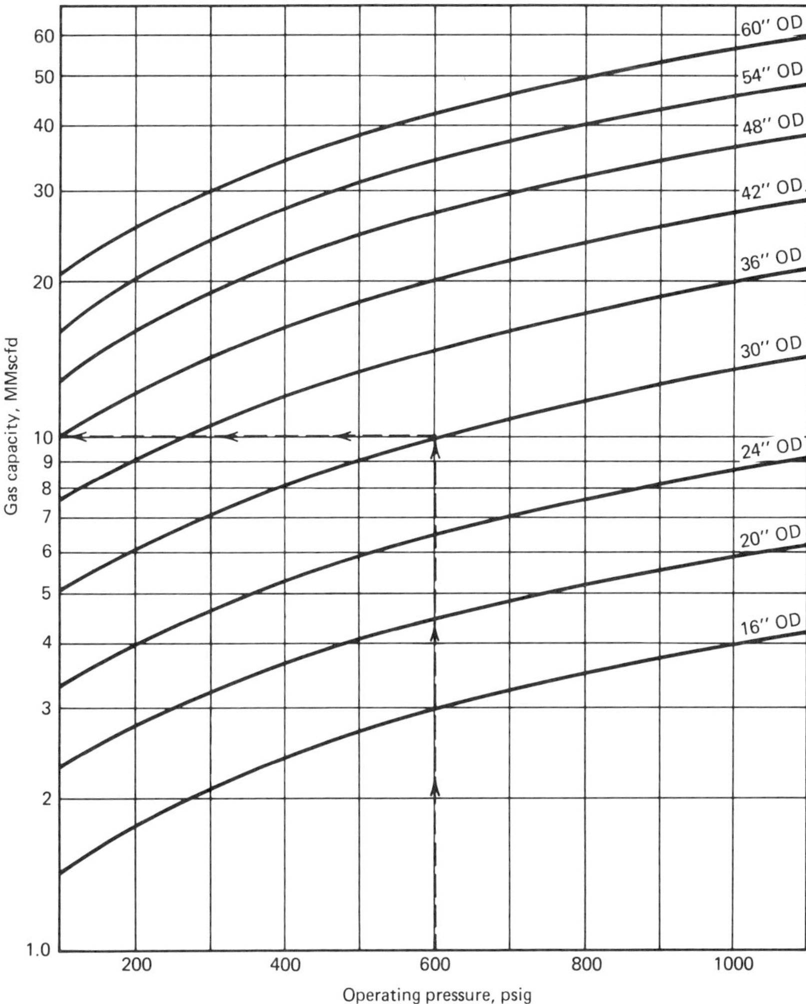

Fig. 4.33 Gas capacity of vertical gas inlet scrubbers based on 0.7 specific gravity, at 100°F. (After Sivalls.)

q_s = gas capacity of contactor at standard conditions (0.7 specific gravity and 100°F) based on operating pressure, MMscfd

C_t = correction factor for operating temperature

C_g = correction factor for gas specific gravity

The temperature and gas specific gravity correction factors for trayed glycol contactors are contained in Tables 4.1 and 4.2, respectively. The temperature and specific gravity factors for packed glycol contactors are contained in Tables 4.3 and 4.4, respectively.

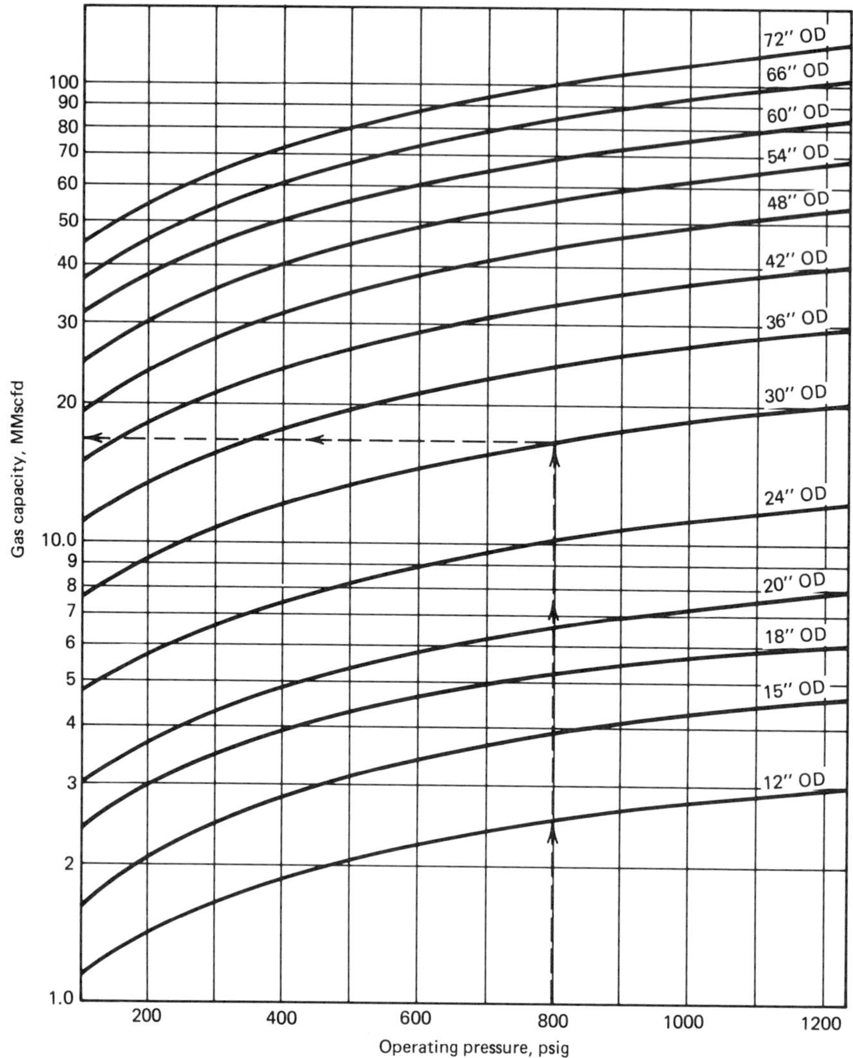

Fig. 4.34 Gas capacity for trayed glycol gas contactors based on 0.7 specific gravity, at 100°F. (After Sivalls.)

Next, determine the required dew point depression and the water removed from the glycol dehydration unit from the following:

Dew point depression

$$°F = \text{inlet gas temp. } °F - \text{outlet dew point temp. } °F$$

$$W_r = \frac{(W_i - W_o)(q)}{24} \tag{4.8}$$

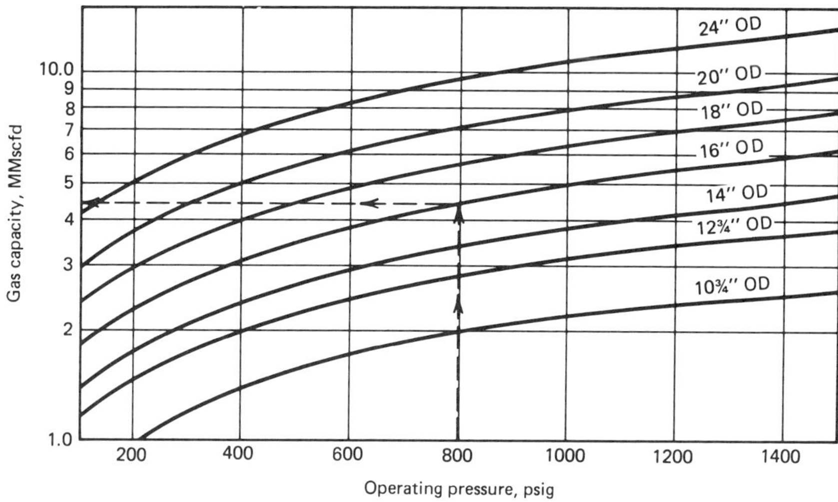

Fig. 4.35 Gas capacity for packed glycol gas contactors based on 0.7 specific gravity, at 100°F. (After Sivalls.)

where

W_r = water removed, lbm/hr
W_i = water content of inlet gas, lb H_2O/MMcf
W_o = water content of outlet gas, lb H_2O/MMcf
q = gas flow rate, MMscfd

The outlet dew point temperature can be found on the water vapor content graph using the outlet gas water content required and the operating pressure. The dew point temperature is the temperature at which the remaining water vapor in the gas will start to condense. The inlet gas temperature is also the inlet dew point temperature since the gas is generally assumed to be water-saturated before it is dehydrated. The water content of the inlet gas can be determined from the same water vapor content graph using the inlet gas temperature and the operating pressure.

If the natural gas stream contains appreciable amounts of either carbon dioxide or hydrogen sulfide, the water content of these sour gases should be taken into account in determining the total water content of the inlet gas stream. Since both carbon dioxide and hydrogen sulfide absorb considerably more water vapor than natural gas, they appreciably increase the total water content and dehydration requirements of the gas stream.

Trayed Contactors

Select the number of actual trays required from Fig. 4.36, using the required dew point depression and the selected glycol to water circulation rate. The data contained in Fig 4.36 will give the approximate number of trays required for rapid

TABLE 4.1
Gas Capacity Correction Factors for Trayed Glycol-Gas Contactors
Temperature Correction Factors, C_t

Operating Temperature, °F	Correction Factor, C_t
40	1.07
50	1.06
60	1.05
70	1.04
80	1.02
90	1.01
100	1.00
110	0.99
120	0.98

Source: After Sivalls.

TABLE 4.2
Gas Capacity Correction Factors for Trayed Glycol-Gas Contactors
Specific Gravity Correction Factors, C_g

Gas Specific Gravity	Correction Factor, C_g
0.55	1.14
0.60	1.08
0.65	1.04
0.70	1.00
0.75	0.97
0.80	0.93
0.85	0.90
0.90	0.88

Source: After Sivalls.

TABLE 4.3
Gas Capacity Correction Factors for Packed Glycol-Gas Contactors
Temperature Correction Factors, C_t

Operating Temperature, °F	Correction Factor, C_t
50	0.93
60	0.94
70	0.96
80	0.97
90	0.99
100	1.00
110	1.01
120	1.02

Source: After Sivalls.

TABLE 4.4
Gas Capacity Correction Factors for Packed Glycol-Gas Contactors
Specific Gravity Correction Factors, C_g

Gas Specific Gravity	Correction Factor, C_g
0.55	1.13
0.60	1.08
0.65	1.04
0.70	1.00
0.75	0.97
0.80	0.94
0.85	0.91
0.90	0.88

Source: After Sivalls.

sizing of field glycol dehydrators. A more detailed consideration of the actual number of trays required will give the accurate results needed for the most economical size contactor.

For a more detailed study, a modified McCabe–Thiele diagram can be constructed to determine the number of theoretical trays for a triethylene glycol dehydrator. This number can be converted to the actual number of trays required by applying the tray efficiency.

First, determine the concentration of the rich TEG leaving the bottom of the glycol-gas contactor.

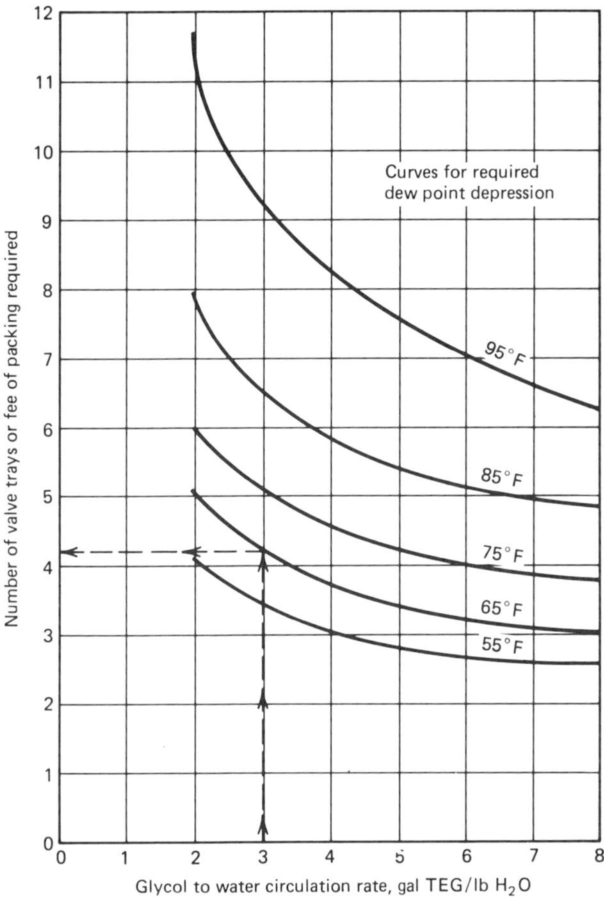

Fig. 4.36 Trays or packing required for glycol dehydrators. (After Sivalls.)

$$\rho_i = \text{Sp gr } (8.34) \tag{4.9}$$

$$\text{Rich TEG} = \frac{(\text{lean TEG})(\rho_i)}{\rho_i + \dfrac{1}{L_w}} \tag{4.10}$$

where

ρ_i = density of lean TEG solution, lbm/gal

Sp gr = specific gravity of lean TEG solution at operating temperature of contactor

Rich TEG = concentration of TEG in rich solution from contactor, %/100

Lean TEG = concentration of TEG in lean solution to contactor, %/100

L_w = glycol to water circulation rate, gal TEG/lb H_2O

The operating line for the McCabe–Thiele diagram is based on connecting a line between a point indicating the top of the column and a point indicating the bottom of the column.

Top of column	lb H_2O/MMcf in outlet gas and lean TEG, %
Bottom of column	lb H_2O/MMcf in inlet gas and rich TEG, %

The equilibrium line on the McCabe–Thiele diagram can be constructed by determining the water content of the gas that would be in equilibrium with various concentrations of triethylene glycol. This can be done by filling in the following table.

Percent TEG	Equilibrium Dew Point Temperature at Contactor Operating Temperature[a]	Water Content of Gas at Dew Point Temperature and Contactor Operating Pressure, lb H_2O/MMcf[b]
99		
98		
97		
96		
95		

[a]Determine from the chart of equilibrium water dew points of glycol solutions at various contact temperature, Fig. 4.37.
[b]Determine from the chart of water vapor content of gas at various temperatures and pressures, Fig. 4.22.

The modified McCabe–Thiele diagram can then be constructed with the operating line and equilibrium line and then stepped off by triangulation to determine the theoretical number of trays required. This procedure is best illustrated by an example (see Example 4.9 and Fig. 4.38).

The actual number of trays then required can be determined using the tray efficiency

$$\text{Number actual trays} = \frac{\text{no. theo. trays}}{\text{tray eff.}} \tag{4.11}$$

where

$$\text{Tray efficiency} = 25\% \text{ for bubble cap trays}$$
$$= 33\tfrac{1}{3}\% \text{ for valve trays}$$

The number of actual trays required as determined from either Fig. 4.36 or by construction of the McCabe–Thiele diagram is based on both theoretical and actual test data using a typical natural gas. Select the next whole number of trays based on the above design procedures after the tray efficiencies have been considered. However, good operation of field dehydrators indicates that a minimum of four trays should be used in any glycol-gas contactor.

Standard field dehydration contactors normally have 24-in. tray spacing. Because of the tendency of glycol to foam in the presence of liquid hydrocarbons,

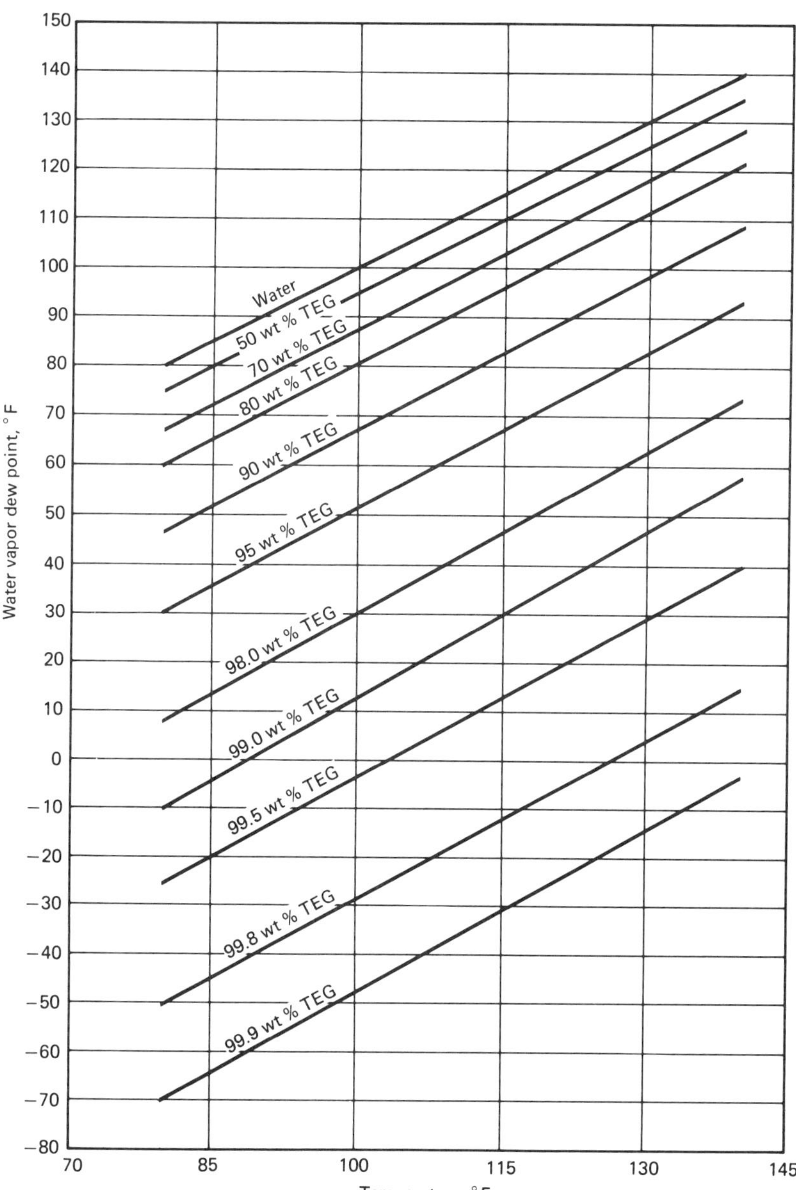

Fig. 4.37 Dew points of aqueous triethylene glycol solutions versus temperature. (After Sivalls.)

it is recommended that no less than 24-in. tray spacing be used to prevent any field problems with the equipment. If any foaming problem does occur, closer tray spacing can result in carryover or entrainment of the glycol in the gas stream and cause excessive glycol losses as well as decreased efficiency in dehydration of the gas.

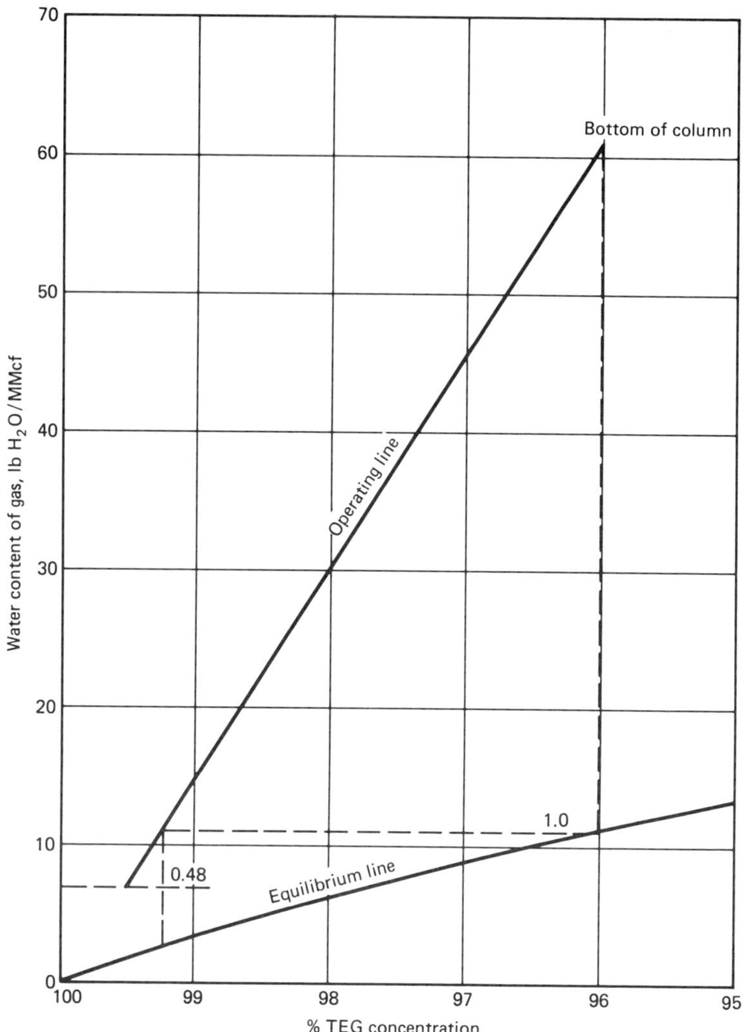

Fig. 4.38 Example of modified McCabe–Thiele diagram. (After Sivalls.)

Packed Contactors

The same procedures can be used for packed column contactors and the depth of packing required can be determined from Fig. 4.36. It is determined in the same manner using the required dew point depression and the selected glycol to water circulation rate. If a more detailed consideration of the depth of packing is required, a modified McCabe–Thiele diagram can be drawn based on the same procedures as described above. The depth of packing required can then be determined from the following empirical relation based on using 1-in. metal pall rings in the contactor.

$$\text{Depth packing, ft} = \text{no. theo. trays} \tag{4.12}$$

Then select the next whole number of feet of packing for use in the contactor. However, good operation indicates that a minimum 4 ft of packing should be used in any gas-glycol contactor.

Additional specifications for standard tray-type glycol-gas contactors are contained in Tables A.16 and A.17. Data on packed column glycol-gas contactors are contained in Tables A.18 and A.19.

Glycol Reconcentrator

For the detailed considerations involved in sizing the various components of the glycol reconcentrator, it is first necessary to calculate the required glycol circulation rate:

$$L = \frac{L_w(W_i)(q)}{24} \tag{4.13}$$

where

L = glycol circulation rate, gas/hr
L_w = glycol to water circulation rate, gal TEG/lb H_2O
W_i = water content of inlet gas, lb H_2O/MMscf
q = gas flow rate, MMscfd

Reboiler

The required heat load for the reboiler can be estimated from the following equation:

$$H_t = 2000L \tag{4.14}$$

where

H_t = Total heat load on reboiler, Btu/h
L = Glycol circulation rate, gph

The above formula for determining the required reboiler heat load is an approximation, which is accurate enough for most high-pressure glycol dehydrator sizing. A more detailed determination of the required reboiler heat load may be made from the following procedure:

$$H_l = L(\rho_i)(C)(T_2 - T_1) \tag{4.15}$$

$$H_w = \frac{970.3(W_i - W_o)(q)}{24} \tag{4.16}$$

$$H_r = 0.25H_w \tag{4.17}$$

$$H_h = 5000 \text{ to } 20{,}000 \text{ Btu/h, depending on boiler size} \tag{4.18}$$

$$H_l = H_l + H_w + H_r + H_h \tag{4.19}$$

where

H_l = sensible heat required for glycol, Btuh
H_w = heat of vaporization required for water, Btuh
H_r = heat to vaporize reflux water in still, Btuh
H_h = heat loss from reboiler and stripping still, Btuh
H_t = total reboiler heat load, Btuh
L = glycol circulation rate, gph
ρ_i = glycol density at average temperature in reboiler, lbm/gal = (sp gr)(8.34)
C = glycol specific heat at average temperature in reboiler, Btu/lbm °F
T_2 = glycol outlet temperature, °F
T_1 = glycol inlet temperature, °F
970.3 = heat of vaporization of water at 212°F 14.7 psia, Btu/lbm
W_i = water content of inlet gas lb H_2O/MMscf
W_o = water content of outlet gas lb H_2O/MMscf
q = gas flow rate, MMscfd

Note: For high-pressure glycol dehydrators, $\rho_i(C)(T_2 - T_1) \approx 1200$.

If the size of the reboiler and stripping still is known or is estimated, the heat loss can be more accurately determined from the following equation:

$$H_h = 0.24(A_s)(T_v - T_a) \tag{4.20}$$

where

H_h = overall heat loss from reboiler and still, Btuh
A_s = total exposed surface area of reboiler and still, sq ft
T_v = temperature of fluid in vessel, °F
T_a = minimum ambient air temperature, °F
0.24 = heat loss from large insulated surfaces, Btuh/sq ft − °F

The actual surface of the firebox required for direct-fired reboilers can be determined from the following equation, which is based on a design heat flux of 7000 Btuh/sq ft. By determining the diameter and overall length of the U-tube firebox required to give the total surface area as calculated, the general overall size of the reboiler can be determined.

$$A = \frac{H_t}{7000} \tag{4.21}$$

where

A = total firebox surface area, sq ft
H_t = total heat load on reboiler, Btuh

Glycol Circulating Pump

The required size of glycol circulating pump can be readily determined using the glycol circulation rate and the maximum operating pressure of the contactor. The most commonly used type of glycol pump for field dehydrators is the glycol powered pump, which uses the rich glycol from the bottom of the contactor to power the pump and pump the lean glycol to the top of the contactor. Sizing data for this type of glycol pump is contained in Table A.21 in the Appendix. For motor-driven positive displacement or centrifugal pumps, the manufacturers of these pumps should be consulted for exact sizing to meet the specific needs of the glycol dehydrator.

Glycol Flash Separator

A flash separator should be installed downstream from the glycol pump (especially when the glycol powered type pump is used) to remove any entrained hydrocarbons from the rich glycol. A small 125 psi WP vertical two-phase separator is adequate for this purpose. The separator should be sized based on a liquid retention time in the vessel of at least five minutes.

$$V = \frac{LT}{60} \tag{4.22}$$

where

V = required settling volume in separator, gal
L = glycol circulation rate, gph
T = retention time $\simeq$ 5.0 min

Liquid hydrocarbon should not be allowed to enter the glycol-gas contactor. Should this be a problem, a three-phase glycol flash separator will keep these liquid hydrocarbons out of the reboiler and stripping still. A liquid retention time of 20 to 30 min should be used in Eq. 4.22 to size a three-phase flash separator.

The hydrocarbon gas released from the flash separator can be piped to the reboiler to use as fuel gas and stripping gas. The amount of gas available from the glycol pump can be determined from the data in Table A.21, based on the glycol circulation rate and the operating pressure of the contactor.

Stripping Still

The size of the packed stripping still for use with the glycol reconcentrator can be determined from Fig. 4.39. The diameter required for the stripping still is normally based on the required diameter at the base of the still using the vapor and liquid loading conditions at that point. The vapor load consists of the water vapor (steam) and stripping gas flowing up through the still. The liquid load consists of the rich glycol stream and reflux flowing downward through the still column. The minimum cross-sectional area or diameter required for the still as read from Fig. 4.39 is based on the glycol to water circulation rate (gal TEG/lb H_2O) and the glycol circulation rate (gph).

Normally, one theoretical tray is sufficient for most stripping still requirements for triethylene glycol dehydration units. For conservative design, the height of packing using 1½-in. ceramic Intalox saddles is held at a minimum of 4 ft. Conservative design and field test data indicate that this height should be gradu-

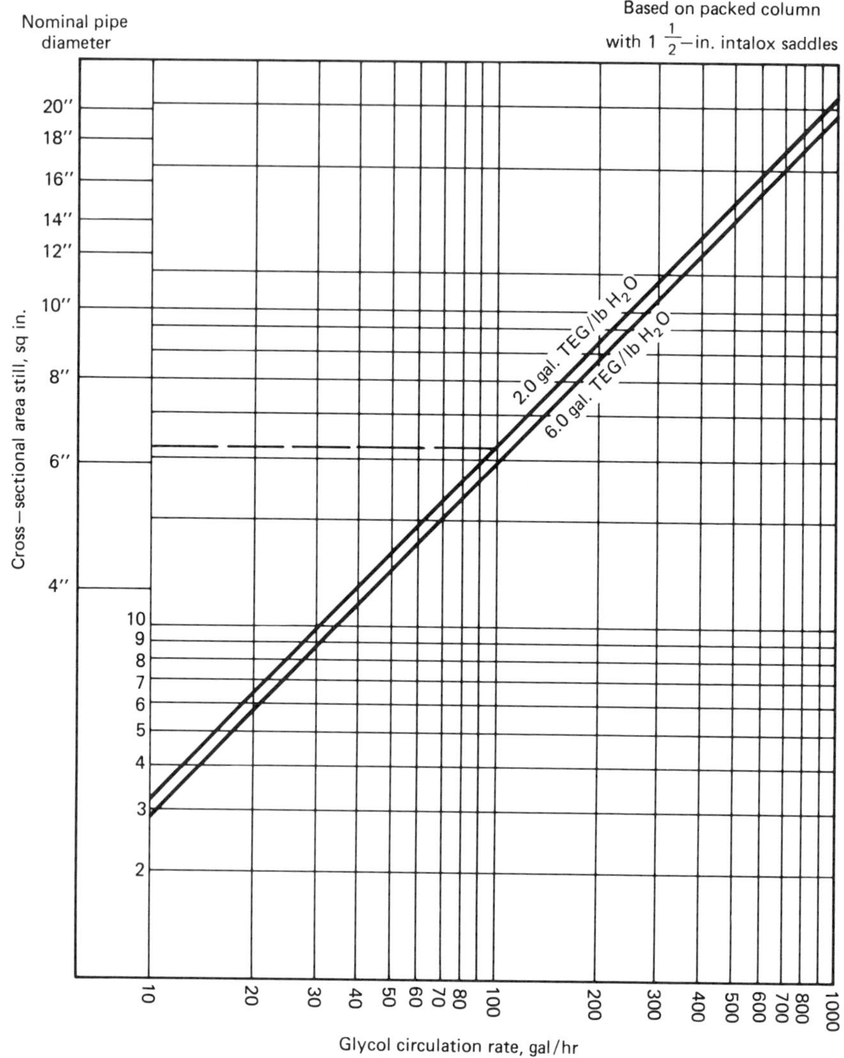

Fig. 4.39 Stripping still size for glycol dehydrators. (After Sivalls.)

ally increased with the size of the glycol reconcentrator to a maximum of approximately 8 ft for a 1,000,000 Btuh unit.

The amount of stripping gas required to reconcentrate the glycol to a high percentage will usually be approximately 2 to10 cu ft/gal of glycol circulated. This stripping gas requirement has been considered in the size of stripping still that is determined using Fig. 4.39.

Specifications for the main components of standard size glycol reconcentrators are contained in Table A.20.

The main design items required for a triethylene glycol dehydration unit such as the glycol-gas contactor, glycol circulation pump, reboiler, and stripping still may be designed using the above described procedures and formulas.

Standard dehydrator units based on manufacturer's catalog data can then be selected based on the minimum design criteria as determined. Also, the above procedures may be used in evaluating the performance or determining the maximum gas capacity of any given specific size glycol dehydration unit.

The following example illustrates the application of the above design procedures to a typical field gas dehydration plant.

Example 4.9. Dehydrator Design. Size a glycol dehydrator for a field installation from standard models to meet the following requirements:

Gas flow rate	10.0 MMscfd
Gas specific gravity	0.70
Operating line pressure	1000 psig
Maximum working pressure of contactor	1440 psig
Gas inlet temperature	100°F
Outlet gas water content	7 lb H_2O/MMscf

Select additional design criteria:

Glycol to water circulation rate	3.0 gal TEG/lb H_2O
Lean glycol concentration	99.5% TEG

Use trayed-type contactor with valve trays

Contactor size

From Fig. 4.34, select a contactor diameter with the approximate gas capacity at operating pressure.

$$q_s \text{ for 24-in. OD contactor at 1000 psig} = 11.3 \text{ MMscfd}$$

Correct for operating conditions from Tables 4.1 and 4.2, using Eq. 4.7,

$$q_o = q_s(C_t)(C_g)$$
$$q_o = 11.3(1.0)(1.0) = 11.3 \text{ MMscfd}$$

Required Dew Point Depression and Water Removed

From the water content chart at 1000 psig:

	Dew Point Temperature	Water content, lb H_2O/MMcf
Inlet	100°F	61
Outlet	33°F	7
	67°F	54 lb H_2O/MMcf

Number of Trays Required

From Fig. 4.36 at 3 gal TEG/lb H_2O) and 67°F dew point depression, No. actual trays = 4.5. For a more detailed study, construct a modified McCabe–Thiele diagram.

Density of lean glycol at 100°F, from Eq. 4.9,

$$\rho_i = (\text{Sp gr})(8.34) = (1.111)(8.34) = 9.266 \text{ lbm/gal}$$

Using Eq. 4.10,

$$\text{Rich TEG} = \frac{(\text{lean TEG})(\rho_i)}{\rho_i + \frac{1}{L_w}} = \frac{(0.995)(9.266)}{9.266 + \frac{1}{3.0}} = 0.960 = 96.0\%$$

Operating line points:

Top of column — 7.0 lb H_2O/MMcf and 99.5% TEG
Bottom of column — 61 lb H_2O/MMcf and 96.0% TEG
Equilibrium line points

Percent TEG	Equilibrium Dew Point Temperature at 100°F	Water Content of Gas at Dew Point Temperature and 1000 psig
99	12	3.2 lb H_2O/MMcf
98	30	6.3
97	40	9.0
96	47	11.7
95	51	13.3

Construct a McCabe–Thiele diagram and determine the number of theoretical trays required. See Fig. 4.38. From Eq. 4.11,

$$\text{No. actual trays} = \frac{\text{no. theo. trays}}{\text{tray eff.}} = \frac{1.48}{0.333} = 4.44$$

The results from the McCabe–Thiele diagram are close to that determined from the approximation curve, Fig. 4.36. In either case, the next whole number of trays should be used.

No. actual trays required = 5

4.3.5 Removal of Acid Gases

Natural gas wellstreams often contain hydrogen sulfide (H_2S) and carbon dioxide (CO_2). These two gases are called acid gases because, in the presence of water, they form acids or acidic solutions. These gases, particularly H_2S, are very undesirable contaminants; unless they are present in very small quantities, they must be removed from a natural gas wellstream.

Hydrogen sulfide must be removed for several reasons, the most important being that it is a toxic, poisonous gas and cannot be tolerated in gases that may be used for domestic fuels. Hydrogen sulfide in the presence of water is extremely corrosive and can cause premature failure of valves, pipeline, and pressure vessels. It can also cause catalyst poisoning in refinery vessels and requires expensive precautionary measures. Most pipeline specifications limit H_2S content to 0.25 g/100 cu ft of gas (about 4 ppm).

Carbon dioxide removal is not always required. However, removal may be required in gas going to cryogenic plants to prevent solidification of the CO_2. Carbon dioxide is corrosive in the presence of water, and as an inert gas has no heating value. Most treating processes that remove H_2S will also remove CO_2; therefore, the volume of CO_2 in the wellstream must be added to the volume of H_2S to arrive at the total acid-gas volume to be removed.

The term sour gas means that the gas contains H_2S in amounts above the acceptable industry limits. Sweet gas means a non-H_2S-bearing gas or gas that has been sweetened by treating. Some of the processes for removing acid gases from natural gas are briefly discussed below.

Iron-sponge Sweetening

The iron-sponge process is a batch process, the sponge being a sensitive, hydrated iron oxide (Fe_2O_3) supported on wood shavings. The reaction between the sponge and H_2S in the gas stream is

$$2\ Fe_2O_3 + 6\ H_2S \longrightarrow 2\ Fe_2S_3 + 6\ H_2O$$

The ferric oxide is present in a hydrated form; without the water of hydration, the reaction will not proceed. Thus, the operating temperature of the vessel must be kept below approximately 120°F or a supplemental water spray must be provided.

Regeneration of the bed is sometimes accomplished by the addition of air (O_2), either continuously or by batch addition. The regeneration reaction is

$$2\ Fe_2S_3 + 3\ O_2 \longrightarrow 2\ Fe_2O_3 + 6\ S$$

Because the sulfur remains in the bed, the number of regeneration steps is limited, and eventually the bed will have to be replaced.

Alkanolamine Sweetening

Alkanolamine encompasses the family of organic compounds of monoethanolamine (MEA), diethanolamine (DEA), and triethanolamine (TEA). The chemicals are used extensively for the removal of hydrogen sulfide or carbon dioxide from other gases and are particularly adapted for obtaining the low acid-gas residuals that are usually specified by pipelines. The alkanolamine process is not selective and must be designed for total acid-gas removal, even though CO_2 removal may not be required or desired to meet market specifications.

Typical reactions of acid gas with MEA are
Absorbing:

$$MEA + H_2S \longrightarrow MEA \text{ hydrosulfide} + heat$$
$$MEA + H_2O + CO2 \longrightarrow MEA \text{ carbonate} + heat$$

Regenerating:

$$MEA \text{ hydrosulfide} + heat \longrightarrow MEA + H_2S$$
$$MEA \text{ carbonate} + heat \longrightarrow MEA + H_2O + CO_2$$

In general, MEA is preferred to either DEA or TEA solutions because it is a stronger base and is more reactive than either DEA or TEA. It also has a lower molecular weight and thus requires less circulation to maintain a given amine to acid-gas mole ratio. In addition, MEA has greater stability and can be readily reclaimed from contaminated solution by semicontinuous distillation.

Glycol/Amine Process

The glycol/amine process uses a solution composed of 10 to 30 weight % MEA, 45 to 85% glycol, and 5 to 25% water for the simultaneous removal of water vapor, H_2S, and CO_2 from gas streams. The combination dehydration and sweetening unit results in lower equipment cost than would be required with the standard MEA unit followed by a separate glycol dehydrator. The main disadvantages of the glycol/amine process are increased vaporization losses of MEA due to high regeneration temperatures, reclaiming must be by vacuum distillation, corrosion problems are present in operating units, and application must be for gas streams that do not require low dew points.

Sulfinol Process

The Shell sulfinol process uses a mixture of solvents, which allows it to behave as both a chemical and physical solvent process. The solvent is composed of sulfolane, diisopropanolamine (DIPA), and water. The sulfolane acts as the physical solvent, while DIPA acts as the chemical solvent.

The main advantages of sulfinol are low solvent circulation rates; smaller equipment and lower plant cost; low heat capacity of the solvent; low utility costs; low degradation rates; low corrosion rates; low foaming tendency; high effectiveness for removal of carbonyl sulfide (COS), carbon disulfide (CS_2), and mercaptans; low vaporization losses of the solvent; low heat-exchanger fouling tendency; and nonexpansion of the solvent when it freezes.

Some of the disadvantages of sulfinol are absorption of heavy hydrocarbons and aromatics, and expense.

REFERENCES

Campbell, J. M. *Gas Conditioning and Processing*. Norman, Okla.: Campbell Petroleum Series, 1976.

Gas Processors Suppliers Assoc. *Engineering Data Book*. 9th ed. 3rd revision. 1977.

Guenther, J. D. "Natural Gas Dehydration." Paper presented at Seminar on Process Equipment and Systems on Treatment Platforms, Taastrup, Denmark. April 26, 1979.

Katz, D. L., et al. *Handbook of Natural Gas Engineering*. New York: McGraw-Hill, 1959.

Kimray, Inc. *Equipment Catalog*. pp. 3–7, section G, 1976.

Kohl, A. L., and F. C. Riesenfeld. *Gas Purification*. New York: McGraw-Hill, pp. 360–362, 1960.

McCarthy, E. L., W. L. Boyd, and L. S. Reid. "The Water Vapor Content of Essentially Nitrogen-Free Natural Gas Saturated at Various Conditions of Temperature and Pressure." *Petroleum Trans. AIME* **189**, pp. 241–43, 1950.

McKetta, J. J., and A. H. Wehe. "Use This Chart for Water Content of Natural Gases." *Petroleum Refiner* **37**, pp. 153–4, 1958.

Petroleum Extension Service. *Field Handling of Natural Gas*. 3rd Ed. Austin: Univ. of Texas Press, 1972.

Selleck, F. T., L. T. Carmichael, and B. H. Sage. "Phase Behavior in the Hydrogen Sulfide-Water System." *Industrial and Engineering Chemistry* **44**, pp. 2219–2226, 1952.

Sivalls, C. R. "Fundamentals of Oil and Gas Separation." Proceedings of the Gas Conditioning Conference, University of Oklahoma, 1977.

Sivalls, C. R. "Glycol Dehydrator Design Manual." Proceedings of the Gas Conditioning Conference, University of Oklahoma, 1976.

Weibe, R., and V. L. Gaddy. "Vapor Phase Composition of Carbon Dioxide-Water Mixtures at Various Temperatures and Pressures to 700 Atmospheres." *Journal of American Chemical Society* **63**, pp. 475–477, 1941.

PROBLEMS

4.1 The analysis of a gas sample obtained from a Panhandle well showed the following composition:

Component	*Mole Fraction*
CH_4	0.8590
C_2H_6	0.0664
C_3H_8	0.0355
i-C_4H_{10}	0.0100
n-C_4H_{10}	0.0120
C_5H_{12}	0.0081
CO_2	0.0000
N_2	0.0090
	1.0000

(a) For pressures of 200 and 400 psia, compute the hydrate-forming temperatures.

(b) For temperatures of 62°F, 70°F, and 80°F, compute the hydrate-forming pressures.

4.2 The sketch illustrates a refrigeration-expansion process to reduce water (and condensate) content of gas well production. The well stream arrives at the wellhead at 2000 psia and 155°F. It is cooled in the separator and heat exchanger and then flashed through the choke into the separator. The separator pressure is controlled at 1000 psia. The separator temperature is controlled at the lowest possible temperature that will not result in hydrates being produced in the separator.

Calculate the following:

(a) Pounds of water per day removed at the free water knockout drum (assume saturated gas at reservoir flowing conditions).

(b) Pounds of water per day removed at the separator.

(c) Separator temperature, °F.

(d) Temperature ahead of the choke, °F.

(e) Heat removed by heat exchange in separator and heat exchanger, Btu/day.

(f) Additional pounds of water per day that must be removed from separator gas if sales specification is for a 15°F water dew point at 1000 psia.

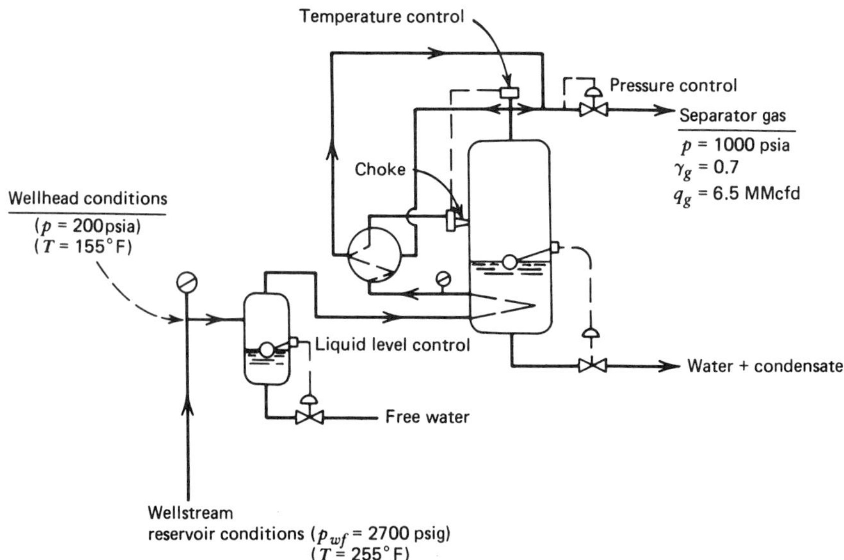

4.3 *Separator Sizing Problem*

Oil production	= 20,000 bopd
Water production	= 15%
Gas rate	= 60 MMscfd at 14.7 psia and 60°F
Maximum working pressure	= 720 psig
Design temperature	= 110°F
Operating pressure	= 500 psig
Flowing temperature	= 120°F
Oil gravity	= 32°API
Gas gravity	= 0.7

Design a proper size horizontal separator to handle the above fluid. *Assume:* No slugs or foaming exists and separator $K = 0.35$.

4.4 *Glycol Dehydration Problem*

Gas flow rate	= 100 MMscfd at 60°F and 14.7 psia
Molecular weight	= 18.85
Operating pressure	= 1000 psig
Operating temperature	= 100°F

Design a glycol dehydration system using trielhylene glycol (TEG) to condition the gas to 7 lb/MMscf.

5

COMPRESSION OF NATURAL GAS

5.1 INTRODUCTION

Until the discovery of natural gas, compressors were mainly used in the mining and metallurgical industries. With the advent of natural gas and its use as a fuel, the necessity arose of transporting natural gas from the gas well to the ultimate consumer. As long as the pressure at the gas well could force the gas through the pipeline to its destination, a compressor was unnecessary. As soon as the pressure dropped, however, some outside means was needed to increase the pressure. Compressors were also essential for gas transmission pipelines extended great distances from the gas field—extended so far that the natural well pressure could not force enough gas through the pipeline to supply the demand.

When a gas has insufficient potential energy for its required movement, a compressor station must be used. Five types of compressor stations are generally utilized:

Field or gathering stations gather gas from wells in which pressure is not sufficient to produce a desired rate of flow into a transmission or distribution system. These stations may handle suction pressures from below atmospheric pressure to 750 psig and volumes from a few thousand to many million cfd.

Relay or main line stations boost pressure in transmission lines. They are generally of large volume and operate with low-compression ratios, usually less than 2. Their pressure range is usually between 200 and 1000 psig, sometimes as high as 1300 psig.

Repressuring or recycling stations are an integral part of a processing or secondary recovery project, generally not involving transportation of natural gas to market. They may discharge at pressures above 6000 psig.

Storage field stations compress trunk line gas for injection into storage wells. These stations may discharge at pressures up to 4000 psig, employing compression ratios as high as 4. Designs of some storage stations permit withdrawing gas from storage and forcing it into high-pressure lines. Storage field stations require precision-design engineering because of their wide range of pressure-volume operating conditions.

Distribution plant stations ordinarily pump gas from holder supply to medium- or high-pressure distribution lines at about 20 to 100 psig, or pump into bottle storage up to 2500 psig.

5.2 TYPES OF COMPRESSORS

The original compressor consisted of a skin bag with a means of expanding and compressing it. A survivor of this ancient compressor is the bellows. The compressors used in the gas industry today fall into three distinct types: jet compressors, rotary compressors, and reciprocating compressors.

5.2.1 Jet Compressors

Motive and suction gas pressures do not vary appreciably with jet compressors. One example is on dually completed gas wells where both high- and low-pressure gases are available and where there is an intermediate pipeline pressure.

5.2.2 Rotary Compressors

Rotary compressors are divided into two classes: the rotary blower and the centrifugal compressor. Rotary blowers (Fig. 5.1) are primarily used in distribution systems where the pressure differential between suction and discharge is not over 15 psi. They are also used for refrigeration and closed regeneration of adsorption plants. The blower is built of a casing in which one or more impellers rotate in opposite directions.

The rotary blower has several advantages. Large quantities of low-pressure gas can be handled at comparatively low horsepower, it has small initial cost and low maintenance cost, it is simple to install and easy to operate and attend, it requires minimum floor space for the quantity of gas removed, and it has almost pulsationless flow. For its disadvantages, it can't withstand high pressures, it has noisy operation due to gear noise and clattering impellers, it improperly seals the clearance between the impellers and the casing, and it overheats if operated above safe pressures. Typically, rotary blowers have a volume up to 17,000 cfm, and have a maximum intake pressure of 10 psig and a differential pressure of 10 psig.

Centrifugal compressors compress gas or air using centrifugal force. In this compressor, work is done on the gas by an impeller. Gas is then discharged at a high velocity into a diffuser where the velocity is reduced and its kinetic energy is

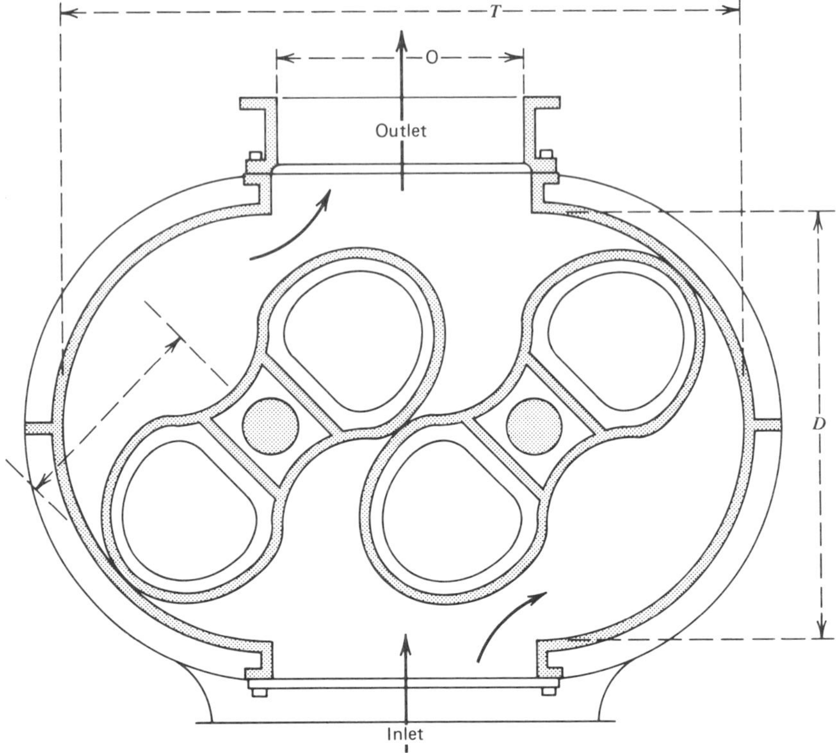

Fig. 5.1 Principles of the rotary blower.

converted to static pressure. All this is done without confinement and physical squeezing, unlike positive-displacement machines.

Essentially, the centrifugal compressor (Figs. 5.2 and 5.3) consists of a housing with flow passages, a rotating shaft on which the impeller is mounted, bearings, and seals to prevent gas from escaping along the shaft. Units differ in shape of volute, impeller, and diffuser.

Centrifugal compressors have few moving parts since only the impeller and shaft rotate. Thus, lubricating oil consumption and maintenance costs are low. They also have continuous delivery without cyclic variations, and cooling water is normally unnecessary because of lower compression ratio and lower friction loss (multistage units for process compression may require some form of cooling). Compression rates are lower because of the absence of positive displacement.

Centrifugal compressors with relatively unrestricted passages and continuous flow are inherently high-capacity, low-pressure ratio machines that adapt easily to series arrangements within a station. In this way, each compressor is required to develop only part of the station compression ratio. Typically, the volume is more than 100,000 cfm and discharge pressure is up to 100 psig.

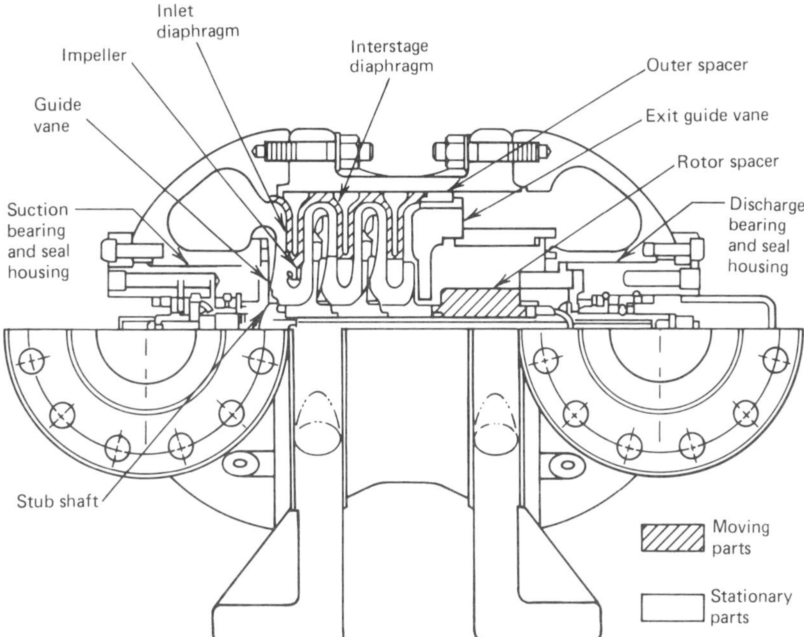

Fig. 5.2 Cross section of a centrifugal compressor. (Courtesy Petroleum Extension Service.)

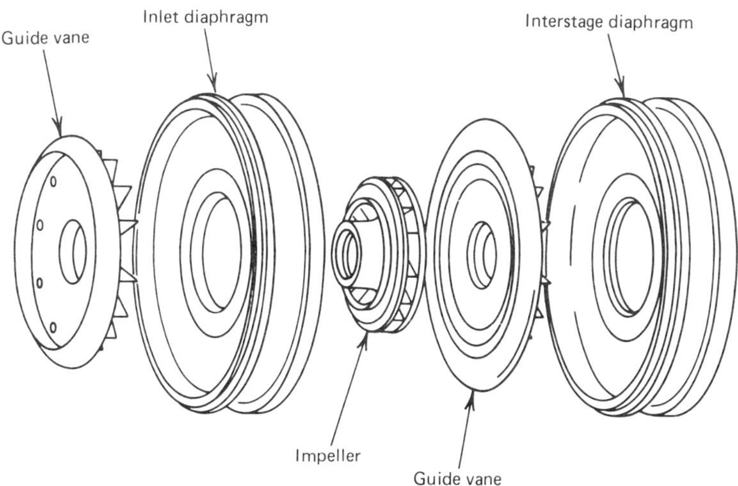

Fig. 5.3 Internal parts of a centrifugal compressor. (Courtesy Petroleum Extension Service.)

5.2.3 Reciprocating Compressors

Reciprocating compressors are most commonly used in the gas industry. These compressors are built for practically all pressures and capacities. Reciprocating compressors (Fig. 5.4) have more moving parts and, therefore, lower mechanical efficiencies than centrifugal machines. Each cylinder assembly of a reciprocating compressor consists of a piston, cylinder, cylinder heads, suction and discharge valves, and the parts necessary to convert rotary motion to reciprocating motion (connecting rod, crosshead, wrist pin, and piston rod). A reciprocating machine is designed for a certain range of compression ratios through the selection of proper piston displacement and clearance volume within the cylinder. This clearance volume can either be fixed or variable, depending on the extent of the operating range and the percent of load variation desired. A typical reciprocating compressor has a volume up to 30,000 cfm and a discharge pressure up to 10,000 psig.

In the selection of a unit, the pressure-volume characteristics and the type of driver must be considered. Small rotary compressors (vane or impeller type) are generally driven by electric motors. Large-volume positive compressors operate at lower speeds and are usually driven by steam or gas engines. They may be driven through reduction gearing by steam turbines or an electric motor. The most widely used reciprocating compressor in the gas industry is the conventional high-speed machine, usually driven by steam turbines or electric motors.

Reciprocating and centrifugal compressors are used extensively in the natural gas industry. Compressors vary in size from small units suitable for individual oil and gas leases (often skid-mounted) to large units for pipeline. The engineer is often required to determine the approximate horsepower needed to compress a given volume of gas from a low pressure to a higher pressure, or estimate the capacity of a given piece of equipment under specific suction and discharge conditions.

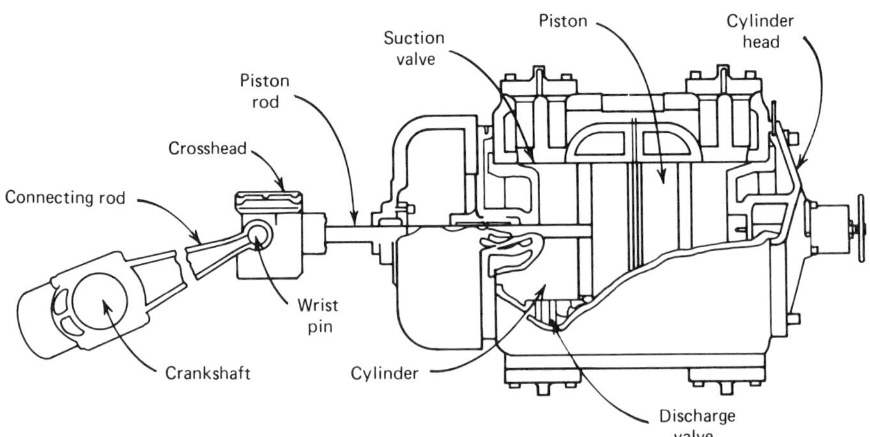

Fig. 5.4 Elements of a typical reciprocating compressor. (Courtesy Petroleum Extension Service.)

5.3 RECIPROCATING COMPRESSORS

There are two basic approaches that can be used to calculate the horsepower theoretically required to compress natural gas. One is analytical expressions. In the case of adiabatic compression, the relationships are complicated and are usually based on the ideal-gas equation. When used for real gases where deviation from ideal-gas law is appreciable, they are empirically modified to take into consideration the gas deviation factor of the gas. The second approach is the enthalpy-entropy or Mollier diagram for real gases. This diagram provides a simple, direct, and rigorous procedure for determining the horsepower theoretically necessary to compress the gas.

5.3.1 The Process

Figure 5.5 illustrates an ideal compression cycle. The compressor cylinder fills with gas at suction pressure, p_1. Gas is compressed to discharge pressure, p_2, along path *C-D* and then displaced from the cylinder at constant discharge pressure. It is impossible to discharge all compressed gas in actual operations. Thus, Fig. 5.6 is obtained. Figure 5.6 shows a typical ideal pressure-volume diagram for a compressor cylinder with corresponding compressor piston locations during reciprocation.

Position 1 is the start of the compression stroke. The cylinder has a full charge of gas at suction pressure. As the piston moves toward position 2, the gas is compressed along line 1-2. At position 2, the pressure in the cylinder becomes greater than the pressure in the discharge line. This causes the discharge valve to

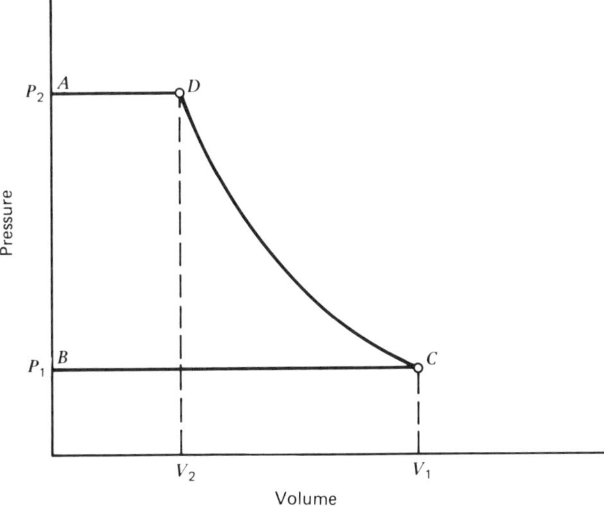

Fig. 5.5 Ideal compression cycle.

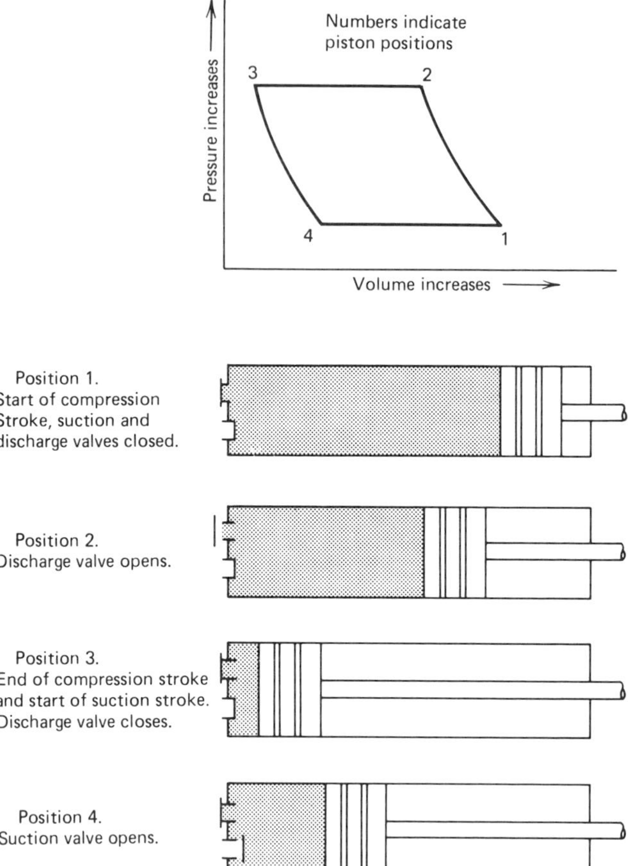

Fig. 5.6 Pressure-volume diagram and piston locations during a reciprocating compressor stroke. (Courtesy Petroleum Extension Service.)

open and allows the original charge of gas to enter the discharge line. This action occurs along line 2-3.

At position 3, the piston has completed its discharge stroke. As soon as it starts its return stroke, the pressure in the cylinder drops, which closes the discharge valve. The gas trapped in the cylinder clearance volume is never discharged but expands along line 3-4. At position 4, the pressure in the cylinder drops below the suction pressure, which causes the suction valve to open. This permits a new charge of gas to enter the cylinder along line 4-1, whereupon the cycle is repeated.

Even though in practice the cylinder may be water-cooled, it is customary to consider the compression process as fundamentally adiabatic; that is, to idealize the compression as one in which there is no cooling of the gas. Furthermore, the process is usually considered to be essentially a perfectly reversible adiabatic,that is, an isentropic process. Thus, in analyzing the performance of a typical reciprocating compressor, one may look upon the compression line 1-2 and the expansion line 3-4 as following the general law

$$pV^k = \text{a constant} \tag{5.1}$$

Where k is an isentropic exponent given by the specific heat ratio

$$k = \frac{C_p}{C_v} \tag{5.2}$$

When a real gas is compressed in a single-stage compression, the compression is polytropic, tending to approach adiabatic or constant-entropy conditions. Adiabatic-compression calculations give the maximum theoretical work or horsepower necessary to compress a gas between any two pressure limits, whereas isothermal-compression calculations give the minimum theoretical work or horsepower necessary to compress a gas. Adiabatic and isothermal work of compression thus give the upper and lower limits, respectively, of work or horsepower requirements to compress a gas. One purpose of intercoolers between multistage compressors is to reduce the horsepower necessary to compress the gas. The more intercoolers and stages, the closer the horsepower requirement approaches the isothermal value.

For the compression process occurring under actual, practical conditions, an equation similar to Eq. 5.1 applies

$$pV^n = \text{a constant} \tag{5.3}$$

In Eq. 5.3, n denotes the polytropic exponent. The isentropic exponent k applies to the ideal frictionless adiabatic process, while the polytropic exponent n applies to the actual process with heat transfer and friction.

In Fig. 5.6, V_3 is called the "clearance volume" and $V_3/(V_1 - V_3)$ is called the "clearance." This volume limits the gas throughout. As it gets larger, more engine horsepower is used in simply recompressing and re-expanding this gas. Area 1234 indicates the compression work done. It depends on the conditions existing during the compression and expansion portions of the cycle, the compression ratio, p_2/p_1, and the clearance volume, V_3.

5.3.2 *Volumetric Efficiency*

The volumetric efficiency E_v represents the efficiency of a compressor cylinder to compress gas. It may be defined as the ratio of the volume of gas actually delivered to the piston displacement, corrected to suction temperature and pressure. The principal reasons that the cylinder will not deliver the piston displacement capacity are wire-drawing, a throttling effect on the valves; heating of the gas during admission to the cylinder; leakage past valves and piston rings; and re-expansion of the gas trapped in the clearance-volume space from the previous stroke. Re-expansion has by far the greatest effect on volumetric efficiency.

The theoretical formula for volumetric efficiency is

$$E_v = 1 - (r^{1/k} - 1)Cl \tag{5.4}$$

where

E_v = volumetric efficiency, fraction
r = cylinder compression ratio
k = C_p/C_v of the gas at atmospheric conditions (isentropic process)
Cl = clearance, fraction

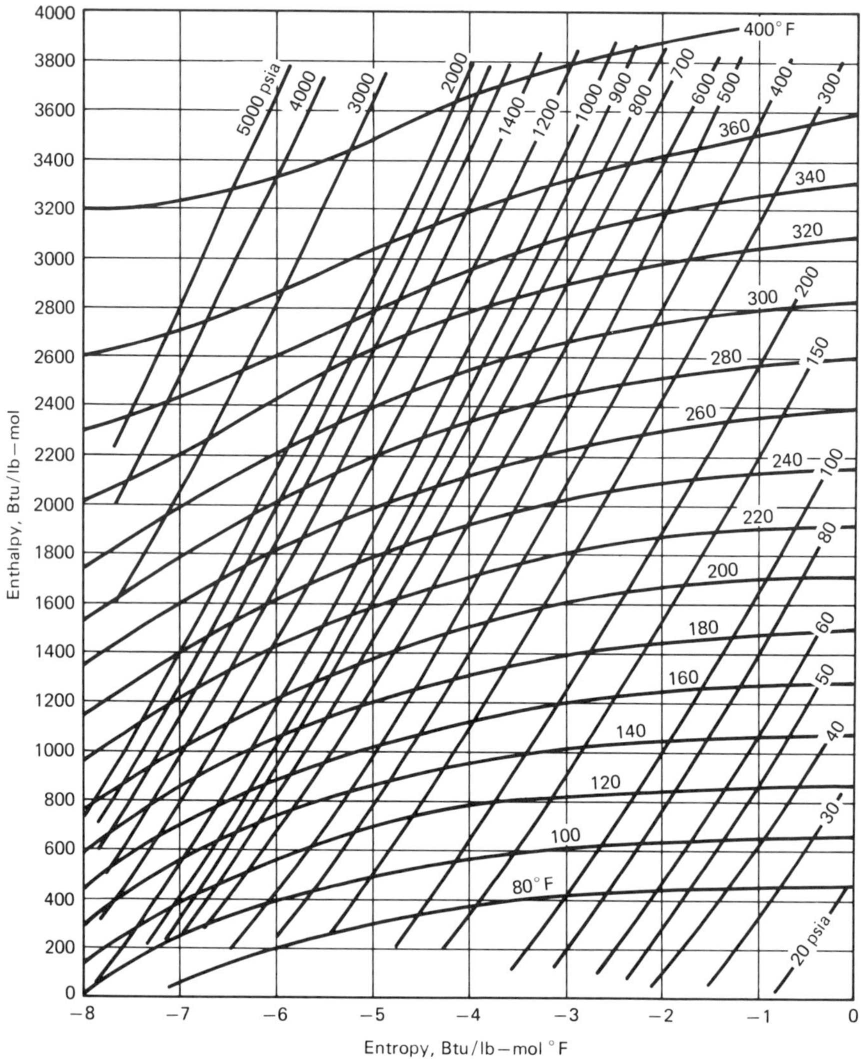

Fig. 5.7 Enthalpy-entropy diagram for a 0.65 to 0.75 specific gravity natural gas. (After Campbell.)

Practically, certain adjustments are made in the theoretical formula, so that a typical equation that might be used in computing compressor performance is

$$E_v = 0.97 - \left[\left(\frac{z_s}{z_d}\right) r^{1/k} - 1\right] Cl - L \tag{5.5}$$

where

z_s = gas deviation factor at suction of the cylinder
z_d = gas deviation factor at discharge of the cylinder

In this practical formula, the values of r, k, and Cl are the same as in the theoretical formula. The constant 1.0 has been reduced to 0.97 to correct for minor inefficiencies such as incomplete filling of the cylinder during the intake stroke. L is a practical correction for the conditions in a particular application that affect the volumetric efficiency and for which the theoretical formula is inadequate.

5.3.3 Ideal Isentropic Horsepower

The basis of the computation is the assumption that the process is ideal isentropic or perfectly reversible adiabatic. The total ideal horsepower for a given compression is the sum of the ideal work computed for each stage of compression. The ideal isentropic work can be determined for each stage of compression in a number of ways.

Mollier Diagram

One simple and rapid way to solve a compression problem is by using the Mollier diagram, if one is available for the gas being compressed. This is done by tracing the increase in enthalpy from the cylinder suction pressure and temperature to its discharge pressure along the path of constant entropy. This involves some care in handling and converting the various units such as cubic feet per minute, pounds of vapor, British thermal units, and horsepower, but it is a simple and straightforward method. All compressor problems for which suitable Mollier diagrams exist should be solved in this manner. The Mollier chart for a typical 0.65 to 0.75 specific gravity natural gas is shown in Fig. 5.7. Other Mollier charts are presented in Figs. 3.9 to 3.14.

For practical purposes, the amount of heat transferred from the gas to the compressor cylinder and piston during a cycle is small compared to the work involved in the compression. Thus, one assumption in compression calculations is

1. $Q = 0$

Other common assumptions made in computing a theoretical work of compression are:

2. Lost work due to friction can be neglected.

3. Kinetic energy effects can be neglected.

The resulting energy balance is

$$w = \Delta H = -n(h_2 - h_1) \tag{5.6}$$

where

n = number of moles being compressed
h_1, h_2 = molar specific enthalpy at intake conditions and discharge conditions, Btu/lb-mol

h_1 and h_2 can be obtained from Brown's enthalpy-entropy charts for gases (Figs. 3.9 to 3.14) for the particular gas gravity or from Fig. 5.7 for average natural gas, or they may be calculated using specific heat and gas deviation factor charts. h_2 is obtained by noting the assumption $Q = 0$, since

$$Q = \int_1^2 T\,dS = T\,\Delta S \tag{5.7}$$

The process from state 1 to state 2 (Fig. 5.8) must be at constant entropy conditions (called isentropic and adiabatic). Start at suction conditions, move up along a constant entropy line to the discharge pressure condition, and read off the resulting temperature and enthalpy.

Example 5.1. What is the theoretical horsepower consumed in compressing 1 MMcfd, measured at 14.73 psia and 60°F, from 65 psia and 80°F to 215 psia? What is the discharge temperature of the gas? Assume a gas gravity of 0.6.

Solution

$$\text{lb-mol/day} = 1(10^6)(14.73)/(10.73 \times 520 \times 1.0) = 2.640(10^3)$$

From Brown's *h-s* chart for 0.6 gravity gas (Fig. 3.9),

h_1 at 65 psia and 80°F = 400 Btu/lb-mol
s_1 at 65 psia and 80°F = −2.3 Btu/lb-mol °F
T_2 at 215 psia and s = −2.3 = 230°F
h_2 at 215 psia and 230°F = 1800 Btu/lb-mol

$$\text{Compression work } w = n(h_2 - h_1) = 2.640(10^3)(1800 - 400)$$
$$= 3.696(10^6)\text{Btu/day}$$

$$\text{Theoretical horsepower required} = \frac{3.696(10^6)(778.2)}{(24)(60)(33{,}000)}$$
$$= 60.5 \text{ hp}$$

Discharge Temperature T_2 = 230°F

In Example 5.1, the gas was compressed from 65 to 215 psia. The ratio of the absolute pressures is called the compression ratio, r. In the example,

$$r = \frac{215}{65} = 3.3$$

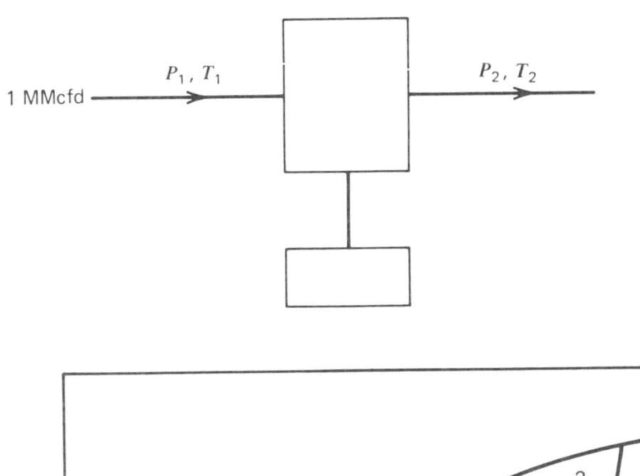

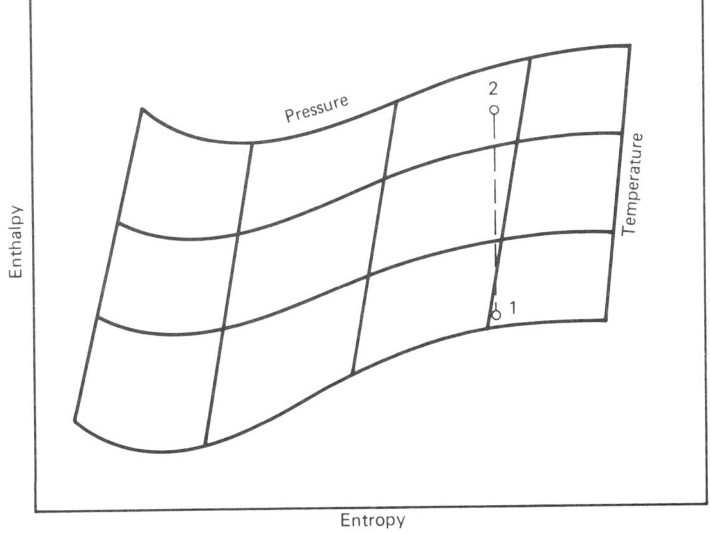

Fig. 5.8 Single-Stage compression.

The volumetric efficiency becomes less, and mechanical stress limitation becomes more pronounced as r increases. As a matter of practical design, r per stage is usually less than 6.0. In actual practice, r seldom exceeds 4.0 when boosting gas from low pressure for processing or sale. When the total compression ratio is greater than this, several stages of compression are used to reach high pressures.

The total power is a minimum when the ratio in each stage is the same. This may be expressed in equation form as

$$\text{Stage } r = \left(\frac{p_d}{p_s}\right)^{1/n} \tag{5.8}$$

where

p_d = final discharge pressure, absolute
p_s = suction pressure, absolute
n = number of stages required

For the two-stage compression (Fig. 5.9),

$$r = \left(\frac{p_3}{p_1}\right)^{1/2}$$

$$\frac{p_2}{p_1} = \frac{p_3}{p_2} = r$$

If $p_1 = 50$ psig and $p_3 = 1000$ psig,

$$r = \left(\frac{1014.73}{64.73}\right)^{1/2} = 3.96$$

and

$$p_2 = 242 \text{ psig}$$

Large compression ratios result in gas being heated to undesirably high temperatures ($T_3 = 450°\text{F}$ when $T_1 = 80°\text{F}$ in above example). Therefore, it is common practice to cool the gas between stages and, if possible, after the final stage of compression (Figs. 5.10 and 5.11). Calculations of the compression work is done as before.

The heat removed in the interstage coolers and aftercoolers is calculated by moving along a constant pressure line on the enthalpy-entropy chart. Figure 5.11 is a qualitative sketch of an enthalpy-entropy diagram, illustrating a two-stage compression with intercoolers and aftercoolers. Point 1 is the initial state of the gas as it enters the compressor. Path 1-2 shows the first stage of compression (constant entropy). The gas is then cooled in the intercoolers at constant pressure (path 2-3); the difference in enthalpy along this path is equal to the heat removed in the intercoolers. Path 3-4 shows the second stage of compression (constant entropy). Path 4-5 shows cooling at constant pressure in aftercoolers. The temperatures at points 2 and 4 are the temperatures of the gas at the end of the first and second stages of compression. The temperatures at points 3 and 5 are the tempuratures to which the gas is cooled in the intercoolers and aftercoolers.

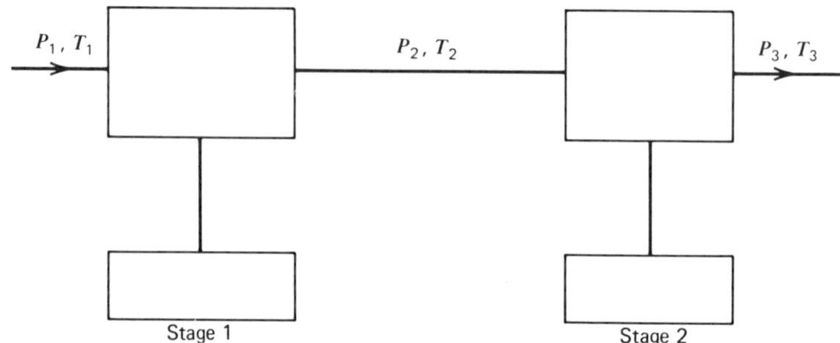

Fig. 5.9 Two-stage compression arrangement.

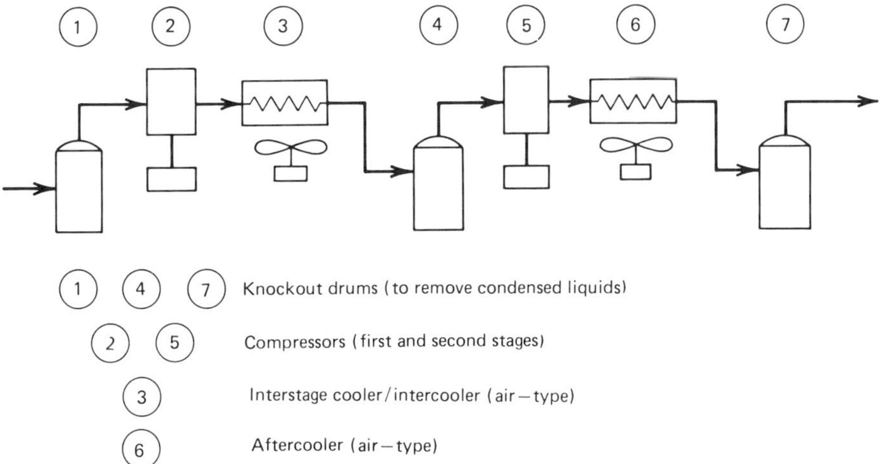

Fig. 5.10 Two-stage compression arrangement with intercoolers and aftercoolers.

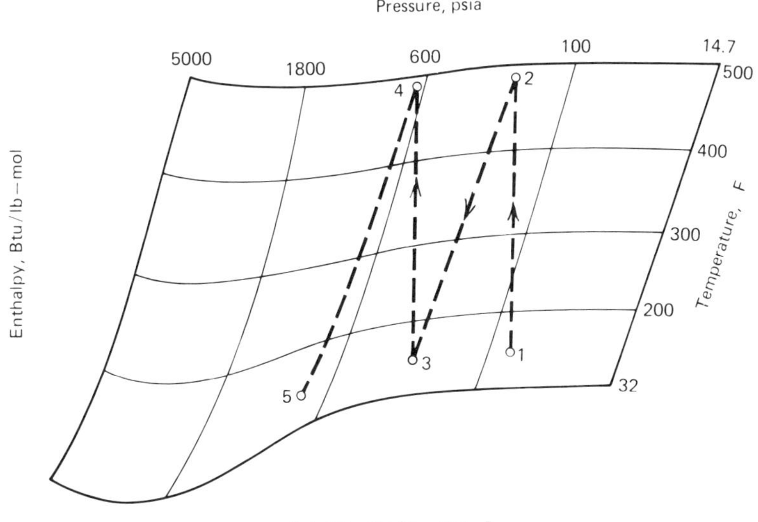

Fig. 5.11 Enthalpy-entropy diagram for a two-stage compression with intercoolers and aftercoolers.

In making rigorous calculations, account for the pressure drop across the inlet and exhaust valves of the compressor and the pressure drop across the coolers and interconnecting pipes.

Analytical Method

Another approach commonly used is to calculate the horsepower for each stage from the isentropic work formula. The basis of the derivation of all analytical expressions for calculating the theoretical work required to compress a quantity of gas is the general energy equation,

$$\int_1^2 V\,dp + \Delta \frac{u^2}{2g_c} + \frac{g}{g_c}\,\Delta Z + lw + w = 0 \tag{5.9}$$

The changes in kinetic energy, potential energy of position, and energy losses are taken as zero. The equation then reduces to

$$-w = \int_{p_1}^{p_2} V\,dp \tag{5.10}$$

Substituting Eqs. 5.1 and 5.2 into Eq. 5.10 and integrating, the theoretical adiabatic work of compression in foot-pound force per pound-mass of gas is obtained:

$$-w\left(\frac{\text{ft-lbf}}{\text{lbm}}\right) = \frac{k}{k-1}\,\frac{53.241T_1}{\gamma_g}\left[\left(\frac{p_2}{p_1}\right)^{(k-1)/k} - 1\right] \tag{5.11}$$

Where deviation from ideal-gas behavior is appreciable, Eq. 5.11 is empirically modified. One such modification is

$$-w\left(\frac{\text{ft-lbf}}{\text{lbm}}\right) = \frac{k}{k-1}\,\frac{53.241T_1}{\gamma_g}\left[\left(\frac{p_2}{p_1}\right)^{Z_1(k-1)/k} - 1\right] \tag{5.12}$$

or

$$-w\left(\frac{\text{Hp}}{\text{MMcfd}}\right) = \frac{k}{k-1}\,\frac{3.027p_b}{T_b}\,T_1\left[\left(\frac{p_2}{p_1}\right)^{Z_1(k-1)/k} - 1\right] \tag{5.13}$$

where

p_b, T_b = base conditions for specifying 1 MMcfd of gas
p_1, T_1 = suction pressure, temperature; psia, °R
p_2 = discharge pressure, psia
z_1 = compressibility factor at suction conditions
k = C_p/C_v at suction conditions

Equation 5.13 gives results that are in reasonably good agreement with those calculated by means of an enthalpy-entropy diagram. For ideal gases,

$$C_v = C_p - R = C_p - 1.986 \tag{5.14}$$

$$k = \frac{C_p}{C_v} = \frac{C_p}{C_p - 1.986} \tag{5.15}$$

$$\frac{k}{k - 1} = \frac{C_p}{1.986} \tag{5.16}$$

The specific heat ratio may also be found, using Kay's rule-type calculation as

$$k = \frac{\Sigma(y_i)(M_i C_{pi})}{\Sigma(y_i)(M_i C_{pi}) - 1.986} \tag{5.17}$$

where

y_i = mole fraction of each component in gas
M_i = component molecular weight
C_{pi} = component specific heat at constant pressure evaluated at average gas temperature in the compressor stage, Btu/lb-°F

For real natural gases in the gravity range $0.55 < \gamma_g < 1$, use the following relationship at approximately 150°F:

$$k^{150°F} \simeq \frac{2.738 - \log \gamma_g}{2.328} \tag{5.18}$$

Use of Eqs. 5.13 and 5.18 with the compression conditions given in the single-stage compression example yields a power requirement of 62.2 hp compared to 60.5 hp obtained using Brown's enthalpy-entropy chart.

From the isentropic (adiabatic) pressure-temperature change for an ideal gas,

$$\frac{T_2}{T_1} = \left(\frac{p_2}{p_1}\right)^{(k-1)/k} \tag{5.19}$$

This may be modified to give the discharge temperature for real gases as

$$\frac{T_2}{T_1} = \left(\frac{p_2}{p_1}\right)^{z_1(k-1)/k} \tag{5.20}$$

Calculation of the heat removed by intercoolers and aftercoolers is accomplished by use of constant pressure specific heat data

$$\Delta H = n\bar{C}_p \, \Delta T \tag{5.21}$$

where

$\bar{C}_p$ = constant pressure molal specific heat at cooler operating pressure and average cooler temperature, Btu/lb-mol-°F

Example 5.2. Calculate the adiabatic horsepower required to compress 1 MMcfd of a 0.6-gravity natural gas from 100 psia and 80°F to 1600 psia. Intercoolers cool the gas to 80°F. What is the heat load on the intercoolers and what is the final gas temperature?

Use:

(a) The enthalpy-entropy diagram.

(b) Analytical expressions.

Solution

$$\text{Overall compression ratio: } r = \frac{1600}{100} = 16$$

This is greater than 6; thus, more than one-stage compression is required. Using two stages of compression,

$$r = \left(\frac{1600}{100}\right)^{1/2} = 4$$

(a) Using the Enthalpy-Entropy Diagram (Fig. 5.12)

First stage: 100 to 400 psia (constant entropy)

$$\Delta h_{1\text{-}2} = h_2 - h_1 = 1990 - 380 = 1610 \text{ Btu/lb-mol}$$

Cooling in intercoolers at constant pressure (400 psia),

$$\Delta h_{2\text{-}3} = h_3 - h_2 = 220 - 1990 = -1770 \text{ Btu/lb-mol}$$

Second stage: 400 to 1600 psia (constant entropy)

$$\Delta h_{3\text{-}4} = h_4 - h_3 = 1920 - 220 = 1700 \text{ Btu/lb-mol}$$

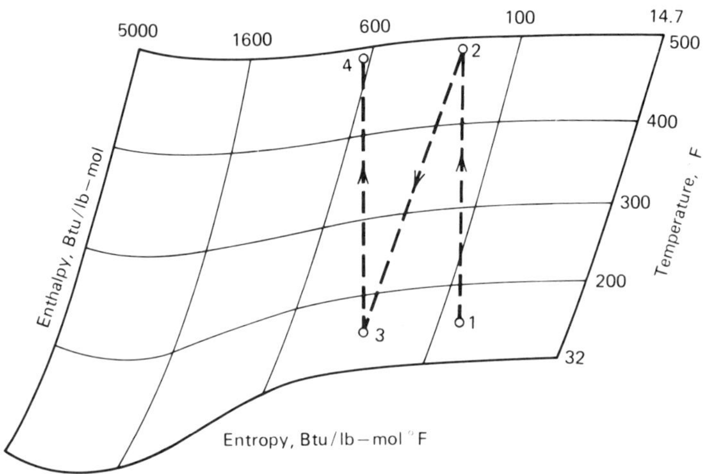

Fig. 5.12 Adiabatic work of compressions from the enthalpy-entropy diagram.

$$\text{Number of moles per day} = \frac{1.0 \times 10^6}{379} = 2.639 \times 10^3$$

$$\text{Adiabatic horsepower} = \Delta h_{1\text{-}2} + \Delta h_{3\text{-}4} = \frac{2.639 \times 10^3 \times 778.2}{24 \times 60 \times 33{,}000}(1610 + 1700)$$
$$= 143 \text{ hp/MMcfd}$$

$$\text{Heat load on intercoolers} = \Delta h_{2\text{-}3} = 2.639 \times 10^3(-1770)$$
$$= -4.67 \times 10^6 Btu/MMcfd$$

The final gas temperature from chart at point 4 is

$$T_4 = 278°F$$

(b) Analytical Method

$$-w\left(\frac{\text{hp}}{\text{MMcfd}}\right) = \frac{k}{k-1} 3.027 \frac{p_b}{T_b} T_1 \left[\left(\frac{p_2}{p_1}\right)^{Z_1(k-1)/k} - 1\right]$$

where

$k = 1.28$, $T_1 = 540°\text{R}$, $p_2/p_1 = 4$, $T_c = 358°\text{R}$
$p_c = 671$ psia, $T_r = 1.51$, $p_{r,1} = 0.149$, $p_{r,2} = 0.595$
$z_1 = 0.985$ at 80°F and 100 psia
$z_2 = 0.940$ at 80°F and 400 psia

First stage

$$-w = \frac{1.28}{0.28}\left(3.027 \times \frac{14.7}{520}\right) 540 \left[(4)^{0.985(0.28/1.28)} - 1\right] = 73.3 \text{ hp}$$

Second stage

$$-w = \frac{1.28}{0.28}\left(3.027 \times \frac{14.7}{520}\right) 540 \left[(4)^{0.940(0.28/1.28)} - 1\right] = 69.5 \text{ hp}$$

$$\text{Total compression work} = 73.3 + 69.5$$
$$= 142.8 \text{ hp/MMcfd}$$

$$\frac{T_4}{T_3} = \left(\frac{p_4}{p_3}\right)^{Z_3(k-1/k)} = (4)^{0.940(0.28/1.28)} = 1.33$$

$$\text{Final gas temperature } T_4 = (1.33)(540)$$
$$= 718°\text{R} = 258°\text{F}$$

$$\frac{T_2}{T_1} = (4)^{0.985(0.28/1.28)} = 1.348$$
$$T_2 = 728°\text{R} = 268°\text{F}$$

$$\text{Average cooler temperature} = \frac{268 + 80}{2} = 174°\text{F}$$

$$\bar{C}_p \text{ at } 174°\text{F and } 400 \text{ psia} = 9.7 + 0.45 = 10.15 \frac{\text{Btu}}{\text{lb-mol °F}}$$

$$\text{Cooler load} = -2.639 \times 10^3 \,(10.15)(268 - 80)$$
$$= -5.036 \times 10^6 \text{ Btu/MMcfd}$$

The results obtained using the analytical expressions compare very well to those obtained from the Mollier diagram.

5.3.4 Actual Horsepower

The theoretical adiabatic horsepower obtained by the above calculations can be converted to brake horsepower (bhp) required at the engine end of the compressor by the use of an efficiency factor, E. The brake horsepower is the horsepower input into the compressor. The efficiency factors include the compression efficiency (compressor-valve losses) and the mechanical efficiency of the compressor. The overall efficiency of a compressor depends on a number of factors, including design details of the compressor, suction pressure, speed of the compressor, compression ratio, loading, and general mechanical condition of the unit.

During the time that the intake valves are open, points 2-3 on the indicator card (Fig. 5.13), and during the time the discharge valves are open, points 4-1, a certain pressure drop occurs through these valves. Furthermore, there are other

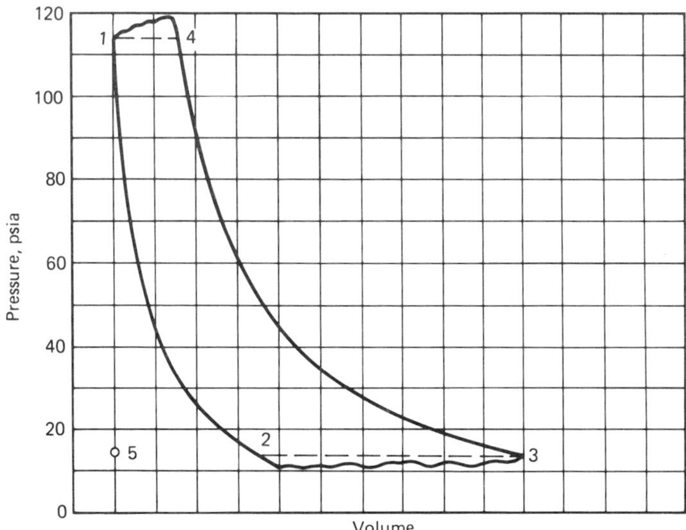

Fig. 5.13 Theoretical indicator card for a positive displacement compressor. (After Rollins.)

thermodynamic deviations from the ideal, perfectly reversible adiabatic process, for example, gas turbulence and heating of the incoming gas. These losses are shown on the indicator card (Fig. 5.13), which might be obtained from a machine in the field. All of these gas losses occurring within the cylinder can be summed up under the general term of compression efficiency. The compression efficiency is the ratio of the isentropic horsepower to the actual horsepower required within the cylinder, the actual horsepower being determined from the actual indicator card. In most modern compressors, the compression efficiency ranges from 83 to 93%.

In addition to thermodynamic losses, there are also the usual mechanical frictional losses in the piston rings, packings, and compressor bearings. These are accounted for by the mechanical efficiency of the compressor. The mechanical efficiency of most modern compressors ranges from 88 to 95%. Thus, most modern compressors have an overall efficiency ranging from 75 to 85%, based on the ideal isentropic compression process as a standard.

The overall efficiencies vary with compression ratio, as illustrated in Fig. 5.14. Actual curves should be obtained from the manufacturer. Applying these factors to the theoretical horsepower gives

$$\text{bhp} = \frac{\text{ideal isentropic hp}}{E} \tag{5.22}$$

and

$$\text{bhp} = \frac{\Delta H(\text{Btuh})}{60} \times \frac{778.2}{33{,}000} \times \frac{1}{E} \tag{5.23}$$

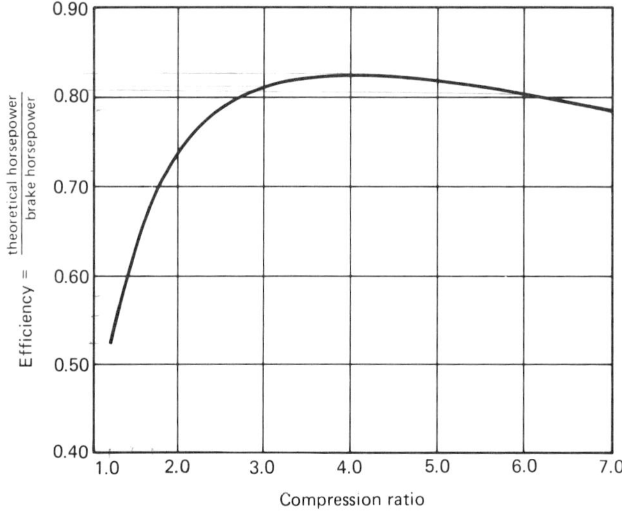

Fig. 5.14 Compressor efficiency. (After Katz et al.)

Using Fig. 5.14 and Eq. 5.22, the 143 hp found for theoretical adiabatic compression at an r of 4 converts to 143/0.827 = 172.9 bhp.

5.3.5 "Quickie" Charts

Figure 5.15 is a monograph that can be used to solve Eq. 5.19. The discharge temperature determined from either Eq. 5.19 or Fig. 5.15 is the theoretical value and is somewhat low since it neglects heat from friction and irreversibilities.

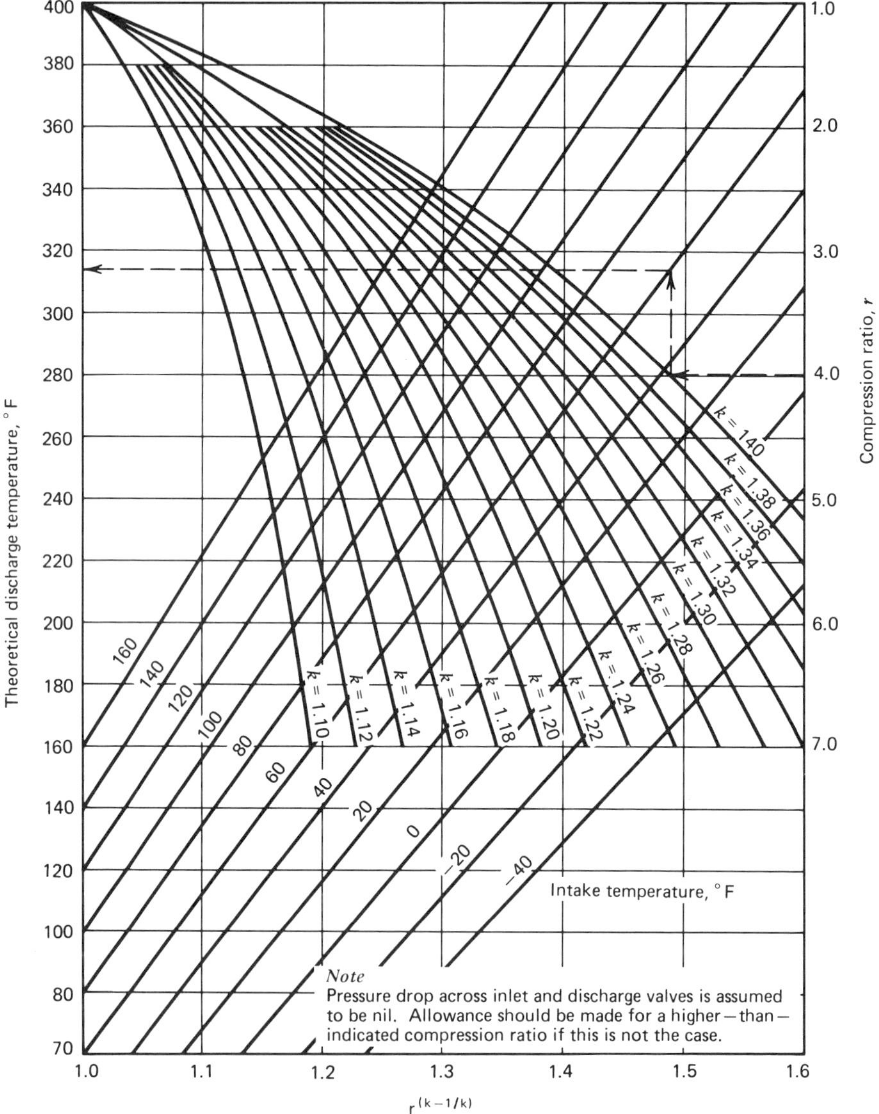

Fig. 5.15 Theoretical discharge temperatures, single-stage compression. Read r to k to T_s to T_d. (Courtesy NGPSA.)

Figures 5.16 to 5.19 are examples of simple correlations that can be used for estimating compressor power. These charts do not replace more accurate methods but are useful for those general planning calculations so often needed. As a general rule, these figures will give slightly higher results than more exact methods. These charts indicate the brake horsepower required per million cubic feet of gas per day with reference to a suction pressure of 14.4 psia plotted against a compression ratio for a number of ratios of specific heats, k. The required brake horsepower is based on a mechanical efficiency of 95% and a compression efficiency of 83.5%. The latter is an assumed value and varies with the compression valve design. Quantity of gas is measured at 14.4 psia and actual intake temperature.

The brake horsepower required for compression may be expressed as

$$\text{bhp} = \frac{Vp_bT_1}{14.4T_b} \times (\text{bhp/MMcfd}) \tag{5.24}$$

or

$$\text{bhp} = \frac{qp_bT_1}{T_b(10)^4} \times (\text{bhp/MMcfd}) \tag{5.25}$$

where

V = inlet capacity of compressor, MMcfd
q = inlet capacity of compressor, cfm
P_b = pressure base at which volume is measured, psia
T_b = temperature base at which volume is measured, °R
14.4 = pressure base of charts, psia
T_1 = inlet temperature of compressor, °F
(bhp/MMcfd) = factor determined from Figs. 5.16 to 5.19

Example 5.3. Find the brake horsepower required to compress 14,000 cfm of gas measured at 14.73 psia and 60°F from 25 psig to 75 psig with an inlet temperature of 90°F. The barometric pressure is 14.4 psia and the ratio of specific heats is 1.35.

Solution

The compression ratio $r = \dfrac{75 + 14.4}{25 + 14.4} = \dfrac{89.4}{39.4} = 2.27$

Therefore, from Fig. 5.16,

$$\text{bhp/MMcfd} = 49.5 \qquad (\text{at } k = 1.35 \text{ and } r = 2.27)$$

The horsepower requirement is

$$\text{bhp} = \frac{14{,}000 \times 14.73 \times 550}{520 \times 10^4} \times 49.5 = 1080 \text{ hp}$$

The horsepower requirement for multistage compression is decreased by intercooling between stages. If intercooling is not employed, the heat of compres-

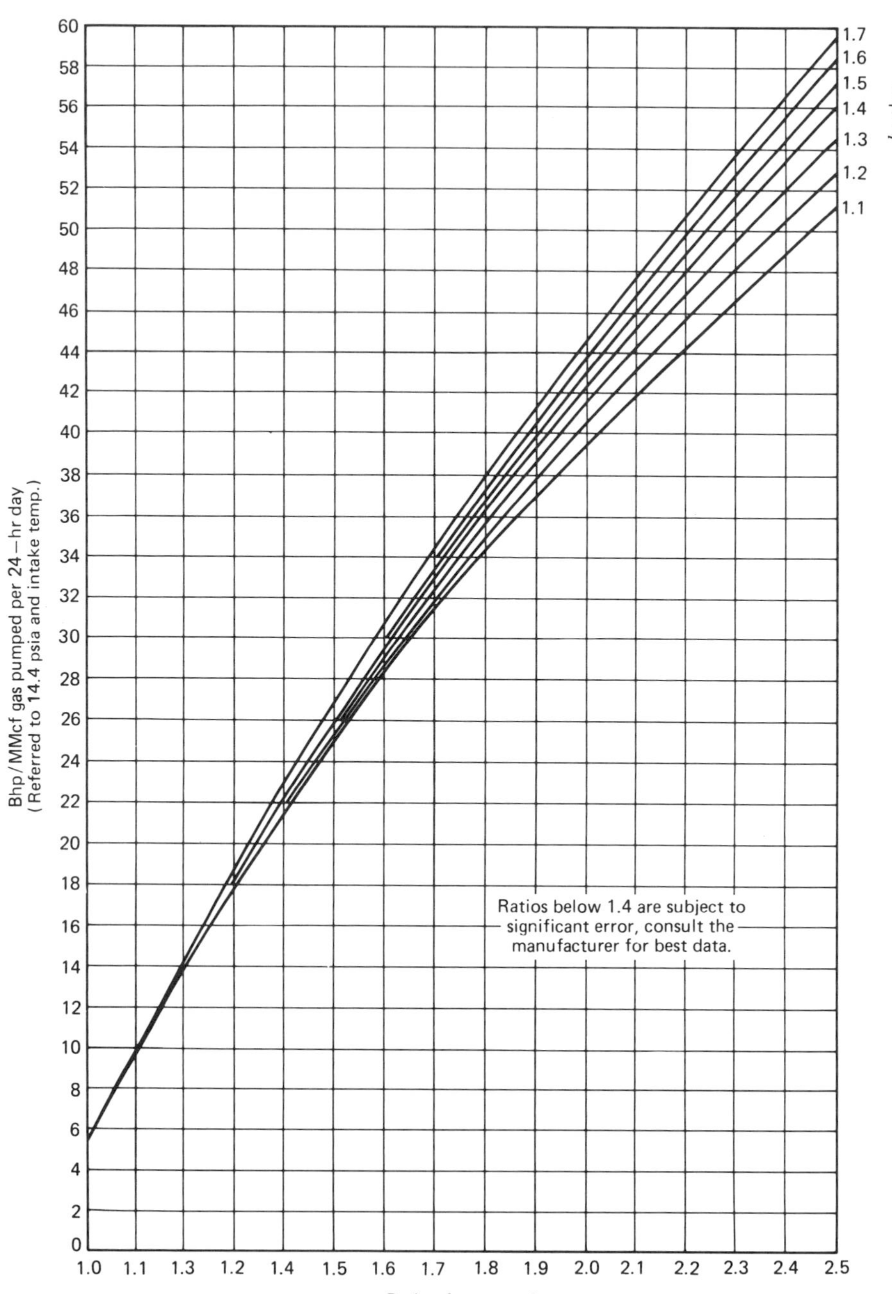

Fig. 5.16 BHP/MM curve. Mechanical efficiency = 95%. Gas velocity through valve = 3000 ft/min (API equation). (Courtesy NGPSA.)

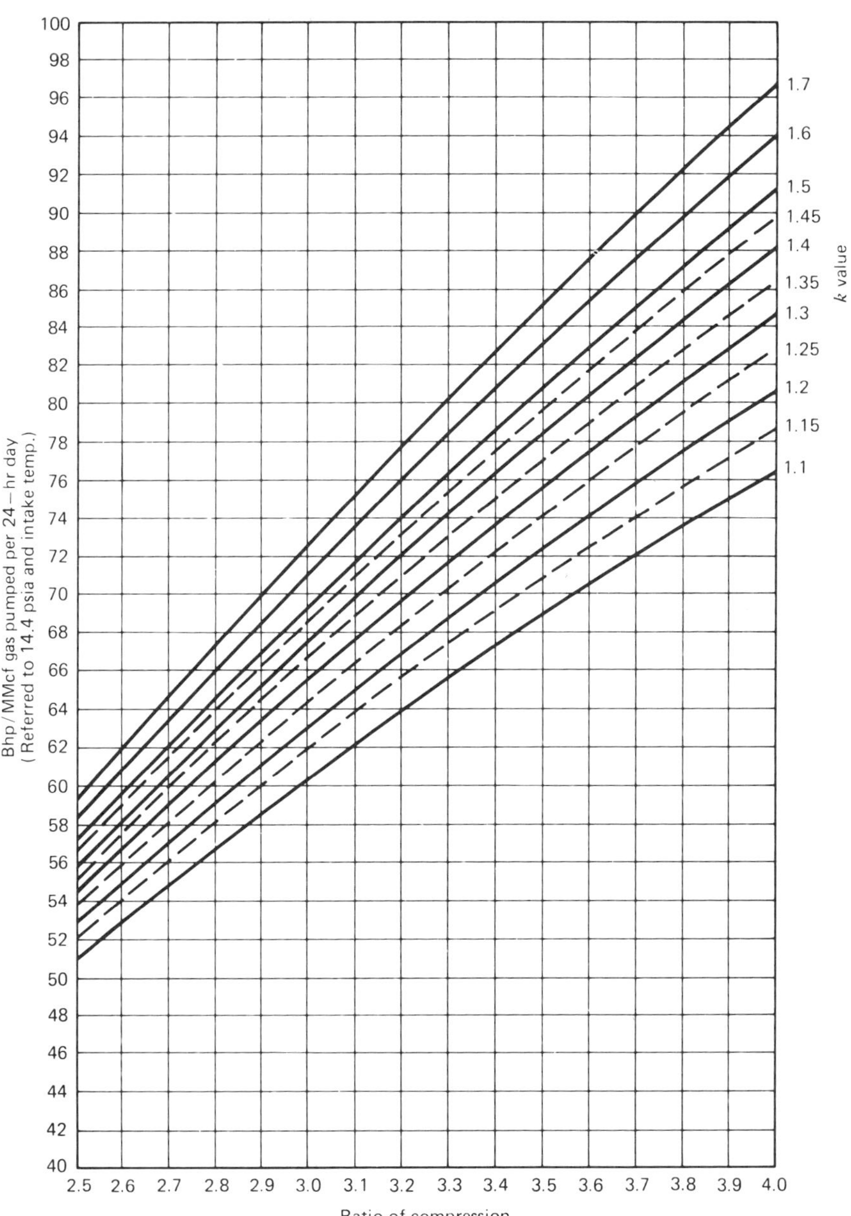

Fig. 5.17 BHP/MM curve. Mechanical efficiency = 95%. Gas velocity through valve = 3000 ft/min (API equation). (Courtesy NGPSA.)

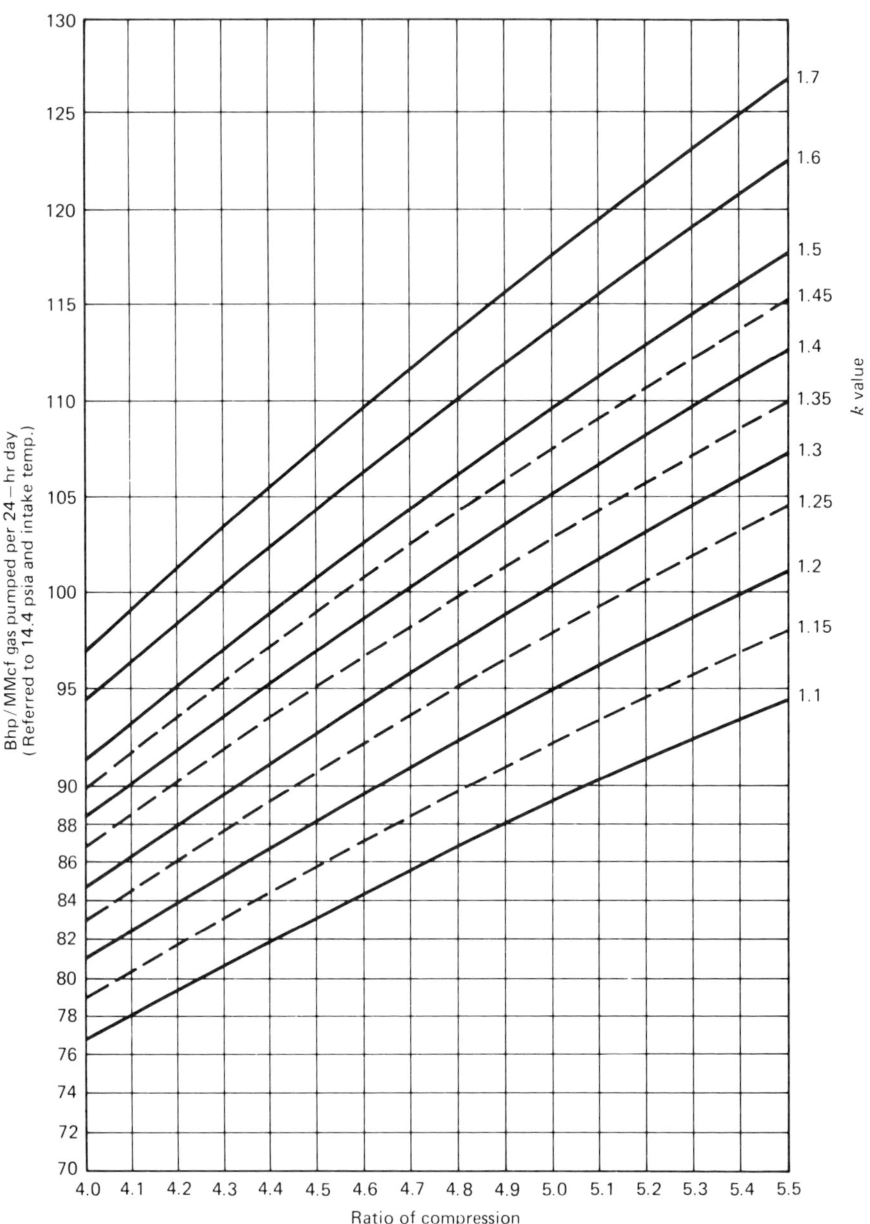

Fig. 5.18 BHP/MM curve. Mechanical efficiency = 95%. Gas velocity through valve = 3000 ft/min (API equation). (Courtesy NGPSA.)

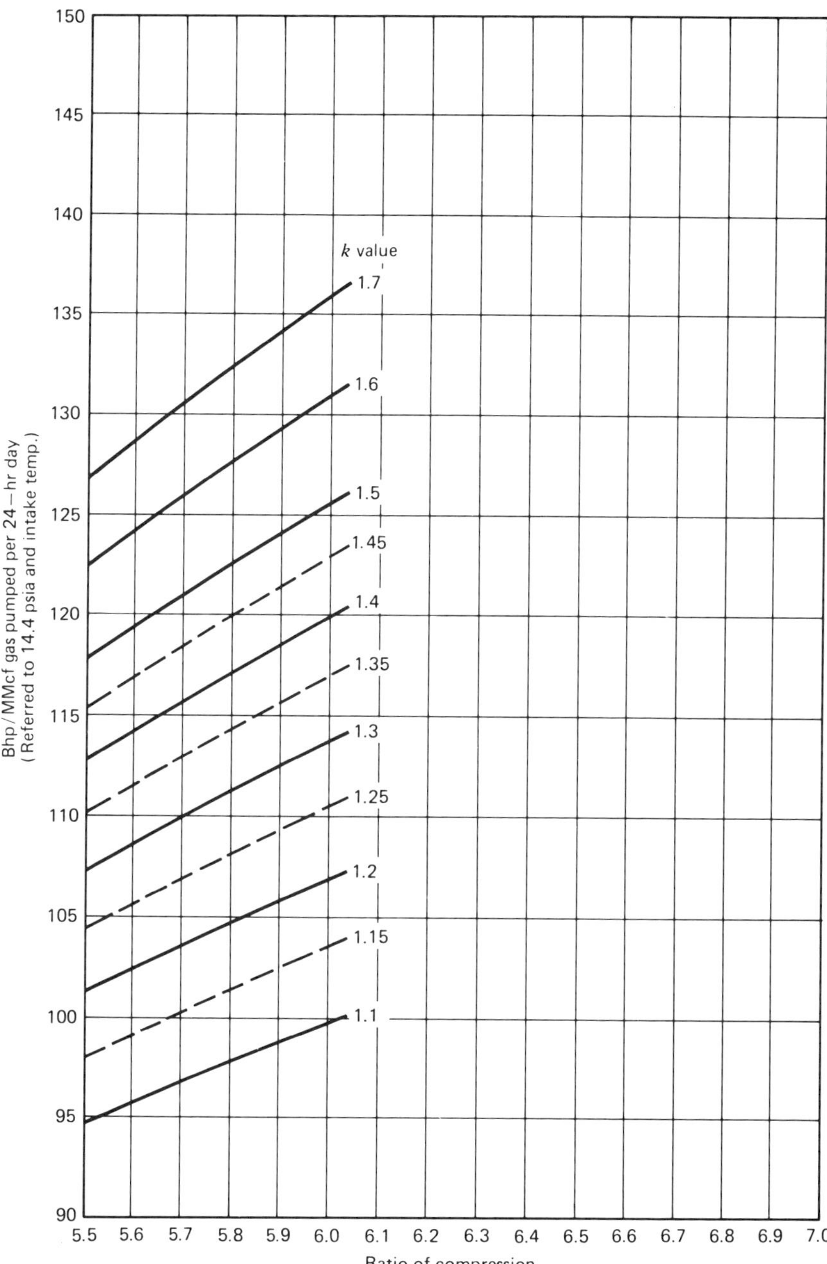

Fig. 5.19 BHP/MM curve. Mechanical efficiency = 95%. Gas velocity through valve = 3000 ft/min (API equation). (Courtesy NGPSA.)

sion in the first stage increases the volume to be compressed in the second stage with resultant increase in horsepower required.The horsepower required by the second stage is in direct proportion to the absolute inlet temperature of the first stage. Hence, the second-stage horsepower requirement will be equal to the first-stage requirement multiplied by the ratio of temperature increase across the first stage. This temperature ratio is given by Eq. 5.19.

The second-stage horsepower may be determined from the following formula, in which subscript 1 refers to the first stage and subscript 2 to the second stage:

$$(\text{bhp/MMcfd})_2 = (\text{bhp/MMcfd})_1(r)^{(k-1)/k} \tag{5.26}$$

If intercooling is used, the total horsepower required is roughly equal to the horsepower requirement of the first stage multiplied by the number of stages (assuming equal compression ratios and the same inlet temperature for all stages). The compression ratio per stage must be corrected for the pressure drop resulting from the intercooler. This pressure drop is generally taken as 4% of the first-stage discharge pressure. Horsepower requirements with and without intercooler may be calculated as shown by the following example.

Example 5.4. Find the horsepower required with and without intercooling when compression is 16,000 cfm of natural gas, $k = 1.28$, measured at 60°F and 14.73 psia, from atmospheric pressure of 14.4 psia to 125 psig. Inlet temperature is 70°F.

Solution

Without Intercooling

$$\text{Overall } r = \frac{p_2}{p_1} = \frac{125 + 14.4}{14.4} = 9.68$$
$$r = (9.68)^{0.5} = 3.11$$

From Fig. 5.17,

$$\text{bhp/MMcfd, First stage (at } k = 1.28,\ r = 3.11) = 67.5$$

From Eq. 5.26,

$$\text{bhp/MMcfd, Second stage} = 67.5(3.11)^{0.2185} = 86.5$$
$$\text{Total bhp/MMcfd} = 154 \text{ hp}$$

The corrected horsepower required (Eq. 5.25) is

$$\text{bhp} = \frac{16{,}000 \times 14.73 \times 530}{520 \times 10^4} \times 154 = 3699 \text{ hp}$$

With Intercooling

Allowing 4% of first-stage discharge pressure or approximately 1.0 psi per stage as a pressure drop between stages, the correct compression ratio r per stage is

$$r = 3.11 + \frac{1.0}{14.4} = 3.18$$

From Fig. 5.17,

$$\begin{aligned} \text{bhp/MMcfd per stage (at } k &= 1.28,\ r = 3.18) \\ &= 68.5 \\ \text{Total bhp/MMcfd} &= 68.5 \times 2 = 137 \text{ hp} \end{aligned}$$

The corrected horsepower requirement is

$$\text{bhp} = \frac{16{,}000 \times 14.73 \times 530}{520 \times 10^4} \times 137 = 3291 \text{ hp}$$

Intercooling shows a saving of approximately 408 hp.

Conversely, Figs. 5.16 to 5.19 may be used to determine the pumping capacity of any given size machine. However, to determine the capacity of a given engine operating as a multistage unit, allowance must be made for the pressure losses between stages. It is customary to assume full horsepower rating for a single-stage unit, 98% of the horsepower rating for a two-stage unit, and 96% of the horsepower rating for a three-stage unit.

The following formulas may be used to determine the pumping capacity of a given unit:

For single-state compression

$$V = \frac{\text{rated bhp} \times 14.4 \times T_b}{(\text{bhp/MMcfd})p_b T_1} \tag{5.27}$$

$$q = \frac{\text{rated bhp} \times T_b \times (10)^4}{(\text{bhp/MMcfd})T_1 p_b} \tag{5.28}$$

For two-stage compression

$$V = \frac{\text{rated bhp} \times 0.98 \times 14.4 \times T_b}{(\text{bhp/MMcfd})nT_1 p_b} \tag{5.29}$$

$$q = \frac{\text{rated bhp} \times 0.98 \times T_b \times (10)^4}{(\text{bhp/MMcfd})nT_1 p_b} \tag{5.30}$$

where

bhp/MMcfd is determined by using the compression ratio per stage

V = inlet capacity of compressor, MMcfd

q = inlet capacity of compressor, cfm

14.4 = pressure base on charts, psia

p_b = pressure base of V or q, psia

T_b = temperature base of V or q, °R

T_1 = inlet temperature, °R

Rated bhp = manufacturer's rating of engine

n = number of stages

Example 5.5. Find the capacity at standard conditions (60°F and 14.73 psia) of an 1100-hp, two-stage gas engine compressor unit compressing natural gas, k =

1.28, from atmospheric pressure of 14.4 psia at 90°F inlet condition to a discharge pressure of 125 psia.

Solution

$$\text{Overall } r = \frac{p_2}{p_1} = \frac{125 + 14.4}{14.4} = 9.68$$

$$r = (9.68)^{0.5} = 3.11 \text{ per stage}$$

From Fig. 5.17,

$$\text{bhp/MMcfd (at } r = 3.11,\ k = 1.28) = 67.5$$

For two-stage compression, Eq. 5.30,

$$q = \frac{1100 \times 0.98 \times 520 \times (10)^4}{67.5 \times 2 \times 550 \times 14.7} = 5{,}136 \text{ cfm}$$

or

$$V = 7.4 \text{ MMcfd}$$

5.4 CENTRIFUGAL COMPRESSORS

Table 5.1 contains formulas for calculating centrifugal compressor head, discharge temperature, and horsepower. These equations are derived from commonly accepted thermodynamic relations of gases. Isothermal compression calculations are generally used only when extensive cooling is accomplished during the compression cycle. Overall isothermal efficiency must be determined by the designer of the compressor. Discharge temperatures cannot be determined directly from the isothermal efficiency, since they depend on the type of cooling and the location of coolers. Isothermal compression calculations are of little use to the estimator of centrifugal machines.

Polytropic relations are commonly used as the basis for comparing centrifugal compressor performance. Adiabatic relations can also be used.

5.4.1 Compression Calculations

Data Required:

1. Physical properties of gas being compressed:

Molecular weight	M
Adiabatic (isentropic) exponent	k
Gas deviation factor	z

TABLE 5.1
Thermodynamic Equations for Compressor Calculations

	Adiabatic	Polytropic	Isothermal
Compression process	$pv^k = C$	$pv^n = C$	$pv = C$
Determination of exponent	$k = \frac{c_p}{c_v}$	$\frac{n-1}{n} = \frac{k-1}{k} \times \frac{1}{E_p}$	
Theoretical discharge temperature (°F abs)	$T_2 = T_1 r^{(k-1)/k}$	$T_2 = T_1 r^{(n-1)/n}$	$T_2 = T_1$
Discharge temperature (°F abs)	$T_2 = T_1 + \frac{T_1[r^{(k-1)/k} - 1]}{E_{ad}}$	$T_2 = T_1 r^{(n-1)/n}$	$T_2 = T_1$
Head (H) (ft-lb/lb)	$H_{ad} = z_{av} R T_1 \frac{[r^{(k-1)/k} - 1]}{\frac{k-1}{k}}$	$H_p = z_{av} R T_1 \frac{[r^{(n-1)/n} - 1]}{\frac{n-1}{n}}$	$H_i = z_{av} R T_1 \ln r$
Gas horsepower (ghp) (using capacity)	$\text{ghp} = \frac{q_1 P_1 \frac{z_1 + z_2}{2z_1} [r^{(k-1)/k} - 1]}{229 E_{ad}(k - 1/k)}$	$\text{ghp} = \frac{q_1 P_1 \frac{z_1 + z_2}{2z_1} [r^{(n-1)/n} - 1]}{229 E_p(n - 1/n)}$	$\text{ghp} = \frac{q_1 P_1 \frac{z_1 + z_2}{2z_1} \ln r}{229 E_i}$
Gas horsepower (ghp) (using weight)	$\text{ghp} = \frac{WH_{ad}}{33{,}000 E_{ad}}$	$\text{ghp} = \frac{WH_p}{33{,}000 E_p}$	$\text{ghp} = \frac{WH_i}{33{,}000 E_i}$
Brake horsepower (bhp)	bhp = ghp + mech. losses	bhp = gph + mech. losses	bhp = ghp + mech. losses

Source: After Rollins.

2. Inlet conditions at compressor flange:

Capacity	cfm, lb/min, MMscfd (at defined conditions), etc.
Inlet temperature	°F
Inlet pressure	p_1 (psia)

3. Discharge pressure at compressor flange p_2 (psia).
4. Water temperature if intercooling is used T_w, °F.
5. Process temperature limitations, if any.

Preliminary Calculations

1. Approximate polytropic efficiency is given by Fig. 5.20, using inlet capacity in cubic feet per minute.
2. Polytropic ratio $(n - 1)/n$ can be obtained from Fig. 5.21, using the specific heat ratio k and the polytropic efficiency E_p; or

$$\frac{n-1}{n} = \frac{k-1}{k} \times \frac{1}{E_p} \tag{5.31}$$

3. Compression ratio is obtained from inlet and discharge pressures:

$$r = p_2/p_1$$

Discharge Temperature

The discharge temperature must be considered in the selection of materials:

$$T_2 = T_1 r^{(n-1)/n} \tag{5.32}$$

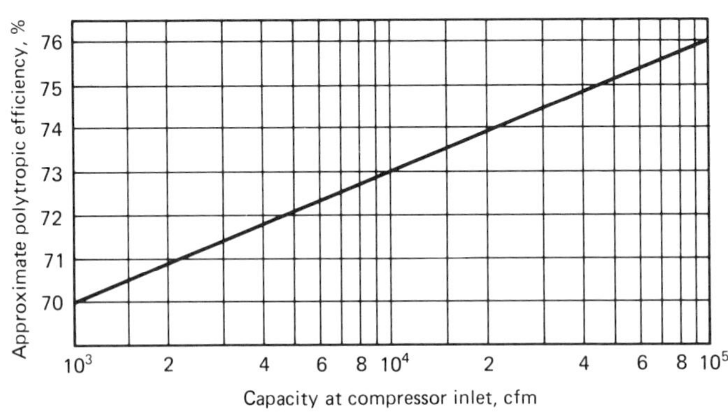

Fig. 5.20 Approximate polytropic efficiency versus compressor inlet capacity. (After Rollins.)

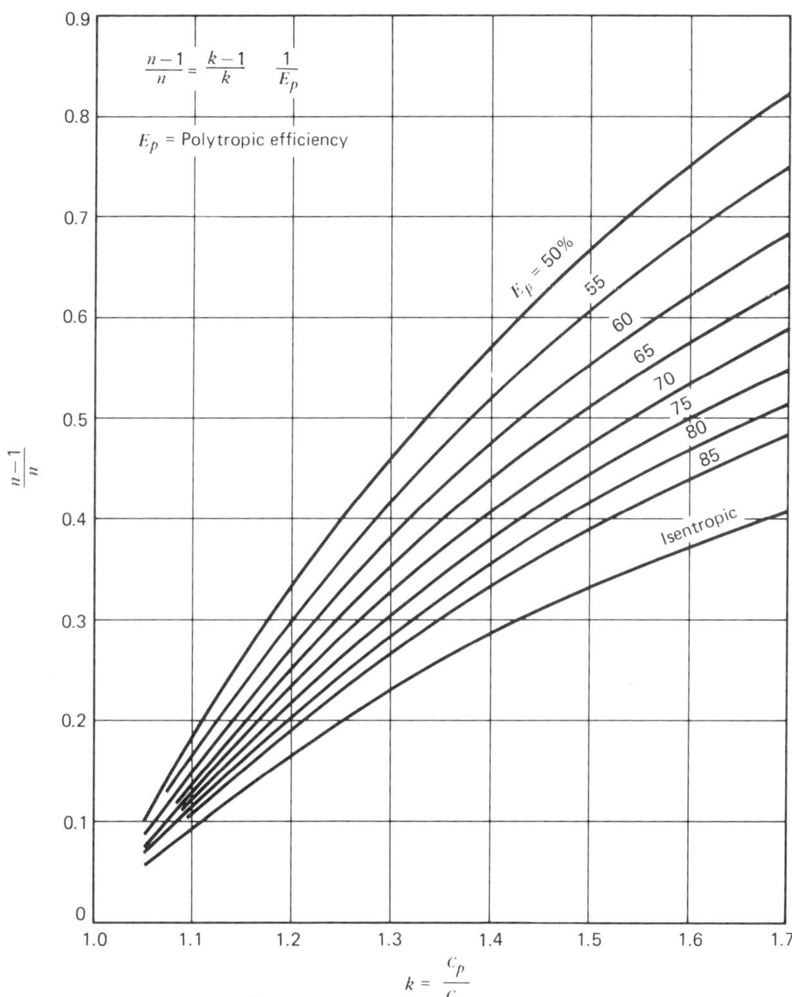

Fig. 5.21 Polytropic ratio $(n - 1)/n$ vs. adiabatic exponent k. (After Rollins.)

Polytropic Head

The polytropic head is an indication of the number of impellers required:

$$H_p \left(\frac{\text{ft-lbf}}{\text{lbm}}\right) = \left(\frac{z_1 + z_2}{2}\right) RT_1 \frac{r^{(n-1)/n} - 1}{(n-1)/n} \tag{5.33}$$

The values of $r^{(n-1)/n}$ may be obtained from Fig. 5.22.

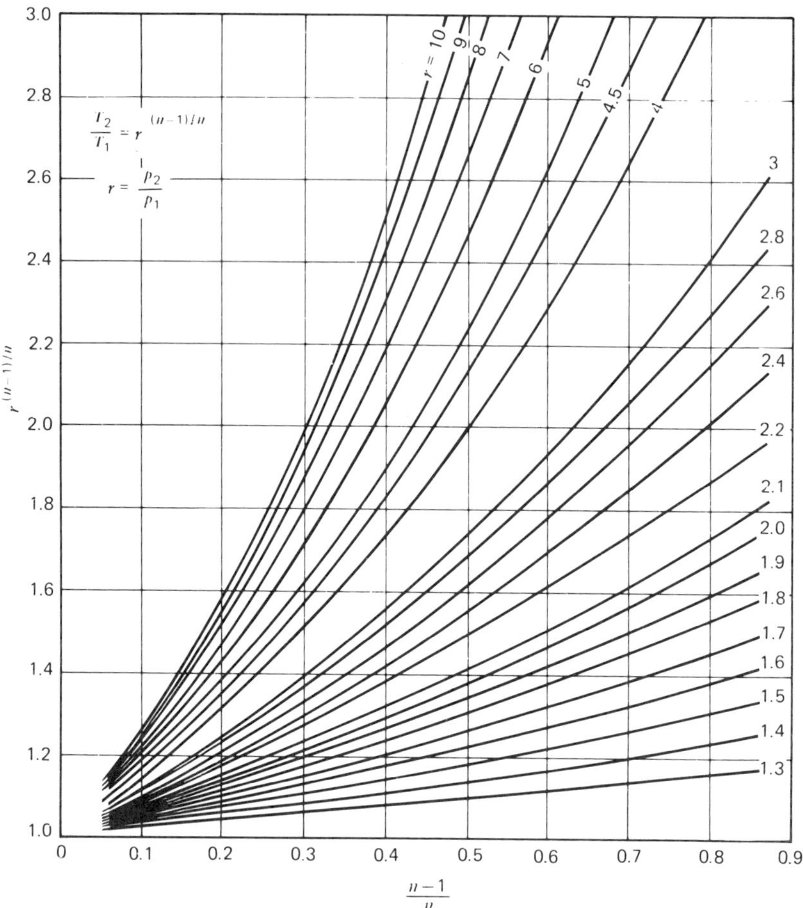

Fig. 5.22 Polytropic temperature ratio versus ratio $(n - 1)/n$. (After Rollins.)

Gas Horsepower (ghp)

1. From mass flow rate,

$$\text{ghp} = \frac{\dot{m} \times H_p}{33{,}000 \times E_p} \tag{5.34}$$

where

$\dot{m}$ = mass flow rate, lbm/min

2. Or, from inlet capacity and inlet pressure,

$$\text{ghp} = \frac{q_1 \times p_1 \times \dfrac{z_1 + z_2}{2z_1} \times \dfrac{r^{(n-1)/n} - 1}{\dfrac{n-1}{n}}}{229 \times E_p} \tag{5.35}$$

Brake Horsepower

The gas horsepower calculated is not the true input horsepower to the compressor. Mechanical and hydraulic losses occur because of bearing losses, seal losses, and other losses which may generally be ignored. In estimating bearing and seal losses, figures of 30 and 20 hp may be used, respectively, as approximations.

Example 5.6. Performing centrifugal compressor calculations using the following data:

Natural gas	$M = 19.7$
	$k = 1.24$
	$R = \dfrac{1544}{M} = 78.3$ psfa cu ft/lbm °R
Inlet pressure	250 psia
Inlet temperature	100°F
Inlet capacity, measured at 14.73 psia, 60°F	100,000 scfm
Discharge pressure	593 psia

Solution

$$r = \frac{p_2}{p_1} = \frac{593}{250} = 2.37$$

To find discharge temperature;

$$\begin{aligned} E_p &= 75\%, \text{ assumed} \\ \frac{n-1}{n} &= \frac{1.24 - 1}{1.24} \times \frac{1}{0.75} = 0.258 \\ T_2 &= 560(2.37)^{0.258} = 560 \times 1.25 \\ &= 700\ °\text{R} = 240°\text{F} \\ z_1 &= 0.97 \text{ (at 250 psia and 100°F)} \\ z_2 &= 0.97 \text{ (at 593 psia and 240°F)} \end{aligned}$$

Mass flow rate

$$\dot{m} = \frac{100{,}000 \times 14.73 \times 144}{520 \times 78.3} = 5210 \text{ lbm/min}$$

Inlet capacity

$$q = \frac{\dot{m}\ z_1 R T_1}{p_1 \times 144}$$

$$q_1 = \frac{5210 \times 0.97 \times 78.3 \times 560}{250 \times 144} = 6155 \text{ cfm}$$

Estimated polytropic efficiency (Fig. 5.20)

$$E_p = 72.4\%$$

$$\frac{n-1}{n} = \frac{1.24 - 1}{1.24} \times \frac{1}{0.724} = 0.268$$

Discharge temperature

$$T_2 = T_1[r^{(n-1)/n}] = 560(2.37^{0.268})$$
$$= 560(1.26) = 706°\text{R} = 246°\text{F}$$

Polytropic head

$$H_p = z_{av}RT_1 \frac{[r^{(n-1)/n} - 1]}{(n-1)/n}$$
$$= 0.97 \times 78.3 \times 560 \times \frac{0.26}{0.268}$$
$$= 41{,}263 \text{ ft-lbf/lbm}$$

Horsepower

$$\text{gph} = \frac{q_1 p_1 \dfrac{z_1 + z_2}{2z_1} \dfrac{[r^{(n-1)/n} - 1]}{\dfrac{n-1}{n}}}{229 \times E_p}$$

$$\text{gph} = \frac{6155 \times 250 \times 1.0 \times \dfrac{0.26}{0.268}}{229 \times 0.724}$$
$$= 9004 \text{ hp}$$
$$\text{bhp} = \text{ghp} + \text{bearing losses} + \text{seal losses}$$
$$= 9004 + 30 + 20$$
$$= 9054 \text{ hp}$$

5.5 ROTARY BLOWERS

Rotary, two-impeller, positive displacement blowers (Fig. 5.1) are constant-volume, variable-pressure machines available in capacities ranging from 5 to 30,000 cfm and pressures up to 12 psig in single stage. In some sizes, two-stage machines are available for pressures up to 20 psig.

The rotary positive blower employs two symmetrical, figure eight-shaped impellers rotating in fixed relationship with each other and in opposite directions within an elongated cylinder. As each lobe of an impeller passes the blower inlet, it traps a quantity of air equal to exactly one-fourth the displacement of the blower. The entrapment occurs four times during each revolution, moving from the air inlet to the outlet. Timing gears position the impellers accurately in relationship to each other, maintaining minute clearances, which allow the rotary positive blower to operate at high volumetric efficiency without internal seal or lubrication. Because of these minute clearances, a certain amount of air escapes past the operating clearances back to the suction side of the blower. This leakage, defined as "slip," is a constant for any given blower at a given pressure. It is expressed in revolutions per minute by dividing the leakage volume per minute by the displacement per revolution.

5.5.1 Design

Total operating speed of a blower, within size range, is determined by the following:

$$\text{Total rpm} = \frac{\text{desired capacity in cfm}}{\text{displacement in cfr}} + \text{slip rpm} \tag{5.36}$$

where

cfr = cubic feet per revolution

Approximate total horsepower consumed equals displacement (cfr) times total rpm times pressure differential (psi) times a constant of 0.005.

$$\text{hp} = 0.005(\text{cfr})(\text{rpm})(\Delta p) \tag{5.37}$$

Approximate temperature rise can be calculated from

$$T_2 = T_1(r)^{(k-1)/k} \tag{5.38}$$

where

r = compression ratio
k = specific heat ratio

Most requirements can be met with a single machine of the required capacity and are suitable to produce the required pressure. The positive displacement blower can be adapted to variable-capacity requirements if provided with a variable-speed transmission or driver. Capacity control can also be provided by installing multiple units of identical or different capacities.

REFERENCES

Campbell, J. M. *Gas Conditioning and Processing*. Norman, Okla.: Campbell Petroleum Series, 1976.

Katz, D. L., et al. *Handbook of Natural Gas Engineering*. New York: McGraw-Hill, 1959.

NGPSA Engineering Data Book. Natural Gas Processors Suppliers Assoc. Tulsa: 1977.

Petroleum Extension Service. *Field Handling of Natural Gas*. Austin: Univ. of Texas, 1972.

Rollins, J. P. *Compressed Air and Gas Handbook*. New York: Compressed Air and Gas Institute, 1973.

PROBLEMS

5.1 A gas reinjection project requires the compression of a 0.6-gravity natural gas from 100 psia and 80°F to 2700 psia. Intercoolers cool the gas to 100°F.

(a) Determine the *minimum* number of compression stages required for optimum compression efficiency.

(b) Carefully sketch an enthalpy-entropy diagram and draw the path taken by the process.

(c) What are the temperatures of the gas at the ends of the first and last stages of compression?

(d) Calculate the total adiabeatic horsepower required to compress 1 MMcfd of this natural gas.

(e) What is the heat load on the intercoolers?

5.2 25 MMcfd (measured at 14.7 psia and 60°F) of a natural gas, with a specific heat ratio of $k = 1.3$, are to be compressed from 1000 to 8000 psia. The inlet temperature is 80°F. Determine the horsepower requirement:

(a) Without intercooling.

(b) With intercooling.

Allow a pressure drop per stage of approximately 4% of first-stage discharge pressure. This is the usual assumption with intercoolers between stages.

(c) A 2100-hp two-stage compressor unit compresses natural gas ($k = 1.3$) from a pressure of 1000 psia at 78°F inlet condition to a discharge pressure of 6000 psia. What is the capacity of this compressor at standard conditions of 60°F and 14.7 psia.

5.3 (a) Figure 5.23 shows the basic flow diagram of a gas-condensate cycling plant. Figure 5.24 shows details of the compression-injection part of the plant. The gas injection rate is 30 MMcfd at 14.7 psia and 60°F. Injection gas gravity may be taken as $\gamma_g = 0.7$. Calculate the brake horsepower required to run the compressors and the heat load for the interstage cooler. Pressures at stations 1 and 4 are 900 and 7000 psia, respectively. Temperatures at stations 1 and 3 are 90 and 130°F, respectively. After all calculations summarize your answers according to the following computational form:

	Station			
	1	**2**	**3**	**4**
Pressure, psia	900			7000
Temperature, °F	90		130	
h, Btu/lb-mol				
s, Btu/1b-mole °R				
ΔH, Btuh				
Brake hp (stage)				

(b) Determine the brake horsepower in problem (a) using "quickie" charts.

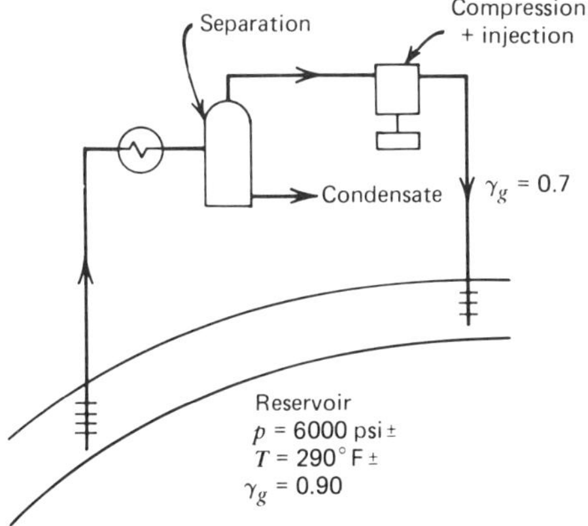

Fig. 5.23 Gas-condensate cycling plant.

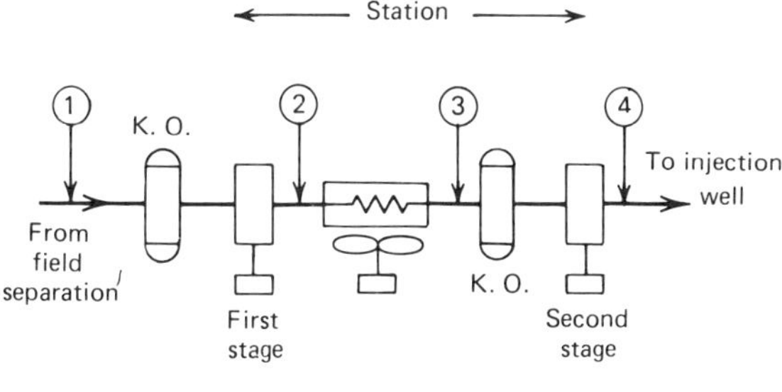

Fig. 5.24 Details of compression-injection part of plant.

5.4 *Reciprocating compressor design*

A remote dry gas well capable of 2.5 MMcfd with 500-psig tubing pressure must be compressed to 825 psig for sales. What is the amount of horsepower required to perform this job? First use a quick estimate, then a calculation.

Additional data: Specific gravity = 0.60, suction temperature = 85°F
Atmospheric pressure (assumed) = 14.7 psia

5.5 *Centrifugal Compressor Design*

Gas Mixture	*Mol %*
Propane	89
n-butane	6
Ethane	5

Calculate the centrifugal compressor required to handle the above gas at the following operating conditions:

Inlet temperature T_1 = 41°F
Inlet pressure P_1 = 20.3 psia
Discharge pressure p_2 = 101.5 psia
Flow rate = 10,540 mol/hr

6

NATURAL GAS MEASUREMENT

6.1 INTRODUCTION

Natural gas is in continuous flow from the time it leaves the reservoir until it reaches its ultimate use, usually in a burner. Unlike other products, it is not packaged or put in warehouses where inventory can be taken, except in underground storage and liquified natural gas storage facilities. Measurement of gas purchases and deliveries is made on a flowing stream of gas; hence, accurate measurement of the total quantity of gas that has passed through a given section of pipe over a period of time is of paramount importance to the gas industry. For example, an error of only 1.0% in the measurement of natural gas in a pipeline delivering 300 MMcfd of gas at $1.00 per thousand cubic feet (Mcf) will amount to a loss of over $1 million/year to either the seller or the purchaser.

At present, the most common method of measuring gas is by volume. As a matter of convenience, most operators account for gas in units of 1000 cu ft, commonly referred to as one Mcf. The total mass of substance in 1 cu ft of gas depends partly on its absolute pressure, which in commercial gas measurement is expressed in pounds per square inch, and its absolute temperature (°F plus 460, also referred to as Rankine temperature). To measure gas in meaningful terms by the volume method, first specify the absolute pressure and temperature of the base or standard cubic foot. In other words, the pressure and temperature of the reference or base cubic foot must be established.

The nationally recognized base temperature for large-volume fuel gas measurement is 520 R (equivalent to 60°F). With regard to a base pressure for large-volume measurement, different states and countries use different base conditions:

State or Location	Pressure Base, psia	Temperature Base, °F
California	14.7	60
Texas, Oklahoma, Arkansas, Kansas, Alberta	14.65	60
Louisiana, Department of the Interior	15.025	60
U.S. Bureau of Standards and Federal Price Commission	14.735	60

Since January 1, 1967, the American Petroleum Institute (API) and the American Gas Association (AGA) have been using 14.73 psia and 60°F as their standard conditions.

Among the presently used pressure bases, one of the most widely used is 14.73 psia. In large-volume fuel gas measurement practices, 14.73 psia is accepted as the equivalent of 30 in. of mercury pressure, which approximates the average atmospheric pressure at sea level. In changing a contract measurement from any other base pressure to "standard base pressure," the effect of the change may be readily offset by adjusting the price of the gas by multiplying it by a factor equal to the standard pressure base divided by the old pressure base. See Table 6.1 for base pressure conversion factors.

6.2 METHODS OF MEASUREMENT

Both gas and liquids may be measured using various measurement techniques, including orifice meters, positive displacement meters, turbine meters, venturi meters, flow nozzles, critical low provers, elbow meters, and variable area meters (rotameters). The selection of the measurement method to be used should be made only after careful analysis of several factors, including the following:

1. Accuracy desired.
2. Expected useful life of the measuring device.
3. Range of flow, temperature.
4. Maintenance requirements.
5. Power availability, if required.
6. Liquid or gas.
7. Cost of operation.
8. Initial cost.
9. Availability of parts.
10. Acceptability by others involved.
11. Purpose for which measurements are to be used.
12. Susceptibility to theft or vandalism.

TABLE 6.1
Base Pressure Conversion Factor for Gas Measurement

Present Pressure Base, psia	Factors to Convert to New Basic Pressures (psia) Indicated Below							
	13.9	14.65	14.7	14.73	14.73 Saturated[a]	14.9	15.025	16.4
13.9	1.0000	0.9488	0.9456	0.9137	0.9603	0.9329	0.9251	0.8476
14.65	1.0540	1.0000	0.9966	0.9946	1.0122	0.9832	0.9750	0.8933
14.7	1.0576	1.0034	1.0000	0.9980	1.0156	0.9866	0.9784	0.8963
14.73	1.0597	1.0055	1.0020	1.0000	1.0177	0.9886	0.9804	0.8982
14.73 saturated (14.4739)*	1.0113	0.9880	0.9846	0.9826	1.0000	0.9714	0.9633	0.8826
14.9	1.0719	1.0171	1.0136	1.0115	1.0294	1.0000	0.9917	0.9085
15.025	1.0809	1.0256	1.0221	1.0200	1.0381	1.0084	1.0000	0.9162
16.4	1.1799	1.1195	1.1156	1.1134	1.1331	1.1007	1.0915	1.0000

$$\text{Factor} = \frac{\text{Present pressure base}}{\text{New pressure base}}$$

Example:

(a) $15.025 \text{ to } 14.73 = \dfrac{15.025}{14.73} = 1.0200$

(b) $14.65 \text{ to } 14.73 = \dfrac{14.65}{14.73} = 0.9916$

(c) $14.73 \text{ sat. to } 14.73 \text{ dry} = \dfrac{14.73 - 0.2561}{14.73} = \dfrac{14.4739}{14.73} = 0.9826$

Note: To convert heating value or to adjust the price to reflect the effect of changing from one pressure base to another, the reciprocals of the above factors apply.

14.73 psia is assumed in commercial gas measurement to be the equivalent to a barometric pressure of 30″ Hg.

*.2561 psia = Water vapor pressure at 60°F (International Critical Tables).

6.2.1 Volumetric Measurement

The simplest method of measuring gas is by the volumetric method. This method determines the cubical content of a container (applying Boyle's and Charles' laws with applicable deviation factors in order to express the quantity of gas in the container) under the existing pressure and temperature in terms of equivalent volume at the desired base pressure and base temperature conditions. Using the simple gas law,

$$\frac{p_1V_1}{T_1} = \frac{p_2V_2}{T_2} = \text{a constant} \tag{6.1}$$

The volumetric formula is

$$q = \frac{pT_bV}{p_bTz} \quad \text{or} \quad \frac{pT_bV_s}{p_bT} = \left(\frac{p}{p_b}\right)\left(\frac{T_b}{T}\right)\frac{1}{z}V \tag{6.2}$$

Equation 6.2 may be written as

$$q = \left(\frac{p}{p_b}\right)\left(\frac{T_b}{T}\right)F_{pv}{}^2V \tag{6.3}$$

where

$$F_{pv} = \text{supercompressibility factor} = \sqrt{\frac{1}{z}}$$

Where calculations are to be made using a particular pressure and temperature base consistently, such as where $p_b = 14.73$ psia and $T_b = 520$ R, the formula may be simplified for more convenient use:

$$q = \frac{520}{14.73}\frac{p}{T}\frac{1}{z}V = 35.3021\left(\frac{p}{T}\right)F_{pv}{}^2V \tag{6.4}$$

A simplified equation for volume of gas in pipelines is derived by C. J. Kribs of Southern Gas Corp:

$$q = D^2(2)p(L) \tag{6.5}$$

where

q = volume of gas in pipeline cubic feet at 14.4 psia pressure base and 60 F
D = internal diameter of pipe, in.
p = mean absolute pressure, psia
L = length of pipeline section, mi

The applicable supercompressibility multiplier $F_{pv}{}^2$ may be included in the pressure or applied to the volume. The 14.4-psia pressure base volume determined by Eq. 6.5 may be converted to another pressure base by applying a factor = 14.4/other pressure base.

6.2.2 Displacement Metering

Displacement metering is essentially a more advanced form of volumetric measurement with an instrument having one or more mobile compartments or diaphragms arranged so that when gas passes through the meter, they are alternately filled or emptied. This motivates an index which registers the summation of the quantity of gas displaced in the compartment over a period of time.

The principle of the diaphragm-type displacement meter may be illustrated by a cylinder and reciprocating piston (Fig. 6.1). As the piston moves in the cylinder from position A to B, a quantity of gas is taken into the cylinder through the inlet port to occupy the space displaced by the piston. On the return stroke, the gas is discharged out of the cylinder through the outlet as the piston returns from B to A.

The volume of space the discharged gas occupied while in the cylinder is equal to the piston displacement. Where the volume of the piston displacement is known, it is a simple matter to connect a counter to the piston rod that will tally the piston displacement for each compression stroke. Since the volume of gas discharged is equal to the total piston displacement, the counter will indicate a measured volume of gas. The pressure and temperature of the gas in the cylinder will be that as supplied to the cylinder through the inlet port. If a thermometer and pressure gauge are added to the cylinder, these conditions may be observed.

From this information, Boyle's and Charles' law formulas can be applied to the volume of gas discharged as indicated on the counter in order to convert this volume to the equivalent quantity of gas at base temperature and pressure conditions. The formula is

$$q = r\left(\frac{pT_b}{p_bTz}\right) \tag{6.6}$$

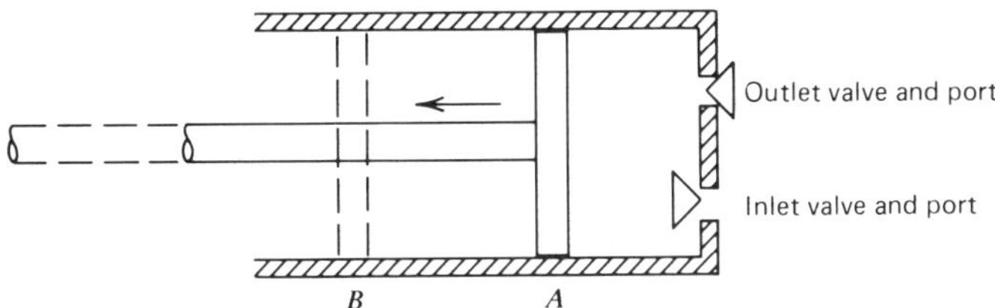

Fig. 6.1 Cylinder and reciprocating piston representing a simple displacement meter.

where

q = quantity of gas at base conditions, cu ft
r = counter registration, cu ft
p = pressure of gas, psia
p_b = pressure base, psia
T = temperature of gas, °R
T_b = temperature base, °R
z = gas deviation factor

The initial reading of the index is subtracted from the final reading to obtain the registration during any period. The displacement meter formula can be rewritten for this procedure:

$$q = (r_2 - r_1)\frac{pT_b}{p_bTz} \tag{6.7}$$

where

r_2 = final index reading, cu ft
r_1 = initial index reading, cu ft

or

$$q = (r_2 - r_1)\left(\frac{p}{p_b}\right)\left(\frac{T_b}{T}\right)F_{pv}{}^2 \tag{6.8}$$

The most common type of displacement meter has diaphragms separating the measuring compartments. These usually have four measuring compartments and two diaphragms. The movement of a diaphragm from one side to the other allows one compartment to fill while the second is discharging (Fig. 6.2).

The rotary displacement meter represents an entirely different mechanical principle than the diaphragm displacement meter. It employs two metal impellers of the same size having a cross-sectional shape resembling the figure 8. These impellers rotate on individual shafts and are designed and spaced to rotate tangentially to each other. They are enclosed in a cylindrical case. Gas flowing through the meter rotates the impellers and, since the close-off volume between an impeller and the case is fixed, a definite volume of gas will pass through the

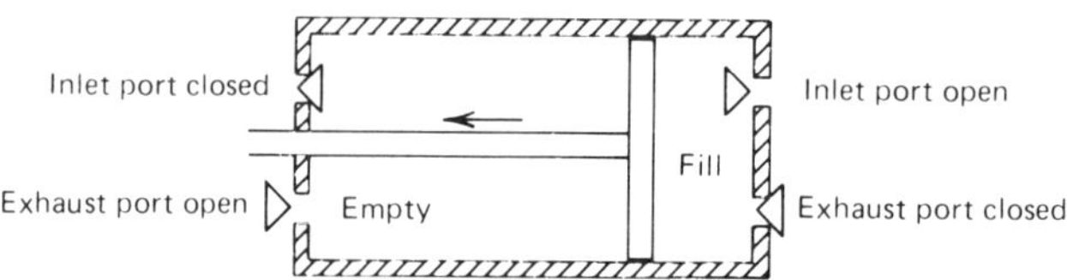

Fig. 6.2 Double-acting and reciprocating piston representation of a two-diaphragm displacement meter.

meter with each revolution of the impellers. By connecting an index to the shaft of an impeller, the volume of gas may be registered by this index.

The wet test meter is a small displacement meter sometimes used for laboratory measurment of gas. This meter measures very small quantities of gas with comparatively high accuracy. The measuring chamber is a drum with four helical partitions that do not extend to the drum center. The drum is mounted within a case filled with water to a height sufficiently above the drum shaft to close the inner edges of the helical partitions. As gas passes through the meter, successive small quantities are sealed off in the chambers formed above the water surface by the helical partitions within the drum. The index is accurately calibrated and may be read to 0.0001 cu ft. The metering pressure may be observed by a water manometer connected to the meter. The metering temperature is obtained from a thermometer inserted in a thermometer well in the meter.

6.2.3 Differential Pressure Methods

Differential pressure methods of gas metering involve the measurement of a pressure difference from which, together with certain other data, the rate of flow is computed on the basis of well-established physical principles. The differential pressure that is measured may be produced by a restriction placed in a pipe (orifice meter), or it may be the difference between kinetic (velocity) pressure and static pressure, as is the case of the pitot tube. There is a direct relationship between the rate of flow and the amount of this pressure drop, or differential. This principle has been widely used and has been developed into a precise and accurate means of measuring fluids (see Section 6.3).

6.2.4 Turbine Meter

The turbine meter uses the flowing gas as a force imparted to a bladed rotor. With appropriate gearing, revolutions of the rotor may be converted to volume. Accuracy curves are usually developed for each turbine meter, and proving or calibration techniques are available. Filters are almost a necessity ahead of turbine meters to permit sustained accuracy and trouble-free operation.

6.2.5 Elbow Meter

Centrifugal force in the curve of a pipe elbow can be used to measure flow. For accuracy, the elbow should be calibrated using some other acceptable measurement as a standard. Accuracy is not usually the objective when elbow meters are used. Relatively little pressure loss or differential pressure is created. Because of this, the meters are used primarily for control or other operations.

6.3 ORIFICE METER

By far the most common type of differential meter used in the fuel gas industry is the orifice meter. This meter consists of a thin flat plate with an accurately machined circular hole that is centered in a pair of flanges or other plate-holding device in a straight section of smooth pipe. Pressure tap connections are provided on the upstream and downstream sides of the plate so that the pressure drop or differential pressure may be measured. Figure 6.3 illustrates the flow pattern through an orifice, how the resulting pressure differential across the orifice is measured, and the change in static pressure that occurs. The advantages of the orifice meter are accuracy, ruggedness, simplicity, ease of installation and maintenance, range of capacity, low cost, acceptance for gas measurement by joint AGA-ASME committee, and availability of standard tables of meter factors.

As its name implies, an orifice meter utilizes an orifice for its basic component in the measurement of natural gas. The typical orifice meter consists of (primarily) a thin stainless steel plate about 3/16 in. thick, with a hole in the center, that is placed in the flow line. Placing an orifice in a pipe in which there is a gas flow causes a pressure difference across the orifice. This pressure difference and the absolute pressure in the line at a specified "tap" location are recorded continuously and are later translated into rate of flow. Two arrangements are commonly used: flange tap and pipe tap (Fig. 6.4).

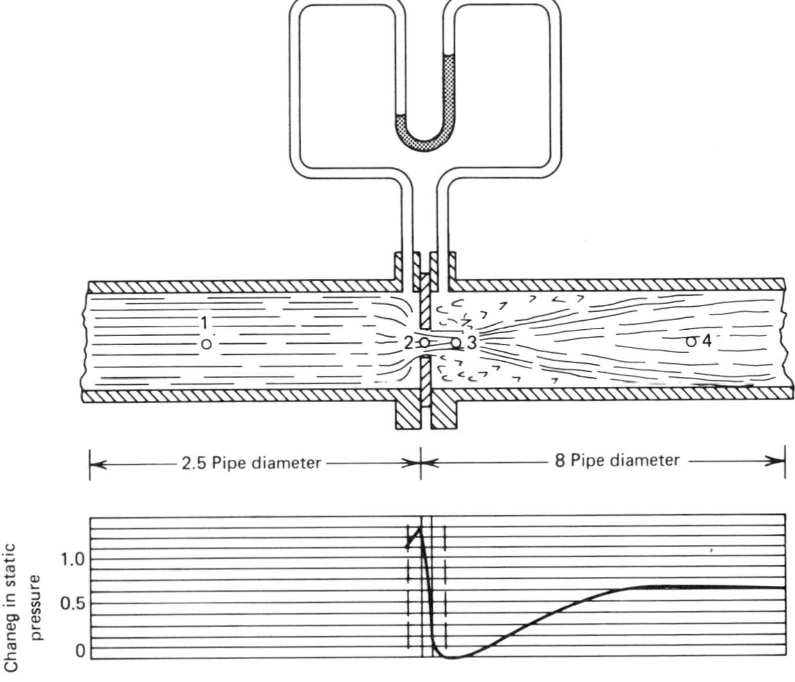

Fig. 6.3 Flow pattern through an orifice and the static pressure gradient.

It is customary to consider a complete orifice meter as composed of two major elements. The first of the two major elements is the differential pressure-producing device called the primary element. This primary element is composed of the following parts (Fig. 6.5):

1. The meter tube—a length of special pipe through which the gas flows.
2. The orifice plate holding and positioning device—an orifice flange or an orifice fitting installed as an integral part of the meter tube to hold the orifice plate in a position perpendicular and concentric to the flow of gas.
3. The orifice plate—a flat circular plate with a centrally bored, sharp-edged orifice machined to an exact, predetermined dimension that forms a calibrated restriction to the flow of gas through the meter tube and is the source of the differential.
4. Pressure taps—precisely located holes through the pipe walls or orifice plate holder from which the gas pressure on each side of the orifice plate may be measured.
5. Straightening vanes—a device that may be inserted in the upstream section of the meter tube to reduce swirling in the gas stream.

The secondary element is called the differential gauge and is the device for measuring the pressures. It is a gauge (or gauges) connected with tubing to the

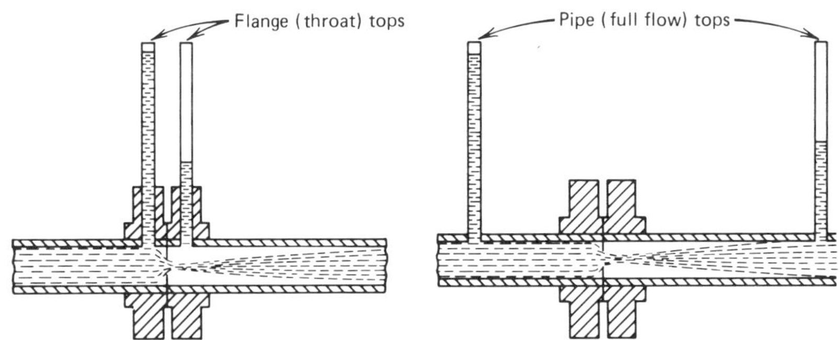

Fig. 6.4 Orifice meters: flange taps, pipe taps.

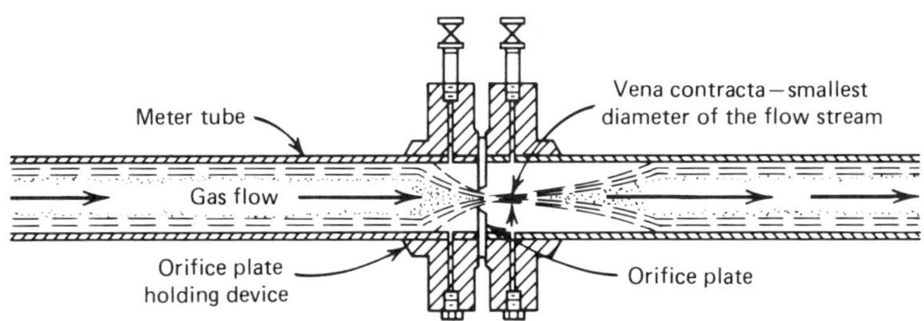

Fig. 6.5 Representation of the primary element of an orifice meter.

upstream and downstream pressure taps of the primary element. One part indicates or records the difference between the pressures on each side of the orifice plate and the other part indicates or records one of these pressures. Recording differential and static pressure gauges, using circular charts with printed scales, are extensively used and they provide a permanent record. Integrating differential gauges are also made, in both the indicating and recording type, that register the flow in uncorrected cubic feet.

The flow of natural gas in the line is calculated by a formula:

$$q = C' \sqrt{h \times p} \tag{6.9}$$

where

q = standard volume per time
C' = a constant
h = pressure drop across the orifice, inches of water
p = static pressure in the line, psia

The constant C' is composed of many other constants that reflect the type of gas being measured, the orifice/pipe diameter ratio, temperature, flow rate, and so on.

6.3.1 Basic Orifice Meter Equation

The basis for the orifice-meter equation is the general energy equation. This equation may be written between any two points in the flowing stream, such as points 1 and 2 on Fig. 6.3:

$$\int_1^2 v\, dp + \int_1^2 \frac{u\, du}{g_c} + \int_1^2 g \frac{dZ}{g_c} = -w - lw \tag{6.10}$$

where

v = specific volume = $\frac{1}{\rho}$ cu ft/lbm
p = pressure, lbf/sq ft
u = average linear-flow velocity, ft/sec
g_c = conversion factor: 32.17 (lbm/lbf)(ft/sec^2)
g = acceleration due to gravity, ft/sec^2
Z = vertical distance above datum, ft
w = work done by flowing fluid, ft-lbf/lbm
lw = work energy lost due to frictional effects, ft-lbf/lbm

For most meters, change in elevation between points 1 and 2, dZ, is zero. Also, no work is done by the flowing fluid, that is, $w = 0$. The lost-work term lw expresses the frictional losses due to viscosity and turbulence of the fluid. These losses can be handled in a manner convenient for meter calculations without reference to friction factor. The basic orifice equation can be written in the form:

$$C^2 \int_1^2 v\,dp + \int_1^2 \frac{u\,du}{g_c} = 0 \tag{6.11}$$

where C^2 is an empirical constant that takes care of friction and other irreversibilities. The point of highest velocity and lowest pressure in the fluid flowing through an orifice is at point 3, called the *vena contracta*. It has the smallest diameter of the flow stream.

To use Eq. 6.11 requires knowledge of the relationship between the specific volume, v and the pressure, p, between points 1 and 2; but for real gases, the integration becomes too complicated for routing use. Since the pressure differential is small as compared with the pressure, an average value ($\bar{v}$ of v or an average value $\bar{\rho}$ of its reciprocal, the density ρ) may be used to simplify the integration. Equation 6.11 thus becomes

$$C^2\bar{v}(144)(p_2 - p_1) + \frac{u_2^2 - u_1^2}{2g_c} = 0 \tag{6.12}$$

or

$$\frac{u_2^2 - u_2^2}{2g_c} = \frac{C^2(144)(p_1 - p_2)}{\bar{\rho}} \tag{6.13}$$

where

p_1, p_2 = pressures, psia
$\bar{\rho}$ = average density, lbm/cu ft

If $\dot{m}$ is the mass flow rate in lbm/sec, then

$$\dot{m} = \rho\, uA \tag{6.14}$$

or

$$u(\text{ft/sec}) = \frac{\dot{m}(\text{lbm/sec})}{\rho(\text{lbm/cu ft})A(\text{sq ft})} \tag{6.15}$$

Substituting Eq. 6.15 in Eq. 6.13,

$$\frac{\dot{m}^2}{\bar{\rho}^2 2g_c}\left[\left(\frac{1}{A_2}\right)^2 - \left(\frac{1}{A_1}\right)^2\right] = \frac{144C^2(p_1 - p_2)}{\bar{\rho}} \tag{6.16}$$

Recognize that u_1 is measured at area A_1, the pipe, and u_2 is measured at area A_2, the orifice. Define the ratio of orifice diameter to pipe diameter as

$$\beta = \frac{D_2}{D_1} \tag{6.17}$$

Equation 6.16 can then be rearranged and solved for $\dot{m}$ to give

$$\dot{m} = CA_2 \sqrt{\frac{2g_c\bar{\rho}(p_1 - p_2)(144)}{(1 - \beta^4)}} \tag{6.18}$$

Equation 6.18 has included the average density $\bar{\rho}$ of the gas. For gases, the average density may be expressed in terms of pressure, temperature, and gas deviation factor, using the gas law:

$$\bar{\rho} = \frac{29\ \gamma_g\ \bar{p}}{\bar{z}\ RT} \tag{6.19}$$

where

γ_g = gas gravity (air = 1.0)
R = gas constant = 10.732

If the pressure difference is expressed in inches of water,

$$(p_1 - p_2)\frac{\text{lbf}}{\text{sq in.}} = \frac{\left(62.43\ \frac{\text{lbm}}{\text{cu ft}}\right)\ \Delta h(\text{in. } H_2O)g\ \frac{\text{ft}}{\text{sec}^2}}{\left(144\ \frac{\text{sq. in}}{\text{sq ft}}\right)\left(12\ \frac{\text{in.}}{\text{ft}}\right)\ \left(g_c\ \frac{\text{lbm}}{\text{lbf}}\ \frac{\text{ft}}{\text{sec}^2}\right)} \tag{6.20}$$

Substitute Eqs. 6.19 and 6.20 into Eq. 6.18:

$$\dot{m}\ \frac{\text{lbm}}{\text{sec}} = CA_2 \sqrt{\frac{2g_c(29\ \gamma_g\ \bar{p})(144)(62.43\ \Delta h)}{(1 - \beta^4)(\bar{z}R\bar{T})(144 \times 12)}} \tag{6.21}$$

Now express Eq. 6.21, which is in terms of mass flow rate, in a form similar to that used for natural gas; that is, in terms of volume flow rate at standard conditions of temperature and pressure, T_b and p_b. Let q_h be the flow rate in cubic feet per hour, and:

$$\rho_s = \frac{29\gamma_g p_b}{z_s RT_b} \tag{6.22}$$

Then

$$\dot{m}(\text{lbm/sec}) = \frac{q_h}{3600}\left[\frac{29\gamma_g p_b}{(1.00)RT_b}\right] \tag{6.23}$$

Substituting Eq. 6.23 into Eq. 6.21 yields

$$q_h = 40{,}077\left(\frac{T_b}{p_b}\right)\frac{CA_2\sqrt{\Delta h.\bar{p}}}{\sqrt{1 - \beta^4}\ \sqrt{\gamma_g T\bar{z}}}$$

$$A_2(\text{sq ft}) = \text{orifice area} = \frac{\pi}{4}\left(\frac{D_2}{12}\right)^2 \tag{6.24}$$

where

D_2 = orifice diameter, in.

Thus

$$q_h = 218.59 \left(\frac{T_b}{p_b}\right) \frac{CD_2^2 \sqrt{\Delta h.\bar{p}}}{\sqrt{1 - \beta^4} \sqrt{\gamma_g T \bar{z}}} \tag{6.25}$$

or

$$q_h = \left(\frac{218.59CD_2^2}{\sqrt{1 - \beta^4}}\right) \frac{T_b}{p_b} \sqrt{\frac{1}{T}} \sqrt{\frac{1}{\gamma_g}} \sqrt{\frac{1}{\bar{z}}} \sqrt{\Delta h.\bar{p}} \tag{6.26}$$

Equation 6.26 is the basic or fundamental orifice-meter equation. For most work in the gas industry,

$$p_b = 14.73 \text{ psia}$$
$$T_b = 60°\text{F} = 520 \text{ R}$$

Assuming the gas to be air, for which $\gamma_g = 1$, and assuming a flowing gas temperature of 520°F, Eq. 6.26 reduces to

$$q_h = 338.40 K_0 \sqrt{\Delta h.\bar{p}} \tag{6.27}$$

where

$$K_0 = \frac{CD_2^2}{\sqrt{1 - \beta^4}}$$

The value $338.40K_0$ becomes the basic orifice flow factor, F_b, which is used in calculating the orifice flow constant, C', in the general orifice-meter equation.

6.3.2 The General Orifice-Meter Equation

In the measurement of most gases, and especially natural gas, it is almost the universal practice to express the flow in cubic feet per hour referred to some specified reference or base condition of temperature and pressure. For the calculation of the quantity of gas, AGA Committee Report No. 3 recommends the formula

$$q_h = C' \sqrt{h_w p_f} \tag{6.28}$$

where

q_h = quantity rate of flow at base conditions, cfh
C' = orifice flow constant
h_w = differential pressure in inches of water at 60°F
p_f = absolute static pressure, psia
$\sqrt{h_w p_f}$ = pressure extension

Because the general orifice-meter equation (Eq. 6.28) appears to be so simple, one may wonder where all these physical laws became involved in the measurement calculations. The orifice flow constant C' may be defined as the rate of flow in cubic feet per hour, at base conditions, when the pressure extension equals unity. It was formerly known as the "flow coefficient." C' is obtained by multiplying a basic orifice factor, F_b, by various correcting factors that are determined by the operating conditions, contract requirements, and physical nature of the installation. This is expressed in the following equation:

$$C' = (F_b)(F_r)(Y)(F_{pb})(F_{tb})(F_{tf})(F_g)(F_{pv})(F_m)(F_l)(F_a) \tag{6.29}$$

where

F_b = basic orifice factor, cfh
F_r = Reynold's number factor (viscosity)
Y = expansion factor
Y_1 = based on upstream static pressure
Y_2 = based on downstream static pressure
Y_m = based on a mean of upstream and downstream static pressures
F_{pb} = pressure base factor (contract)
F_{tb} = temperature base factor (contract)
F_{tf} = flowing temperature factor
F_g = specific gravity factor
F_{pv} = supercompressibility factor
F_m = manometer factor for mercury meter
F_l = gauge location factor
F_a = orifice thermal expansion factor

The derivation of some of these factors is very complex. Actually, several factors can be determined only by very extensive tests and experimentation, from which tables of data have been accumulated so that a value may be obtained. Tables for these factors are available (Tables A.22 to A.37 in the Appendix) and should be referred to for actual values when making calculations. There are two sets of tables—one for flange taps and one for pipe taps—and care should be taken to use appropriate tables for the type of taps in the installation.

Basic Orifice Factor, F_b

This is dependent on the location of the taps, the internal diameter of the run, and the size of the orifice. F_b can be obtained from Table A.22 for flange taps and Table A.27 for pipe taps for published inside diameters. For sizes not listed in these tables and sizes outside the tolerance limits, the exact value of F_b should be calculated for the particular value of β based on the actual value of orifice diameter by using the equations given in AGA Report No. 3. Interpolation should not be relied on.

Reynolds Number Factor, F_r

This factor is dependent on the pipe diameter and the viscosity, density, and velocity of the gas. F_r is obtained by using Table A.23 for flange taps and Table

A.28 for pipe taps. For its determination, the average extension at which the meter operates must be known in addition to the orifice and pipe size. The extension $\sqrt{h_w p_f}$ used in the determination of F_r may be chosen in several ways. The accuracy of each method will depend on individual conditions of flow. The method selected should give as accurate an average $\sqrt{h_w p_f}$ as possible consistent with the tolerances of the rest of the factors. For pressures below 100 psig, it may be necessary to calculate F_r daily, where some estimated average pressure extension may be sufficient for pressure above 100 psig.

Several methods, any one of which may be mutally agreed on, are used for arriving at these averages. Some examples are:

1. Daily average of $\sqrt{h_w p_f}$.
2. Average $\sqrt{h_w p_f}$ after plate installation and periodic average of $\sqrt{h_w p_f}$ under normal conditions of flow for season.
3. One-half of the square root of the range of the chart (100 in., 1000# chart at 14.7 psia would be $1/2\sqrt{100 \times 1014.7} = 159.27$).
4. One-half of the differential and three-quarters of the pressure range (100 in., 1000# chart at 14.7 psia would be $\sqrt{50 \times 764.7} = 195.54$).

Tables A.23 and A.28 have been calculated by using average values of viscosity, 0.000,006,9 lbm/ft-sec, of temperature, 60°F, and of specific gravity, 0.65, applying particularly to natural gas. If the gas being metered has a viscosity, temperature, or specific gravity quite different from these, the value of F_r in Tables A.23 and A.28 may not be applicable. However, for variations in viscosity of from 0.000,005,9 to 0.000,007,9 lbm/ft-sec, in temperature of from 30 to 90°F, or in specific gravity of from 0.55 to 0.75, the variations in the factor F_r would be well within the recommended tolerances.

Expansion Factor, *Y*

Unlike liquids, when a gas flows through an orifice, the change in velocity and pressure is accompanied by a change in the density. The expansion of the gas through the orifice is essentially adiabatic. Under these conditions, the density of the stream changes because of the pressure drop and the adiabatic temperature change. An expansion factor Y, computed for the adiabatic and reversible case, is included in the orifice-meter formula to correct for this variation in density. It is a function of the differential pressure, the absolute pressure, the diameter of the pipe, the diameter of the orifice, and the type of taps. Refer to Tables A.24, A.25, and A.26 for flange taps and Tables A.29 and A.30 for pipe taps for the expansion factors. Care should be taken to select the correct table for the pressure tap from which the static pressure is measured (upstream, Y_1, or downstream, Y_2). The same procedures for selecting h_w/p_f are used in this case as with the F_r factor.

Pressure Base Factor, F_{pb}

Most locations use 14.73 psia as a standard base. This is the pressure base adopted by the American Gas Association for its standard, which represents atmospheric pressure at sea level. This factor is a direct application of Boyle's law in the correction for the difference in base from 14.73 psia. The pressure base is set by contract or tariff. For this reason, a close check of each contract should be made to see that the pressure base factor table (Table A.31) and the basic factor tables (Tables A.22 and A.27) are computed on the same base pressure. The basic orifice flow factor is determined under a defined standard condition. Variation of pressure from this defined pressure base (14.73 psia) will affect the calculated volume of the gas. So, in order to correct the volume of the gas to contract pressure, it is necessary to apply a multiplier determined by dividing the absolute base pressure by the absolute contracted base pressure:

$$F_{bp} = \frac{14.73}{\text{contracted pressure base}}$$

F_{bp} values are given in Table A.31.

Temperature Base Factor, F_{tb}

Sixty degrees Fahrenheit is almost universally used as the base temperature in calculating gas measured by orifice meters. If it was desired to calculate the measurement, however, on some other contract temperature base, this factor would be used in a direct application of Charles' law to correct for this change from 60°F. F_{tb} values may be taken from Table A.32. Gas measured at one base temperature will have a different calculated volume if it is sold to a customer on a different base. That is, if the gas is measured at a base temperature of 60°F and sold at a base temperature of 70°F, the company must correct the volume to the contract temperature or, in this case, lose money. It is clear that the absolute temperature of the base (60°F) divided by the absolute temperature of the contract will give a factor that should be applied to correct the meter reading to the terms of the contract temperature. Thus

$$F_{tb} = \frac{T_b}{520}$$

Flowing Temperature Factor, F_{tf}

The flowing temperature has two effects on the volume. A higher temperature means a lighter gas so that flow will increase. Also, a higher temperature causes the gas to expand, which reduces the flow. The combined effect is to cause the quantity of flow of a gas to vary inversely as the square root of the absolute flow temperature. Thus

$$F_{tf} = \sqrt{\frac{520}{460 + \text{actual flow temperature}}}$$

F_{tf} value is given in Table A.33 and should be based on the actual flowing temperature of the gas. It should be used to correct for flowing temperature. This is usually applied to the average temperature during the time gas is passing. The temperature may be taken by recording charts or by periodic indicating thermometer readings.

Specific Gravity Factor, F_g

F_g (see Table A.34) is used to correct for changes in the specific gravity and should be based on the actual flowing specific gravity of the gas as determined by test. The specific gravity may be determined continuously by a recording gravitometer or by gravity balance on a daily, weekly, or monthly schedule, or as often as necessary to meet conditions of the contract. The basic orifice factor is determined by air with a specific gravity of 1.00. With a given force applied on a gas, a larger quantity of lightweight gas can be pushed through an orifice than that of a heavier gas. To make the basic orifice factor usable for any gas, the proper correction for the specific gravity of the gas being measured must be applied. This factor varies inversely as the square root of specific gravity. Thus

$$F_g = \frac{1}{\sqrt{\gamma_g}}$$

Supercompressibility Factor, F_{pv}

This factor corrects for the fact that gases do not follow the ideal gas laws. It varies with temperature, pressure, and specific gravity. Supercompressibility factors may be obtained from AGA tables. The development of the general hydraulic flow equation involves the actual density of the fluid at the point of measurement. In the measurement of gas, this depends on the flowing pressure and temperature. To translate the calculated volume at the flowing pressure and temperature to base pressure and temperature, it is necessary to apply the law for an ideal gas. All gases deviate from this ideal gas law to a greater or lesser extent. The actual density of a gas under high pressure is usually greater than the theoretical density obtained by calculation of the ideal gas law. This deviation has been termed *supercompressibility*. A factor to account for this supercompressibility is necessary in the measurement of some gases. This factor is particularly appreciable at high line pressures.

The following equation is used to determine the supercompressibility factor:

$$F_{pv} = \frac{z_b}{\sqrt{z}}$$

where

z_b = gas deviation factor for base conditions
z = gas deviation factor for operating conditions

z_b and z may be calculated from gas deviation factor charts. z_b, for many applications measuring gas in terms of standard conditions, will be equal to unity. However, z_b should be calculated for any gas that will deviate from ideal behavior at standard conditions. If unobtainable from actual determinations, the supercompressibility factor, F_{pv}, may be taken from Tables A.35(a), A.35(b), A.35(c), A.35(d), and A.35(e). This is an empirical method of evaluating the supercompessibility factor for normal natural gas mixtures. The accuracy of determining the factor from this method will be within reasonable limits if a specific gravity of 0.75 and diluent contents of 12 mole % nitrogen and/or 5 mole % carbon dioxide are not exceeded. Supercompressibile tests with approved apparatus may be used to establish the suitability of employing the tables for various gas mixtures.

Manometer Factor, F_m

This is used with mercury differential gauges and compensates for the column of compressed gas opposite the mercury leg. Usually, this is not considered for pressures below 500 psia, nor is it required for mercuryless differential gauges. The weight of the gas column over the mercury reservoir of orifice meter gauges introduces an error in determining the differential pressure across the orifice, unless some adjustment is made. This error is consistently in one direction and becomes increasingly important with increasing pressure. The application of the manometer factor from Table A.36 will compensate for this error. The correction varies with ambient temperature, static pressure, and specific gravity. Since the correction is very small, usually some average conditions are selected and a factor is agreed on. One method is to obtain the average ambient temperature from the weather bureau and, from previous recordings, arrive at an average pressure and specific gravity.

Other applicable factors that may affect the total flow of gas as recorded by the orifice meter are discussed below.

Gauge Location Factor, F_l

F_l is used where orifice meters are installed at locations other than 45° latitude and sea level elevation. This is given in Table A.37. F_l may affect the total flow of gas as recorded by the orifice meter.

Orifice Thermal Expansion Factor, F_a

This is introduced to correct for the error resulting from expansion or contraction of the orifice operating at temperatures appreciably different from the temperature at which the orifice was bored. The factor may be calculated from the following equation:

304 and 316 stainless steel

$$F_a = 1 + [0.000{,}018{,}5\ (°F-68)]$$

Monel

$$F_a = 1 + [0.000{,}015{,}9\ (°F-68)]$$

where

°F = gas flowing temperature at orifice

These formulas assume that orifice bore diameter has been measured at a temperature of 68°F. Like gauge location factor, this may affect the orifice meter's record of total gas flow.

6.3.3 Computation of Volumes

After the differential pressure, static pressure, and temperature data at the field location have been recorded on charts, the latter must be picked up and taken to some location for processing. For standard gauges, this requires trips to the field location once a day, every other day, every third day, or once a week, depending on the chart rotation.

With the advent of automatic changers, this is no longer necessary. Charts for several days may be loaded at one time. At the completion of recording, the chart automatically changes, and several fully recorded charts may be picked up at one time. This saves much chart-changing time and allows more accurate chart recording because faster rotating charts are economically feasible. At the central chart processing locations, the charts are integrated or scanned to obtain chart units per period of operation (usually 24 hours). These chart units must then be converted to volume by use of the proper basic orifice factor and all the related factors. The most proficient manner of doing this is by programming the rather complex calculations on a computer.

For those who do not have enough charts to make this economical or because of unavailability of a computer, a manual calculation of the various factors utilizing tables must be made.

Example 6.1. ***(After AGA Committee Report)***

Given

Meter equipped with flange taps, with static pressure from downstream tap:

D_1 = line size = 8.071 in. actual ID

D_2 = orifice size = 1.000 in.
flowing temperature = 65°F
ambient temperature = 70°F

p_b = contract pressure base = 14.65 psia
temperature base = 50°F = 510°R

γ_g = specific gravity = 0.570

H_w = total heating value = 999.1 Btu/cu ft

X_n = mole fraction nitrogen content = 0.011

X_c = mole fraction carbon dioxide content = 0.000

h_w = average differential head = 50 in. water

p_f = average downstream gauge pressure
= 370 psig

Required

The orifice flow constant and the quantity rate of flow for 1 hour at base conditions.

Solution

$$\beta = \frac{D_2}{D_1} = \frac{1.000}{8.071} = 0.1239$$

$$\text{Average } \sqrt{h_w p_f} = \sqrt{50 \times 384.4} = 138.64$$

$$\text{Average } \frac{h_w}{p_f} = \frac{50}{384.4} = 0.1301$$

From Table A.22, for a 1-in. plate in an 8.071-in. ID line,

$$F_b = 200.38$$

From Table A.23, for a 1-in. plate in an 8.071-in. ID line,

$$b = 0.0680$$

$$F_r = 1 + \frac{b}{\sqrt{h_w p_f}} = 1 + \frac{0.0680}{138.64}$$
$$= 1 + 0.000,49 = 1.0005$$

From Table A.25 (Y_2 for downstream static pressure), interpolating, for h_w/p_f = 0.1301 and $\beta = 0.1239$,

$$Y_2 = 1.0008$$

From Table A.31, for $p_b = 14.65$ psia,

$$F_{pb} = 1.0055$$

From Table A.32, for temperature base = 50°F,

$$F_{tb} = 0.9808$$

From Table A.33, for flowing temperature = 65°F,

$$F_{tf} = 0.9952$$

From Table A.34, for specific gravity = 0.570,

$$F_g = 1.3245$$

(a) Using the specific gravity method of supercompressibility factor evaluation, the pressure adjustment index [Table A.35(b)] is

$$f_{pg} = \gamma_g - 13.84X_c + 5.420X_n$$
$$= 0.570 - 13.84(0) + 5.420(0.011)$$
$$= 0.630$$

for p_f = 370.0 psig and f_{pg} = 0.630, by interpolation,

$$\begin{aligned}\text{Pressure adjustment} &= \Delta p = 0.52 \text{ psi} \\ \text{Adjusted pressure} &= p_f = p_f + \Delta p \\ &= 370.5 \text{ psig}\end{aligned}$$

The temperature adjustment index [Table A.35(c)] is

$$\begin{aligned}f_{tg} &= \gamma_g - 0.472X_c - 0.793X_n \\ &= 0.570 - 0.472(0) - 0.793(0.011) \\ &= 0.561\end{aligned}$$

for flowing temperature = 65°F and f_{tg} = 0.561, by interpolation,

$$\begin{aligned}\text{Temperature adjustment} &= \Delta T = 19.90^\circ \\ \text{Adjusted temperature} &= T_f + \Delta T \\ &= 65 + 19.90 \\ &= 84.9^\circ\text{F}\end{aligned}$$

for p_f = 370.5 psig and T_f = 84.9°F, by interpolation [Table A.35(a)],

$$F_{pv} = 1.0254$$

(b) Using the heating value method of supercompressibility factor evaluation, the pressure adjustment index [Table A.35(d)] is

$$\begin{aligned}f_{ph} &= \gamma_g - 0.5688\frac{H_w}{1000} - 3.690X_c \\ &= 0.570 - 0.5688(0.9991) - 3.690(0) \\ &= 0.0017\end{aligned}$$

for p_f = 370.0 psig and f_{ph} = 0.0017, by interpolation,

$$\begin{aligned}\text{Pressure adjustment} &= \Delta p = 0.52 \text{ psi} \\ \text{Adjusted pressure} &= p_f = p_f + \Delta p \\ &= 370.0 + 0.52 \\ &= 370.5 \text{ psig}\end{aligned}$$

The temperature adjustment index [Table A.35(e)] is

$$\begin{aligned}f_{th} &= \gamma_g + 1.814\frac{H_w}{1000} + 2.641X_c \\ &= 0.570 + 1.814(0.9991) + 2.641(0) \\ &= 2.382\end{aligned}$$

for flowing temperature = 65°F and f_{th} = 2.382, by interpolation,

$$\begin{aligned}\text{Temperature adjustment} &= \Delta T = 19.92^\circ \\ \text{Adjustment temperature} &= T_f + \Delta T \\ &= 84.9^\circ\text{F}\end{aligned}$$

for p_f = 370.5 psig and T_f = 84.9°F, by interpolation [Table A.35(a)],

$$F_{pv} = 1.0254$$

for p_f = 370 psig, ambient temperature = 70°F, and γ_g = 0.570, by interpolation (Table A.36),

$$F_m = 0.9993$$

Then the orifice flow constant is

$$\begin{aligned}C' &= (200.38)(1.0005)(1.0008)(1.0055)(0.9808)(0.9952) \\ &\quad \times (1.3245)(1.0254)(0.9993) \\ &= 267.25\end{aligned}$$

and the rate of flow for 1 hour at base conditions is

$$q_h = 267.25\sqrt{50 \times 384.4} = 37{,}052 \text{ cfh}$$

Example 6.2 (After AGA Committee Report)

Given

Meter equipped with pipe taps, using the upstream static pressure connection:

- D_1 = 6 in. = line size = 6.065 in. actual ID
- D_2 = 2.000 in. = orifice size
- T_f = 50°F = flowing temperature
- p_b = 8 oz above 14.4 psia = 14.9 psia = pressure base
- T_b = 50°F = 510°R = temperature base
- γ_g = 0.650 = specific gravity
- Average differential head, h_w = 60 in. water
- Average upstream static pressure = p_f = 90 psia

Required

The orifice flow constant and the quantity rate of flow at average conditions for 1 hour.

Solution

$$\beta = \frac{D_2}{D_1} = \frac{2.000}{6.065} = 0.330$$

$$\text{Average } \sqrt{h_w p_f} = \sqrt{60 \times 90} = 73.49$$

$$\text{Average } \frac{h_w}{p_f} = \frac{60}{90} = 0.67$$

From Table A.27 for 2-in. plate in 6-in. line,

$$F_b = 870.93$$

From Table A.28 for 2-in. plate in 6-in. line,

$$b = 0.0273$$

$$F_r = 1 + \frac{b}{\sqrt{h_w p_f}} = 1 + \frac{0.0273}{73.49} = 1.000,37$$

From Table A.29 (Y_1 for upstream static pressure), interpolating, for $h_w/p_f = 0.67$ and $\beta = 0.330$

$$Y_1 = 0.990,96$$

From Table A.31, for $p_b = 14.9$ psia,

$$F_{pb} = 0.9886$$

From Table A.32, for temperature base = 50°F,

$$F_{tb} = 0.9808$$

From Table A.33, for flowing temperature = 50°F,

$$F_{tf} = 1.0098$$

From Table A.34, for specific gravity = 0.650,

$$F_g = 1.2403$$
$$\gamma_g = 0.650$$
$$p_{pc} = 670 \text{ psia}$$
$$T_{pc} = 375\ °\text{R}$$
$$p_{pr} = \frac{90}{670} = 0.1343$$
$$T_{pr} = \frac{510}{375} = 1.36$$
$$z = 0.985$$
$$F_{pv} = \frac{1}{\sqrt{z}} = \frac{1}{\sqrt{0.985}} = 1.0076$$

Then, the orifice flow constant is

$$C' = (870.93)(1.0037)(0.990,96)(0.9886)(0.9808)(1.0098)(1.2403) \times (1.0076) = 1059.97$$

The average rate of flow for 1 hour is

$$q_h = 1059.97\sqrt{60 \times 90} = 77,892 \text{ cfh}$$

6.3.4 *Recording Charts*

Round charts have been used extensively on all kinds of recording instruments associated with gas measurement. Circular charts for recording differential and static pressure gauges are usually 12 in. in diameter. The chart scale ranges generally used in fuel gas metering are given below.

Common Differential Pressure Ranges, in. of water	Common Static Pressure Ranges, psig
0 to 10	0 to 100
0 to 20	0 to 250
0 to 50	0 to 500
0 to 100	0 to 1000
0 to 200	0 to 2500

There are two principal types of meter charts: the uniform scale direct-reading chart (Fig. 6.6) for the differential pressure in inches of water and the static pressure in pounds per square inch, and a chart that reads the square root (Fig. 6.7). Clocks turn the charts at desired speed, one turn each time period.

Direct-reading Charts

In this type of scale, the lines are spaced an equal distance apart. The scale value of each line, in terms of the full range of the instrument with which it is used, should be 1, 2, or 5 units, or some multiple of these. In many cases, the differential pressure and static pressure are recorded on a chart with a common spacing. To illustrate this, assume Fig. 6.6 is used on a gauge having a differential pressure range of 100 in. of water and a static pressure range of 500 psig. Then each circular line on the chart represents 2 in. of water pressure and 10 psig.

Square Root Charts

This scale shows the square root of the percentage of the full-scale range of the gauge, or as reprcsented by the full scale of the chart. A reading at full scale or full range of the gauge will be 10, the square root of 100. Using the 100-in., 500-lb gauge, a chart reading of 5 would represent a differential pressure of 25 in. or a static pressure of 125 psia.

For square root charts, a chart factor may be defined as

$$\text{Chart factor} = \sqrt{\frac{\text{meter range}}{100}}$$

$$\text{actual pressure} = (\text{chart reading} \times \text{chart factor})^2$$

Thus,

$$\text{Actual pressure} = \left(\frac{\text{chart reading}}{10}\right)^2 \times \text{meter range}$$

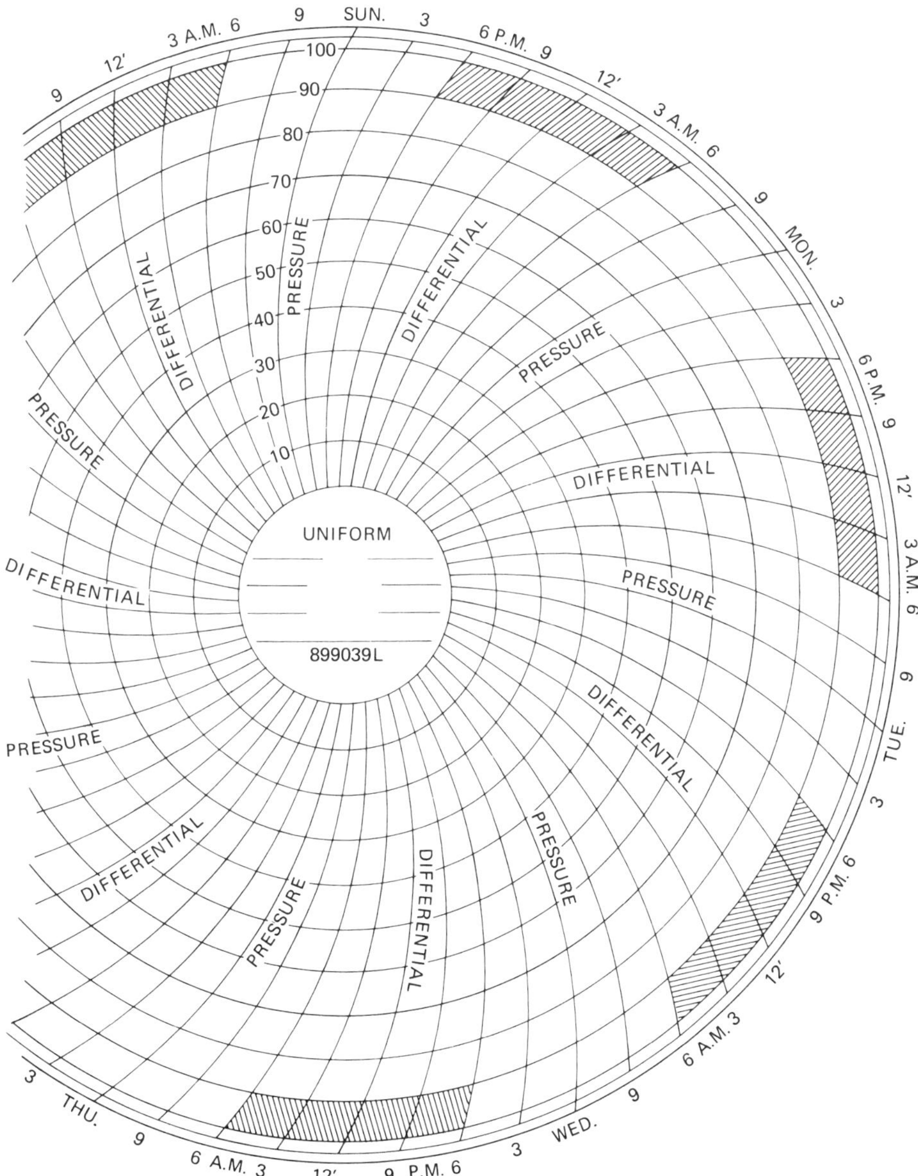

Fig. 6.6 Typical uniform scale direct-reading chart.

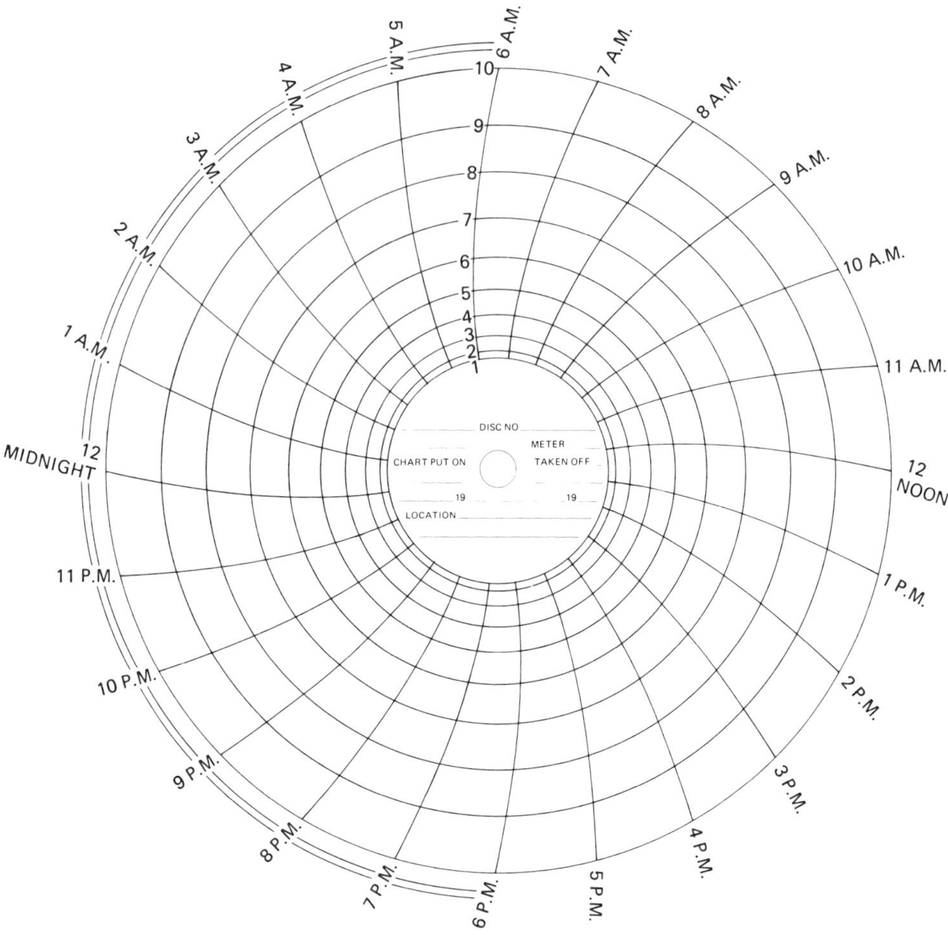

Fig. 6.7 Square root chart.

Example 6.3

50 in. × 100 lb gauge
Differential pressure range, R_h = 50 in.
Static pressure range, R_p = 100 psi

Assume square root chart readings are

$$\text{Differential} = 7.2$$
$$\text{Static} = 9.4$$

$$\begin{aligned}\text{Differential pressure, } h_w &= \left(\frac{\text{chart reading}}{10}\right)^2 \times R_h \\ &= (0.72)^2 \times 50 \\ &= 25.92 \text{ in. water}\end{aligned}$$

$$\text{Static pressure, } p_f = \left(\frac{\text{chart reading}}{10}\right)^2 \times R_p$$
$$= (0.94)^2 \times 100$$
$$= 88.36 \text{ psia}$$

The static pressure to be used in all gas computation is the absolute pressure. Hence, when a square root chart is used, the static pen is set so that, theoretically, it should read zero only if subjected to a pressure of absolute zero. Figure 6.8 is a typical orifice-meter chart recording. The recorded differential of a typical flow pattern is shown as a weaving line on the chart. The smoother line on the chart represents the static pressure.

6.3.5 Orifice Meter Selection

Primary to proper meter selection, it is necessary to know the following about the characteristics and conditions of the flow to be metered: maximum peak hourly

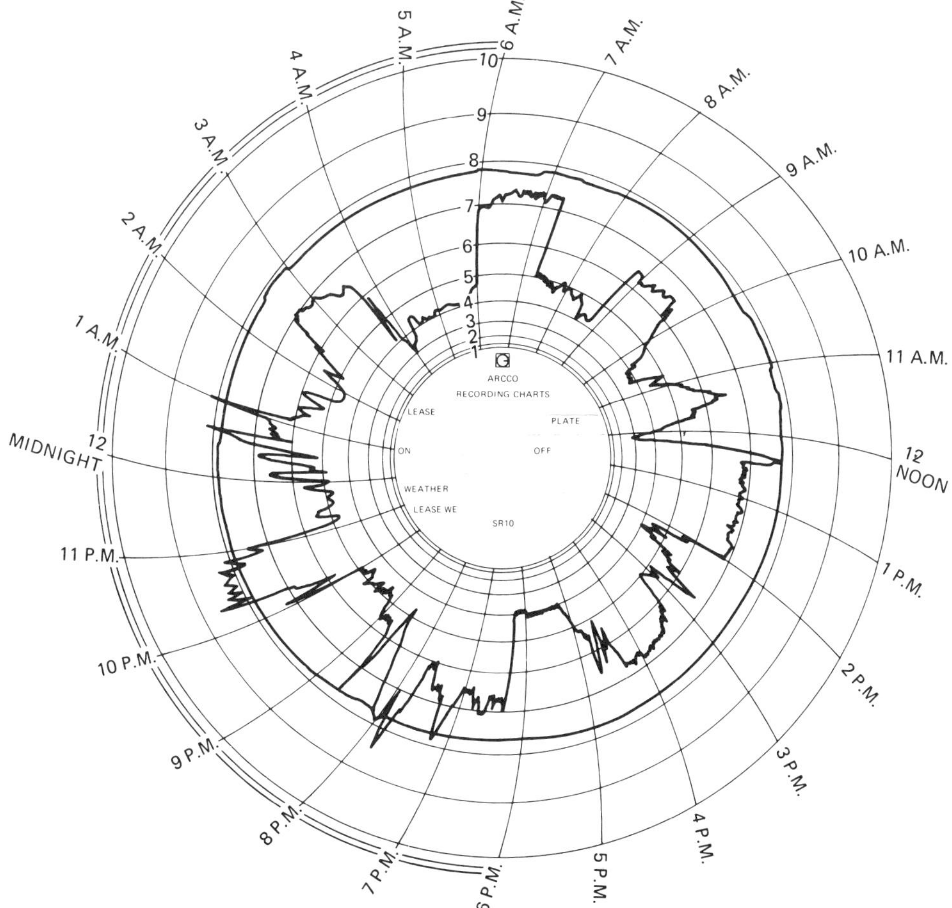

Fig. 6.8 Typical orifice meter chart and recording.

rate, duration of maximum peak or uniformity of flow, minimum hourly rate or its uniformity, metering gauge pressure required and available, and permissible pressure variations.

The quantity of gas flowing through an orifice at constant pressure varies as the square root of the differential pressure. Accordingly, for half of a given rate of flow, the differential pressure will be one-fourth of that for the given rate. Because of mechanical and installation limitations, it has been considered impractical to construct a differential gauge that will continuously record pressures with acceptable accuracy below about one-sixteenth of its maximum range. Therefore, the working range of one orifice plate and one differential gauge is from maximum capacity to about one-sixteenth of maximum. The maximum and minimum capacity can be changed by changing the orifice size.

6.3.6 Uncertainties in Flow Measurements

Table 6.2 may be used to estimate the reliability of measurement data. The AGA procedures discussed in this chapter are representative of practical means that may be used with reasonable accuracy. Other methods may be equally practical.

Flow measurement is often used as a basis for control only. As a general rule, the more accurate measuring installations will give more accurate control. However, in most cases, satisfactory control can be expected as long as the errors remain constant.

Several constant and variable errors exist (courtesy Petroleum Extension Service):

Constant Errors

1. Incorrect information about the bore of the orifice plate.
2. Contour of the orifice plate (convex or concave).
3. Dullness of the orifice edge.
4. Thickness of the orifice edge.
5. Eccentricity of the orifice bore in relation to the pipe bore.
6. Incorrect information about pipe bore.
7. Excessive recess between the end of the pipe and the face of the orifice plate.
8. Excessive pipe roughness.

Variable Errors

1. Flow disturbances caused by insufficient length of meter tube or irregularities in the pipe, welding, and so on.
2. Incorrect locations of differential taps in relation to the orifice plate.
3. Pulsating flow.
4. Progressive buildup of solids, dirt, and sediment on the upstream side of orifice plate.
5. Improper check-valve operation.
6. Accumulation of liquid in the bottom of a horizontal run.

TABLE 6.2

Estimated Uncertainties in the Measurement or Evaluation of Items Involved in Fluid Metering

Item or Factor	Uncertainty Range[a] "Good" Laboratories	"Field Practice"
Throat diameter of orifice or flow nozzle: <5 in.	±0.0001 in.	to ±0.002 in.
>5 in.	±0.0002 in.	to ±0.005 in.
Pipe diameter: 1-in. pipe to 30-in. OD pipe	±0.001 in.	to ±0.10 in.
Static pressure: dead weight gauge, single reading	±0.05 lb.	to ±1.0 lb
average 10 readings	±0.01 lb	to ±0.2 lb
test gauge, single reading	±0.1 lb	to ±2.0 lb
average 10 readings	±0.02 lb	to ±0.5 lb
indicating gauge, single reading	±0.5 lb	to ±10. lb
average 10 readings	±0.1 lb	to ±2. lb
recording chart, regardless of chart scale	±0.01 in.	to ±0.05 in.
Differential pressure: mercury or water manometer, single reading	±0.001 in.	to ±0.25 in.
average 10 readings	±0.0002 in.	to ±0.05 in.
recording chart, regardless of chart scale	±0.01 in.	to ±0.05 in.
Temperature: mercury-in-glass, thermocouple, resistance thermo.	±0.01°F	to ±5.°F
temperature recorder, regardless of chart scale	±0.01 in.	to ±0.05 in.
Specific weight (or density): water at <140°F	±1 in 60,000	to ±1 in 6000
at >140°F	±1 in 12,000	to ±1 in 3000
other liquid, direct determination	±1 in 50,000	to ±1 in 200
air	±1 in 5,000	to ±1 in 200
steam, superheated	±1 in 2,000	to ±1 in 200
wet	±1 in 400	to ±1 in 50
other gases, direct determination	±1 in 2,000	to ±1 in 50
Specific gravity: liquids	±1 in 5,000	to ±1 in 50
gases	±1 in 2,000	to ±1 in 50
Weighings, commercial scales	±1 in 10,000	to ±1 in 100
Coefficient of discharge or of flow: by calibration	±0.05%	to ±1.0%
from table or curve	±0.5%	to ±5.0%
Compressibility or supercompressibility of gases	±0.1%	to ±5.0%
Expansion factor, compressible fluids $\left[x = \frac{p_1 - p_2}{p_1} 100\right]$	±0.5% of x	to ±2% of x

Source: Courtesy AGA, *Gas Engineers Handbook*.

[a]Signifies an estimate of the uncertainty: that is, difference to be expected between observed or determined values of the measurement and the time value, if such were known or could be determined exactly.

7. Liquids in the piping or meter body.
8. Changes in operating conditions from those used in the coefficient calculations (i.e., specific gravity, atmospheric pressure, temperature, etc.).
9. Incorrect zero adjustment of the meter.
10. Nonuniform calibration characteristic of the meter.
11. Corrosion or deposits in the meter range tube or float chamber.
12. Emulsification of liquids with mercury.
13. Dirty mercury.
14. Incorrect arc for meter pens.
15. Formation of hydrates in meter piping or meter body.
16. Leakage around the orifice plate (applies to orifice fittings).
17. Wrong range of chart.
18. Incorrect time for rotation of chart.
19. Excessive friction in the meter's stuffing box.
20. Meter not level (mercury-type only).
21. Excessive friction between the pen and chart.
22. Overdampening of the meter response.

6.3.7 Mass-Flow Meter

The orifice meter may be used to measure gas on a mass-flow basis. The density of the flowing gas is measured with a densitometer. The densitometer is substituted for the static-pressure element and makes determination of specific gravity and supercompressibility corrections unnecessary.

The following equation from AGA Report No. 3 may be used for mass flow computations:

$$W = 1.0618 F_b F_r Y \sqrt{h_w \gamma_g} \qquad (6.30)$$

where

W = flow rate, lbm/hr
F_b = basic orifice factor
F_r = Reynold's number factor
Y = expansion factor
h_w = differential pressure
γ_g = specific weight of gas at flowing conditions, lb/cu ft

6.4 NATURAL GAS LIQUID MEASUREMENT

Field measurement of natural gas liquids is accomplished by the conventional gauging of tanks and by use of various metering techniques. The orifice meter is sometimes used. Installation and operation requirements are about the same as for gas.

For measurement in gallons, the following equation may be used:

$$q_h = C' \sqrt{h_w} \tag{6.31}$$

where

q_h = rate of liquid flow, gph
C' = orifice constant ($F_b \times F_g \times F_r$)
h_w = differential pressure, inches of water
F_b = basic orifice factor
F_g = specific gravity factor
F_r = Reynold's number factor

If pound units are desired, the formula is

$$W = SND^2 F_a F_m F_c F_{pv} \sqrt{\gamma_f h_w} \tag{6.32}$$

where

W = rate of flow, lbm/day
S = a value determined from the bore of the orifice and internal diameter of the metering tube
N = combined constant for weight-flow measurement (=68,045 when W is in pounds/day)
D = ID of tube, inches
F_a = orifice thermal expansion factor
F_m = manometer factor (= 1.000 for bellows-type meter)
F_c = viscosity factor (usually assumed equal to 1.000)
F_{pv} = supercompressibility factor
γ_f = specific gravity of liquid stream at flowing temperature and pressure as determined by gravitometer readings
h_w = differential pressure, inches of water

Equation 6.32 (after Foxboro), for simplicity, may be written as:

$$W = 68{,}045 SD^2 \sqrt{\gamma_f h_w} \tag{6.33}$$

It is frequently necessary to make measurements for operation and allocation purposes when the fluid is two-phase, that is, both gas and liquid. Accuracy suffers if the fluid is not in single-phase. Certain precautions should be taken to arrive at acceptable measurements of a two-phase stream:

1. Keep pressure and temperature as high as possible at the meter.
2. Use a free-water knockout ahead of the meter.
3. A vertical meter run may sometimes improve the differential-pressure relationship to the volume.
4. Use test data from periodic full-scale separator tests to determine coefficient or meter factor.
5. Connect manifold lead lines to bottom of bellows-type meter with self-draining pots installed above orifice fitting.

REFERENCES

American Gas Association. *Gas Engineer's Handbook*. New York: Industrial Press, 1965.

American Gas Association. "Orifice Metering of Natural Gas." Gas Measurement Committee Report No. 3. New York: January 1956.

Foxboro. *Principles and Practices of Flow Meter Engineering*. 8 ed. Foxboro, Mass.: Foxboro Co., 1961.

Katz, D. L., et al. *Handbook of Natural Gas Engineering*. New York: McGraw-Hill, 1959.

NGPSA Engineering Data Book. Tulsa: Gas Processors Suppliers Assoc., 1972.

Petroleum Extension Service. *Field Handling of Natural Gas*. Austin: Univ. of Texas Press, 1972.

Terrell, C. E., and H. S. Bean. *AGA Gas Measurement Manual*. New York: AGA, 1963.

PROBLEMS

6.1 A bellow-type meter is used to meter the gas flowing to a plant. The meter reading is 1,250,000 cu ft at meter conditions. Meter pressure is 150 psig, atmospheric pressure is 14.5 psia, T_f = 90°F, gas gravity = 0.79, and gas price is $1.00/Mscf. The percent error is +4.5%. Compute the gas bill assuming standard conditions of 14.73 psia and 60°F.

6.2 Calculate the gas flow rate, Q, in MMscfd for the following conditions:

Pipe ID = 8.071	p_b = 15.4 psia
Orifice ID = 4.00	h_w = 64 in.
T_f = 80°F	p_f = 625 psig
T_b = 65°F	P_{atm} = 14.5 psia
γ_g = 0.72	

Flange taps, static pressure measured upstream.

6.3 An orifice meter is equipped with an L-10 square root chart. The maximum range of the static element is 900 psia, and the maximum range of the differential element is 225 in. water. Compute the following:

(a) Actual static pressure when the chart reading is 8.5.

(b) Chart reading when the actual differential is 100 in. of water.

(c) Meter constant, M, where for the square root chart,

$q_h = C'Mh_uP_u$

h_u = differential pen reading from square root chart

P_u = static pen reading from square root chart

(d) Atmospheric pressure setting of the static pen.

6.4 You are asked to interpret an orifice meter chart where the static and differential readings are varying with time. Set up the column headings that you would use to calculate the static and differential pressures for computation of volumes for:

(a) Normal linear chart.
(b) L-10 square root chart.

6.5 The following data are given:

$T_b = T_f = 60°F$
$P_b = 14.73$ psia
$q = 1$ MMcfd at base conditions (60°F, 14.73 psia)
S.G. = 0.6 (air = 1.0)
$P_r = 100$ psig (downstream tap)
4-in. nominal meter tube (SCH 40)
Orifice is 1.625 in. ID with flange taps
100-in. differential meter (bellows type)
250-psi static element

Assuming that the above data are correct, find the percent error that would be caused by incorrectly "zeroing" the static pen on the lowest line on the L-10 chart.

6.6 A 2-in. orifice is used for measuring the flow of gas in a 4-in. nominal diameter (3.438-in. internal diameter) pipeline. The differential pressure is 15 in. of water. The static pressure taken from the upstream tap is 50 psig. Pipe taps are used. The temperature of the gas is 80°F, and the specific gravity is 0.7 (air = 1). What is the hourly rate of flow measured at base conditions of 15.15 psia and 70°F? Assume $F_a = F_1 = F_m = 1$.

6.7 What is the casing gas production from a well with the following orifice meter information?

Readings from Square-Root Chart

Differential	= 5.6
Static	= 6.5
Chart range	= 50 in. × 100#
Pipe diameter	= 2.067 in.
Orifice diameter	= 0.5 in.

Static Pressure Taken at Downstream Tap

Flowing temperature	= 80°F
Gas gravity	= 0.65 (air = 1)
Base conditions	= 14.7 psia and 60°F

Assume $F_m = F_l = F_a$ = unity. Flange taps are used.

6.8 An orifice meter having a 1.5-in. orifice in a 4.026-in. internal-diameter flange is located in a gas line through which gas is flowing. If the water manometer indicates a differential in pressure of 10 in. of water, at what rate is the gas flowing?

Specific gravity of gas	= 0.6 (air = 1)
Static pressure taken at downstream tap	= 100 psig
Flowing temperature	= 60°F
Base conditions	= 14.7 psia and 60°F

Assume expansion factor, supercompressibility factor, F_m, F_l, and F_a each equals unity.

6.9 You are attempting to determine the input gas volume to an intermitting gas lift well. There is a 4 in. Sch 40 (4.026 in. ID) meter run installed at the well, with a 1½-in. orifice plate. To this you hook up a portable flow recorder (100-in. H_2O, 1000 psig). The differential trace is impossible to read because of the cylcing of the injection: six minutes injection every 30 minutes. To improve the readability of the chart, you use a 24-minute hub and determine that the average differential reading during several kicks is 60-in. and that the average static pressure is 685 psig.

The following data are known or assumed:

(a) Flange taps, static pressure from upstream tap.
(b) Pressure base, 14.65 psia.
(c) Temperature base, 60°F.
(d) Flowing temperature, 90°F.
(e) Gas specific gravity, 0.625.

Calculate the meter coefficient and the daily rate of gas injection.

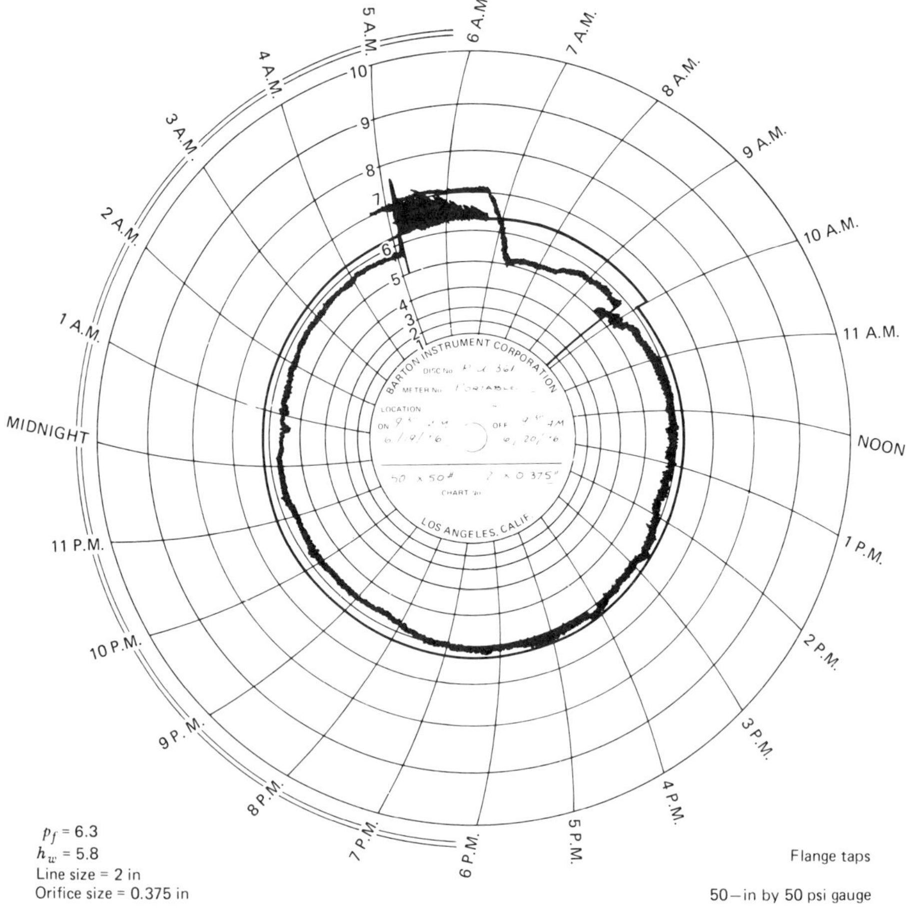

6.10 What is the casing gas production for Well RU 361? Use the following parameters in addition to those shown on the chart in your calculations:
(a) Static pressure taken at downstream tap.
(b) Flowing temperature = 76°F.
(c) Gas gravity = 0.63.
(d) Base conditions = 14.7 psia and 60°F.
Assume F_m, F_l, and F_a equal 1.000. *Note:* Smooth line is p_f.

7

GAS GATHERING AND TRANSPORTATION

7.1 INTRODUCTION

The transmission of gas to the consumer may be divided into four distinct units: the gathering system, the compression station, the main trunk line, and the distribution lines.

Pipelines, which comprise the gathering system, main trunk line, and distribution lines, provide an economical method of transporting fluids over great distances. After the initial capital investment required for their construction, they show low operating costs and unit costs that decline with large volumes of throughput. Many factors must be considered in the design of long-distance gas pipelines. These include the nature and volume of the gas to be transmitted, the length of the line, the type of terrain to be crossed, and the maximum elevation of the route.

After the compression station is located and its size is determined by the quantity of gas to be handled, the gathering system is designed. This involves the location of the wells, the availability of right of way, the amount of gas to be handled, the distance to be transported, and the pressure difference between the field and the main transmission line. The gas wells are generally located in groups around a geological structure or within the defined limits of a pool or gas reservoir. The problem is to get the gas to the compression station. In a new field, the gathering system must be large enough to handle the production of additional leases. The gathering system is made up of branches that lead into trunk lines. The trunk line is small at the most distant well and, as more wells along the line are attached to it, the line must be larger to accommodate the greater volume of gas.

In addition to the gathering system and major trunk pipelines, there is also a network of smaller-diameter feeder and transmission mains that may carry gas to centers of consumption. In addition, complex systems of still smaller-diameter distribution piping run to individual homes, shops, and factors.

The design of the transmission system calls for the services of a well-trained and experienced engineering and legal staff. Complex engineering studies are needed to decide on the diameter, yield strength, and pumping horsepower required to give the optimum results for any particular pipeline transmission system. Computer programs that enable high-pressure gas transmission networks to be dynamically simulated on a digital computer have been developed and are commonly used by gas pipeline companies. Several designs are usually made so that the most economical one can be selected. The maximum carrying capacity of a pipeline is limited by its initial parameters of construction. In general, the tendency is to use higher transmission pressures and strong materials of construction. For economic operation, it is important to preserve full pipeline utilization.

Studies of the flow conditions of natural gases in pipelines have led to the development of complex equations (e.g., the Weymouth equation, the Panhandle equation, and the Modified-Pandhandle equation) for relating the volume transmitted through a gas pipeline to the various factors involved, thus deciding the optimum pressures and pipe dimensions to be used. From equations of this type, various combinations of pipe diameter and wall thickness for a desired rate of gas throughout can be calculated. An optimum balance is sought between pipe tonnage and pumping horsepower.

7.2 REYNOLDS NUMBER AND FRICTION FACTOR

Flow of natural gas in pipelines always results in some mechanical energy being converted into heat. The so-called "lost work," lw, represents all energy losses resulting from irreversibilities of the flowing stream. In the case of single-phase flow, such as flow of gas in pipe, these irreversibilities consist primarily of friction losses: internal losses due to viscosity effects and losses due to the roughness of the wall of the confining flow string.

With the exception of completely laminar flow, the energy losses lw of actual systems cannot be predicted theoretically; they must be determined by actual experiment and then correlated as some function of the flow variables.The lost work is usually calculated using a friction factor, f. By dimensional analysis, it can be shown that the friction factor is a function of the Reynolds number, N_{Re} and of the relative roughness, e/D.

The theoretical basis of most fluid flow equations is the general energy equation. Considering a steady-state system, the energy balance may be written as

$$U_1 + p_1V_1 + \frac{mu_1^2}{2g_c} + \frac{mgZ_1}{g_c} + Q - w = U_2 + p_2V_2 + \frac{mu_2^2}{2g_c} + \frac{mgZ_2}{g_c} \tag{7.1}$$

where

U = internal energy
pV = energy of expansion or compression
$\frac{mu^2}{2g_c}$ = kinetic energy
$\frac{mgZ}{g_c}$ = potential energy
Q = heat energy added to fluid
w = shaft work done by the fluid on the surroundings

Dividing Eq. 7.1 through by m to obtain an energy per unit mass balance and writing the resulting equation in differential form yields

$$dU + d\left(\frac{p}{\rho}\right) + \frac{u\,du}{g_c} + \frac{g}{g_c}\,dZ_1 + dQ - dw = 0 \tag{7.2}$$

Equation 7.2 can be converted to a mechanical energy balance using the following thermodynamic relations:

$$dh = T\,ds + \frac{dp}{\rho} \tag{7.3}$$

and

$$dU = dh - d\left(\frac{p}{\rho}\right) = T\,ds + \frac{dp}{\rho} - d\left(\frac{p}{\rho}\right) \tag{7.4}$$

where

h = enthalpy
s = entropy
T = temperature

Using Eq. 7.4 in Eq. 7.2 results in

$$T\,ds + \frac{dp}{\rho} + \frac{u\,du}{g_c} + \frac{g}{g_c}\,dZ_1 + dQ - dw = 0 \tag{7.5}$$

Clausis inequality for an irreversible process states that

$$ds \geqslant \frac{-dQ}{T} \tag{7.6}$$

or

$$T\,ds = -dQ + d(lw) \tag{7.7}$$

where lw is usually called lost work and represents losses due to irreversibilities, such as friction. Using Eq. 7.7 in Eq. 7.5 gives

$$\frac{dp}{\rho} + \frac{u\,du}{g_c} + \frac{g}{g_c}\,dZ_1 + d(lw) - dw = 0 \tag{7.8}$$

If no work is done by or on the fluid, $dw = 0$, and we obtain

$$\frac{dp}{\rho} + \frac{u\,du}{g_c} + \frac{g}{g_c}\,dZ_1 + d(lw) = 0 \tag{7.9}$$

We may consider a general case where the pipe under consideration is inclined at some angle θ to the horizontal. Since $dZ = dL \sin\theta$, the energy equation becomes

$$\frac{dp}{\rho} + \frac{u\,du}{g_c} + \frac{g}{g_c}\,dL \sin\theta + d(lw) = 0 \tag{7.10}$$

Equation 7.10 may be written in terms of pressure gradient by multiplying through by ρ/dL:

$$\frac{dp}{dL} + \frac{\rho u\,du}{g_c\,dL} + \frac{g}{g_c}\,\rho \sin\theta + \rho\,\frac{d(lw)}{dL} = 0 \tag{7.11}$$

Considering pressure drop as being positive in the direction of flow, Eq. 7.11 can be written as

$$\frac{dp}{dL} = \frac{g}{g_c}\,\rho \sin\theta + \left(\frac{dp}{dL}\right)_f + \frac{\rho u\,du}{g_c\,dL} \tag{7.12}$$

where the pressure gradient due to viscous shear or frictional losses has been expressed as

$$\left(\frac{dp}{dL}\right)_f \equiv \rho\,\frac{d(lw)}{dL} \tag{7.13}$$

We can define a friction factor

$$f' = \frac{\tau_w}{\rho u^2/2g_c} = \frac{\text{wall shear stress}}{\text{kinetic energy per unit volume}} \tag{7.14}$$

Equation 7.14 defines a dimensionless group, which reflects the relative importance of wall shear stress to the total losses. The wall shear stress can be easily evaluated from a force balance between pressure forces and viscous forces:

$$\tau_w = \frac{D}{4}\left(\frac{dp}{dL}\right)_f \tag{7.15}$$

Using this in Eq. 7.14 yields

$$\left(\frac{dp}{dL}\right)_f = \frac{2f'\,\rho u^2}{g_c D} \tag{7.16}$$

This is the Fanning equation and f' is called the Fanning friction factor. In terms of the Darcy–Weisbach or Moody friction factor, $f = 4f'$, Eq. 7.16 becomes

$$\left(\frac{dp}{dL}\right)_f = \frac{f\,\rho u^2}{2g_c D} \tag{7.17}$$

The equation that relates lost work per unit length of pipe and the flow variables is

$$\frac{d(lw)}{dL} = \frac{fu^2}{2g_c D} \tag{7.18}$$

where

lw = mechanical energy converted to heat, ft-lbf/lbm
u = flow velocity, ft/sec
g_c = gravitational conversion factor = 32.17 lbm ft/lbf sec^2
D = pipe diameter, ft
f = Moody friction factor

Integration of Eq. 7.18 results in

$$lw = \frac{fu^2}{2g_c}\,\frac{L}{D} \tag{7.19}$$

A similar equation, using the Fanning fraction factor f' is

$$lw = \frac{f'2u^2L}{g_c D} \tag{7.20}$$

Figure 7.1 is a Moody friction factor chart. It is a log-log graph of (log f) versus (log N_{Re}). Four general conditions of flow are evident: laminar, critical, transition, and turbulent.

7.2.1 The Reynolds Number, N_{Re}

The Reynolds number (N_{Re}) is a dimensionless group defined as

$$N_{Re} = \frac{D(\text{ft})\ u(\text{ft/sec})\ \rho(\text{lbm/cu ft})}{\mu(\text{lbm/ft sec})} \tag{7.21}$$

The dimensionless group, $N_{Re} = Du\,\rho/\mu$ is the ratio of fluid momentum forces to viscous shear forces. It is used as a parameter to distinguish between

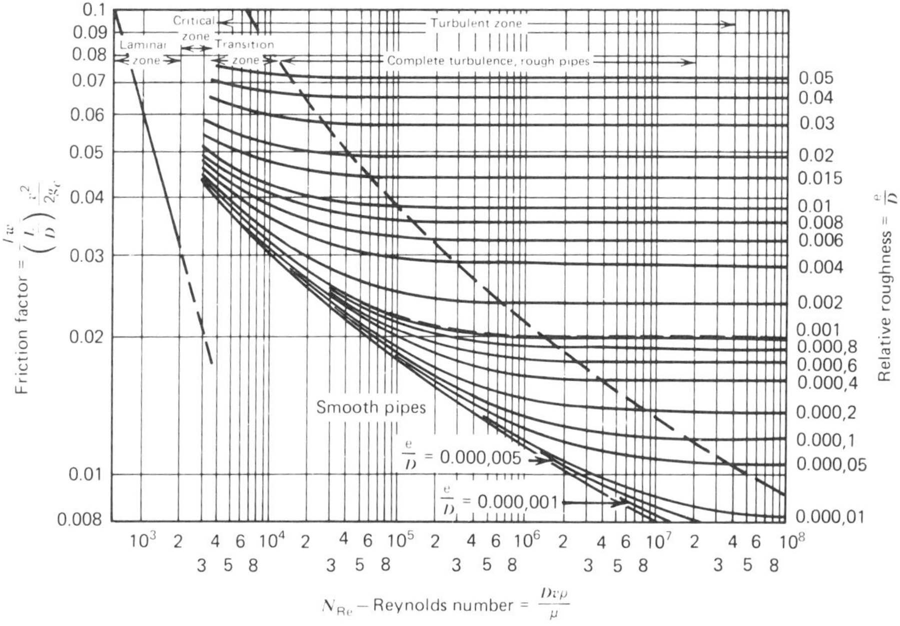

Fig. 7.1 Friction factors for any type of commercial pipe.

laminar and turbulent fluid flow. The change from laminar to turbulent flow is usually assumed to occur at a Reynolds number of 2100 for flow in a circular pipe. If units of ft, ft/sec, lbm/cu ft, and centipoise are used, the Reynolds number equation becomes

$$N_{Re} = 1488 \frac{Du\, \rho}{\mu} \tag{7.22}$$

The mass flux is given by

$$\mu(\text{ft/sec})\ \rho(\text{lbm/cu ft}) \equiv \left[\frac{\text{lbm}}{\text{sec}} \times \frac{1}{\text{ft}^2}\right] = \frac{\text{mass flow rate}}{\text{area}}$$

If gas is flowing at q (Mcfd) measured at base conditions of T_b (°R) and p_b (psia), the mass flow rate in lbm/sec is given by

$$\left[\frac{\text{lbm}}{\text{sec}}\right] \frac{(1000)q\, Mp_b}{(24)(3600)z_b R T_b} = \frac{(3.128 \times 10^{-2})q\, \gamma_g\, p_b}{z_b T_b}$$

$$\text{Cross-sectional area of pipe} = \frac{D^2 \pi}{(4)(144)}$$

Using these in Eq. 7.21 and noting that 1 cp = 6.7197 × 10^{-4} lbm/ft sec,

$$N_{Re} = \frac{\frac{D}{12}\left(\frac{3.128 \times 10^{-2} q \gamma_g p_b}{z_b T_b}\right)\left(\frac{4 \times 144}{\pi D^2}\right)}{[\mu(\text{cp}) \times 6.7197 \times 10^{-4}]}$$

or

$$N_{Re} = \frac{711 p_b(\text{psia}) q(\text{Mcfd}) \gamma_g}{T_b(°\text{R}) D(\text{in.}) \mu(\text{cp})} \quad (7.23)$$

Consider some common base conditions:

p_b(psia)	T_b(°R)	$711 \frac{p_b}{T_b}$
14.4	520 (60°F)	19.69
14.65	520 (60°F)	20.03
14.73	520 (60°F)	20.14
15.025	520 (60°F)	20.54

Thus, for all practical purposes, the Reynolds number for natural gas flow problems may be expressed as

$$N_{Re} \simeq \frac{20 q \gamma_g}{\mu D} \quad (7.24)$$

where

q = gas flow rate at 60°F and 14.73 psia, Mcfd
γ_g = gas gravity(air = 1)
μ = gas viscosity at flow conditions (temperature and pressure), cp
D = pipe diameter, in.

7.2.2 Relative Roughness, e/D

The inside wall of a pipe is not normally smooth. Wall roughness is a function of pipe material, method of manufacture, and environment to which it has been exposed. From a microscopic sense, wall roughness is not uniform, and thus the distance from the peaks to valleys on the wall surface will vary greatly. The absolute roughness, e, of a pipe wall is defined as the mean protruding height of relatively uniformly distributed and sized, tightly packed sand grains that would give the same pressure gradient behavior as the actual pipe wall.

Dimensional analysis suggests that the effect of roughness is not due to its absolute dimensions, but to its dimensions relative to the inside diameter of the pipe. Relative roughness, e/D, is the ratio of the absolute roughness to the pipe internal diameter:

$$\text{Relative roughness} = \frac{e\ (\text{ft})}{D\ (\text{ft})} \quad \text{or} \quad \frac{e\ (\text{in.})}{D\ (\text{in.})} \tag{7.25}$$

The selection of value of pipe wall roughness is sometimes difficult since the absolute roughness is not a directly measurable property for a pipe. The way to evaluate the absolute roughness is to compare the pressure gradients obtained from the pipe of interest with a pipe that is sand-roughened. Typical results have been presented by Moody in Fig. 7.2. Thus, if measured pressure gradients are available, the friction factor and Reynolds number can be calculated and an effective e/D obtained from the Moody diagram. This value of e/D should then be used for future predictions until updated. If no information is available on roughness, a value of $e = 0.0006$ in. is recommended for tubing and line pipe. Typical values of absolute roughness of interest in natural gas flow problems are:

	e (in.)
Drawn tubing	0.00006
Well tubing	0.0006
Line pipe	0.0007
Galvanized pipe	0.006
Cement-lined pipe	0.01–0.1

7.2.3 Equations for the Friction Factor, $f(N_{Re}, e/D)$

Fluid flow ranges in nature between two extremes: laminar or streamline flow and turbulent flow (Fig. 7.1). Within this range are four distinct regions. The equation for the friction factor in terms of Reynolds number and relative roughness varies for each of the four regions.

Laminar Single-Phase Flow

The friction factor for laminar flow can be determined analytically. The Hagen–Poiseuille equation for laminar flow is

$$\left(\frac{dp}{dL}\right)_f = \frac{32\ \mu u}{g_c D^2} \tag{7.26}$$

Equating the frictional pressure gradients given by Eqs. 7.17 and 7.26 gives

$$\frac{f\ \rho u^2}{2g_c D} = \frac{32\ \mu u}{g_c D^2}$$

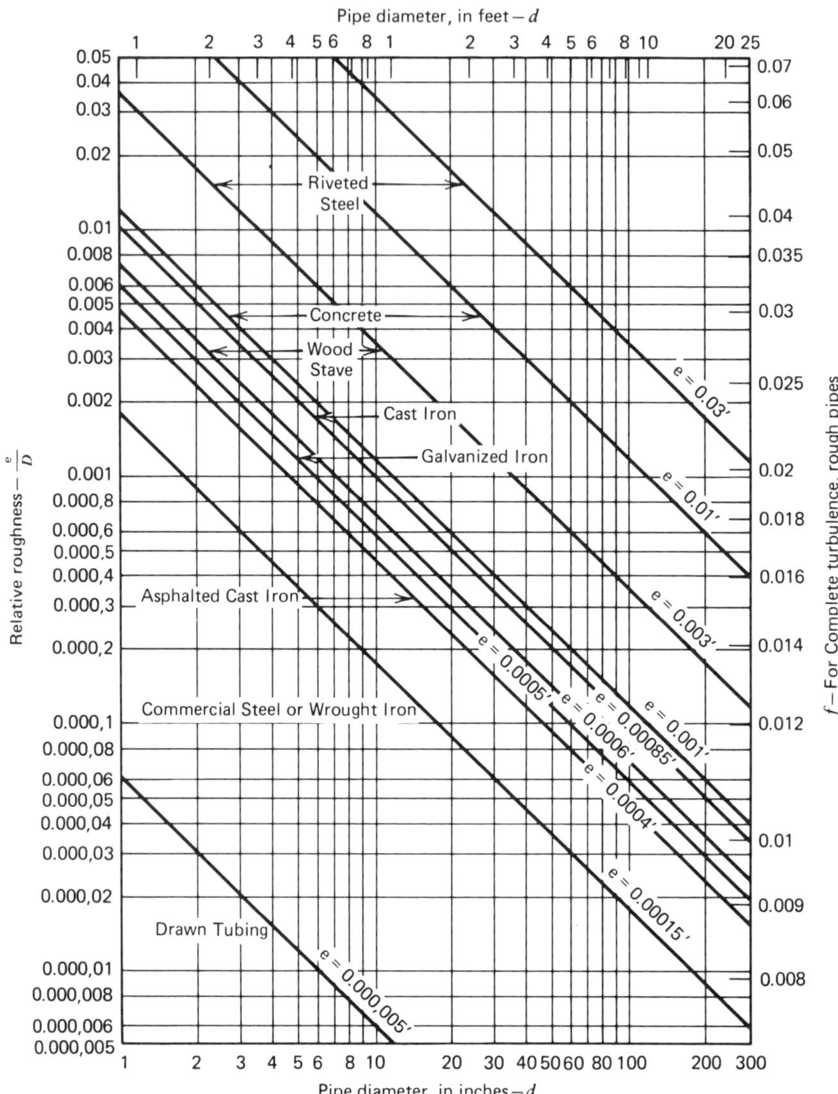

Fig. 7.2 Relative roughness of pipe materials and friction factors for complete turbulence.

or

$$f = \frac{64\mu}{du\,\rho} = \frac{64}{N_{Re}} \tag{7.28}$$

Turbulent Single-Phase Flow

Experimental studies of turbulent flow have shown that the velocity profile and pressure gradient are very sensitive to the characteristics of the pipe wall, that is, the smoothness of the wall. Only the most accurate empirical correlations for friction factors are presented.

Smooth Wall Pipe. Several correlations, each valid over different ranges of Reynolds number, are available. Drew, Koo, and McAdams presented the most commonly used correlation in 1930:

$$f = 0.0056 + 0.5N_{\mathrm{Re}}^{-0.32} \tag{7.29}$$

Equation 7.29 is explicit in f and covers a wide range of Reynolds numbers, $3 \times 10^3 < N_{\mathrm{Re}} < 3 \times 10^6$.

Blasius also developed a correlation that may be used for N_{Re} up to 10^5 for smooth pipes:

$$f = 0.316N_{\mathrm{Re}}^{-0.25} \tag{7.30}$$

Rough Wall Pipe. In turbulent flow, the effect of wall roughness on friction factor depends on the relative roughness and Reynolds number. When the thickness of the laminar sublayer that exists within the boundary layer is large enough, the behavior approximates that of smooth pipe. The thickness of the laminar sublayer is a function of the Reynolds number.

Nikuradse's friction factor correlation is still the best one available for fully developed turbulent flow in rough pipes:

$$\frac{1}{\sqrt{f}} = 1.74 - 2\log\left(\frac{2e}{D}\right) \tag{7.31}$$

His equation is valid for large values of the Reynolds number where the effect of relative roughness is dominant (Fig. 7.1).

The correlation that is used as the basis for modern friction factor charts was proposed by Colebrook and White in 1939:

$$\frac{1}{\sqrt{f}} = 1.74 - 2\log\left(\frac{2e}{D} + \frac{18.7}{N_{\mathrm{Re}}\sqrt{f}}\right) \tag{7.32}$$

Equation 7.32 is applicable to smooth pipes and to flow in transition and fully rough zones of turbulent flow. It degenerates to the Nikuradse correlation (Eq. 7.31) at large values of the Reynolds number.

Equation 7.32 is not explicit in f. However, values of f can be obtained by an iterative procedure using the following form of the equation:

$$f_{\mathrm{calculated}} = \left[1.74 - 2\log\left(\frac{2e}{D} + \frac{18.7}{N_{\mathrm{Re}}\sqrt{\dfrac{1}{f_{\mathrm{guess}}}}}\right)\right]^{-2} \tag{7.33}$$

An explicit correlation for friction factor was presented by Jain in 1976:

$$\frac{1}{\sqrt{f}} = 1.14 - 2 \log \left(\frac{e}{D} + \frac{21.25}{N_{\mathrm{Re}}^{0.9}} \right) \tag{7.34}$$

This correlation is comparable to the Colebrook and White correlation. For relative roughness between 10^{-6} and 10^{-2} and Reynolds number between 5×10^3 and 10^8, the errors were within $\pm 1.0\%$ when compared with the Colebrook and White correlation. Equation 7.34 is recommended for all calculations requiring friction factor determination for turbulent flow

7.2.4 Total Pressure Drop

The pressure gradient equation for a pipeline at any angle of inclination can be written as

$$\frac{dp}{dL} = \frac{g}{g_c} \rho \sin \theta + \frac{f \rho u^2}{2 g_c D} + \frac{\rho u \, du}{g_c \, dL} \tag{7.35}$$

The total pressure gradient is made up of three distinct components:

$$\frac{dp}{dL} = \left(\frac{dp}{dL} \right)_{el} + \left(\frac{dp}{dL} \right)_{f} + \left(\frac{dp}{dL} \right)_{acc} \tag{7.36}$$

where

$\left(\frac{dP}{dL} \right)_{el} \equiv \frac{g}{g_c} \rho \sin \theta$ is the component due to elevation or potential energy change.

$\left(\frac{dp}{dL} \right)_{f} \equiv \frac{f \rho u^2}{2 g_c D}$ is the component due to frictional losses.

$\left(\frac{dp}{dL} \right)_{acc} \equiv \frac{\rho u \, du}{g_c \, dL}$ is the component due to convective acceleration or kinetic energy change.

The elevation component applies for compressible and incompressible, steady-state and transient flows, in vertical and inclined systems. It is zero for horizontal flow. The friction loss component applies to any type of flow at any pipe angle and causes a pressure drop in the direction of flow. The acceleration component causes a pressure drop in the direction of velocity increase in any flow condition in which velocity change occurs. It is zero for constant-area, incompressible flow.

Equation 7.35 applies for any fluid in steady-state, one-dimensional flow for which ρ, f, and u can be defined. It is in differential equation form and would have to be integrated to yield pressure drop as a function of flow rate, pipe diameter, and fluid properties.

7.3 PIPELINE-FLOW CALCULATIONS

Engineering of long-distance transportation of natural gas by pipeline requires a knowledge of flow formulas for calculating capacity and pressure requirements. There are several equations in the petroleum industry for calculating the flow of gases in pipelines. In the early development of the natural gas transmission industry, pressures were low and the equations used for design purposes were simple and adequate. However, as pressure increased to meet higher capacity demands, equations were developed to meet the new requirements. Probably the most common pipeline flow equation is the Weymouth equation, which is generally preferred for smaller-diameter lines ($D \leqslant 15$ in. $\pm$). The Panhandle equation and the Modified Panhandle equation are usually better for larger-sized transmission lines.

7.3.1 Pipeline Equation

Consider steady-state flow of dry gas in a constant-diameter, horizontal pipeline. The mechanical energy equation (Eq. 7.35) becomes

$$\frac{dp}{dL} = \frac{f\,\rho u^2}{2g_c D} = \frac{pM}{ZRT}\frac{fu^2}{2g_c D} \tag{7.37}$$

Many pipeline equations have been developed by integrating Eq. 7.37. The difference in these equations originated from the methods used in handling the z-factor and friction factor.

Integrating Eq. 7.37 gives

$$\int dp = \frac{Mfu^2}{2Rg_c D}\int \frac{p}{zT}\,dL \tag{7.38}$$

If temperature is assumed constant at average value in pipeline, $\bar{T}$, and gas deviation factor, $\bar{z}$, is evaluated at average temperature and average pressure, $\bar{p}$, Eq.7.38 can be evaluated over a distance L between upstream pressure, p_1, and downstream pressure, p_2:

$$p_1^2 - p_2^2 = \frac{25\gamma_g q^2\,\bar{T}\,\bar{z}fL}{D^5} \tag{7.39}$$

where

p = pressure, psia
γ_g = gas gravity (air = 1)
q = gas flow rate, MMscfd (at 14.7 psia, 60°F)
$\bar{T}$ = average temperature, °R
$\bar{z}$ = gas deviation factor at $\bar{T}$ and $\bar{p}$
$\bar{p}$ = $(p_1 + p_2)/2$
L = pipe length, ft
D = pipe internal diameter, in.
f = Moody friction factor = $f(N_{\text{Re}}, e/D)$

Equation 7.39 may be written in terms of flow rate measured at arbitrary base conditions (T_b and p_b):

$$q = \frac{CT_b}{p_b}\left[\frac{(p_1^2 - p_2^2)D^5}{\gamma_g \bar{T}\bar{z} f L}\right]^{0.5} \tag{7.40}$$

C is a constant with a numerical value that depends on the units used in the pipeline equation. These are summarized in Table 7.1.

The use of Eq. 7.40 involves an iterative procedure. The gas deviation factor depends on pressure and the friction factor depends on flow rate or diameter. This problem prompted several investigators to develop pipeline flow equations that are noniterative or explicit. This has involved substitutions for the friction factor f. The specific substitution used may be diameter dependent only (Weymouth equation) or Reynolds number dependent only (Panhandle equations).

7.3.2 Weymouth Equation—Horizontal Flow

The basis for the Weymouth equation is the usual energy balance between points 1 and 2 (Fig. 7.3). In the horizontal flow case, points 1 and 2 are at the same elevation, but it is not necessary that the line connecting them be horizontal. The assumptions made in this flow situation are:

1. The kinetic-energy change is negligible and can be taken as zero.
2. The flow is steady state and isothermal.
3. Flow is horizontal.
4. Heat is not transferred to or from the gas to the surroundings.

TABLE 7.1
Value of C for Various Units

P	*T*	*D*	*L*	*q*	*C*
psia	°R	in.	mi	scfd	77.54
psia	°R	in.	ft	scfd	5634
psia	°R	in.	ft	MMscfd	5.634×10^{-3}
kPa	K	m	m	m³/day	1.149×10^{6}

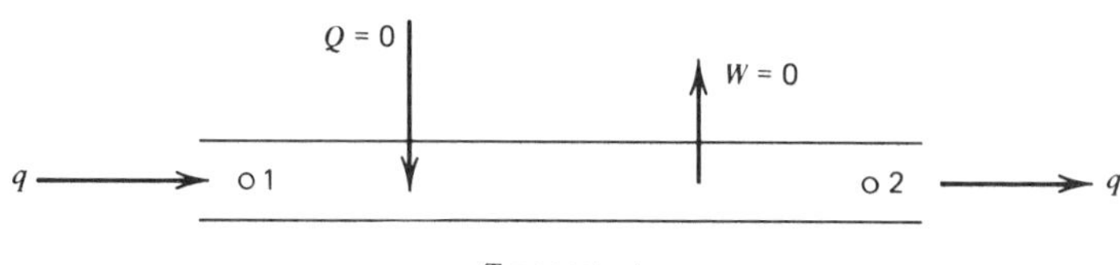

Fig. 7.3 Isothermal gas flow in horizontal lines.

5. There is no work done by the gas during flow.

With these assumptions, the energy balance is left with the expansion work and lost work terms. Thus,

$$v\,dp + lw = 0 \tag{7.41}$$

or

$$(144)\,v\,dp + \frac{fu^2}{2g_cD}\,dL = 0 \tag{7.42}$$

where

v = specific volume, cu ft/lbm
p = pressure, psia
f = Moody friction factor, dimensionless
u = velocity, ft/sec
D = pipe diameter, ft
L = length of pipe, ft
g_c = conversion factor = 32.17 lbm-ft/lbf-sec^2

The velocity u in Eq. 7.42 can be expressed in terms of the volume flow rate and the cross-sectional area of the pipe:

$$u\text{ (ft/sec)} = \frac{\text{cu ft}}{\text{sec}} \times \frac{1}{\text{sq ft}}$$

Let q_h be the volume flow rate of gas, cfh, measured at base conditions, T_b (°R) and p_b (psia). Then

$$u = \left(\frac{q_h}{3600}\right)\left(\frac{T}{T_b}\right)\left(\frac{p_b}{p}\right)\left(\frac{z}{1.00}\right)\left(\frac{4}{\pi}\right)\left(\frac{1}{D^2}\right)$$

For real gases,

$$v\text{(cu ft/lbm)} = \frac{zRT}{pM} = \frac{10.732z\,T}{29\gamma_g p}$$

Using these in Eq. 7.42

$$(144)\left(\frac{10.732zT}{29\gamma_g p}\right)dp + \frac{f}{2 \times 32.17}\left(\frac{4q_hTp_b\,z}{3600T_b\pi pD^2}\right)^2\frac{dL}{D} = 0$$

or

$$53.29\,\frac{zT}{\gamma_g p}\,dp + 1.9444 \times 10^{-9}\,\frac{f}{D^5}\left(\frac{q_hTp_b\,z}{T_bp}\right)^2 dL = 0$$

Integrating and using an average value of z,

$$1.9444 \times 10^{-9} \frac{f}{D^5}\left(\frac{q_h T p_b \bar{z}}{T_b}\right)^2 \int_0^L dL = -53.29 \frac{\bar{z}T}{\gamma_g} \int_{p_1}^{p_2} p\, dp$$

or

$$q_h^2 = \frac{53.29}{1.9444 \times 10^{-9}} \left(\frac{T_b}{p_b}\right)^2 \frac{(p_1^2 - p_2^2)\, D^5}{2f\bar{z}T\gamma_g L}$$

If L is in miles and D in inches, then

$$q_h^2 = \frac{53.29}{1.9444 \times 10^{-9}} \frac{(p_1^2 - p_2^2)(D/12)^5}{2f\bar{z}T\gamma_g(5280L)}$$

or

$$q_h = 3.23 \frac{T_b}{p_b}\left[\frac{(p_1^2 - p_2^2)\, D^5}{\gamma_g \bar{z} T f L}\right]^{0.5} \tag{7.43}$$

Equation 7.43 is the general steady-flow equation for isothermal gas flow over a pipeline. It is generally attributed to Weymouth. The terms are defined as follows:

q_h = gas flow rate, cfh at p_b and T_b
T_b = base temperature, °R
p_b = base pressure, psia
p_1 = inlet pressure, psia
p_2 = outlet pressure, psia
D = inside diameter of pipe, in.
γ_g = gas specific gravity (air = 1)
T = average flowing temperature, °R
f = Moody friction factor
L = length of pipe, miles
$\bar{z}$ = gas deviation factor at average flowing temperature and average pressure

Equation 7.43 may be written as

$$q_h = 3.23 \frac{T_b}{p_b}\left(\frac{1}{f}\right)^{0.5} \left[\frac{p_1^2 - p_2^2}{\gamma_g T L \bar{z}}\right]^{0.5} D^{2.5} \tag{7.44}$$

where

$$\left(\frac{1}{f}\right)^{0.5} = \text{transmission factor}$$

The Moody friction factor, which occurs in Eqs. 7.43 and 7.44, may itself be a function of flow rate, q_h, and pipe roughness, e. If flow conditions are in the fully turbulent region, f can be calculated from the relationship

$$f_{\text{turb}} = \frac{1}{\left(1.14 - 2\log\dfrac{e}{D}\right)^2} \tag{7.45}$$

where f depends only on the relative roughness, e/D. When flow conditions are not completely turbulent, f depends on the Reynolds number, also:

$$N_{\text{Re}} \simeq \frac{20q(\text{Mcfd})\gamma_g}{\mu D} \simeq \frac{0.48q_h\gamma_g}{\mu D} \tag{7.46}$$

Therefore, to make use of the Weymouth equation in the form of Eq. 7.43 or Eq. 7.44 requires a trial-and-error procedure to calculate q_h.

To eliminate the trial-and-error procedure, Weymouth proposed that f vary as a function of diameter in inches as follows:

$$f = \frac{0.032}{D^{1/3}} \tag{7.47}$$

With this simplification, Eq.7.43 reduces to

$$q_h = 18.062\,\frac{T_b}{pb}\left[\frac{(p_1^2 - p_2^2)\,D^{16/3}}{\gamma_g\bar{T}L\bar{z}}\right]^{0.5} \tag{7.48}$$

Equation 7.48 is the form of Weymouth equation commonly used in industry.

7.3.3 Effects of Assumptions

The use of Eqs. 7.43, 7.44, or 7.48 to calculate transmission factors for an existing transmission line or for the design of a new transmission line involves a few assumptions that were mentioned earlier. The effects of these assumptions on work on long commercial pipelines under normal operating conditions are now discussed.

Mechanical Work

No mechanical work is done on the fluid between the points at which the pressures are measured. In the study of an existing pipeline, the pressure-measuring stations should be placed so that no mechanical energy is added to the system between stations. Thus, the conditions of this assumption can be fulfilled.

Steady Flow

The flow is steady; that is, the same mass of gas passes each cross section of the pipe in a given interval of time. Steady flow in pipeline operation seldom, if ever, exists in actual practice because pulsations, liquid in the pipeline, and variations in input or outlet gas volumes cause deviations from steady-state conditions. Deviations from steady-state flow are the major cause of difficulties experienced in pipeline flow studies.

Isothermal Flow

The flow is isothermal or can be considered isothermal at an average effective temperature. The heat of compression is usually dissipated into the ground along a pipeline within a few miles downstream from the compressor station. Otherwise, the temperature of the gas is very near that of the containing pipe and, as pipelines usually are buried, the temperature of the flowing gas is not influenced appreciably by rapid changes in atmospheric temperature.

Constant Compressibility

The compressibility of the fluid can be considered constant, and an average effective gas deviation factor may be used. Equations 7.43, 7.44, and 7.48 contain an average gas deviation factor, $\bar{z}$. This comes about because the equation that was integrated was

$$q_h = \text{const.} \int_1^2 \frac{p}{z}\, dp \tag{7.49}$$

It was elected to take $\bar{z}$ outside the integral sign. Consider Fig. 7.4. When the two pressures p_1 and p_2 lie in a region where z is essentially linear with pressure, then it is accurate enough to evaluate $\bar{z}$ at the average pressure $\bar{p} = (p_1 + p_2)/2$. One can also use the arithmetic average of the z's with the same result: $\bar{z} = (z_1 + z_2)/2$. On the other hand, should p_1 and p_2 lie in the range illustrated by the double-hatched lines, the proper average would result from determining the area under the z-curve and dividing it by the difference in pressure:

$$\bar{z} = \frac{\displaystyle\int_{p_1}^{p_2} z\, dp}{(p_1 - p_2)} = \frac{z \text{ area}}{(p_1 - p_2)} \tag{7.50}$$

Also, $\bar{z}$ may be evaluated at an average pressure given by

$$p_m = \frac{2}{3}\left(\frac{p_1^3 - p_2^3}{p_1^2 - p_2^2}\right) \tag{7.51}$$

Horizontal Pipeline

The pipeline is horizontal. In actual practice, transmission lines seldom, if ever are horizontal, so that factors are needed in Eqs. 7.43, 7.44, and 7.48 to compensate for changes in elevation. With the trend to higher operating pressures in transmission lines, the need for these factors is greater than is generally realized. This correction for change in elevation is discussed in the next section.

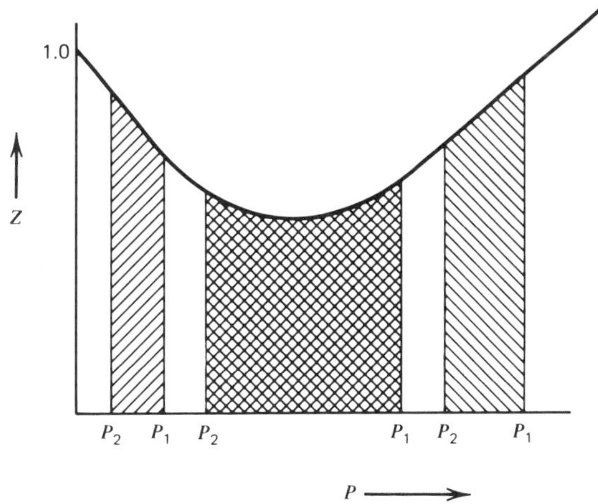

Fig. 7.4 Determination of average z-factors.

Kinetic-energy Term

The pipeline is long enough so that changes in the kinetic-energy term can be neglected. The assumption is justified for work with commercial transmission lines.

Example 7.1.

Given

$T_b = 520\ °R$
$p_b = 14.7$ psia
$p_1 = 400$ psia
$p_2 = 200$ psia
$D = 12.09$ in.
$\gamma_g = 0.60$
$T = 520\ °R$
$L = 100$ mi
$e = 0.0006$ in.

Required

Flow rate in cu ft/hr through pipeline.

Solution

A. Gas properties

For $\gamma_g = 0.6$,

$$p_{pc} = 672 \text{ psia}$$
$$T_{pc} = 358\ °R$$
$$\bar{p}_{pr} = \frac{300}{672} = 0.446$$

$$T_{pr} = \frac{520}{358} = 1.453$$

$$\bar{z} = 0.950$$

$$\mu_1 \text{ at } 60°\text{F} = 0.0103 \text{ cp}$$

$$\frac{\mu}{\mu_1} \text{ at 300 psia and } 60°\text{F} = 1.05$$

$$\bar{\mu} \text{ at 300 psia and } 60°\text{F} = (1.05)(0.0103) = 0.010{,}82 \text{ cp}$$

$$\bar{N}_{\text{Re}} \simeq \frac{0.48 q_h \gamma_g}{\mu D} = \frac{(0.48)(0.6) q_h}{(0.010{,}82)(12.09)} = 2.2016 q_h$$

$$\frac{e}{D} = \frac{0.0006}{12.09} = 0.000{,}05$$

B. Trial-and-Error Calculation of q_h

First trial:

$$q_h = 100{,}000 \text{ cfh}$$

$$N_{\text{Re}} = 2.2 \times 10^5, \qquad f = 0.0158$$

$$q_h = 3.23 \frac{T_b}{p_b}\left(\frac{1}{f}\right)^{0.5} \left[\frac{(p_1^2 - p_2^2\, D^5}{\gamma_g \bar{T} L \bar{z}}\right]^{0.5}$$

$$= 3.23 \left(\frac{520}{14.7}\right) \left(\frac{1}{f}\right)^{0.5} \left[\frac{(160{,}000 - 40{,}000)(12.09)^5}{(0.6)(0.95)(520)(100)}\right]^{0.5}$$

$$= 116{,}843{,}84 \quad \left(\frac{1}{f}\right)^{0.5} \quad = 929{,}560 \text{ cu ft/hr}$$

Second trial:

$$q_h = 500{,}000 \text{ cfh}$$

$$N_{\text{Re}} = 1.1 \times 10^6$$

$$f = 0.0125$$

$$q_h = 1{,}045{,}083 \text{ cu ft/hr}$$

Third trial:

$$q_h = 1{,}000{,}000 \text{ cfh}$$

$$N_{\text{Re}} = 2.2 \times 10^6$$

$$f = 0.012$$

$$q_h = 1{,}066{,}633 \text{ cfh}$$

C. Calculation of q_h Using the Weymouth Equation Without f

$$q_h = 18.062 \frac{T_b}{p_b}\left[\frac{(p_1^2 - p_2^2)\, D^{16/3}}{\gamma_g \bar{T} L \bar{z}}\right]^{0.5}$$

$$= 18.062\left(\frac{520}{14.7}\right)\left[\frac{(160{,}000 - 40{,}000)(12.09)^{16/3}}{(0.6)(520)(100)(0.950)}\right]^{0.5}$$

$$= 989{,}859 \text{ cfh}$$

7.3.4 Weymouth Equation—Nonhorizontal Flow

The three forms of the Weymouth equation developed above (Eqs. 7.43, 7.44, and 7.48) were on the basis that points 1 and 2 were at equal elevations above a datum. In actual practice, transmission lines often deviate considerably from the

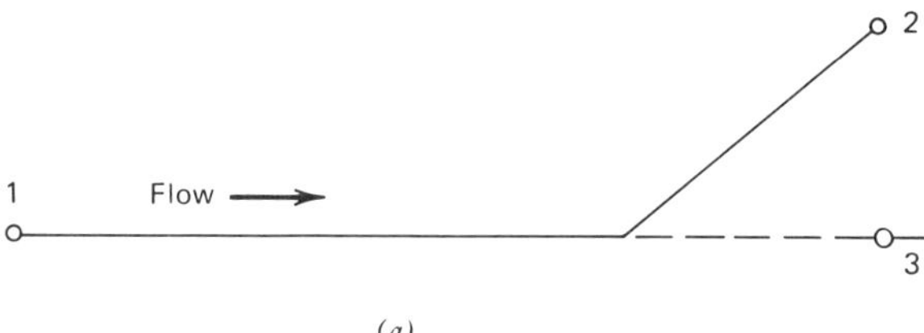

(a)

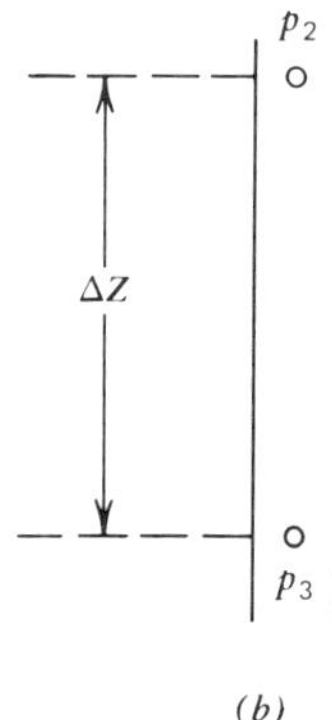

(b)

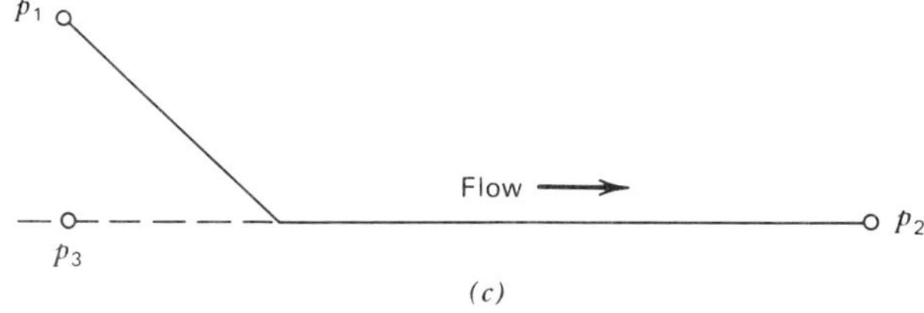

(c)

Fig. 7.5 Compensation for difference in elevation.

horizontal. Account should be taken of substantial pipeline elevation changes. Given all the previous assumptions, with the exception of horizontal flow, the energy balance reduces to

$$\int_1^2 v\,dp + \frac{g}{g_c}\Delta Z + \int_1^2 \frac{fu^2}{2g_cD}\,dL = 0 \tag{7.52}$$

Equation 7.52 is the starting point for any flow calculations that take into consideration differences in elevation.

Figure 7.5*a* shows flow in nonhorizontal pipeline; point 2 is at a higher elevation than point 1, points 1 and 3 are at equal elevations. The simplest way to handle the calculation is to add, or subtract, from one of the points the amount of pressure equivalent to the weight of the gas column caused by the elevation difference, and then proceed with the calculation as in the case of horizontal flow. Of course, the length term L must be the actual length and not the horizontal length only. However, this is usually a minor difference.

Consider the gas column illustration in Fig. 7.5*b*. The pressure gradient is given by

$$\frac{dp}{dz} = \frac{\rho_g}{144} \tag{7.53}$$

Using $\rho_g = \dfrac{pM}{zRT} = \dfrac{29\gamma_g p}{zRT}$ and an average gas deviation factor $\bar{z}$,

$$\int_{p_2}^{p_3} \frac{dp}{p} = \frac{29\gamma_g}{(144)(10.732)\bar{z}\bar{T}} \int_{Z_2}^{Z_3} dz \tag{7.54}$$

or

$$\ln\frac{p_3}{p_2} = \frac{0.01875\gamma_g}{\bar{z}T}\Delta Z \tag{7.55}$$

The natural logarithm of y may be expanded as

$$\ln y = (y-1) - 1/2(y-1)^2 + 1/3(y-1)^3 \ldots \qquad 2 > y > 0$$

For all practical purposes, $p_3/p_2 \leqslant 1.05$. Thus, all but the first term may be neglected, leaving

$$\left(\frac{p_3}{p_2} - 1\right) = \frac{0.018{,}75\gamma_g\,\Delta Z}{\bar{z}T} \tag{7.56}$$

or

$$(p_3 - p_2) = \Delta p = p_2\left(\frac{0.018{,}75\gamma_g\,\Delta Z}{\bar{z}T}\right) \tag{7.57}$$

Equation 7.57 may be written in terms of gas formation volume factor, B_g(cu ft/scf), as

$$\Delta p = \frac{0.000,529\gamma_g}{B_g} \tag{7.58}$$

Incorporating Eq. 7.55 into the Weymouth equation, Eq. 7.43, gives

$$q_h = 3.23\,\frac{T_b}{p_b}\left[\frac{(p_1^2 - e^s p_2^2)\,D^5}{\gamma_g \bar{T} f L \bar{z}}\right]^{0.5} \tag{7.59}$$

where

e = base of natural logarithm = 2.718
s = $0.0375\gamma_g\,\Delta Z/T\bar{z}$
ΔZ = outlet elevation minus inlet elevation (note that ΔZ is positive when outlet is higher than inlet).

Consider the configuration in Fig. 7.5*c*. Here the correction is made on p_1 and Eq. 7.15 becomes

$$q_h = 3.23\,\frac{T_b}{p_b}\left[\frac{(p_1^2/e^s - p_2^2)D^5}{\gamma_g \bar{T} f L \bar{z}}\right]^{0.5} \tag{7.60}$$

Equations 7.59 and 7.60 are adequate for most purposes. A general and more rigorous form of Weymouth equation with compensation for elevation is

$$q_h = 3.23\,\frac{T_b}{p_b}\left(\frac{1}{f}\right)^{0.5}\left[\frac{(p_1^2 - e^s p_2^2)\,D^5}{\gamma_g T \bar{z} L_e}\right]^{0.5} \tag{7.61}$$

where L_e is the effective length of the pipeline.

The effective length of the pipeline is based on the profile of the line between pressure-measuring stations. For a uniform slope,

$$L_e = \frac{(e^s - 1)}{s}\,L \tag{7.62}$$

For a nonuniform slope (where elevation change cannot be simplified to a single section of constant gradient), an approach in steps to any number of sections, n, will yield

$$L_e = \frac{(e^{s_1} - 1)}{s_1}L_1 + \frac{e^{s_1}(e^{s_2} - 1)}{s_2}L_2 + \frac{e^{s_1+s_2}(e^{s_3} - 1)}{s_3}L_3 + \cdots + \frac{e^{\Sigma s_{n-1}}(e^{s_n} - 1)L_n}{s_n} \tag{7.63}$$

where

$$s_1 = 0.0375\gamma_g \Delta Z_1/T\bar{z}$$
$$s_2 = 0.0375\gamma_g \Delta Z_2/T\bar{z}$$
$$s_n = 0.0375\gamma_g \Delta Z_n/T\bar{z}$$

Note: $\Sigma s_n = s_1 + s_2 + s_3 + \cdots + s_n$. The numerical subscripts refer to the individual sections of the overall line, which is operating under pressure differential $(p_1 - p_2)$. For example, if the line is divided into four sections, n would equal 4 and subscript 2 would pertain to the properties of the second section.

7.3.5 Panhandle A Equation—Horizontal Flow

The Panhandle A pipeline flow equation assumes that f varies as follows:

$$f = \frac{0.085}{N_{\mathrm{Re}}^{0.147}} \tag{7.64}$$

The pipeline flow equation is thus

$$q = 435.87 \left(\frac{T_b}{p_b}\right)^{1.07881} \left(\frac{p_1^2 - p_2^2}{\bar{T}L\bar{z}}\right)^{0.5394} \left(\frac{1}{\gamma_g}\right)^{0.4604} D^{2.6182} \tag{7.65}$$

where

q is the gas flow rate, cfd measured at T_b and p_b.
Other terms are as in Weymouth equation.

7.3.6 Modified Panhandle (Panhandle B) Equation—Horizontal Flow

This is probably the most widely used equation for long lines (transmission and delivery). The modified Panhandle equation assumes that f varies as

$$f = \frac{0.015}{N_{\mathrm{Re}}^{0.0392}} \tag{7.66}$$

and results in

$$q = 737 \left(\frac{T_b}{p_b}\right)^{1.02} \left[\frac{p_1^2 - p_2^2}{\bar{T}L\bar{z}\gamma_g{}^{0.961}}\right]^{0.510} D^{2.530} \tag{7.67}$$

where the units are the same as in Eq. 7.65.

7.3.7 Clinedinst Equation—Horizontal Flow

The pipeline flow equation of Clinedinst rigorously considers the deviation of natural gas from ideal behavior. This equation is a rigorous integration of Eq.

7.42. The only assumptions are those made in arriving at Eq. 7.42. The equation is

$$q = 3973.0 \frac{z_b T_b p_{pc}}{p_b}\left[\frac{D^5}{\gamma_g \bar{T} L f}\left(\int_0^{p_{r,1}} \frac{p_r}{z}\,dp_r - \int_0^{p_{r,2}} \frac{p_r}{z}\,dp_r\right)\right]^{0.5} \tag{7.68}$$

where

q = volumetric flow rate, Mcfd
p_{pc} = pseudocritical pressure, psia
D = pipe internal diameter, in.
L = pipe length, ft
p_r = pseudoreduced pressure
$\bar{T}$ = average flowing temperature, °R
γ_g = gas gravity
z_b = gas deviation factor at T_b and p_b, normally accepted as 1.0

Values of the integral functions $\int_0^{p_r} \frac{p_r}{z}\,\mathrm{dp_r}$ are tabulated in Table A-6 of Katz et al, *Handbook of Natural Gas Engineering*.

7.3.8 Pipeline Efficiency

All the pipeline flow equations developed above are for 100% efficient conditions. In actual pipelines, water, condensates, and sometimes crude oil accumulate in low spots in the line. There are often scales and "junk" (dead rabbits, etc.) left in the line. The net result is that the flow rates calculated for the 100% efficient cases are often modified by multiplying them by an efficiency factor E. The efficiency factor expresses the actual flow rate as a fraction of the theoretical flow rate. An efficiency factor ranging from 0.85 to 0.95 would represent a "clean" line. A few values of efficiency factors are given below.

Type of Line	Liquid Content of Gas (gal/MMcf)	*E*
Dry-gas field	0.1	0.92
Casing-head gas	7.2	0.77
Gas and condensate	800	0.60

7.3.9 Transmission Factors

The transmission factor is defined as $(1/f)^{0.5}$. Of the many factors in Eq. 7.44, the transmission factor has long been the most difficult to evaluate. Thus, the literature contains many different empirical transmission factors that have been used to meet the needs of pipeline engineers. Table 7.2 presents some transmission factors that are the most significant and have either best stood the test of usage or have strong foundations in basic flow theories.

TABLE 7.2
Transmission Factors for Pipeline Flow Equations
(for Use in Eq. 7.44)

Flow Equation	Transmission Factor, $(1/f)^{0.5}$	Remarks
Smooth pipe (laminar)	$2\log_{10}(f^{0.5}N_{Re}) + 0.3$	Seldom applicable to large-diameter natural gas transmission lines
Rough pipe (fully turbulent)	$2\log_{10}(3.7D/e)$	Characterizes most natural gas transmission operating conditions. Table 7.3 gives a number of solutions.
Weymouth	$1.10^{a} \times 5.6D^{0.167}$	Reasonably good approximation of preceding (rough pipe) formula for $D = 10$ in. and $e = 0.002$ in.
Panhandle	$0.92^{a} \times 3.44N_{Re}^{0.073}$	Large-diameter transmission piping where Re varies from 5×10^{6} to 20×10^{6}.
Modified panhandle	$0.90^{a} \times 8.25N_{Re}^{0.0196}$	

[a]Average steel pipeline efficiency; aluminum varies from 0.92 to 0.96.

TABLE 7.3
Selected Rough Pipe Transmission Factors

$(1/f)^{0.5} = 1\log_{10}(3.7D/e)$
where $e = 0.0007$ in.[a]

D, pipe ID, in.	$(1/f)^{0.5}$
10.00	9.45
13.50	9.70
15.44	9.82
19.38	10.02
23.25	10.18
25.31	10.25
29.25	10.38

[a]Average effective roughness for clean steel pipe in transmission service.

7.3.10 Summary of Pipeline Equations

A general pipeline flow equation that is noniterative may be written as

$$q = a_1 E \left(\frac{T_b}{p_b}\right)^{a_2} \left(\frac{p_1^2 - p_2^2}{\bar{T}\bar{z}L}\right)^{a_3} \left(\frac{1}{\gamma_g}\right)^{a_4} D^{a_5} \tag{7.69}$$

where E is the efficiency factor. The units to be used in Eq. 7.69 are

q = cfd measured at T_b and p_b

T = °R

p = psia

L = miles

D = inches

The values of the constants a are given below for the different pipeline flow equations.

Equation	a_1	a_2	a_3	a_4	a_5
Weymouth	433.5	1.0	0.5	0.5	2.667
Panhandle A	435.87	1.0788	0.5394	0.4604	2.618
Panhandle B	737.0	1.02	0.510	0.490	2.530

7.4 GAS FLOW IN SERIES, PARALLEL, AND LOOPED PIPELINES

It is often desirable to increase the throughput of a pipeline while maintaining the same pressure drop and level. This need may occur when new gas wells are developed in an area serviced by an existing pipeline. A similar type of problem may arise when an existing pipeline must be "pressure derated" because of age (corrosion, etc.) but it is desired to maintain the same throughput.

A common economical solution to the above problems is to place one or more lines in parallel, either partially or throughout the whole length, or to replace a portion of the line with a larger one. This requires calculations involving flow in series, parallel, and series-parallel (looped) lines.

The philosophy involved in deriving the special relationships used in the solution of complex transmission systems is to express the various lengths and diameters of the pipe in the system as equivalent lengths of a common diameter or equivalent diameters of a common length—equivalent meaning that both lines will have the same capacity with the same total pressure drop.

All the examples that follow will be based on the Weymouth equation. At the end of the discussion, a summary of the relationships based on the Panhandle and Clinedinst equations will be given.

7.4.1 Series Pipelines

Consider an L-mile long, D_A-in. internal diameter pipeline operating with a total pressure drop of $p_1 - p_2$ psi. This pipeline is altered by replacing the first L_B miles with a D_B-in. internal diameter line (Figs. 7.6*a* and 7.6*b*). If the total pressure drop in the new system is $p_1 - p_2$, what is the capacity of the new series pipeline? Also, what is the pressure p_3 at the junction of the D_B-in. line and the D_A-in. line?

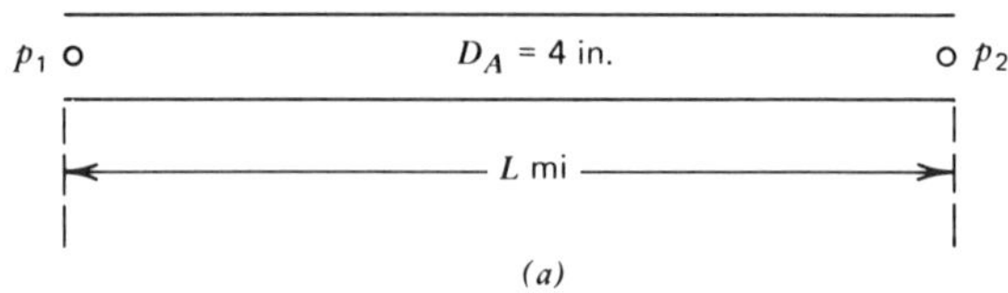

(a)

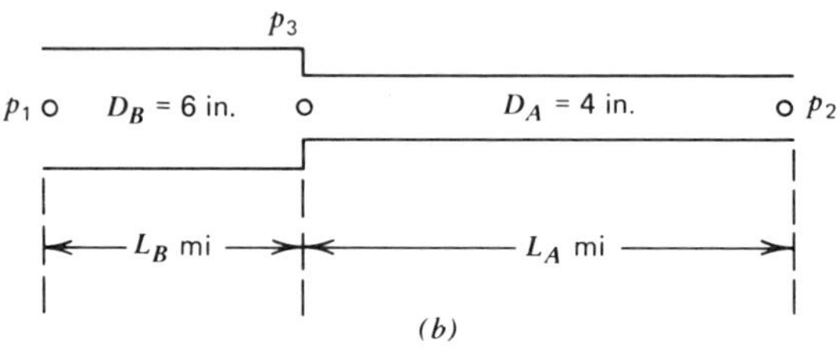

(b)

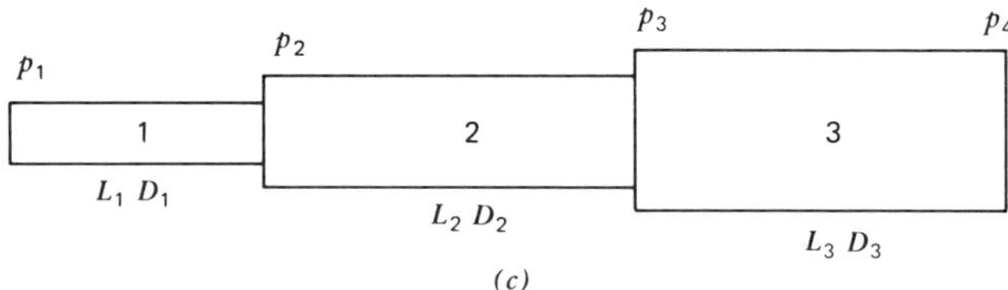

(c)

Fig. 7.6 Pipes in series.

Using Eq. 7.48,

$$q_h = 18.062 \frac{T_b}{p_b}\left[\frac{(p_1^2 - p_2^2)\, D^{16/3}}{\gamma_g T L \bar{z}}\right]^{0.5} \tag{7.48}$$

Since $p_1 - p_2$ is constant, Eq. 7.48 can be written as

$$q_h = K_1 \left(\frac{D^{16/3}}{L}\right)^{0.5} \tag{7.70}$$

or

$$L = \frac{KD^{16/3}}{q_h^{\,2}} \tag{7.71}$$

Thus, for transporting a given quantity of gas at a given pressure drop, length is proportional to diameter raised to the power 16/3. Therefore, the equivalent length of a D_A-in. line, L'_A, that would have the same pressure drop as the L_B miles of D_B-in. line is

$$\frac{L'_A}{L_B} = \left(\frac{D_A}{D_B}\right)^{16/3}$$

or

$$L'_A = L_B\left(\frac{D_A}{D_B}\right)^{16/3} \tag{7.72}$$

Therefore, the series line shown has a total equivalent length of

$$L_{Aeq} = L_A + L'_A = L_A + L_B\left(\frac{D_A}{D_B}\right)^{16/3} \tag{7.73}$$

The volume flow rate q_h of gas that will flow for given values of p_1, p_2, γ_g, T, and $\bar{z}$ can be calculated using Eq. 7.48 with $D_A L_{Aeq}$. However, from Eq. 7.70, the flow rate is proportional to $(1/L)^{0.5}$. Thus, percent change in flow rate is

$$\Delta q_h = \left[\frac{\left(\frac{1}{L_{Aeq}}\right)^{0.5} - \left(\frac{1}{L}\right)^{0.5}}{\left(\frac{1}{L}\right)^{0.5}}\right] \tag{7.74}$$

Example 7.2

$$\begin{aligned} L &= 10 \text{ mi} \\ L_A &= 7 \text{ mi} \\ L_B &= 3 \text{ mi} \\ D_A &= 4 \text{ in.} \\ D_B &= 6 \text{ in.} \end{aligned}$$

Then,

$$L'_A = 3\left(\frac{4}{6}\right)^{16/3} = 0.345 \text{ mi}$$

$$L_{Aeq} = 7 + 0.345 = 7.345 \text{ mi}$$

$$\Delta q_h\% = \left[\frac{\left(\frac{1}{7.345}\right)^{0.5} - \left(\frac{1}{10}\right)^{0.5}}{\left(\frac{1}{10}\right)^{0.5}}\right] \times 100 = 16.7\%$$

The equivalent lengths and diameters can then be expressed as

$$L_1' = L_2 \left(\frac{D_1}{D_2}\right)^{16/3} \tag{7.75}$$

or

$$D_1' = D_2 \left(\frac{L_1}{L_2}\right)^{3/16} \tag{7.76}$$

If the Weymouth equation with friction factor f is used where the flow of gas, pressure differential, temperature, gas gravity, and deviation factor are the same for two different pipelines, the relationship of equivalent lengths and diameters is expressed as

$$L_1' = L_2 \left(\frac{f_2}{f_1}\right) \left(\frac{D_1}{D_2}\right)^5 \tag{7.77}$$

or

$$D_1' = D_2 \left(\frac{f_1}{f_2}\right)^{1/5} \left(\frac{L_1}{L_2}\right)^{1/5} \tag{7.78}$$

where L_1' is the length of a pipe of diameter D_1 and friction factor f_1 equivalent to the length L_2 of a pipe of diameter D_2 and friction factor f_2, or where D_1' is the diameter of a pipe of length L_1 and friction factor f_1 equivalent to the diameter D_2 of a pipe of length L_2 and factor f_2.

The calculation of the pressure at the junction point of the $D_A - D_B$ line is accomplished by noting that the flow rates in the two sections are equal and calculating in terms of the equivalent D_A-in. system. Using lengths of L_A' miles for the B-section and L_A miles for the A-section, Eq. 7.48 is

$$\frac{q_A p_b}{18.062 T_b} \left(\frac{\gamma_g T}{D_A{}^{16/3}}\right)^{0.5} = \text{constant} = \left[\frac{p_1^2 - p_3^2}{L_A' \bar{z}_B}\right]^{0.5} = \left[\frac{p_3^2 - p_2^2}{L_A \bar{z}_A}\right]^{0.5}$$

or

$$(p_1^2 - p_3^2)\, L_A \bar{z}_A = (p_3^2 - p_2^2)\, L_A' \bar{z}_B \tag{7.79}$$

L_A, L_A', p_1, and p_2 are known. Equation 7.79 contains three unknowns: p_3, $\bar{z}_A$, and $\bar{z}_B$. It is best solved by trial and error using p_3 as the variable. To start, assume

$$\bar{z}_A = \bar{z}_B \simeq \frac{z_{p_1} + z_{p_2}}{2}$$

Equation 7.75 may be extended to three or more pipes in series (Fig. 7.6*c*). The flow rates are same in all sections:

$$q_t = q_1 = q_2 = q_3$$

Total pressure drop equals sum of pressure drops:

$$\Delta p_t = \Delta p_1 + \Delta p_2 + \Delta p_3$$

or

$$(p_1^2 - p_4^2) = \frac{\gamma_g \bar{z} T q_t^2 p_b{}^2 L_e}{(18.062)^2 T_b{}^2 D^{16/3}}$$
$$= \frac{\gamma_g \bar{z} T p_b{}^2}{(18.062)^2 T_b{}^2} \left[\frac{q_1{}^2 L_1}{D_1{}^{16/3}} + \frac{q_2{}^2 L_2}{D_2{}^{16/3}} + \frac{q_3{}^2 L_3}{D_3{}^{16/3}} \right]$$

Therefore,

$$\frac{L_e}{D^{16/3}} = \frac{L_1}{D_1{}^{16/3}} + \frac{L_2}{D_2{}^{16/3}} + \frac{L_3}{D_3{}^{16/3}} \tag{7.80}$$

Using the Weymouth equation, which contains f,

$$\frac{fL_e}{D^5} = \frac{f_1 L_1}{D_1{}^5} + \frac{f_2 L_2}{D_2{}^5} + \frac{f_3 L_3}{D_3{}^5} + \cdots + \frac{f_n L_n}{D_n{}^5} \tag{7.81}$$

7.4.2 Parallel Pipelines

Consider once more the L-mile, D_A-in. internal diameter pipeline. Suppose the full length is paralleled with a new D_B-in. internal diameter line (Fig. 7.7*a*). What

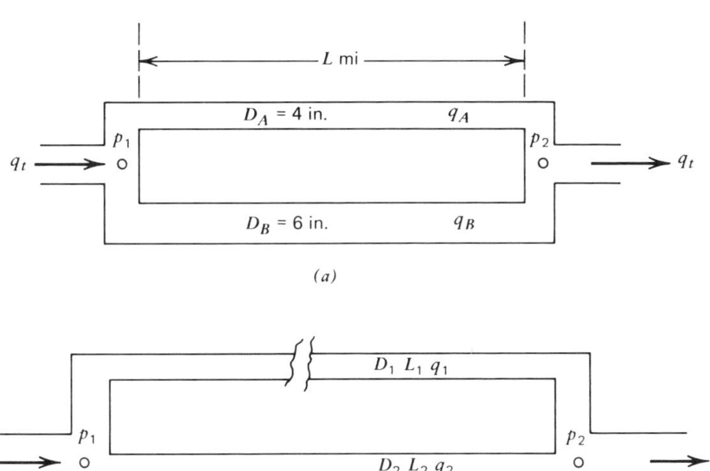

Fig. 7.7 Pipes in parallel.

would be the resulting increase in capacity? The old flow rate using only D_A-in. line is q_A. The new flow rate with both lines is $q_t = q_A + q_B$. The length, L, is constant. Using the Weymouth equation (without f),

$$q_h = \text{const}\ (D^{16/3})^{0.5} = \text{const}\ (D^{8/3}) \tag{7.82}$$

The ratio of new to old flow rates is

$$\frac{q_t}{q_A} = \frac{q_A + q_B}{q_A} = \left(1 + \frac{q_B}{q_A}\right) = \left[1 + \left(\frac{D_B}{D_A}\right)^{8/3}\right] \tag{7.83}$$

Percent increase in capacity is

$$\%\ \text{increase in}\ q_h = \frac{q_B}{q_A} \times 100 = 100\left(\frac{D_B}{D_A}\right)^{8/3} \tag{7.84}$$

Example 7.3

$$L = 10\ \text{mi}$$
$$D_A = 4\ \text{in.}$$
$$D_B = 6\ \text{in.}$$
$$\frac{q_t}{q_A} = \left[1 + \left(\frac{6}{4}\right)^{8/3}\right] = 3.95$$
$$\%\ \text{increase in}\ q_h = 100\left(\frac{6}{4}\right)^{8/3} = 295\%$$

Equations 7.82 to 7.84 are for equal lengths of parallel pipelines. Where the lengths of the two parallel lines are not equal, use

$$q_h = \text{const}\left(\frac{D^{16/3}}{L}\right)^{0.5} \tag{7.85}$$

The ratio of the flow rates becomes

$$\frac{q_t}{q_A} = \frac{q_A + q_B}{q_A} = \left[1 + \left(\frac{D_B}{D_A}\right)^{8/3}\left(\frac{L_A}{L_B}\right)^{1/2}\right] \tag{7.86}$$

$$\%\ \text{increase in}\ q_h = 100\left[\left(\frac{D_B}{D_A}\right)^{8/3}\left(\frac{L_A}{L_B}\right)^{1/2}\right] \tag{7.87}$$

The above expressions for two parallel lines may be extended to three or more lines in parallel (Fig. 7.7*b*).

$$q_t = q_1 + q_2 + q_3$$
$$\Delta p_t = \Delta p_1 = \Delta p_2 = \Delta p_3$$

$$q_t = \frac{18.062}{\sqrt{\gamma_g T \bar{z}}} \frac{T_b}{p_b} \left[\frac{(p_1^2 - p_2^2) D^{16/3}}{L_e} \right]^{0.5}$$
$$= \frac{18.062}{\sqrt{\gamma_g T \bar{z}}} \frac{T_b}{p_b} (p_1^2 - p_2^2)^{0.5} \left[\left(\frac{D_1^{16/3}}{L_1} \right)^{0.5} + \left(\frac{D_2^{16/3}}{L_2} \right)^{0.5} + \left(\frac{D_3^{16/3}}{L_3} \right)^{0.5} \right]$$

Therefore,

$$\left(\frac{D^{16/3}}{L_e} \right)^{0.5} = \left(\frac{D_1^{16/3}}{L_1} \right)^{0.5} + \left(\frac{D_2^{16/3}}{L_2} \right)^{0.5} + \left(\frac{D_3^{16/3}}{L_3} \right)^{0.5} \tag{7.88}$$

If the Weymouth equation, which contains the friction factor, f, is used, Eq. 7.88 becomes

$$\left(\frac{D^5}{fL_e} \right)^{0.5} = \left(\frac{D_1^{5}}{f_1 L_1} \right)^{0.5} + \left(\frac{D_2^{5}}{f_2 L_2} \right)^{0.5} + \left(\frac{D_3^{5}}{f_3 L_3} \right)^{0.5} + \cdots + \left(\frac{D_n^{5}}{f_n L_n} \right)^{0.5} \tag{7.89}$$

7.4.3 Looped Pipelines

Most often in the design of complex transmission systems, only a part of the line is parallel. This is known as looping. In the looped pipeline illustrated (Fig. 7.8*a*), the original line consisted of segments A and C, of the same diameter. In the existing system, a looping segment (B) has been added to increase the capacity of the pipeline system.

The looped pipeline system (Fig. 7.8*a*) may be represented by a series system (Fig. 7.8*b*). To obtain the total flow rate, the looped section and unlooped section are solved as a series flow situation.

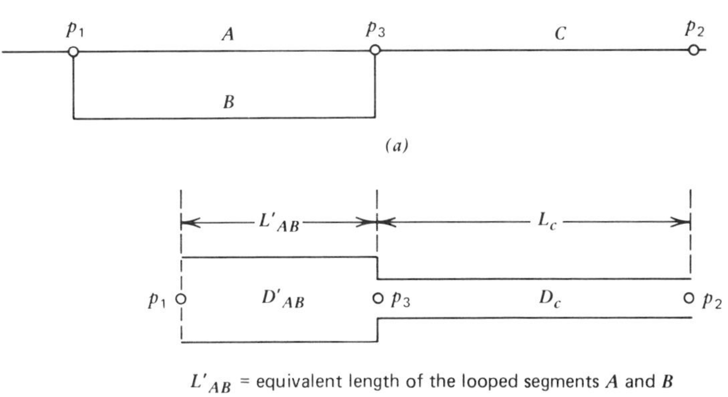

Fig. 7.8 Looped pipelines.

Consider the parallel flow situation in the looped section, using the Weymouth equation (without f):

$$q_h = \text{const}\left(\frac{D^{16/3}}{L}\right)^{0.5} \tag{7.85}$$

From Eq. 7.88,

$$\frac{(D'_{AB})^{8/3}}{(L'_{AB})^{1/2}} = \frac{D_A^{8/3}}{L_A^{1/2}} + \frac{D_B^{8/3}}{L_B^{1/2}} \tag{7.90}$$

Solving Eq. 7.90 for the equivalent length of the looped segment yields

$$L'_{AB} = \left[\frac{1}{\left(\frac{D_A}{D'_{AB}}\right)^{8/3}\left(\frac{1}{L_A}\right)^{1/2} + \left(\frac{D_B}{D'_{AB}}\right)^{8/3}\left(\frac{1}{L_B}\right)^{1/2}}\right]^2 \tag{7.91}$$

If the equivalent diameter of the looped segment D'_{AB} is selected to be the same as the diameter of the unlooped segment,

$$D'_{AB} = D_C = D_A$$

Equation 7.91 reduces to

$$L'_{AB} = \left[\frac{1}{\left(\frac{1}{L_A}\right)^{1/2} + \left(\frac{D_B}{D_C}\right)^{8/3}\left(\frac{1}{L_B}\right)^{1/2}}\right]^2 \tag{7.92}$$

Also, if the lengths of the paralleled line are equal, $L_A = L_B$, then

$$L'_{AB} = \left[\frac{1}{\left(\frac{1}{L_A}\right)^{1/2}\left\{1 + \left(\frac{D_B}{D_C}\right)^{8/3}\right\}}\right]^2 \tag{7.93}$$

The ratio of flow rate after looping to original flow rate is given by

$$\frac{q}{q_o} = \left(\frac{L_o}{L'}\right)^{0.5} \tag{7.94}$$

$$\%\text{ increase in } q_h = 100\left(\frac{q - q_0}{q_0}\right) = 100\left[\left(\frac{L_o}{L'}\right)^{0.5} - 1\right] \tag{7.95}$$

where

q_0 = original flow rate before looping
q = flow rate after looping
L_o = original length of pipeline
L' = equivalent of pipeline after looping = $L'_{AB} + L_c$

Example 7.4

Section	Diameter *D*, in.	Length *L*, mi
A	4 ID	3
B	6 ID	3
C	4 ID	7

$$L'_{AB} = \left[\frac{1}{\left(\frac{1}{3}\right)^{1/2} \left\{1 + \left(\frac{6}{4}\right)^{8/3}\right\}} \right]^2 = 0.192 \text{ mi}$$

$$L' = L'_{AB} + L_C = 7.192 \text{ mi}$$

$$\frac{q}{q_0} = \left(\frac{10}{7.192}\right)^{0.5} = 1.18$$

$$\% \text{ increase in } q_h = 100 \left[\left(\frac{10}{7.192}\right)^{0.5} - 1 \right] = 18\%$$

The above equations and calculational procedure can be generalized to yield several useful equations and a looping chart. These equations help determine pipe size for completely or partially paralleling an original line in order to increase the system capacity by a fixed amount. The equations are derived on the basis that, after a line has been paralleled, the temperature, gas gravity, mean z-factor, and pressures at the inlet and outlet of the original lines are as they were before paralleling.

The following expression gives the fraction of the total length of original line that must be paralleled (looped) in order to increase the flow rate by a fixed amount. It assumes that the lengths of the lines in the looped section are equal:

$$Y = \frac{1 - (q_0/q)^2}{1 - 1/(1 + W)^2} \tag{7.96}$$

where

Y = fraction of original line paralleled (looped), starting downstream
q_0 = original flow rate before looping
q = flow rate after looping
$W = (D_B/D_o)^{8/3}$ (Weymouth without f)

or

$W = (D_B/D_o)^{2.5} (f_o/f_B)^{0.5}$ (Weymouth with f)
D_o = diameter of original line
D_B = diameter of added parallel line
f_o = friction factor of original line
f_B = friction factor of added parallel line

Equation 7.96 can be rearranged in several different ways, for example,

$$q = \frac{q_0}{\{Y[1/(1 + W)^2 - 1] + 1\}^{0.5}} \tag{7.97}$$

or

$$W = \left[\frac{Y}{(q_o/q)^2 - 1 + Y}\right]^{0.5} - 1 \tag{7.98}$$

When $Y = 1$ and the entire length of the original line has been paralleled,

$$W = \frac{q}{q_0} - 1 \tag{7.99}$$

or

$$\frac{q}{q_0} = 1 + W = \left[1 + \left(\frac{D_B}{D_o}\right)^{2.5}\left(\frac{f_o}{f_B}\right)^{0.5}\right] \tag{7.100}$$

If the diameter and friction factor of the parallel line are the same as those of the original line, then $W = 1$, and

$$Y = \frac{4}{3}[1 - (q_0/q)^2] \tag{7.101}$$

and

$$q/q_0 = \frac{2}{(4 - 3Y)^{0.5}} \tag{7.102}$$

The chart of Fig. 7.9 yields rapid solutions for these equations.

7.4.4 Extensions to Other Pipeline Equations

The equivalent lengths may be calculated by the following expressions:

$$L_e = \left(\frac{D}{D_1}\right)^A L_1 \tag{7.103}$$

where

L_e = equivalent of selected pipe size (e.g., Table 7.4)
D = ID of selected pipe size
L_1 = length of actual pipe of size D_1
D_1 = ID of actual pipe section
A = exponent which depends on pipeline equation used (Table 7.5).

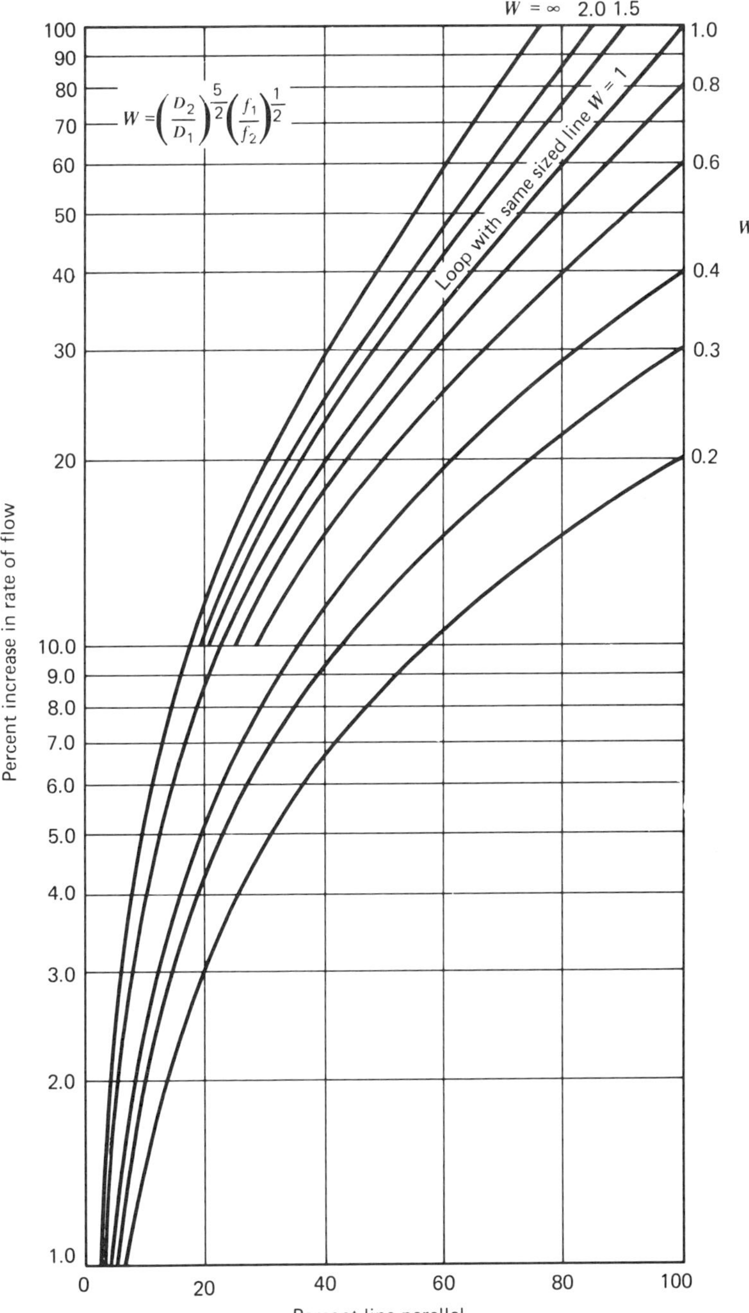

Fig. 7.9 Design of parallel lines. (After Katz et al.)

TABLE 7.4
Miles of Pipe of Various Diameters Having Delivery Capacity Equivalent to One Mile of a Diameter Given in Extreme Left Column

Internal Diameter, in.	2.067	3.068	4.026	6.065	8.071	10.025	12.125
2.067	1.000	8.218	35.01	311.40	1429.4	—	—
3.068	0.1217	1.000	4.250	37.89	173.9	622.2	—
4.026	0.1286	0.2347	1.000	8.895	40.83	146.1	357.8
6.065	0.0032	0.0264	0.1124	1.000	4.590	16.42	40.23
8.071			0.0245	0.2178	1.000	3.577	8.763
10.250				0.0609	0.2795	1.000	2.450
12.125				0.0249	0.1141	0.4082	1.000
13.375				0.0147	0.0676	0.2419	0.5925
15.375					0.0322	0.1151	0.2818
17.375						0.0599	0.1468

Example: Capacity of 1 mile of 8.071-in. pipe is equivalent to 0.0245 miles of 4.026-in or 3.577 miles of 10.025-in. pipe for same pressure conditions.
Source: Courtesy AGA.

TABLE 7.5
Exponents for Equations 7.103, 7.104, and 7.105

	A	*B*	*C*
Weymouth equation	5.333	0.50	2.667
Panhandle equation	4.854	0.5394	2.618
Modified Panhandle equation	4.961	0.51	2.53

For series flow, the equivalent length is the sum of the individual equivalent lengths. When the lines are connected in parallel, their combined equivalent length is derived from

$$L_{eq} = \left[\frac{1}{(1/L_{e1})^B + (1/L_{e2})^B + (1/L_{e3})^B + \cdots}\right]^{1/B} \tag{7.104}$$

If each of the parallel lines is of a single pipe size and of equal length, Eq. 7.104 may be expressed as

$$L_{eq} = \left[\frac{1}{D_1{}^c + D_2{}^c + D_3{}^c + \cdots}\right]^{1/B} D^A L \tag{7.105}$$

Terms are defined under Eq. 7.103 and exponents are given in Table 7.5.

Other relationships equivalent to those derived using Weymouth equations are summarized in Table 7.6 for Panhandle and Clinedinst equations.

TABLE 7.6
Comparison of Complex Flow Formulas

Quantity	Weymouth	Clinedinst	Panhandle
		Lines in Series	
Equivalent diameter	$D_1 = D_2 \left(\frac{L_1}{L_2}\right)^{3/16}$	$D_1 = D_2 \left(\frac{L_1}{L_2}\right)^{1/5}$	$D_1 = D_2 \left(\frac{L_1}{L_2}\right)^{0.2060}$
Equivalent length	$L_1 = L_2 \left(\frac{D_1}{D_2}\right)^{16/3}$	$L_1 = L_2 \left(\frac{D_1}{D_3}\right)^{5}$	$L_1 = L_2 \left(\frac{D_1}{D_2}\right)^{4.854}$
		Lines in Parallel	
Equivalent diameter	$\frac{D_0^{8/3}}{L_0^{1/2}} = \frac{D_1^{8/3}}{L_1^{1/2}} + \frac{D_2^{8/3}}{L_2^{1/2}}$	$\left(\frac{D_0^3}{L_0}\right)^{1/2} = \left(\frac{D_1^3}{L_1}\right)^{1/2} + \left(\frac{D_2^3}{L_2}\right)^{1/2}$	$\frac{D_0^{2.618}}{L_0^{0.5394}} = \frac{D_1^{2.618}}{L_1^{0.5394}} + \frac{D_2^{2.618}}{L_2^{0.5394}}$
Looping requirements[a]	$Y = \frac{1 - \left(\frac{q}{q_1}\right)^2}{1 - \left(\frac{D^{8/3}}{D^{8/3} + D_1^{8/3}}\right)^2}$	$Y = \frac{1 - \left(\frac{q}{q_1}\right)^2}{1 - \left(\frac{D^{5/2}}{D^{5/2} + D_1^{5/2}}\right)^2}$	
Entire line looped[a]	$\frac{q_1}{q} = \left[1 + \left(\frac{D_1}{D}\right)^{8/3}\right]$	$\frac{q_1}{q} = \left[1 + \left(\frac{D_1}{D}\right)^{5/2}\right]$	
Diameter of original and parallel lines the same[a]	$Y = \frac{4}{3}\left[1 - \left(\frac{q}{q_1}\right)^2\right]$	$Y = \frac{4}{3}\left[1 - \left(\frac{q}{q_1}\right)^2\right]$	

Source: After Campbell.

[a]Assumes that the length of all lines in loop section are the same.

7.5 GAS-LIQUID FLOW IN PIPELINES

The use of a single pipeline for simultaneous flow of natural gas and oil is increasing. Such systems frequently offer substantial economic advantages for offshore applications as well as for many onland installations. In some cases, simultaneous flow of gas and oil occurs naturally in gathering systems of oil with associated solution gas or in gas-condensate flow to central processing equipment.

Multiphase flow conditions cover a broad spectrum of operating situations. On the one hand is the condition of gas flow in a pipeline accompanied by small amounts of liquid due either to carryover from separators or due to condensation in the pipeline. At the other extreme is the flow of crude or condensate, which enters the pipe as one phase but, because of the drop in pressure along the line, releases small amounts of gas, which then travel with the liquid in multiphase flow. These extremes of gas-liquid ratios and all conditions in between are situations of two-phase flow. Common to all is the fact that the pressure drop is usually substantially greater than in single-phase flow and, in some cases, the flow can be quite unsteady.

This condition of two-phase flow is exceedingly complex. Early attempts to use design methods equivalent to those used for single-phase flow frequently resulted in a line that was either inadequate or highly overdesigned. In two-phase flow, an overdesigned line draws a penalty not only in excessive cost but in very unstable operation with slugging of liquid and fluctuating pressures.

Three methods for handling pipeline flowing both liquid and gas will be discussed. Dukler Case II is considered one of the most accurate correlations for horizontal flow. Flanigan correlation accounts for the added pressure drop caused by lifting the liquid up hills in a hilly terrain pipeline. It ignores any pressure recovery in the downhill section and the angle of the hill is not taken into consideration. Beggs and Brill correlation takes angle of inclination into account. It can be used for downward two-phase flow such as might occur in offshore gathering lines.

7.5.1 Dukler Case II Correlation

This correlation published by Dukler in 1969 is the most widely used method for a wide range of conditions in horizontal flow. The AGA-API Design Manual by Baker et al. gives full details about this procedure and example calculations. Further examples may be found in Brown, *The Technology of Artificial Lift Methods, Volume 1*.

The procedure for calculating pressure loss in a horizontal flow line from known upstream pressure by method of Dukler, Case II is outlined below.

1. Assume a downstream pressure and calculate an arithmetic average pressure, $\bar{p} = (p_1 + p_2)/2$.
2. Obtain average values of solution gas-oil ratio, $\bar{R}_s$, oil formation volume factor, $\bar{B}_o$, and gas deviation factor, $\bar{z}$.

3. Calculate the volumetric flow rates of liquid and gas in the pipeline in cfs:

$$q_{LPL} = \frac{q_L \bar{B}_o(5.615)}{86{,}400} \tag{7.106}$$

$$q_{gPL} = \frac{q_L(GOR - \bar{R}_s)}{86{,}400} \frac{p_b}{\bar{p}} \frac{\bar{T}}{T_b} \bar{z} \tag{7.107}$$

4. Calculate λ, the volumetric liquid fraction input to the pipeline. This is the no-slip holdup.

$$\lambda = \frac{q_{LPL}}{q_{LPL} + q_{gPL}} = \frac{W_L/\rho_L}{W_L/\rho_L + W_g/\rho_g} \tag{7.108}$$

5. Calculate the liquid density in pipeline:

$$\bar{\rho}_L = \frac{\gamma_L(62.4) + \gamma_g(0.0764)\bar{R}_s/5.615}{\bar{B}_o} \tag{7.109}$$

6. Calculate the gas density:

$$\bar{\rho}_g = \gamma_g(0.0764) \frac{\bar{p}}{p_b} \frac{T_b}{\bar{T}} \frac{1}{\bar{z}} = 2.701 \, \gamma_g \frac{\bar{p}}{\bar{T}\bar{z}} \tag{7.110}$$

7. Calculate the velocity of the mixture, in feet per second:

$$\bar{u}_m = \frac{q_{LPL} + q_{gPL}}{\pi D^2/576} \tag{7.111}$$

D is pipeline ID in inches.

8. Calculate the viscosity of a two-phase mixture:

$$\mu_{TP} = \mu_L \lambda + \mu_g(1 - \lambda) \tag{7.112}$$

9. Estimate a value of the liquid holdup in the pipeline ($\bar{H}_L$).

10. Calculate the two-phase density:

$$\rho_{TP} = \bar{\rho}_L \left(\frac{\lambda^2}{\bar{H}_L}\right) + \bar{\rho}_g \left[\frac{(1 - \lambda)^2}{1 - \bar{H}_L}\right] \tag{7.113}$$

11. Calculate an approximate value of the two-phase Reynolds number:

$$(N_{\text{Re}})_{TP} = \frac{D\bar{u}_m \, \rho_{TP}}{\mu_{TP}} \tag{7.114}$$

12. With the no slip holdup λ from step 4 and the approximate Reynolds number, $(N_{\text{Re}})_{TP}$, from step 11, go to Fig. 7.10 and read a value of $\bar{H}_L$.

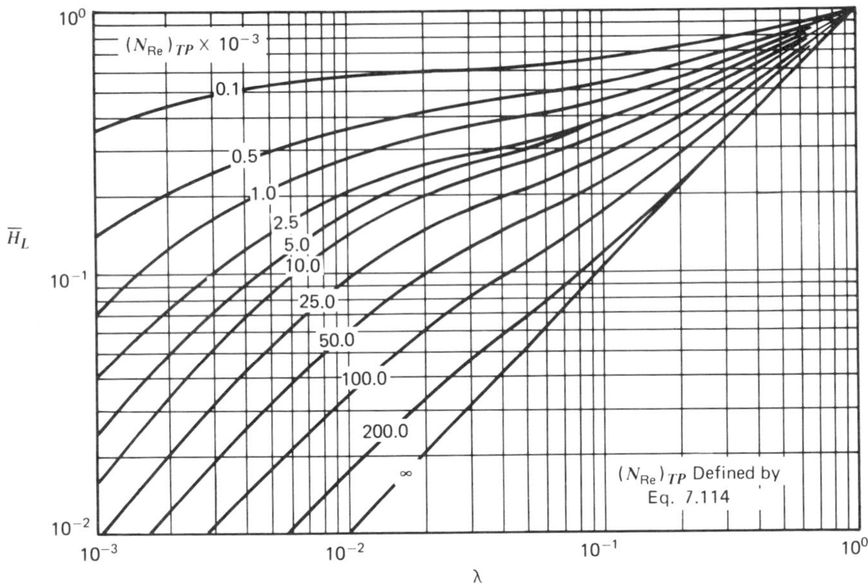

Fig. 7.10 Dukler liquid holdup correlation. (After Baker et al.)

13. Check the value of $\bar{H}_L$ of step 12 with the assumed value of step 9. If they agree within 5%, the accuracy is sufficient. If they do not agree, repeat steps 9 to 13 until they agree within 5%.

14. When the assumed value and calculated value of $\bar{H}_L$ agree within 5%, use the calculated value of $\bar{H}_L$ to determine the two-phase Reynolds number.

15. Determine the value f_{Tp}/f_o from Fig. 7.11.

16. Calculate f_o, the single-phase friction factor. There are several correlations available for calculating the friction factor of a smooth pipe. The equation used for developing the correlation shown in Fig. 7.11 is

$$f_o = 0.001{,}40 + \frac{0.125}{(N_{Re})_{TP}^{0.32}} \tag{7.115}$$

17. Calculate f_{TP}, the two-phase friction factor:

$$f_{TP} = \frac{f_{TP}}{f_o} \times f_0 \tag{7.116}$$

18. Calculate the pressure drop due to friction:

$$\Delta p_f = \frac{2 f_{TP} L \bar{u}_m^{\ 2}\, \rho_{TP}}{12 g_c D} \tag{7.117}$$

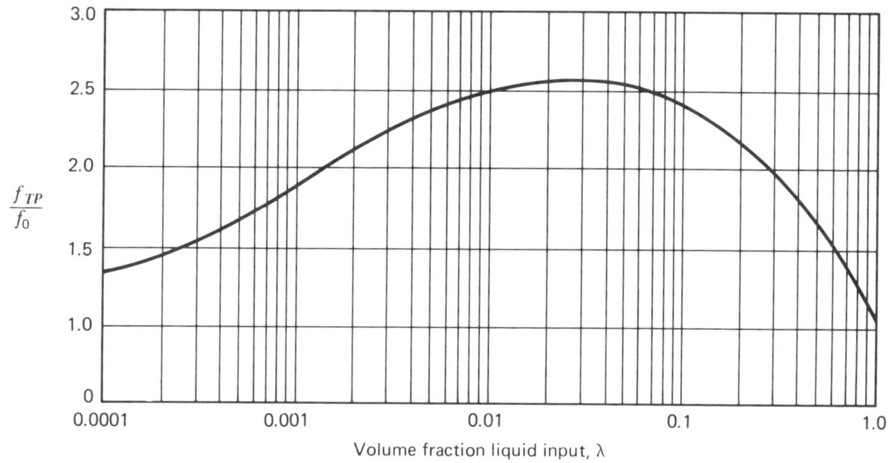

Fig. 7.11 Normalized friction factor curve. (After Baker et al.)

where

$$L = \text{ft}$$
$$u_m = \text{ft/sec}$$
$$\rho_{TP} = \text{lbm/cu ft}$$
$$D = \text{in.}$$

19. Acceleration pressure drop may also be calculated. This is usually very small in pipelines but may be significant in process piping.

$$\Delta p_{acc} = \frac{1}{144 g_c A}\left\{\left[\frac{\bar{\rho}_g q_{gPL}^2}{1 - \bar{H}_L} + \frac{\bar{\rho}_L q_{LPL}^2}{\bar{H}_L}\right]_{\text{downstream}} - \left[\frac{\bar{\rho}_g q_{gPL}^2}{1 - \bar{H}_L} + \frac{\bar{\rho}_L q_{LPL}^2}{\bar{H}_L}\right]_{\text{upstream}}\right\} \cos\theta \tag{7.118}$$

where θ is the angle of pipe bend. For a horizontal pipe $\theta = 0$ and cosine $\theta = 1$.

20. Elevation pressure drop—if any elevation changes take place, the pressure drop due to elevation changes may be calculated in the following manner:

(a) Calculate the superficial gas velocity

$$u_{sg} = \frac{q_{gPL}(144)}{\pi/4\ D^2} \tag{7.119}$$

(b) From Flanigan's correlation (Fig. 7.12), read the value of H_{LF}, the elevation factor.

(c) Calculate the pressure drop due to elevation changes

$$\Delta p_{el} = \frac{\bar{\rho}_L H_{LF} \Sigma z}{144} \tag{7.120}$$

where Σz is the summation of uphill rises in the direction of flow, feet.

21. Calculate total pressure drop:

$$\Delta p_{\text{total}} = \Delta p_f + \Delta p_{el} + \Delta p_{acc} \tag{7.121}$$

7.5.2 The Flanigan Correlation

Flanigan proposed a method of calculating the effect of hills and pressure drop in flowlines and pipelines. His correlation ignores pressure recovery in the downhill section. He treated the uphill sections as though they affected the pressure drop in the same manner as would a vertical column containing an equivalent amount of liquid. Since a two-phase line is not completely filled with liquid. Flanigan used a liquid holdup term, H_{LF}, to represent that fraction of the total static pressure drop that exists as the elevation component in two-phase flow. A correlation for H_{LF} as a function of gas superficial velocity, u_{sg}, is presented in Fig. 7.12.

The Flanigan holdup factor is calculated from the equation

$$H_{LF} = \frac{1}{1 + 0.3264 u_{sg}^{1.006}} \tag{7.122}$$

The gas superficial velocity is calculated at average pressure and temperature in the line. This requires an iterative procedure since $\bar{p} = (p_1 + p_2)/2$ is unknown.

The additional pressure drop due to hills is calculated from Eq. 7.120. Flanigan correlation, along with Dukler friction factor correlation, is recommended in the AGA-API Design Manual by Baker, et al.

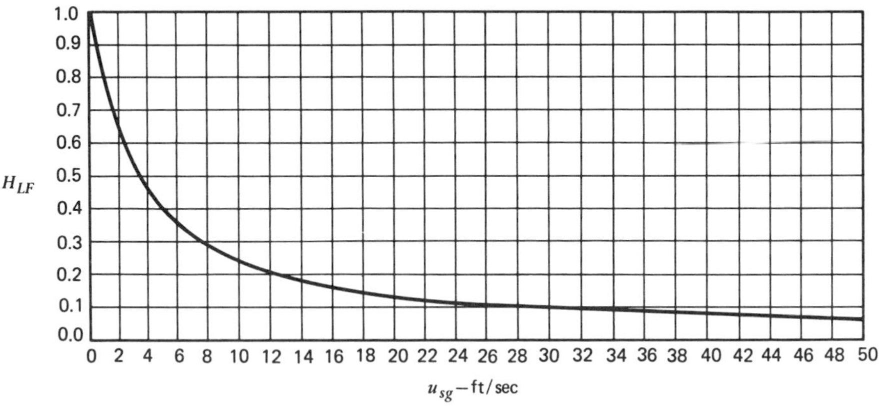

Fig. 7.12 Flanigan holdup factor.

Flanigan proposed using the Panhandle A equation to calculate the pressure drop due to friction, based on the gas flow rate. A correlation for pipeline efficiency factor as a function of superficial gas velocity and liquid loading is presented in Fig. 7.13. The gas velocity is in feet per second and liquid-gas ratio in bbl/MMscf.

Flanigan's procedure for calculating the total pressure drop is as follows:

1. Calculate the superficial velocity in feet per second:

$$u_{sg} = 31{,}194 \frac{q_g \bar{z} \bar{T}}{D^2 \bar{p} T_b} \tag{7.123}$$

where

q_b = gas flow rate, MMscfd
p_b = 14.7 psia

2. Calculate the liquid to gas ratio R in barrels per million standard cubic feet of gas.

3. Calculate $u_{sg}/R^{0.32}$ and determine the Panhandle pipeline efficiency factor E from Fig. 7.13.

4. Calculate the frictional pressure drop using the Panhandle A equation:

$$q_g = 4.3587 \times 10^4 E \left(\frac{T_b}{p_b}\right)^{1.078,81} \left(\frac{p_1^2 - p_2^2}{\bar{T} L \bar{z}}\right)^{0.5394} \left(\frac{1}{\gamma_g}\right)^{0.4604} D^{2.6182} \tag{7.124}$$

5. Determine the Flanigan holdup factor H_{LF} from Fig. 7.12.

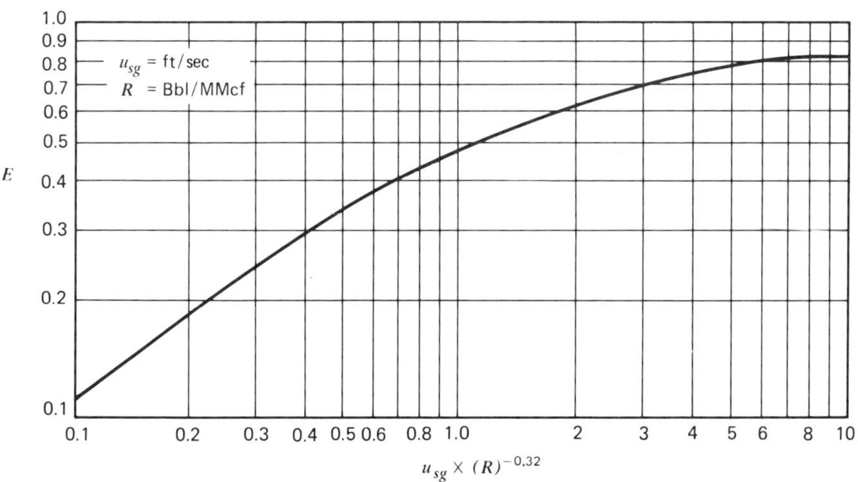

Fig. 7.13 Flanigan efficiency factor.

6. Calculate the pressure drop due to elevation using Eq. 7.120.

7. The total pressure drop

$$\Delta p_{\text{total}} = \Delta p_f + \Delta p_{el} \tag{7.125}$$

Example 7.5 (after Baker et al.).

Problem:

Two gas-condensate wells feed into a 4-in. gathering line 2.10 mi long. Well A will flow at the rate of 3 MMcfd, and well B will flow at the rate of 1 MMcfd. The following data are available on each well:

	Well *A*	**Well** *B*
Liquid GPM[a] at 1000 psig	4.10	3.20
Liquid GPM at 500 psig	3.10	2.30
Gas specific gravity	0.68	0.62

[a]Gallons per thousand cubic feet of gas.

The summation of the uphill rises in the line is 143 ft. The initial pressure at the wells is 900 psig. What is the pressure drop in the line?

Solution

$$\text{Line diameter} = \frac{4.026}{12} = 0.3355 \text{ ft}$$

$$\text{Line length} = (2.10))(5280) = 11{,}088 \text{ ft}$$

$$q_{sg} = \frac{4000}{24} = 166.7 \text{ Mcfh} = 46.3 \text{ ft/sec}$$

Assume an average pressure in the pipeline of 850 psig or 865 psia. Assume an average temperature in the pipeline of 60°F or 520°R. Calculate the volume of pipeline liquid from the product of the gas volumes and the GPM. From a plot of GPM versus pressure (Fig.7.14), the GPM at 850 psig is 3.85 for well A and 2.95 for well B:

$$\begin{aligned} q_{LPL} &= (3000)(3.85) + (1000)(2.95) \\ &= 14{,}500 \text{ gal/day} = 80.8 \text{ cfh} = 0.0224 \text{ cu ft/sec} \end{aligned}$$

Calculate the liquid density. In this particular example, the condensate is typical and has a specific gravity of approximately 0.6:

$$\bar{\rho}_L = (0.6)(62.4) = 37.4 \text{ lbm/cu ft}$$

Calculate the weighted average specific gravity of the commingled gas stream:

$$\gamma_g = \frac{(3)(0.68) + (1)(0.62)}{3 + 1} = 0.665$$

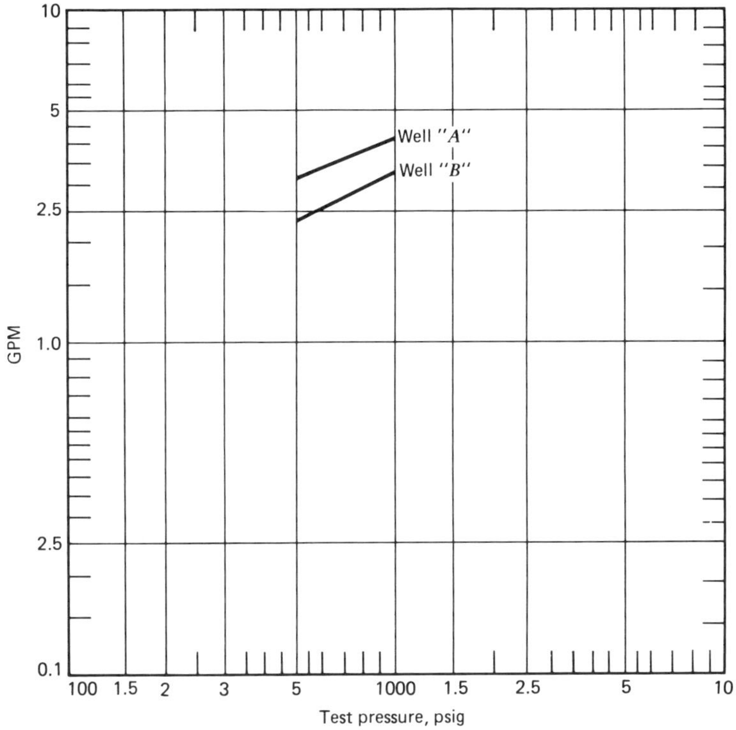

Fig. 7.14 Liquid content of wellstreams. (After Baker et al.)

Calculate the gas viscosity. The molecular weight of the gas is

$$M = (0.665)(28.97) = 19.00$$

From Fig. 2.10, the viscosity of the gas at atmospheric pressure is 0.010,25 cp. From Fig. 2.5, the critical temperature of the gas is 372 R and the critical pressure is 669 psia. Calculate the reduced temperature and pressure:

$$p_R = \frac{865}{669} = 1.29$$

$$T_R = \frac{520}{372} = 1.40$$

From Fig. 2.11, the viscosity ratio is 1.2. Calculate the gas viscosity at pipeline conditions:

$$\mu_g = \mu_1 \frac{\mu}{\mu_1} = (0.010,25)(1.2) = 0.0123 \text{ cp}$$

From the value of p_R and T_R developed above and by referring to Fig. 2.4,

$$\bar{z} = 0.84$$

Calculate the gas volume at pipeline conditions:

$$q_{gPL} = q_{gs}\left(\frac{14.7}{\bar{p}}\right)\left(\frac{\bar{T}}{520}\right)\bar{z} = (166.7)\left(\frac{14.7}{865}\right)\left(\frac{520}{520}\right)(0.84)$$

$$= 2.38 \text{ Mcf/hr} = 2380 \text{ cfh}$$

$$= 0.661 \text{ ft}^3\text{/sec}$$

Calculate the density of the gas at pipeline conditions from Eq. 7.110:

$$\rho_g = 2.701\ \gamma_g \frac{\bar{p}}{\bar{T}\bar{z}} = \frac{(2.701)(0.665)(865)}{(520)(0.84)} = 3.56 \text{ lbm/cu ft}$$

Calculate the liquid viscosity. Assume that the average composition of the condensate is normal octane. From a table of physical properties, the critical temperature is 564 R, the critical pressure is 362 psia, and the molecular weight is 128.3. Calculate the reduced pressure and temperature:

$$p_R = \frac{865}{362} = 2.39$$

$$T_R = \frac{520}{564} = 0.92$$

From Fig. 7.15,

$$\mu_L/\sqrt{M} = 0.012$$

$$\mu_L = 0.012\sqrt{M} = 0.012\sqrt{128.3} = 0.136 \text{ cp}$$

Calculate λ, the input liquid-volume fraction from Eq. 7.108:

$$\lambda = \frac{q_{LPL}}{q_{LPL} + q_{gPL}} = \frac{0.0224}{0.0224 + 0.661} = 0.0328$$

Calculate $\bar{u}_m$, the mixture velocity from Eq. 7.111:

$$\bar{u}_m = \frac{q_{LPL} + q_{gPL}}{\pi D^2/576} = \frac{0.0224 + 0.661}{\pi(4.026)^2/576} = 7.74 \text{ ft/sec}$$

Calculate μ_{TP}, the mixture viscosity from Eq. 7.112:

$$\mu_{TP} = \mu_L\lambda + \mu_g(1 - \lambda) = (0.136)(0.0328) + (0.0123)(0.9672) = 0.0164 \text{ cp}$$

Converting to English units,

$$\mu_{TP} = (0.0164)(0.000{,}672) = 0.000{,}011{,}0 \text{ lbm/ft sec}$$

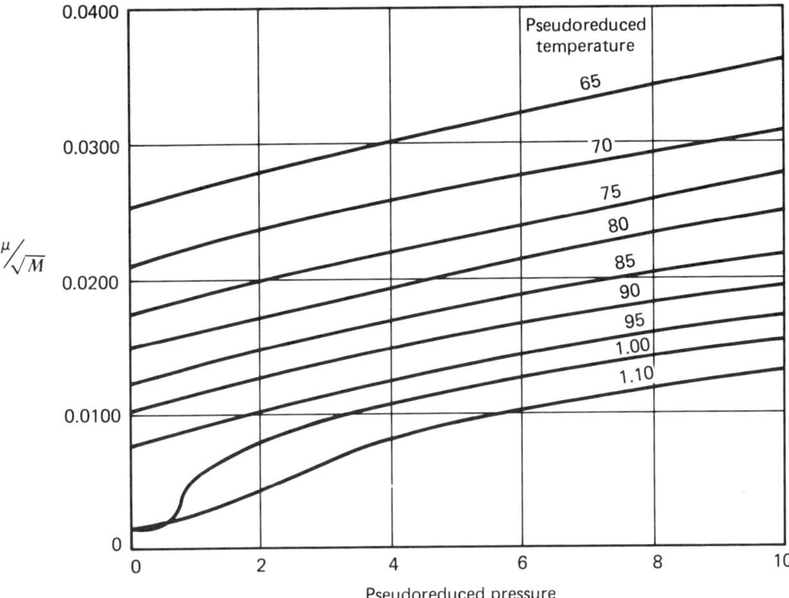

Fig. 7.15 Viscosity of light liquid hydrocarbons. (After Baker et al.)

Determine the friction factor ratio from Fig. 7.11:

$$f_{TP}/f_o = 2.532$$

Now, calculate the two-phase Reynolds number. This is a trial-and-error calculation. Assume a value for $\bar{H}_L$, the liquid holdup. Assume $\bar{H}_L = 0.03$. Calculate the two-phase density from Eq. 7.113:

$$\rho_{TP} = \frac{\bar{\rho}_L \lambda^2}{\bar{H}_L} + \frac{\bar{\rho}_G(1-\lambda)^2}{1-\bar{H}_L} = \frac{(37.4)(0.0328)^2}{(0.03)} + \frac{(3.56)(0.9672)^2}{(0.97)} = 4.77 \text{ lbm/cu ft}$$

$$(N_{\text{Re}})_{TP} = \frac{D\bar{u}_m \rho_{TP}}{\mu_{TP}} = \frac{(0.3355)(7.74)(4.77)}{(0.000{,}011{,}0)} = 1{,}126{,}000$$

From Fig. 7.10, $\bar{H}_L = 0.033$. This is not a close enough check, and the calculation must be repeated with the new value of $\bar{H}_L$:

$$\rho_{TP} = \frac{(3.74)(0.328)^2}{(0.033)} + \frac{(3.56)(0.9672)^2}{(0.967)} = 4.66 \text{ lbm/cu ft}$$

$$(N_{\text{Re}})_{Tp} = \frac{(0.3355)(7.74)(4.66)}{(0.000{,}011{,}0)} = 1{,}100{,}000$$

From Fig. 7.10, $\bar{H}_L = 0.033$. This checks. With this value of Reynolds number, calculate the single-phase friction factor from Eq. 7.115:

$$f_o = 0.0014 + \frac{0.125}{(N_{\text{Re}})^{0.32}} = 0.0014 + \frac{0.125}{(1{,}100{,}000)^{0.32}} = 0.002{,}86$$

Calculate the two-phase friction factor (Eq. 7.116):

$$f_{TP} = f_o \times \frac{f_{TP}}{f_o} = (0.002{,}86)(2.532) = 0.007{,}24$$

Calculate the pressure drop due to friction from Eq. 7.117:

$$\Delta p_F = \frac{2 f_{TP}\, L\, u_m^{\,2}\, \rho_{TP}}{12 g_c D} = \frac{(2)(0.007{,}24)(11{,}088)(7.74)^2(4.66)}{(12)(32.2)(4.026)} = 28.8 \text{ psi}$$

Next, the pressure drop due to elevation changes must be considered. Calculate u_{sg}, the superficial gas velocity from Eq. 7.119:

$$u_{sg} = \frac{q_{gPL}(144)}{\pi/4D^2} = \frac{(0.661)(144)}{\pi/4(4.026)^2} = 7.48 \text{ ft/sec}$$

From Fig. 7.12, $H_{LF} = 0.30$. Calculate the elevation pressure drop from Eq. 7.120:

$$\Delta p_{el} = \frac{H_{LF}\bar{\rho}_L \Sigma Z}{144} = \frac{(0.30)(37.4)(143)}{(144)} = 11.1 \text{ psi}$$

Calculate the total pressure drop from Eq. 7.121:

$$\Delta p_{\text{total}} = \Delta p_f + \Delta p_{el} + \Delta p_{acc}$$

The downstream pressure would be 900 minus 39.9 or 860.1 psig. The originally assumed average pressure was 850 psig. For greater accuracy, the entire calculation may be repeated, using as an average line pressure the average of 900 and 860.1, or 880 psig.

7.5.3 Beggs and Brill Correlation

The Beggs and Brill correlation was developed especially for inclined or directional flow and can be used for pipe at any angle of inclination, including downhill flow. Data for the correlation were taken in a small-scale test facility consisting of 1 in. and 1½-in. pipes using air and water as fluids. Three flow regimes were considered and correlations were given for liquid holdup and friction factor. The flow regimes used were those that would exist if the pipeline were horizontal. The liquid holdup is first calculated for horizontal flow and then corrected for actual inclination angle.

This correlation is accurate for directional and inclined flow and can be used for horizontal and vertical flow. It tends to predict too much pressure recovery in downhill flow, if the flow pattern is stratified.

The horizontal flow patterns or regimes used in this study are illustrated in Fig. 7.16. The original flow pattern map was slightly modified to include a transition zone between the segregated and intermittent flow regimes. The modified flow pattern map is superimposed on the original in Fig. 7.17.

Determination of the correct flow regime requires calculating several dimensionless numbers, including a two-phase Froude number. The following variables are used to determine which flow regime would exist if the pipe were in a *horizontal* position. This will be the actual flow regime only in horizontal flow and is used as a correlating parameter for other pipe inclination angles.

Froude number (N_{FR})

$$N_{FR} = \frac{\bar{u}_m{}^2}{gD}$$

where

$\bar{u}_m$ is the mixture velocity in feet per second.

Volumetric liquid fraction input (λ_L):

$$\lambda_L = \frac{u_{SL}}{u_m}$$

where

u_{SL} is the superficial liquid velocity in feet per second.

Limiting parameters

$$\begin{aligned} L_1 &= 316\lambda_L^{0.302} \\ L_2 &= 0.000{,}925{,}2\lambda_L^{-2.4682} \\ L_3 &= 0.10\lambda_L^{-1.4516} \\ L_4 &= 0.5\lambda_L^{-6.738} \end{aligned}$$

The horizontal flow regime limits are

Segregated

$$\lambda_L < 0.01 \quad \text{and} \quad N_{FR} < L_1$$

or

$$\lambda_L \geqslant 0.01 \quad \text{and} \quad N_{FR} < L_2$$

Transition

$$\lambda_L \geqslant 0.01 \quad \text{and} \quad L_2 < N_{FR} \leqslant L_3$$

Intermittent

$$0.01 \leqslant \lambda_L < 0.4 \quad \text{and} \quad L_3 < N_{FR} \leqslant L_1$$

or

$$\lambda_L \geqslant 0.4 \quad \text{and} \quad L_3 < N_{FR} \leqslant L_4$$

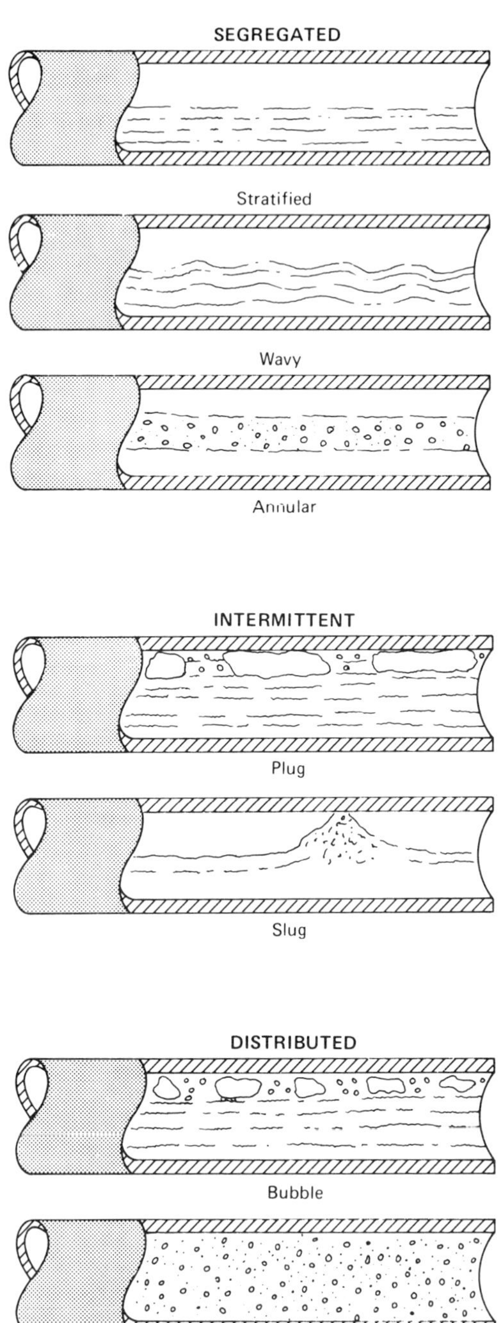

Fig. 7.16 Horizontal flow patterns. (After Beggs and Brill.)

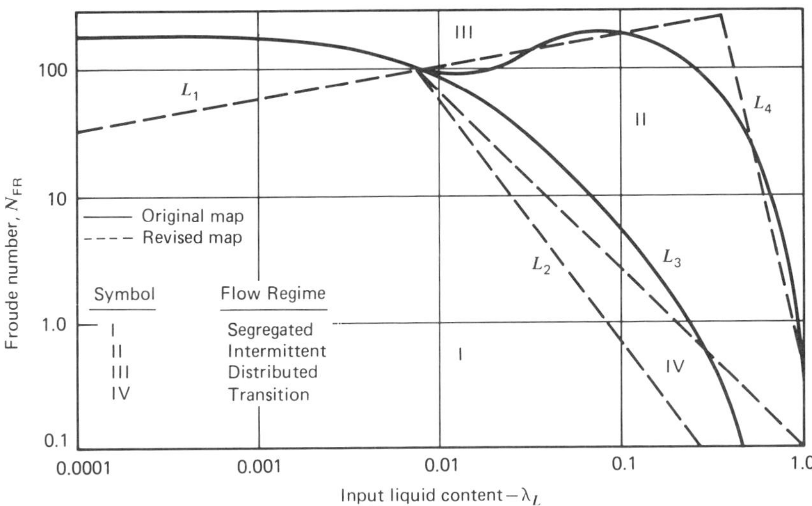

Fig. 7.17 Horizontal flow pattern map. (After Beggs and Brill.)

Distributed

$$\lambda_L < 0.4 \quad \text{and} \quad N_{FR} \geqslant L_1$$

or

$$\lambda_L \geqslant 0.4 \quad \text{and} \quad N_{FR} > L_4$$

The liquid holdup depends on flow regime and is given by

$$H_{L(\theta)} = H_{L(0)}\,\psi$$

where $H_{L(0)}$ is the holdup that would exist at the same conditions in a horizontal pipe and ψ is a correction factor for pipe inclination.

$$H_{L(0)} = \frac{a\lambda_L^b}{N_{FR}^c}$$

with the constraint, $H_{L(0)} \geqslant \lambda_L$. The values of the constants a, b, and c for each flow regime are given below.

Flow Regime	a	b	c
Segregated	0.98	0.4846	0.0868
Intermittent	0.845	0.5351	0.0173
Distributed	1.065	0.5824	0.0609

When the flow falls in the transition regime, the liquid holdup is a weighted average of the segregated and intermittent values:

$$H_L \text{ (transition)} = A \times H_L \text{ (segregated)} + B \times H_L \text{ (intermittent)}$$

where

$$A = \frac{L_3 - N_{FR}}{L_3 - L_2}$$

$$B = \frac{N_{FR} - L_2}{L_3 - L_2} = 1 - A$$

The pipe inclination correction factor is given by

$$\psi = 1 + C[\sin(1.8\theta) - 0.333 \sin^3(1.8\theta)]$$

where θ is the pipe inclination angle to horizontal, and

$$C = (1 - \lambda_L) \ln (d\lambda_L^e N_{LV}^f N_{FR}^g)$$

with the constraint $C \geq 0$. The liquid velocity number is calculated using

$$N_{LV} = 1.938 \left(\frac{u_{SL}}{\sigma_L}\right)^{0.25}$$

where σ_L is the liquid surface tension in dynes per centimeter. The liquid surface tension may be calculated from the values for oil and water:

$$\sigma_L = \sigma_o f_o + \sigma_w f_w = \sigma_o \left(\frac{1}{1 + \text{WOR}}\right) + \sigma_w \left(\frac{\text{WOR}}{1 + \text{WOR}}\right)$$

The constants d, e, f, and g are given for each flow regime below.

Flow Regime	d	e	f	g
Segregated uphill	0.011	−3.768	3.539	−1.614
Intermittent uphill	2.96	0.305	−0.4473	0.0978
Distributed uphill	No correction		$C = 0$, $\psi = 1$, $H_L \neq f(\theta)$	
All flow regimes downhill	4.70	−0.3692	0.1244	−0.056

Once $H_{L(\theta)}$ is determined, the two-phase density is calculated from

$$\rho_{TP} = \rho_L H_L + \rho_g(1 - H_L)$$

The pressure gradient due to elevation change is then

$$\left(\frac{dp}{dL}\right)_{el} = \frac{g}{g_c} \rho_{TP} \sin\theta$$

The pressure gradient due to friction is given by

$$\left(\frac{dp}{dL}\right)_f = \frac{f_{TP}\,\rho_o\,\bar{u}_m^{\,2}}{2g_cD}$$

where the no-slip density is calculated from

$$\rho_o = \rho_L\lambda_L + \rho_g(1 - \lambda_L)$$

and the two-phase friction factor is given by

$$f_{TP} = \frac{f_{TP}}{f_o} \times f_o$$

The single-phase or no-slip friction factor f_o is determined by the Moody diagram or Jain's correlation (Eq. 7.34), using the following Reynolds number:

$$N_{\mathrm{Re}} = \frac{D\bar{u}_m\,\rho_o}{\mu_o}$$

where μ_o is a no-slip viscosity:

$$\mu_o = \mu_L\lambda_L + \mu_g(1 - \lambda_L)$$

The ratio f_{TP}/f_o is calculated from

$$\frac{f_{TP}}{f_o} = e^s$$

where

$$s = [\ln(y)]/\{-0.0523 + 3.182\ln(y) - 0.8725\,[\ln(y)]^2 + 0.018{,}53\,[\ln(y)]^4\}$$

and

$$y = \frac{\lambda_L}{[H_{L(\theta)}]^2}$$

The value of s is unbounded in the interval $1 < y < 1.2$ and should be calculated, in this interval, from

$$s = \ln(2.2y - 1.2)$$

The acceleration pressure gradient is

$$\left(\frac{dp}{dL}\right)_{acc} = \frac{\rho_{TP}\bar{u}_m u_{sg}}{g_c\bar{p}}\left(\frac{dp}{dL}\right)$$

where u_{sg} is the superficial gas velocity in feet per second.

The total pressure gradient can then be calculated using

$$\left(\frac{dp}{dL}\right)_{total} = \frac{\left(\frac{dp}{dL}\right)_{el} + \left(\frac{dp}{dL}\right)_{f}}{1 - E_k}$$

where E_k is an acceleration term:

$$E_k = \frac{\rho_{TP}\bar{u}_m u_{sg}}{g_c \bar{p}}$$

Detailed procedure and example calculations on the use of Beggs and Brill correlation are given in Brown, *The Technology of Artificial Lift Methods*, Volume 1.

7.5.4 Application of Multiphase Flow Correlations

The correlations previously described can be programmed for computer calculation. When such programs are available, pressure traverses can be calculated accurately by computer because the pipeline or flowline can be divided into small increments. Also the exact conditions of pressure, temperature, and fluid properties for each case can be used. If time permits, computer calculations are recommended.

Pressure traverse curves or "working curves" have been prepared using some of the correlations. Although computer solutions are more accurate, a computer is not always readily available. A set of working curves becomes a necessity if the practicing engineer should have immediate access to multiphase flow correlations. When using these working curves one has no control over pressure, temperature, and fluid properties. However, prepared pressure traverse curves may be calculated using conditions in a particular field, if time permits. Solutions derived from working curves are sufficiently accurate for most field problems.

Pressure traverse curves are available from Kermit E. Brown, *The Technology of Artificial Lift Methods*, Volumes 3A and 3B. Figure 7.18 is a set of curves from Brown's book and will be used to illustrate the use of pressure traverse curves.

Example 7.6

Given

Pipeline size = 8.00 in. ID

Liquid flow rate = 20,000 bbl/day

Gas-liquid ratio = 5000 scf/STB

Gas specific gravity = 0.65

Oil gravity = 35°API

Average flowing temperature = 120°F

Length of flowline = 10,000 ft

Outlet pressure = 500 psig

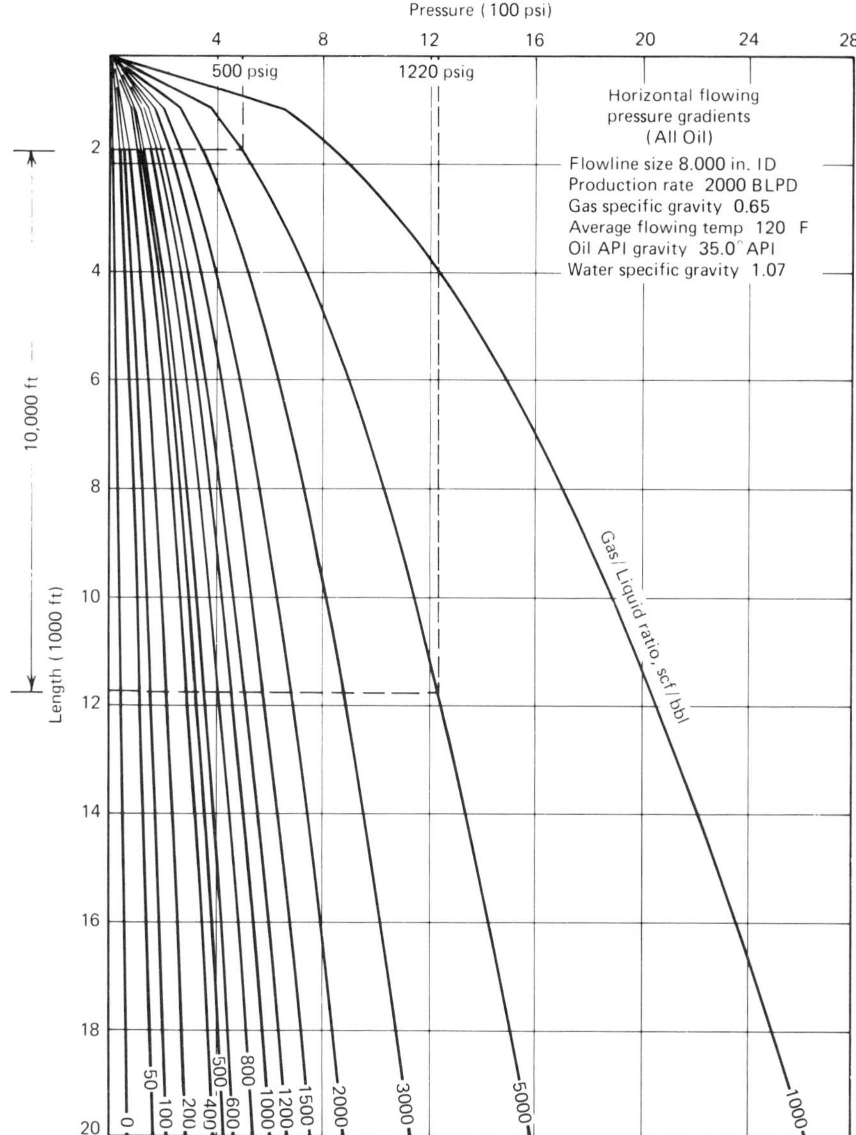

Fig. 7.18 Horizontal flowing pressure gradients. (After Brown).

Required

Find the pipeline inlet pressure.

Solution

(a) Select a set of curves for the correct line size, liquid flow rate, flowing temperature, and fluid properties. Figure 7.18 is selected.

(b) Select a curve for the gas-liquid ratio of 5000 scf/STB.

(c) Locate the known 500-psig pressure on the pressure axis and select a length equivalent to this pressure by proceeding vertically downward from 500 psig at zero length until intersecting the 5000 scf/STB gas-oil ratio line. This is found to be 1800 ft.

(d) Correct the length for pipeline length. If known the pressure is the outlet pressure, add the pipeline length to the value found in step (c). If the known pressure is the inlet pressure, subtract the pipeline length from the value found in step (c). Thus, the corrected graph length is 10,000 + 1800 = 11,800 ft.

(e) The unknown pressure is the pressure corresponding to the corrected length. At the intersection of the 11,800-ft-length line and the 5000 scf/STB gas-oil ratio line draw a vertical line to intersect the pressure axis at 1220 psig. This is the pipeline inlet pressure.

7.6 PIPELINE ECONOMICS

In planning a pipeline system, bear in mind that the scale of operation of a pipeline has considerable effect on the unit costs. By doubling the diameter of the pipe, other factors remaining constant, the capacity increases more than sixfold. On the other hand, the cost approximately doubles, so that cost per unit delivered decreases to one-third of the original unit cost. It is this scale effect that justifies multiproduct lines. Whether it is, in fact, economical to install a large-diameter pipeline at the outset depends on scale and the following factors:

1. Rate of growth in demand (it may be uneconomical to operate at low-capacity factors during initial years). (Capacity factor is the ratio of actual average discharge to design capacity.)
2. Operating factor (the ratio of average throughput at any time to maximum throughput during the same time period), which will depend on the rate of draw-off and can be improved by installing storage at the consumer's end.
3. Reduced power costs due to low-friction losses while the pipeline is not operating at full capacity.
4. Certainty of future demands.
5. Varying costs with time (both capital and operating).
6. Rates of interest and capital availability.
7. Physical dificulties in the construction of a second pipcline if required.

Figure 7.19 illustrates some important points to be considered in planning a pipeline. At any particular throughput q_1, there is a certain diameter at which overall costs will be a minimum (in this case D_2). At this diameter, the cost per million cubic feet per day of throughput could be reduced further if throughput was increased. Costs would be a minimum at some throughput q_2. Thus, a pipeline's optimum throughput is not the same as the throughput for which it is the optimum diameter. If q_1 were increased by an amount q_3, so that total throughput $q_4 = q_1 + q_3$, it might be more economical to increase the flow through the pipe with diameter D_2 than to install a second pipeline, for example, q_4C_4 is less than $q_1C_1 + q_3C_3$.

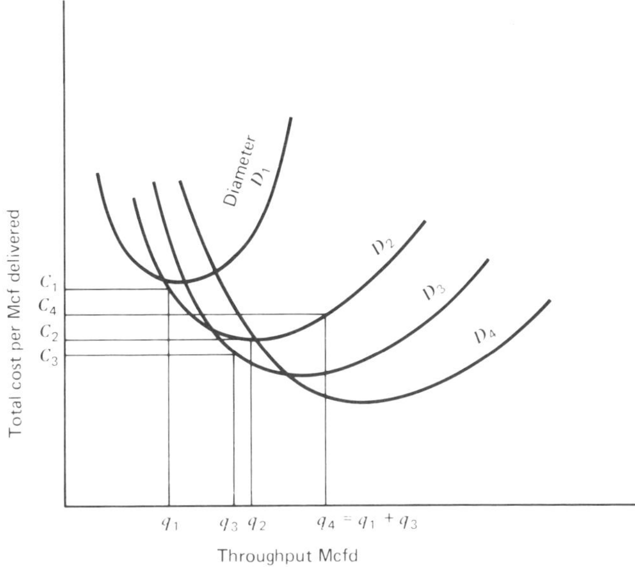

Fig. 7.19 Cost-throughput curves for different diameter pipes. (After Stephenson.)

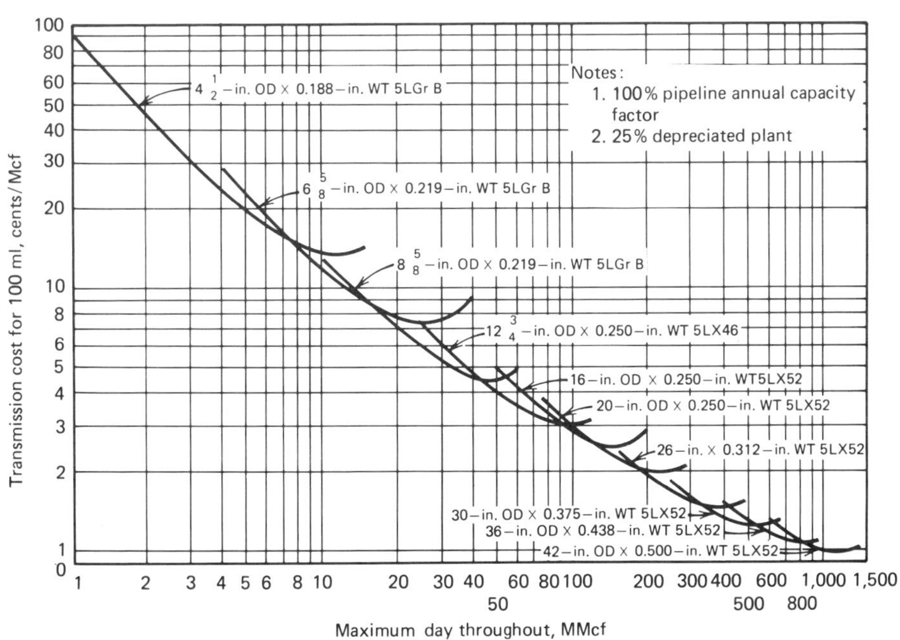

Fig. 7.20 Effect of volume throughput on gas transmission costs (*Source:* Testimony, Fred A. Hough, Chief Gas Engineer, Bechtel Corp. Docket no. AR61-2, et al. Hearing exhibit no. 59. Date identified, 12-10-63. Date admitted, 5-26-64. pp. 2456–2462.)

The power cost per unit of additional throughput decreases with increasing pipe diameter, so the corresponding likelihood of it being most economic to increase throughput through an existing line increases. Fig. 7.20 is a typical chart illustrating the effect of volume throughput on 1963 gas transmission costs.

REFERENCES

American Gas Assoc. *Gas Engineer's Handbook*. New York: The Industrial Press, 1965.

Baker, O. "Multiphase Flow in Pipelines." *Oil and Gas Journal*. November 10,1958.

Baker, O., et al. *Gas-Liquid Flow in Pipelines II, Design Manual*. AGA and API Monograph, Project NX-28. October 1970.

Beggs, H. D., and J. P. Brill. "A Study of Two-phase Flow in Inclined Pipes." *Journal of Petroleum Technology* **25**, p. 607, 1973.

Brown, K. E. *The Technology of Artificial Lift Methods*, Volume 1. Tulsa: Petroleum Publishing Company, 1977.

Brown, K. E. *The Technology of Artificial Lift Methods*, Volumes 3a and 3b. Tulsa: PennWell Publishing Company, 1980.

Campbell, J. M. "Elements of Field Processing." Tulsa: *Oil and Gas Journal*, reprint series.

Colebrook, C. F. *J. Inst. Civil Eng.* **11**, p. 133, 1938.

Drew, T. B., E. C. Koo, and W. H. McAdams. *Trans. Am. Inst. Chem. Eng.* **28**, 56, 1930.

Duckler, A. E. *Gas-Liquid Flow in Pipelines I, Research Results*. AGA and API Monograph, Project NX-28, 1969.

Flanigan, O. "Effect of Uphill Flow on Pressure Drop in Design of Two-Phase Gathering Systems," *Oil and Gas Journal*, March 10, 1958.

Jain, A. K. "An Accurate Explicit Equation for Friction Factor," *J. Hydraulics Div. ASCE* **102**, No. HY5, May 1976.

Katz, D. L., et al. *Handbook of Natural Gas Engineering*. New York: McGraw-Hill, 1959.

Nikuradse, J. *Forschungshelf*, p. 301, 1933.

NGPSA Engineering Data Book. Tulsa: Gas Processors Suppliers Assoc., 1977.

Standing, M. B. *Volumetric and Phase Behavior of Oil Field Hydrocarbon Systems*. La Habra, Cal.: Chevron Oil Field Research, September 1970.

Stephenson, D. *Pipeline Design for Water Engineers*. New York: American Elsevier Publishing Co., 1976.

PROBLEMS

7.1 It is desired that 24,000,000 cfd of gas measured at standard conditions of 14.7 psia and 60°F be delivered to a city 50 miles from the field. The gas is delivered to the pipeline at 300-psia pressure, and it is to be transmitted through a pipeline of sufficient size so that the pressure at the city is not less than 50 psia. Assume the specific gravity of the gas to be 0.65 and the temperature of flow 60°F. What is the minimum size of pipeline required for the job?

Fig. 7.21 Series pipeline.

7.2 A series pipeline (see Fig. 7.21) consists of three lengths: (1) 30 miles of 6-in. pipe, (2) 40 miles of 8-in. pipe, and (3) 50 miles of 10-in. pipe. Determine the equivalent lengths of 6-, 8-, and 10-in. pipe.

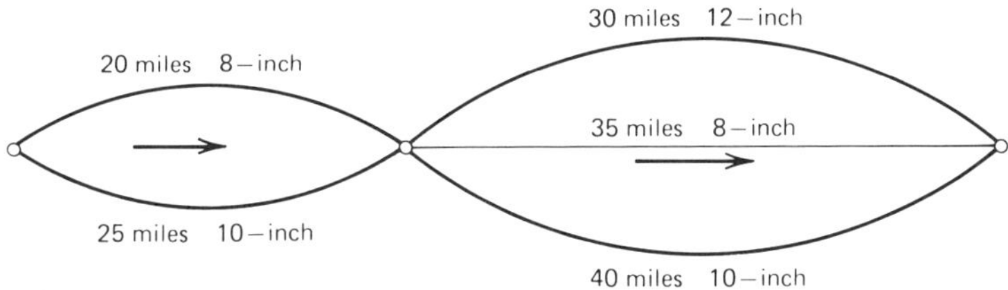

Fig. 7.22 Series-loop system.

7.3 A pipeline system is composed of two sections, *AB* and *BC* (see Fig. 7.22). The former consists of two parallel lines and the latter of three parallel lines, of sizes and lengths as indicated in the figure. 50 MMcfd measured at standard conditions of 14.7 psia and 60°F are to be transmitted through this system. The pressure at *C* is to be maintained at 50 psia. The specific gravity of the gas is 0.60 and its temperature is 60°F.

(a) Determine the pressure at *A*.

(b) What would be the pressure at *B*?

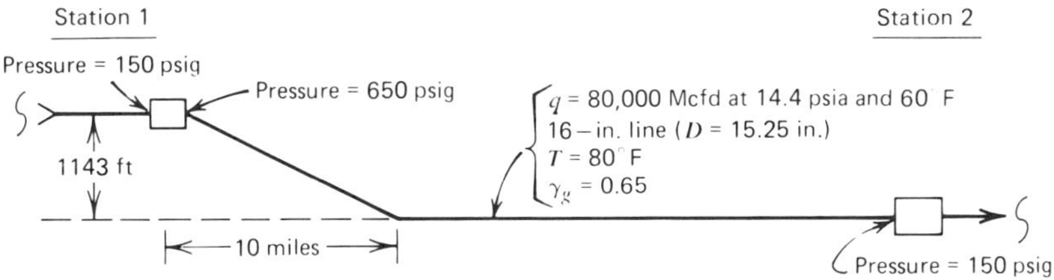

Fig. 7.23 Pipeline system.

7.4 Figure 7.23 illustrates a situation in which gas will be flowing from Station 1 to Station 2. Compressors in both stations are to operate with 150-psig intake pressure and 650-psig discharge pressure. The pipeline drops 1143 ft in elevation in the first 10 miles after which it is horizontal. Average temperature in the line during flow is 80°F. The gas gravity is 0.65 (air = 1) and contains no nonhydrocarbon components. The pipeline will be 16 in. (inside diam. = 15.25 in.). The absolute roughness of the pipe is 0.0006 in. The quantity of gas to be transported is 80,000 Mcfd at standard conditions of 14.4 psia and 60°F. How many miles should Station 2 be from Station 1?

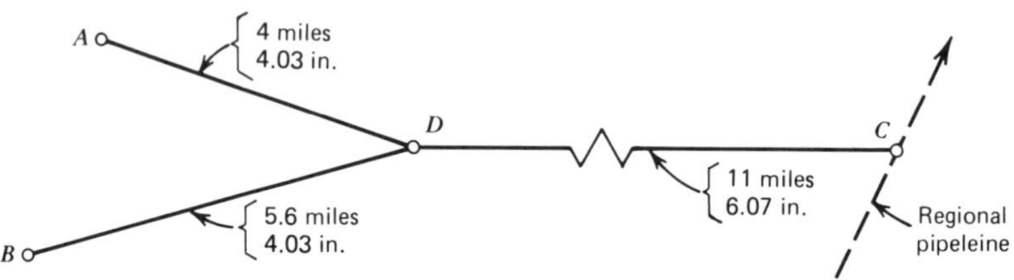

Fig. 7.24 Gas-gathering system.

7.5 Figure 7.24 shows a gas-gathering system that delivers gas from lease *A* and lease *B* to the terminal location *C*. At the terminal location the total gas is sold to a regional gas transmission line operating at 450 psig.

Gas produced at lease *B* is gas from primary separators operating at 225 psig and 85°F. Thus, it is obvious that some compression must be required at lease *B*. The question is: To what pressure must the gas be compressed at lease *B*?

Additional data and assumptions	Lease *A*	Lease *B*
Gas production, MMcfd	3.5	1.5
Gas gravity (air = 1)	0.65	0.80

No elevation differences (flat land)
Line temperature = 85°F
Standard conditions = 14.7 psia and 60°F
Absolute roughness = 0.0006 in. (all pipes)
Use Weymouth equation that contains f.

Calculate the pressures at *A* and *B*.

7.6 What is the result of looping an operating line with 6.626-in. OD pipe? The present operating conditions are as follows:

8.625 in. O.D. Sch 40 line

15,000 ft long

13,000 Mcfd flow

100-psig inlet pressure

0.70 specific gravity

80°F flowing temperature

1.000 compressibility factor

Adding the 6.625-in. OD line permits:

(a) An increase in q if outlet pressure is unchanged.
(b) An increase in outlet pressure if q remains constant.
(c) A combination in which q and P_2 are both increased.

Estimate the resulting increase in output pressure if q remains constant.

8

GAS WELL PERFORMANCE

The surveillance of gas wells is of increasing interest to the petroleum industry as the economic importance of natural gas continues to grow. Referring to Fig. 8.1, it can be seen that the ability of a gas reservoir to produce for a given set of reservoir conditions depends directly on the flowing bottom-hole pressure, P_{wf}. The ability of this reservoir to deliver a certain quantity of gas depends both on the inflow performance relationship and the flowing bottom-hole pressure. The flowing bottom-hole pressure on its part depends on the separator pressure and the configuration of the piping system.

These conditions can be mathematically expressed as

$$q = C(\bar{p}_R{}^2 - p_{wf}{}^2)^n \tag{8.1}$$

and

$$p_{wf} = p_{\text{sep}} + \Delta p_{fl} + \Delta p_{ch} + \Delta p_{tb} + \Delta p_{\text{res}} \tag{8.2}$$

where

q = gas flow rate
C = a numerical coefficient, characteristic of the particular well
$\bar{p}_R$ = shut-in reservoir pressure
p_{wf} = flowing bottom-hole pressure
n = a numerical exponent, characteristic of the particular well
p_{sep} = separator pressure
Δp_{fl} = pressure drop in flowline
Δp_{ch} = pressure drop in surface choke
Δp_{tb} = pressure drop in well tubing
Δp_{res} = pressure drop in other restrictions, such as subsurface safety valves, valves, fittings, and so on.

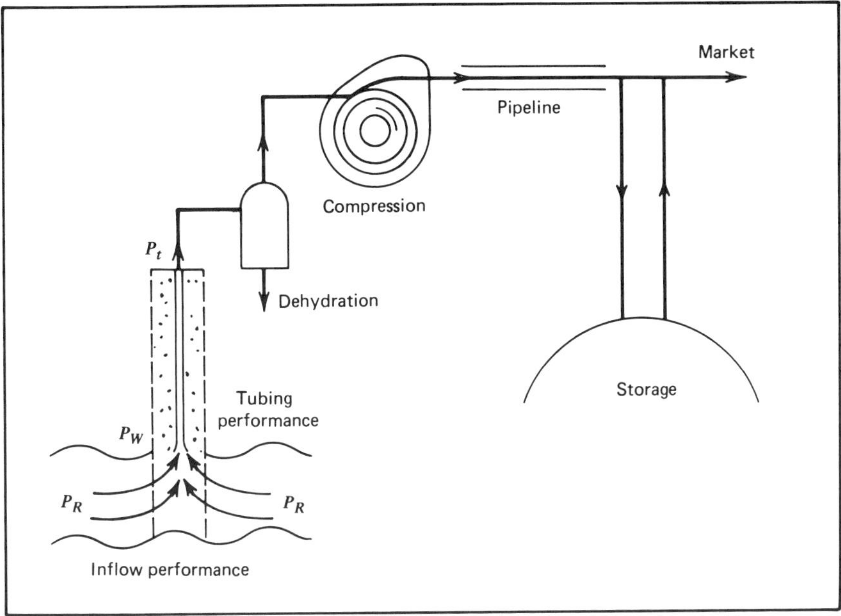

Fig. 8.1 Gas production schematic.

In order to determine the deliverability of the total well system, it is necessary to calculate all the parameters and pressure drops listed in Eqs. 8.1 and 8.2. This chapter will discuss static and flowing bottom-hole pressures in gas wells and methods of determining pressure drops in tubing for single-phase gas flow and for multiphase gas-liquid flow. Inflow and outflow performance curves will be discussed. Equations for calculating pressure drops across restrictions will be presented. The problems of liquid loading in gas wells will also be examined.

8.1 STATIC AND FLOWING BOTTOM-HOLE PRESSURES

Often, the static or flowing pressure at the formation must be known in order to predict the productivity or absolute open flow potential of gas wells. The preferred method is to measure the pressure with a bottom-hole pressure gauge. It is often impractical or too expensive to measure static or flowing bottom-hole pressures with bottom-hole gauges. However, for many problems, a sufficiently precise value can be estimated from wellhead data (gas specific gravity, surface pressure and temperature, formation temperature, and well depth). Calculation of static (or shut-in) pressure amounts to evaluating the pressure difference equal to the weight of the column of gas. In the case of flowing wells, the gas column weight and friction effects must be evaluated and summed up.

8.1.1 Basic Energy Equation

Numerous equations have been developed for flow of gas through pipes. These equations result from the different assumptions used in integrating the fundamental mechanical energy equation. In the case of steady-state flow, this energy balance can be expressed as follows (Fig. 8.2):

$$144\ v\ dp + \frac{u\ du}{2\alpha g_c} + \frac{g\ dZ}{g_c} + \frac{fu^2}{2g_c D}\ dL + w_s = 0 \tag{8.3}$$

or

$$\frac{144}{\rho}\ dp + \frac{u\ du}{2\alpha g_c} + \frac{g\ dZ}{g_c} + \frac{fu^2}{2g_c D}\ dL + w_s = 0 \tag{8.4}$$

where

v = specific volume of fluid, cu ft/lbm
ρ = density of the fluid, lbm/cu ft
p = pressure, psia
u = average velocity of the fluid, ft/sec
α = correction factor to compensate for the variation of velocity over the tube cross section (α ranges from 0.5 for laminar flow to 1.0 for fully developed turbulent flow. A value of 0.90 is usually satisfactory for practical gas flow problems.)
Z = distance in the vertical direction, ft
f = Moody friction factor, dimensionless
D = inside diameter of the pipe, ft
L = length of the flow string, ft (for a vertical flow string, $L = Z$)
$u\ du/2\alpha g_c$ = pressure drop due to kinetic energy effects
$fu^2\ dL/2g_c D$ = pressure drop due to friction effects
w_s = mechanical work done on or by the gas ($w_s = 0$)

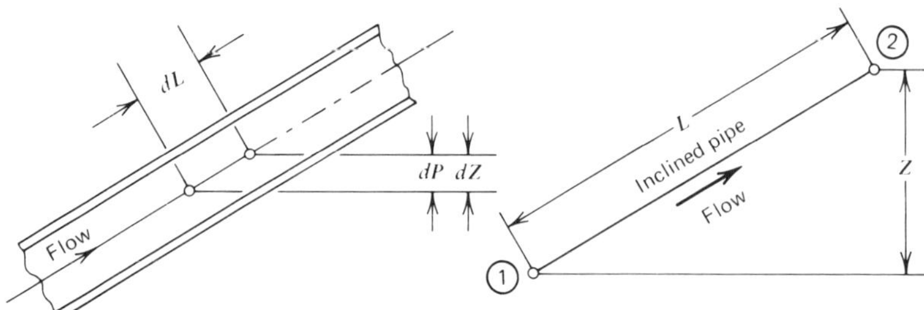

Fig. 8.2 Flow in pipe. (After Aziz.)

g = acceleration due to gravity, ft/sec^2
g_c = 32.17 = conversion factor, lbm ft/lbf sec^2

The second term in Eq. 8.3 or 8.4 expresses kinetic energy. This term usually is neglected in pipeline flow calculations. The magnitude of error caused by neglecting the change in kinetic energy will be discussed later. If no mechanical work is done on the gas (compression) or by the gas (expansion through a turbine or engine), the term w_s is zero.

The reduced form of the mechanical energy equation may be written as

$$\frac{144}{\rho}\,dp + \frac{g}{g_c}\,dZ + \frac{fu^2\,dL}{2g_cD} = 0 \tag{8.5}$$

or

$$144\int_1^2 \frac{dp}{\rho} + \frac{g}{g_c}\int_1^2 dZ + \frac{1}{2g_cD}\int_1^2 fu^2\,dL = 0 \tag{8.6}$$

All equations now in use for gas flow and static head calculations are various forms of Eq. 8.6.

The density of a gas at a point in a vertical pipe at pressure p and temperature T may be written as

$$\rho_g = \frac{28.97\gamma_g p}{zRT} \tag{8.7}$$

where

ρ_g = density of gas, lbm/cu ft
28.97 = molecular weight of air
γ_g = specific gravity of gas at standard conditions (air = 1)
p = pressure, lbf/sq in.
z = gas deviation factor
R = gas constant, 10.732 cu ft psia/lb-mole °R
T = absolute temperature, °R

The velocity of gas flow u_g at a cross section of a vertical pipe may be defined as

$$u_g = \frac{4\dot{m}}{\pi D^2\rho_g} = \frac{4mzRT}{\pi D^2(28.97)\gamma_g p} \tag{8.8}$$

where

u_g = gas velocity, ft/sec
$\dot{m}$ = mass flow rate, lbm/sec
D = inside diameter of flow channel, ft

Combining Eqs. 8.6, 8.7, and 8.8 and employing petroleum engineering units, the general vertical flow equation becomes

$$\int_1^2 \frac{\dfrac{z\,dp}{p}}{1+\dfrac{667fq^2T^2z^2}{D^5p^2}} = \int_1^2 \frac{28.97\gamma_g\,dL}{10.732(144)T} = \int_1^2 \frac{0.01875\gamma_g\,dL}{T} \tag{8.9}$$

Assuming a constant average temperature in the interval of interest, Eq. 8.9 becomes

$$\int_1^2 \frac{\dfrac{z\,dp}{p}}{1+\dfrac{667fq^2\overline{T}^2z^2}{D^5p^2}} = \frac{0.01875\gamma_g L}{\overline{T}} \tag{8.10}$$

where

p = pressure, psia
q = gas flow rate, MMscfd
D = inside diameter of pipe, in.
$\overline{T}$ = average temperature through distance L, °R
$L = Z$ = vertical distance of interest, ft

Equations 8.9 and 8.10 cannot be easily integrated because of the interdependence of variables z, p, and T. Thus, various investigators have made different simplifying assumptions, such as substituting average values of z and/or T, that have resulted in several final equations of varying degrees of accuracy.

Rzasa and Katz and Vitter integrated Eq. 8.10 by assuming the gas deviation factor constant for the entire flow column at assumed average conditions of temperature and pressure. Sukker & Cornell and Poettmann integrated the equation on the assumption that temperature is constant at the average value but gas deviation factor varies with pressure. Although these assumptions are fairly accurate in relatively shallow wells, they become more erroneous as depth increases. A more realistic approach is that of Cullender & Smith and Crawford & Fancher, who treated gas deviation factor as a function of both temperature and pressure.

The various methods of estimating bottom-hole pressure from surface measurements are summarized in the following discussion. All of these methods apply to both gas and condensate wells, provided that the flow rate q_{wg}, specific gravity γ_{wg}, and gas deviation factor z_{wg} of the wet gas or total well effluent (gas plus vapor equivalent of condensate liquids) are substituted for q, γ_g, and z in the basic equations.

Example 8.1. Calculate the change in kinetic energy during flow in a deep natural gas well producing under the following conditions:

Flow rate = 11.716 MMscfd

Gas gravity = 0.98

Flow string ID = 2.441 in.

Wellhead pressure = 2685 psia

Bottom-hole pressure = 4563 psia

Pseudocritical pressure = 768 psia

Pseudocritical temperature = 477 °R

Well depth = 10,471 ft

Wellhead temperature = 606 °R

Bottom-hole temperature = 702 °R

Solution

Eq. 8.4 may be written as

$$53.34\,\frac{Tz\,dp}{\gamma_g p} + \frac{(0.4152q)^2}{g_c D^4} \times \frac{Tz}{p} \times d\left(\frac{Tz}{p}\right) + \frac{g}{g_c}\,dZ + \frac{0.0862f}{g_c D^5} \times dL\left(\frac{Tz}{p}\right)^2 q^2 = 0 \qquad (8.11)$$

where

$Z/L = 1$ for vertical flow
$Z = 0$ for horizontal flow

From Eq. 8.11, change in kinetic energy expressed in terms of pressure change is

$$\Delta KE = \int_1^2 \frac{\rho u\,du}{2\alpha g_c}$$

$$\begin{aligned}\Delta KE &= \frac{(0.4152)^2}{(53.34)(2\alpha g_c)} \times \frac{\gamma_g q^2}{D^4}\left[\left(\frac{Tz}{p}\right)_2 - \left(\frac{Tz}{p}\right)_1\right] \\ &= \frac{(0.4152 \times 144)^2}{(53.34)(2)(0.9)(32.17)} \times \frac{(0.98)(11.716)^2}{(2.441)^4}\,[0.1388 - 0.1308] \\ &= 0.035 \text{ psia}\end{aligned}$$

This is very small compared to a total pressure drop of 1878 psia and can be neglected.

8.1.2 Adjusting for Liquid Production

The produced liquid or condensate can be converted to its gas equivalent assuming that it behaves as an ideal gas when vaporized in the produced gas. Using the ideal gas law, the gas equivalent of liquid hydrocarbons production is given by

$$GE_{hc} = V = \frac{nRT_b}{P_b} \tag{8.12}$$

or

$$GE_{hc} = \left(\frac{350.5\gamma_o}{M_o}\right)\frac{RT_b}{p_b} \tag{8.13}$$

Using base or standard conditions of 14.7 psia and 520°F and a universal gas constant of 10.73,

$$GE_{hc} = 133{,}000\gamma_o/M_o \qquad \text{scf/STB} \tag{8.14}$$

The molecular weight of the condensate is given by

$$M_o = \frac{44.29\gamma_o}{1.03 - \gamma_o} = \frac{6084}{°\text{API} - 5.9} \tag{8.15}$$

It is common for some water to be produced as condensate from the gas phase. This water will be fresh water and should be added to the gas production. The gas equivalent of water production is calculated using

$$GE_w = \left(\frac{350.5\gamma_w}{M_w}\right)\frac{RT_b}{p_b} \tag{8.16}$$

At $p_b = 14.7$ psia and $T_b = 520$ °R,

$$\begin{aligned} GE_w &= \frac{350.5 \times 1.00 \times 10.73 \times 520}{18 \times 14.7} \\ &= 7390 \text{ scf/STB of water} \end{aligned}$$

Example 8.2. Determine the total daily gas production from a reservoir, including the gas equivalents of liquid hydrocarbons (condensate) and water. The following data are available:

Separator gas production = 2.0 MMscfd

Condensate production = 60.0 STB/day

Stock tank gas production = 5.0 Mscfd

Fresh water production = 4.0 STB/day

Reservoir pressure = 1000 psia

Reservoir temperature = 250°F

Condensate gravity= 55°API

Solution

Condensate specific gravity:

$$\gamma_o = \frac{141.5}{131.5 + 55} = 0.759$$

Condensate molecular weight

$$\begin{aligned} M_o &= (44.29)(0.759)/(1.03 - 0.759) \\ &= 124 \text{ lb/lb-mole} \end{aligned}$$

Gas equivalent of condensate

$$\begin{aligned} GE_{hc} &= \frac{(350.5)(0.759)}{124} \times \frac{(10.73)(520)}{14.7} \\ &= 814 \text{ scf/STB} \\ &= 814 \times 60 = 48{,}840 \text{ scfd} \end{aligned}$$

From Fig. 2.13, water content at 1000 psia and 240°F is 1380 lb/MMscf or 3.94 bbl/MMscf. The gas equivalent of water production is

$$GE_w = (7390)(3.94) = 29{,}117 \text{ scfd}$$

Total daily gas production is

$$q = 2000 + 5.0 + 48.8 + 29.1 = 2082.9 \text{ Mcfd}$$

8.2 STATIC BOTTOM-HOLE PRESSURE

The estimation of static bottom-hole pressure from surface measurements only involves calculating the additive pressure exerted by the weight of the static fluid column. Figure 8.3 illustrates conditions in the tubing or casing of a nonflowing gas well; p_{ts} is the pressure at the surface and p_{ws} is the pressure at depth feet below the surface. The temperature profile in the shut-in gas well is shown in Fig. 8.4. The temperature profile is not quite a straight line but, because of circulation within the well, it tends to be higher than indicated by a straight line connecting the surface and reservoir temperatures.

Since the flow rate (q or q_{wg}) is equal to zero, Eq. 8.9 simplifies to

$$\int_1^2 \frac{z\,dp}{p} = \int_1^2 \frac{0.018{,}75\,\gamma_g\,dL}{T} \tag{8.17}$$

Thus, the static condition is a special case of the general vertical flow equation.

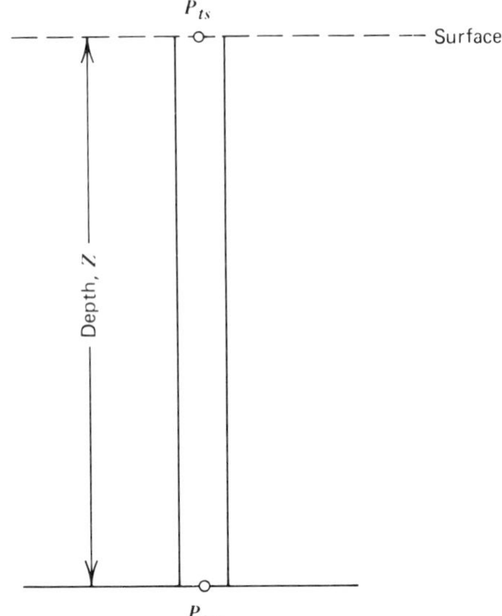

Fig. 8.3 Static gas well.

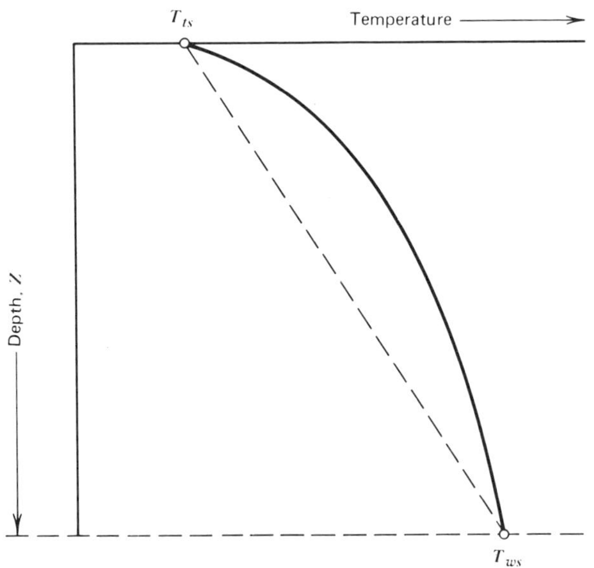

Fig. 8.4 Temperature profile in a static well.

8.2.1 Average Temperature and Deviation Factor Method

The average temperature and average gas deviation method is frequently used because of its simplicity. Equation 8.17 may be written as

$$\int_{p_{ts}}^{p_{ws}} \frac{z\, dp}{p} = \int_0^Z \frac{0.018{,}75\ \gamma_g\, dL}{T} \tag{8.18}$$

If T and z are taken outside the integrals as average values $\bar{T}$ and $\bar{z}$, Eq. 8.18 becomes

$$\ln \frac{p_{ws}}{p_{ts}} = \frac{0.018{,}75\ \gamma_g Z}{\bar{z}\bar{T}} \tag{8.19}$$

or

$$p_{ws} = p_{ts}\ e^{0.01875\gamma_g Z/\bar{z}\bar{T}} \tag{8.20}$$

where

p_{ws} = static bottom-hole pressure, psia
p_{ts} = static wellhead pressure, psia
γ_g = gas specific gravity (air = 1)
Z = well depth from surface, ft
$\bar{T}$ = average temperature, °R (usually arithmetic mean of bottom and wellhead temperatures, assuming linear temperature profile)
$\bar{z}$ = z value at arithmetic mean temperature and arithmetic mean pressure

The pressure difference, $p_{ws} - p_{ts}$, equivalent to the weight of the gas column can be obtained by subtracting p_{ts} from both sides of Eq. 8.20:

$$p_{ws} - p_{ts} = p_{ts}\ e^{0.01875\gamma_g Z/\bar{z}\bar{T}} - 1 \tag{8.21}$$

The Problem of $\bar{z}$ and $\bar{T}$

The solution of Eq. 8.20 for the static bottom-hole pressure at some given depth below the surface is best handled by trial-and-error methods. This is because p_{ws} is a function of T_{ts}, T_{ws}, p_{ts}, and p_{ws}. The calculation procedure involves guessing a value of p_{ws} and obtaining a value for $\bar{z}$ at $(p_{ts} + p_{ws})/2$ and $(T_{ts} + T_{ws})/2$. Then Eq. 8.20 is used to obtain a value for p_{ws}. If the calculated value of p_{ws} is within the desired initial guess, the procedure ends. If not, make another guess and continue the trial-and-error process.

The procedure may be applied as a one-step calculation from the wellhead to the sand face or as a multistep (usually two) calculation.

Example 8.3. Calculate the static bottom-hole pressure of a gas well having a depth of 5790 ft. The gas gravity is 0.60 and the pressure at the wellhead is 2300 psia. The average temperature of the flow string is 117°F.

Solution

$$Z = 5790 \text{ ft}$$
$$\gamma_g = 0.6$$
$$p_{ts} = 2300 \text{ psia}$$
$$\bar{T} = \frac{T_{ts} + T_{ws}}{2} = 117°\text{F} = 577°\text{R}$$
$$p_{ws} = p_{ts}\, e^{(0.01875\times 0.60\times 5790)/(577\times \bar{z})} = 2300e^{0.113/z}$$

First trial

$$p_{ws} = 2500 \text{ psia}$$
$$\bar{p} = 2400$$
$$p_{pr} = \frac{2400}{672} = 3.57$$
$$T_{pr} = \frac{577}{358} = 1.61$$
$$\bar{z} = 0.822$$
$$p_{ws} = 2300\ e^{0.113/0.822} = 2639 \text{ psia}$$

Second trial

$$\bar{p} = \frac{2639 + 2300}{2} = 2470 \text{ psia}$$
$$p_{pr} = 3.67$$
$$\bar{z} = 0.822$$

There is no appreciable change in $\bar{z}$ for this trial, so the first trial is sufficient.

$$p_{ws} = 2639 \text{ psia}$$

8.2.2 Sukkar and Cornell Method

One of the easiest methods to use for estimating bottom-hole pressure is that of Sukkar and Cornell, who integrated the entire left-hand side of Eq. 8.10 numerically at several constant (average) temperatures. This work has the advantage of improved accuracy and also permits calculations of bottom-hole pressures without a trial-and-error procedure.

Converting Eq.8.10 to the Sukkar and Cornell form and employing pseudo-reduced parameters gives

$$\int_{p_{pr1}}^{p_{pr2}} \frac{z\ dp_{pr}/p_{pr}}{1 + \dfrac{Bz^2}{{p_{pr}}^2}} = \frac{0.018,75\gamma_g Z}{\bar{T}} \tag{8.22}$$

For the static case, $B = 0$, and Eq. 8.22 reduces to

$$\int_{p_{pr1}}^{p_{pr2}} \frac{z}{p_{pr}}\ dP_{pr} = \frac{0.018,75\gamma_g Z}{\bar{T}} \tag{8.23}$$

where

p_{pr1} = pseudoreduced wellhead pressure

p_{pr2} = pseudoreduced bottom-hole pressure

z = gas deviation factor of gas (well effluent) at $\bar{T}$ and at the varying pressures within the flow string

$$B = \frac{667 f q^2 \bar{T}^2}{D^5 p_{pc}^{\ 2}} \tag{8.24}$$

p_{pc} = pseudocritical pressure of gas (well effluent), psia

The integral on the left-hand side of Eq. 8.15 has been solved for various values of pseudoreduced temperature, pseudoreduced pressure, and the factor B. The results are presented in tabular form (Tables 8.1 and 8.2). Table 8.1 was prepared for the low-pressure range of reduced pressures from 1.0 to 5.0, or pressures from about 600 psia to 3200 psia. Table 8.2 was prepared for the high-pressure range of reduced pressures of about 3.0 to 12.0, or pressures above 2000 psia. Adequate overlap has been provided so that most problems can be worked entirely on one chart. In order to avoid negative values of the integrals on Table 8.1, a constant value of 0.5 was added to all of the integrals. The constant disappears when integrating between two limits.

Because of its accuracy and simplicity, the Sukkar and Cornell method is recommended for general engineering use on gas and condensate wells with bottom-hole pressures less than 10,000 psia (the approximate pressure limit of Table 8.2).

Example 8.4. Calculate the static bottom-hole pressure for the gas well of Example 8.3 using the Sukkar and Cornell method.

Solution

The steps involved in the Sukkar and Cornell method are as follows.

(a) The right-hand side of Eq. 8.23 is computed:

$$\frac{0.018{,}75\gamma_g Z}{\bar{T}} = \frac{(0.018{,}75)(0.6)(5790)}{577} = 0.1129$$

(b) Calculate $\bar{T}_{pr}$ and p_{pr1}:

$$\bar{T}_{pr} = \frac{577}{359} = 1.610$$

Pseudoreduced wellhead pressure:

$$p_{pr1} = \frac{2300}{672} = 3.423$$

(c) The value of integral on the left-hand side of Eq. 8.23 is obtained from Table 8.2, using $B = 0$ under static conditions. By linear interpolation, the integral is 1.1996 for $p_{pr} = 3.423$ and $\bar{T}_{pr} = 1.610$.

TABLE 8.1

Sukkar–Cornel Integral

Table of $0.5 + \int_3^{P_r} \frac{(z/P_r)\, dP_r}{1 + B(z/P_r)^2}$

	$T_r = 1.5$				$T_r = 1.6$				$T_r = 1.7$			
P_r	$B = 0$	$B = 5$	$B = 10$	$B = 20$	$B = 0$	$B = 5$	$B = 10$	$B = 20$	$B = 0$	$B = 5$	$B = 10$	$B = 20$
1.0	1.4236	0.9215	0.7876	0.6790	1.4613	0.9233	0.7852	0.6775	1.4927	0.9242	0.7828	0.6719
1.1	1.3380	0.9032	0.7773	0.6735	1.3733	0.9052	0.7751	0.6702	1.4031	0.9063	0.7728	0.6666
1.2	1.2606	0.8838	0.7662	0.6675	1.2939	0.8861	0.7642	0.6643	1.3219	0.8874	0.7621	0.6609
1.3	1.1902	0.8636	0.7545	0.6611	1.2213	0.8661	0.7527	0.6580	1.2476	0.8676	0.7507	0.6547
1.4	1.1255	0.8427	0.7420	0.6562	1.1547	0.8454	0.7404	0.6513	1.1792	0.8471	0.7387	0.6481
1.5	1.0659	0.8212	0.7289	0.6468	1.0931	0.8242	0.7276	0.6440	1.1159	0.8261	0.7260	0.6410
1.6	1.0107	0.7993	0.7153	0.6390	1.0359	0.8025	0.7142	0.6364	1.0570	0.8045	0.7129	0.6336
1.7	0.9603	0.7772	0.7012	0.6308	0.9825	0.7705	0.7003	0.6284	1.0020	0.7827	0.6992	0.6258
1.8	0.9113	0.7549	0.6866	0.6223	0.9325	0.7583	0.6860	0.6201	0.9503	0.7605	0.6851	0.6177
1.9	0.8662	0.7326	0.6718	0.6134	0.8856	0.7360	0.6714	0.6113	0.9018	0.7383	0.6707	0.6092
2.0	0.8238	0.7102	0.6566	0.6042	0.8413	0.7136	0.6564	0.6023	0.8559	0.7159	0.6559	0.6004
2.1	0.7838	0.6880	0.6413	0.5946	0.7994	0.6913	0.6412	0.5930	0.8124	0.6936	0.6408	0.5914
2.2	0.7459	0.6660	0.6257	0.5849	0.7597	0.6691	0.6258	0.5834	0.7711	0.6713	0.6256	0.5821
2.3	0.7099	0.6442	0.6100	0.5748	0.7220	0.6471	0.6102	0.5736	0.7318	0.6491	0.6101	0.5725
2.4	0.6758	0.6226	0.5943	0.5646	0.6860	0.6252	0.5946	0.5636	0.6944	0.6271	0.5946	0.5627
2.5	0.6432	0.6013	0.5785	0.5542	0.6516	0.6036	0.5788	0.5534	0.6585	0.6053	0.5789	0.5526
2.6	0.6121	0.6804	0.5627	0.5435	0.6188	0.5823	0.5630	0.5430	0.6242	0.5837	0.5631	0.5424
2.7	0.5823	0.5597	0.5469	0.5328	0.5873	0.5612	0.5472	0.5324	0.5913	0.5623	0.5473	0.5320
2.8	0.5537	0.5395	0.5312	0.5220	0.5570	0.5405	0.5314	0.5217	0.5597	0.5413	0.5315	0.5214
2.9	0.5263	0.5195	0.5155	0.5110	0.5280	0.5201	0.5156	0.5109	0.5293	0.5205	0.5157	0.5108
3.0	0.5000	0.5000	0.5000	0.5000	0.5000	0.5000	0.5000	0.5000	0.5000	0.5000	0.5000	0.5000

3.1	0.4145	0.4807	0.4845	0.4890	0.4729	0.4802	0.4843	0.4890	0.4717	0.4797	0.4842	0.4891
3.2	0.4500	0.4619	0.4692	0.4780	0.4469	0.4607	0.4688	0.4780	0.4443	0.4598	0.4686	0.4781
3.3	0.4262	0.4433	0.4540	0.4666	0.4217	0.4416	0.4534	0.4669	0.4179	0.4402	0.4530	0.4671
3.4	0.4031	0.4251	0.4390	0.4555	0.3973	0.4228	0.4381	0.4557	0.3922	0.4210	0.4376	0.4561
3.5	0.3808	0.4074	0.4241	0.4443	0.3736	0.4043	0.4229	0.4446	0.3674	0.4020	0.4222	0.4450
3.6	0.3590	0.3897	0.4093	0.4331	0.3506	0.3861	0.4079	0.4334	0.3432	0.3833	0.4070	0.4338
3.7	0.3379	0.3724	0.3947	0.4219	0.3283	0.3682	0.3930	0.4222	0.3198	0.3649	0.3918	0.4227
3.8	0.3173	0.3554	0.3802	0.4108	0.3065	0.3506	0.3782	0.4110	0.2970	0.3468	0.3768	0.4115
3.9	0.2972	0.3387	0.3659	0.3997	0.2853	0.3333	0.3636	0.3999	0.2747	0.3289	0.3619	0.4003
4.0	0.2776	0.3222	0.3517	0.3886	0.2646	0.3163	0.3491	0.3887	0.2531	0.3114	0.3472	0.3891
4.1	0.2584	0.3060	0.3377	0.3776	0.2444	0.2995	0.3348	0.3776	0.2319	0.2941	0.3326	0.3780
4.2	0.2396	0.2900	0.3238	0.3665	0.2246	0.2830	0.3206	0.3665	0.2112	0.2771	0.3181	0.3668
4.3	0.2211	0.2743	0.3101	0.3556	0.2053	0.2667	0.3065	0.3554	0.1910	0.2603	0.3037	0.3557
4.4	0.2031	0.2588	0.2965	0.3446	0.1864	0.2506	0.2925	0.3444	0.1711	0.2437	0.2895	0.3446
4.5	0.1853	0.2434	0.2830	0.3337	0.1678	0.2348	0.2787	0.3334	0.1517	0.2274	0.2754	0.3335
4.6	0.1679	0.2283	0.2696	0.3229	0.1495	0.2191	0.2650	0.3224	0.1327	0.2112	0.2614	0.3225
4.7	0.1508	0.2134	0.2564	0.3121	0.1316	0.2037	0.2515	0.3115	0.1140	0.1954	0.2476	0.3115
4.8	0.1340	0.1986	0.2432	0.3014	0.1140	0.1884	0.2380	0.3006	0.0957	0.1797	0.2339	0.3005
4.9	0.1174	0.1841	0.2302	0.2907	0.0966	0.1734	0.2247	0.2898	0.0777	0.1642	0.2203	0.2896
5.0	0.1011	0.1697	0.2173	0.2800	0.0796	0.1585	0.2115	0.2791	0.0600	0.1488	0.2068	0.2787

TABLE 8.2

Sukkar–Cornel Integral

Table of $\int_{12}^{P_r} \frac{(z/P_r)\,dP_r}{1 + B(z/P_r)^2}$

	$T_r = 1.5$			$T_r = 1.6$			$T_r = 1.7$		
P_r	$B = 0$	$B = 5$	$B = 10$	$B = 0$	$B = 5$	$B = 10$	$B = 0$	$B = 5$	$B = 10$
2.0	1.5946	1.3471	1.1904	1.6458	1.3708	1.2027	1.6925	1.3926	1.2134
2.1	1.5546	1.3249	1.1751	1.6040	1.3485	1.1875	1.6490	1.3703	1.1983
2.2	1.5167	1.3029	1.1595	1.5643	1.3263	1.1721	1.6077	1.3480	1.1831
2.3	1.4808	1.2811	1.1438	1.5275	1.3042	1.1565	1.5684	1.3258	1.1676
2.4	1.4466	1.2595	1.1281	1.4906	1.2824	1.1408	1.5310	1.3038	1.1521
2.5	1.4140	1.2382	1.1123	1.4562	1.2608	1.1251	1.4951	1.2820	1.1364
2.6	1.3829	1.2173	1.0965	1.4234	1.2395	1.1093	1.4608	1.2604	1.1206
2.7	1.3531	1.1966	1.0807	1.3919	1.2184	1.0935	1.4297	1.2497	1.1048
2.8	1.3246	1.1764	1.0650	1.3616	1.1977	1.0777	1.3963	1.2180	1.0890
2.9	1.2972	1.1564	1.0493	1.3325	1.1772	1.0619	1.3659	1.1972	1.0732
3.0	1.2708	1.1369	1.0338	1.3045	1.1571	1.0462	1.3366	1.1767	1.0575
3.1	1.2453	1.1176	1.0187	1.2775	1.1373	1.0306	1.3082	1.1565	1.0417
3.2	1.2208	1.0988	1.0030	1.2515	1.1179	1.0151	1.2809	1.1366	1.0261
3.3	1.1970	1.0803	0.9878	1.2262	1.0988	0.9997	1.2544	1.1170	1.0105
3.4	1.1740	1.0621	0.9728	1.2018	1.0800	0.9844	1.2288	1.0977	0.9951
3.5	1.1516	1.0442	0.9579	1.1782	1.0615	0.9692	1.2039	1.0787	0.9797
3.6	1.1299	1.0266	0.9431	1.1552	1.0433	0.9542	1.1798	1.0600	0.9645
3.7	1.1087	1.0093	0.9285	1.1329	1.0254	0.9393	1.1563	1.0416	0.9493
3.8	1.0881	0.9923	0.9140	1.1112	1.0078	0.9245	1.1336	1.0235	0.9343
3.9	1.0680	0.9756	0.8997	1.0899	0.9905	0.9099	1.1113	1.0057	0.9194
4.0	1.0484	0.9591	0.8856	1.0692	0.9735	0.8954	1.0897	0.9881	0.9047
4.1	1.0292	0.9429	0.8715	1.0490	0.9567	0.8810	1.0685	0.9708	0.8901
4.2	1.0104	0.9269	0.8576	1.0292	0.9401	0.8668	1.0478	0.9538	0.8756
4.3	0.9920	0.9112	0.8439	1.0099	0.9239	0.8528	1.0276	0.9370	0.8612
4.4	0.9739	0.8957	0.8303	0.9909	0.9078	0.8388	1.0077	0.9204	0.8470
4.5	0.9562	0.8804	0.8168	0.9723	0.8919	0.8250	0.9883	0.9041	0.8239
4.6	0.9388	0.8652	0.8034	0.9541	0.8763	0.8113	0.9693	0.8880	0.8189
4.7	0.9216	0.8503	0.7902	0.9361	0.8608	0.7977	0.9506	0.8721	0.8051
4.8	0.9048	0.8356	0.7770	0.9185	0.8456	0.7843	0.9323	0.8564	0.7914
4.9	0.8882	0.8210	0.7640	0.9012	0.8305	0.7710	0.9143	0.8409	0.7778
5.0	0.8719	0.8066	0.7511	0.8842	0.8156	0.7578	0.8966	0.8256	0.7643
5.1	0.8558	0.7923	0.7384	0.8674	0.8009	0.7447	0.8791	0.8104	0.7509
5.2	0.8400	0.7783	0.7257	0.8508	0.7864	0.7317	0.8620	0.8645	0.7377
5.3	0.8243	0.7643	0.7131	0.8345	0.7720	0.7188	0.8451	0.7797	0.7246

TABLE 8.2 (Continued)

Sukkar–Cornel Integral

Table of $\int_{12}^{P_r} \frac{(z/P_r)\, dP_r}{1 + B(z/P_r)^2}$

	$T_r = 1.5$			$T_r = 1.6$			$T_r = 1.7$		
P_r	$B = 0$	$B = 5$	$B = 10$	$B = 0$	$B = 5$	$B = 10$	$B = 0$	$B = 5$	$B = 10$
5.4	0.8089	0.7505	0.7007	0.8184	0.7577	0.7060	0.8285	0.7651	0.7115
5.5	0.7936	0.7369	0.6883	0.8026	0.7436	0.6933	0.8120	0.7506	0.6986
5.6	0.7785	0.7233	0.6760	0.7869	0.7297	0.6807	0.7958	0.7363	0.6858
5.7	0.7636	0.7090	0.6638	0.7714	0.7158	0.6683	0.7799	0.7221	0.6730
5.8	0.7488	0.6966	0.6517	0.7561	0.7021	0.6559	0.7641	0.7081	0.6604
5.9	0.7342	0.6834	0.6396	0.7410	0.6886	0.6435	0.7485	0.6942	0.6479
6.0	0.7198	0.6703	0.6277	0.7260	0.6751	0.6313	0.7331	0.6805	0.6354
6.1	0.7055	0.6573	0.6158	0.7112	0.6618	0.6192	0.7179	0.6668	0.6231
6.2	0.6913	0.6445	0.6040	0.6966	0.6486	0.6071	0.7039	0.6533	0.6108
6.3	0.6773	0.6317	0.5923	0.6821	0.6355	0.5952	0.6880	0.6400	0.5986
6.4	0.6634	0.6190	0.5807	0.6678	0.6225	0.5833	0.6733	0.6267	0.5866
6.5	0.6496	0.6065	0.5691	0.6536	0.6096	0.5715	0.6588	0.6135	0.5746
6.6	0.6360	0.5940	0.5576	0.6396	0.5968	0.5597	0.6444	0.6005	0.5626
6.7	0.6224	0.5816	0.5461	0.6257	0.5841	0.5481	0.6301	0.5875	0.5508
6.8	0.6090	0.5692	0.5347	0.6119	0.5715	0.5365	0.6160	0.5747	0.5390
6.9	0.5957	0.5570	0.5234	0.5982	0.5590	0.5250	0.6020	0.5620	0.5273
7.0	0.5824	0.5448	0.5122	0.5847	0.5466	0.5135	0.5882	0.5493	0.5157
7.1	0.5693	0.5327	0.5010	0.5712	0.5343	0.5021	0.5745	0.5368	0.5041
7.2	0.5562	0.5207	0.4898	0.5579	0.5221	0.4908	0.5609	0.5243	0.4927
7.3	0.5433	0.5088	0.4787	0.5447	0.5099	0.4796	0.5474	0.5120	0.4811
7.4	0.5304	0.4969	0.4677	0.5315	0.4978	0.4684	0.5340	0.4997	0.4699
7.5	0.5176	0.4851	0.4567	0.5185	0.4858	0.4572	0.5207	0.4875	0.4586
7.6	0.5049	0.4733	0.4457	0.5056	0.4739	0.4461	0.5076	0.4754	0.4474
7.7	0.4923	0.4616	0.4349	0.4928	0.4620	0.4351	0.4945	0.4634	0.4363
7.8	0.4797	0.4500	0.4240	0.4800	0.4503	0.4242	0.4816	0.4514	0.4252
7.9	0.4673	0.4384	0.4132	0.4674	0.4385	0.4133	0.4687	0.4396	0.4141
8.0	0.4549	0.4269	0.4025	0.4548	0.4269	0.4024	0.4560	0.4278	0.4032
8.1	0.4425	0.4155	0.3918	0.4423	0.4153	0.3916	0.4433	0.4161	0.3922
8.2	0.4303	0.4041	0.3811	0.4299	0.4038	0.3809	0.4307	0.4044	0.3814
8.3	0.4181	0.3927	0.3705	0.4176	0.3924	0.3702	0.4182	0.3928	0.3706
8.4	0.4059	0.3814	0.3599	0.4053	0.3810	0.3595	0.4058	0.3813	0.3598
8.5	0.3939	0.3702	0.3494	0.3931	0.3696	0.3489	0.3935	0.3698	0.3491
8.6	0.3818	0.3589	0.3389	0.3810	0.3583	0.3383	0.3812	0.3585	0.3385
8.7	0.3699	0.3478	0.3284	0.3690	0.3491	0.3278	0.3691	0.3471	0.3279

TABLE 8.2 (Continued)

Sukkar–Cornel Integral

Table of $\int_{12}^{P_r} \frac{(z/P_r)\, dP_r}{1 + B(z/P_r)^2}$

	$T_r = 1.5$			$T_r = 1.6$			$T_r = 1.7$		
P_r	$B = 0$	$B = 5$	$B = 10$	$B = 0$	$B = 5$	$B = 10$	$B = 0$	$B = 5$	$B = 10$
8.8	0.3579	0.3367	0.3180	0.3570	0.3359	0.3173	0.3570	0.3359	0.3173
8.9	0.3461	0.3256	0.3076	0.3451	0.3248	0.3069	0.3450	0.3247	0.3068
9.0	0.3343	0.3146	0.2972	0.3332	0.3137	0.2965	0.3330	0.3135	0.2964
9.1	0.3225	0.3036	0.2869	0.3214	0.3027	0.2862	0.3211	0.3024	0.2860
9.2	0.3108	0.2926	0.2766	0.3097	0.2917	0.2758	0.3093	0.2914	0.2756
9.3	0.2992	0.2817	0.2664	0.2980	0.2808	0.2656	0.2976	0.2804	0.2653
9.4	0.2876	0.2709	0.2561	0.2864	0.2699	0.2553	0.2859	0.2695	0.2550
9.5	0.2760	0.2600	0.2460	0.2748	0.2591	0.2451	0.2743	0.2586	0.2448
9.6	0.2645	0.2492	0.2358	0.2663	0.2483	0.2350	0.2627	0.2477	0.2345
9.7	0.2531	0.2385	0.2257	0.2519	0.2375	0.2248	0.2512	0.2370	0.2244
9.8	0.2417	0.2278	0.2156	0.2405	0.2268	0.2147	0.2397	0.2262	0.2143
9.9	0.2303	0.2171	0.2055	0.2291	0.2162	0.2047	0.2284	0.2155	0.2042
10.0	0.2190	0.2065	0.1955	0.2178	0.2055	0.1947	0.2170	0.2049	0.1941
10.1	0.2077	0.1959	0.1855	0.2065	0.1949	0.1847	0.2057	0.1943	0.1841
10.2	0.1964	0.1853	0.1755	0.1953	0.1844	0.1747	0.1945	0.1837	0.1742
10.3	0.1852	0.1748	0.1655	0.1841	0.1739	0.1648	0.1833	0.1732	0.1642
10.4	0.1741	0.1643	0.1556	0.1730	0.1634	0.1549	0.1722	0.1627	0.1543
10.5	0.1629	0.1538	0.1457	0.1619	0.1529	0.1450	0.1611	0.1523	0.1454
10.6	0.1519	0.1433	0.1358	0.1508	0.1385	0.1351	0.1501	0.1419	0.1346
10.7	0.1408	0.1329	0.1260	0.1398	0.1322	0.1253	0.1391	0.1315	0.1248
10.8	0.1298	0.1225	0.1131	0.1289	0.1218	0.1155	0.1281	0.1212	0.1150
10.9	0.1188	0.1122	0.1063	0.1179	0.1115	0.1058	0.1172	0.1109	0.1053
11.0	0.1078	0.1018	0.0966	0.1070	0.1012	0.0960	0.1064	0.1007	0.0956
11.1	0.0969	0.0915	0.0868	0.0962	0.0910	0.0863	0.0956	0.0904	0.0859
11.2	0.0860	0.0813	0.0771	0.0853	0.0807	0.0766	0.0848	0.0803	0.0762
11.3	0.0752	0.0710	0.0673	0.0746	0.0705	0.0670	0.0741	0.0701	0.0666
11.4	0.0648	0.0608	0.0577	0.0638	0.0604	0.0573	0.0634	0.0600	0.0570
11.5	0.0535	0.0506	0.0480	0.0531	0.0502	0.0477	0.0527	0.0499	0.0474
11.6	0.0428	0.0404	0.0383	0.0424	0.0401	0.0381	0.0421	0.0399	0.0379
11.7	0.0320	0.0303	0.0287	0.0317	0.0301	0.0285	0.0315	0.0298	0.0284
11.8	0.0213	0.0201	0.0191	0.0211	0.0200	0.0190	0.0210	0.0198	0.0189
11.9	0.0106	0.0100	0.0095	0.0105	0.0100	0.0095	0.0104	0.0099	0.0094

(d) The value obtained in step 1 is subtracted from this integral:

$$1.1996 - 0.1129 = 1.0867$$

(e) The pseudoreduced bottom-hole pressure, corresponding to the integral of 1.0867 and the average pseudoreduced temperature of 1.610 for $B = 0$, is $p_{pr} \simeq 3.92$, from Table 8.2.

(f) Multiply pseudoreduced bottom-hole pressure obtained in step 5 by the pseudocritical pressure to obtain the required static bottom-hole pressure:

$$p_{ws} = (3.92)(672) = 2634 \text{ psia}$$

8.2.3 Cullender and Smith Method

Starting with a more realistic approach that gas deviation factor is a function of both temperature and pressure, Cullender and Smith changed the units in Eq. 8.9 and rearranged it in the following form:

$$\int_{p_t}^{p_w} \frac{\dfrac{p}{Tz}\,dp}{\dfrac{2.6665(f/4)q^2}{D^5} + \dfrac{1}{1000}\left(\dfrac{p}{Tz}\right)^2} = \frac{1000\ \gamma_g\ Z}{53.34} \tag{8.25}$$

where

f = Moody friction factor
p_w = bottom-hole pressure, psia
p_t = wellhead (tubing) pressure, psia
γ_g = gas gravity (air = 1.00)
L = length of pipe, ft
q = gas flow rate MMscfd
T = absolute temperature, °R
z = gas deviation factor
D = internal diameter, in.
Z = depth of well, ft

Define

$$I = \frac{p/Tz}{2.6665(f/4)q^2/D^5 + 1/1000(p/Tz)^2} \tag{8.26}$$

which, for the static case, reduces to

$$I = 1000\left(\frac{Tz}{p}\right) \tag{8.27}$$

Equation 8.25 can be solved by numerical techniques, but because this is tedious and time-consuming, we suggest that acceptable accuracy can be obtained from a compromise two-step solution employing the trapezoidal and Simpson's rules. Selecting depths 0, $Z/2$, and Z, for the static case the integral may be expressed as

$$\int_{p_{ts}}^{p_{ws}} \left(1000 \frac{Tz}{p}\right) dp \simeq \frac{(p_{ms} - p_{ts})(I_{ms} + I_{ts})}{2} + \frac{(p_{ws} - p_{ms})(I_{ws} + I_{ms})}{2} \quad (8.28)$$

and Eq. 8.25 becomes

$$(p_{ms} - p_{ts})(I_{ms} + I_{ts}) + (p_{ws} - p_{ms})(I_{ws} + I_{ms}) = 37.5\ \gamma_g Z \quad (8.29)$$

Equation 8.29 may be separated into two expressions, one for each half of the flow string. For the upper half,

$$(p_{ms} - p_{ts})(I_{ms} + I_{ts}) = 37.5\ \gamma_g \frac{Z}{2} \quad (8.30)$$

For the lower half,

$$(p_{ws} - p_{ms})(I_{ws} + I_{ms}) = 37.5\ \gamma_g \frac{Z}{2} \quad (8.31)$$

The static bottom-hole pressure at depth Z in the well is finally given by

$$p_{ws} = p_{ts} + \frac{(3)(2)(1000)\gamma_g Z}{53.34(I_{ts} + 4I_{ms} + I_{ws})}$$

or

$$p_{ws} = p_{ts} + \frac{112.5\gamma_g Z}{I_{ts} + 4I_{ms} + I_{ws}} \quad (8.32)$$

where

I_{ts} is evaluated at $H = 0$, I_{ms} at $Z/2$, and I_{ws} at Z.

In applying this technique a trial-and-error procedure is required. The trapezoidal rule is used to evaluate I_{ms} and then I_{ws} in a two-step calculation.

The procedure is to solve first for an intermediate temperature and pressure condition at the midpoint of the vertical column and then repeat the calculations for bottom-hole conditions. A value of I_{ts} is first calculated from Eq. 8.27 at surface conditions. Then I_{ms} is assumed ($I_{ts} = I_{ms}$ as a first approximation) and p_{ms} is calculated for the midpoint conditions. Using this value of I_{ms}, a new value of I_{ms} is computed. The new value of I_{ms} is then used to recalculate p_{ms}. This procedure is repeated until successive calculations of p_{ms} are within the desired accuracy (usually within 1 psi difference). The entire procedure is repeated for the second interval to calculate I_{ws} and p_{ws}.

The Cullender and Smith method, which does not make any of the simplifying assumptions of other methods, is the most accurate method for calculating bottom-hole pressures. This method is generally applicable to shallow and deep wells, sour gases, and digital computation.

Example 8.5. Calculate the static bottom-hole pressure for the gas well of Example 8.3 using the Cullender and Smith method.

Solution

(a) Determine the value of z at wellhead conditions and compute I_{ts}:

$$p_{pr} = \frac{p_{ts}}{p_{pc}} = \frac{2300}{672} = 3.42$$

$$T_{pr} = \frac{T_{ts}}{T_{pc}} = \frac{74 + 460}{358} = \frac{534}{358} = 1.49$$

$$z = 0.765$$

$$I_{ts} = \frac{1000Tz}{p} = \frac{(1000)(534)(0.765)}{2300} = 178$$

(b) Calculate I_{ms} for intermediate conditions at a depth of 5790/2 or 2895 ft, assuming a straightline temperature gradient. As a first approximation, assume

$$I_{ms} = I_{ts} = 178$$

Then, from Eq. 8.30,

$$\Delta p = \frac{37.5\gamma_g(Z/2)}{I_{ts} + I_{ms}} = \frac{(37.5)(0.6)(2895)}{178 + 178} = 183 \text{ psi}$$

$$p_{ms} = p_{ts} + \Delta p = 2300 + 183 = 2483 \text{ psia}$$

$$p_{pr} = \frac{2483}{672} = 3.69$$

$$\bar{T} = \frac{T_{ts} + T_{ws}}{2} = \frac{74 + 160}{2} = 117°\text{F} = 577°\text{R}$$

$$\bar{T}_{pr} = \frac{577}{358} = 1.61$$

$$z = 0.820$$

$$I_{ms} = \frac{(1000)(577)(0.820)}{2483} = 191$$

$$\Delta p = \frac{(37.5)(0.6)(2895)}{178 + 191} = 177 \text{ psi}$$

$$p_{ms} = 2300 + 177 = 2477 \text{ psia}$$

Since the two values of p_{ms} are not equal, the calculations are repeated with $p_{ms} = 2477$ psia.

$$
\begin{aligned}
p_{pr} &= 3.69 \\
z &= 0.820 \\
I_{ms} &= 191 \\
p_{ms} &= 2477 \text{ psia}
\end{aligned}
$$

This is a check of the pressure at 2895 ft.

(c) Calculate I_{ws} at bottom-hole conditions assuming, for the first trial, $I_{ws} = I_{ms} = 191$. Then, from Eq. 8.31,

$$
\begin{aligned}
\Delta p &= \frac{37.5\gamma_g(Z/2)}{I_{ms} + I_{ws}} = \frac{(37.5)(0.6)(2895)}{191 + 191} = 171 \text{ psi} \\
p_{ws} &= p_{ms} + \Delta p = 2477 + 171 = 2648 \text{ psia} \\
p_{pr} &= \frac{2648}{672} = 3.94 \\
T_{pr} &= \frac{T_{ws}}{T_{pc}} = \frac{620}{358} = 1.73 \\
z &= 0.865 \\
I_{ws} &= \frac{(1000)(620)(0.865)}{2648} = 203 \\
\Delta p &= \frac{(37.5)(0.6)(2895)}{191 + 203} = 166 \text{ psi} \\
p_{ws} &= 2477 + 166 = 2643 \text{ psia}
\end{aligned}
$$

Repeating the calculation,

$$
\begin{aligned}
p_{pr} &= 3.93 \\
z &= 0.865 \\
I_{ws} &= 203 \\
p_{ws} &= 2643 \text{ psia}
\end{aligned}
$$

(d) Finally, using Eq. 8.32,

$$
\begin{aligned}
p_{ws} &= 2300 + \frac{(112.5)(0.6)(5790)}{178 + 4(191) + 203} \\
&= 2641 \text{ psia}
\end{aligned}
$$

8.3 FLOWING BOTTOM-HOLE PRESSURE

The flowing bottom-hole pressure of a gas well is the sum of the flowing wellhead pressure, the pressure exerted by the weight of the gas column, the kinetic energy change, and the energy losses resulting from friction. As the kinetic energy change is usually very small (about 0.1%) compared to the other energies, it is usually omitted in the calculations. This leaves the general mechanical energy equation (Eq. 8.3) in the form presented in Eq. 8.5, Eq. 8.6, or Eq. 8.9, for the situation of no heat loss from gas to surroundings and no work performed by the system.

Equation 8.9 may be written as

$$\frac{53.34}{\gamma_g}\frac{Tz}{p}\,dp + dZ + 0.002{,}68\,\frac{f}{D^5}\left(\frac{Tz}{p}\right)^2 q^2 dL = 0 \tag{8.33}$$

This equation is the basis for all methods of calculating flowing bottom-hole pressures from wellhead observations. The only assumptions made so far are single-phase gas flow and negligible kinetic energy change.

8.3.1 Average Temperature and Deviation Factor Method

This method is not as accurate as other methods for calculating flowing bottom-hole pressures. It is frequently used to obtain an approximate value of flowing bottom-hole pressure because of its simplicity.

Consider a slanting well of length L at an angle θ from the horizontal (Fig. 8.5):

$$L = \frac{Z}{\sin\theta} \qquad \text{thus} \qquad dL = \frac{dZ}{\sin\theta} \qquad \sin\theta = \frac{Z}{L}$$

Thus

$$dL = \frac{L}{Z}\,dZ \tag{8.34}$$

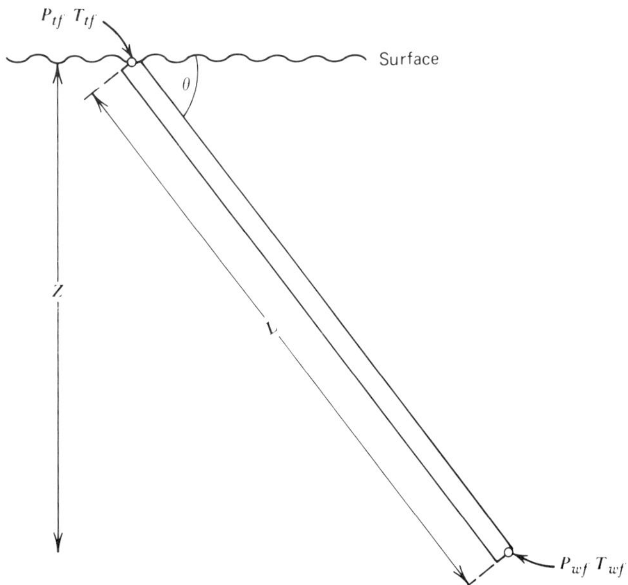

Fig. 8.5 Flowing gas well.

Using Eq. 8.34 in Eq. 8.33,

$$\frac{53.34}{\gamma_g}\frac{Tz}{p}\,dp + \left[1 + 0.002{,}68\,\frac{f}{D^5}\left(\frac{Tz}{p}\right)^2 q^2 \frac{L}{Z}\right] dZ = 0 \tag{8.35}$$

Using average values and integrating yields,

$$\frac{53.34}{\gamma_g}\,\bar{T}\bar{z}\int_{p_{tf}}^{p_{wf}} \frac{dp}{\left[p + 0.002{,}68\,\dfrac{\bar{f}}{D^5}(\bar{T}\bar{z}q)^2 \dfrac{L}{Z}\dfrac{1}{p}\right]} = -\int_0^{-Z} dZ \tag{8.36}$$

From calculus,

$$\int \frac{p\,dp}{C^2 + p^2} = \frac{dp}{p + \dfrac{C^2}{p}} = \frac{1}{2}\ln\,(C^2 + p^2)$$

Therefore Eq.8.36 becomes

$$\ln\left[\frac{C^2 + p_{wf}{}^2}{C^2 + p_{tf}{}^2}\right] = \frac{2\gamma_g Z}{53.34\bar{T}\bar{z}} \tag{8.37}$$

That is,

$$\frac{C^2 + p_{wf}{}^2}{C^2 + p_{tf}{}^2} = e^{2\gamma_g Z/53.34\bar{T}\bar{z}} \tag{8.38}$$

Substituting for C in Eq. 8.38,

$$p_{wf}{}^2 = p_{tf}{}^2 e^{2\gamma_g Z/53.34\bar{T}\bar{z}} + \frac{0.002{,}68}{D^5}\,\bar{f}(\bar{T}\bar{z}q)^2\,\frac{L}{Z}\,(e^{2\gamma_g Z/53.34\bar{T}\bar{z}} - 1)$$

or

$$p_{wf}{}^2 = p_{tf}{}^2\,e^s + \frac{25\gamma_g\bar{T}\bar{z}\bar{f}L(e^s - 1)q^2}{sD^5} \tag{8.39}$$

where

p_{wf} = flowing bottom hole pressure, psia
p_{tf} = flowing wellhead pressure,psia
s = $2\gamma_g Z/53.34\ \bar{T}\bar{z}$
$\bar{T}$ = arithmetic average of bottom hole and wellhead temperatures, °R
$\bar{z}$ = gas deviation factor at the arithmetic average temperature and arithmetic average pressure
$\bar{f}$ = Moody friction factor at arithmetic average temperature and pressure

L = length of flow string, ft
Z = vertical distance of reservoir from surface, ft
q = gas flow rate, MMcfd at 14.65 psia and 60°F
D = flow string diameter, in.

If the Fanning friction factor is used Eq. 8.39 becomes

$$p_{wf}^2 = p_{tf}^2 \, e^s + \frac{100\gamma_g \bar{T} \bar{z} \bar{f}_F L(e^s - 1)q^2}{sD^5} \tag{8.40}$$

Equation 8.39 may be applied as a one-step calculation from the wellhead to the sandface or as a multistep (usually two) calculation. The average value of gas deviation factor for each step may be obtained by estimation or iteration.

Example 8.6. Calculate the sandface pressure of a flowing gas well from the following surface measurements:

q = 5.153 MMscfd
D = 0.1663 ft = 1.9956 in. ≃ 2 in.
γ_g = 0.60
Depth = 5790 ft (bottom of casing)
T_{wf} = 160°F
T_{tf} = 83°F
p_{tf} = 2122 psia
e = roughness of tubing = 0.0006 in.
L = length of tubing = 5700 ft (well is vertical, $L = Z$)

Solution

Using Eq.8.39,

$$p_{wf}^2 = p_{tf}^2 \, e^s + \frac{25\gamma_g \bar{T} \bar{z} \bar{f} L(e^s - 1)q^2}{sD^5}$$

First Trial

Guess, p_{wf} = 2500 psia

$$\bar{p} = \frac{2500 + 2122}{2} = 2311 \text{ psia}$$

$$p_{pr} = \frac{2311}{672} = 3.44$$

$$\bar{T} = \frac{160 + 83}{2} = 121.5°\text{F} = 581.5 \text{ R}$$

$$\bar{T}_{pr} = \frac{581.5}{358} = 1.62$$

$$\bar{z} = 0.825; \qquad \frac{\mu}{\mu_1} = 1.45$$

At 1.0 atm and 121.5°F,

$$\mu_1 = 0.0115 \text{ cp}$$

Viscosity at average pressure:

$$\bar{\mu} = 0.0115(1.45) = 0.0167 \text{ cp}$$

The Reynold's number is given by

$$\bar{N}_{\text{Re}} \simeq \frac{20(\text{Mcfd})\gamma_g}{\mu D} = \frac{(20)(5153)(0.6)}{(0.0167)(2)} = 1.85(10^6)$$

$$\frac{e}{D} = \frac{0.0006}{2} = 0.0003$$

From the Moody friction chart,

$$\bar{f} = 0.015$$

$$s = \frac{2\gamma_g Z}{53.34\bar{T}\bar{z}} = \frac{(2)(0.6)(5700)}{(53.34)(581.5)(0.825)} = 0.2673$$

$$e^s = 1.3064$$

$$p_{wf}^2 = (2122)^2(1.3064) + \frac{(25)(0.6)(581.5)(0.825)(0.015)(5700)(0.3064)(5.153)}{(0.2673)(2)^5}$$

$$= 6{,}467{,}790 \text{ psia}^2$$

or

$$p_{wf} = 2543 \text{ psia}$$

Second Trial

$$\bar{p} = \frac{2543 + 2122}{2} = 2332.6 \text{ psia}$$

$$p_{pr} = \frac{2332.6}{672} = 3.47$$

$$\bar{z} \simeq 0.825$$

There is no appreciable change in $\bar{z}$ for this trial; so the first trial is sufficiently accurate.

$$p_{wf} = 2543 \text{ psia}$$

In summary, the assumptions made in the average temperature and average gas deviation factor method (or the Rzasa–Katz method) are:

1. Steady-state flow

2. Single-phase gas flow, although it may be used for condensate flow if proper adjustments are made in the flow rate, gas gravity, and Z-factor
3. Change in kinetic energy is small and may be neglected
4. Constant temperature at some average value
5. Constant gas deviation factor at some average value
6. Constant friction factor over the length of the conduit

8.3.2 The Sukkar and Cornell Method

Sukkar and Cornell published tabular data (Tables 8.1 and 8.2) for solving bottom-hole pressure calculation problems within the reduced temperature-reduced pressure range of $1.5 < T_{pr} < 1.7$; $1 < p_{pr} < 12$. The original Sukkar and Cornell method applies only to vertical wells. The assumptions made in this method include:

1. Steady-state flow.
2. Single-phase flow.
3. Change in kinetic energy is small and may be neglected.
4. Temperature is constant at some average value.
5. Friction is constant over the length of the conduit.

All equations in the Sukkar and Cornell Method are based on the assumption that the temperature is constant at some average value. Katz, in *Handbook of Natural Gas Engineering*, proposes the use of the log mean average temperature rather than the arithmetic average. The assumption of constant temperature also implies that the gas deviation factor is only dependent on a constant temperature. This assumption usually is not justified for gas flow with large temperature gradients.

In January 1974, Messer, Raghavan, and Ramey published an extension of the Sukkar and Cornell method that would apply to slanted wells at reduced temperatures to 3.0 and reduced pressures to 30.0. A more complete version of the extended Sukkar–Cornell integral for bottom-hole pressure calculations is contained in preprint paper SPE 3913. [see Tables A.38(a) to A.38(m)]. The Sukkar–Cornell integral may be written as

$$\frac{\gamma_g L \cos\theta}{53.34\bar{T}} = \int_{(p_{tf})_r}^{(p_{wf})_r} \frac{z/p_{pr}}{1 + B(z/p_{pr})^2}\, dp_{pr} \tag{8.41}$$

where

$$\bar{T} = \frac{T_{res} - T_{surf}}{\ln \dfrac{T_{res}}{T_{surf}}} = \text{log average temperature} \tag{8.42}$$

$$B = \frac{667 f q^2 \overline{T}^2}{D^5 p_{pc}{}^2 \cos\theta} \tag{8.43}$$

D is in inches

$$\cos\theta = Z/L$$
$$\theta = \text{angle of drift from vertical} \tag{8.44}$$

The integral on the right-hand side of Eq. 8.41 may be evaluated from any arbitrary lower limit, say $p_{pr} = 0.2$:

$$\int_{(p_{tf})_r}^{(p_{wf})_r} I(p_r)dp_r = \int_{0.2}^{(p_{wf})_r} I(p_r)dp_r - \int_{0.2}^{(p_{tf})_r} I(p_r)dp_r \tag{8.45}$$

Therefore,

$$\int_{0.2}^{(p_{wf})_r} I(p_r)dp_r = \int_{0.2}^{(p_{tf})_r} I(p_r)dp_r + \frac{\gamma_g Z}{53.34\overline{T}} \tag{8.46}$$

The extended Sukkar–Cornell integral is based on the Standing and Katz z-factor chart (Fig. 2.4) developed for natural gases containing small amounts of contaminants (hydrogen sulfide, nitrogen, and carbon dioxide). The z-factor may be adjusted for the presence of contaminants by using pseudocritical temperature and pseudocritical pressure corrections presented by Wichert and Aziz (Fig. 2.6).

Example 8.7 (after Messer, Raghavan, and Ramey). The bottom-hole pressure in a flowing gas well is to be calculated for the following conditions:

Depth, $Z = 20{,}500$ ft

Diameter of tubing, $D = 2.00$ in.

Gas gravity, $\gamma_g = 0.75$ (air = 1.00)

Gas flow rate, $q = 8.82$ MMscfd (at 60°F and 14.65 psia)

Moody friction factor, $f = 0.016$ (Chapter 7)

Reservoir temperature, $T_{res} = 900°R$

Surface temperature, $T_{surf} = 540°R$

Wellhead pressure, $p_{tf} = 14{,}575$ psia

$\theta = 0°$, $\cos\theta = 1$ (hole is vertical)

Analysis of gas indicates that the major contaminants are CO_2 (24%) and H_2S (38%), and the remainder is essentially methane.

Solution

Calculate the pseudocritical properties:

$$p_{pc} = \sum_{i=1}^{N} (y_i p_{ci})$$
$$= (0.24)(1071) + (0.38)(1306) + (0.38)(668) = 1007.2 \text{ psia}$$
$$T_{pc} = (0.24)(548) + (0.38)(672) + (0.38)(343) = 517.2°R$$

Entering Fig. 2.6 for ε: ε = 31, the adjusted pseudocritical temperature is

$$T'_{pc} = T_{pc} - \varepsilon = 517.2 - 31 = 486°\text{R}$$

The adjusted pseudocritical pressure is

$$\begin{aligned} p'_{pc} &= p_{pc}\left(\frac{T_{pc}}{T_{pc} + B(1 - B)\varepsilon_3}\right) \\ &= (1007.2)\left(\frac{486}{517.2 + (0.38)(1 - 0.38)(31)}\right) \\ &= 934 \text{ psia} \end{aligned}$$

The log average temperature (Eq. 8.35) is

$$\bar{T} = \frac{900 - 540}{\ln \dfrac{900}{540}} = 705 \text{ R}$$

$$T_{pr} = \frac{705}{486} = 1.45$$

From Eq. 8.36,

$$B = \frac{(667)(0.016)(8.82)^2(705)^2}{(2.00)^5(934)^2(1)} = 14.78 \simeq 15.0$$

$$\frac{\gamma_g L \cos\theta}{53.34\bar{T}} = \frac{(0.75)(20{,}500)(1)}{(53.34)(705)} = 0.4089$$

The tubing head calculation for $I(p_{pr})$ is

$$p_{tf} = 14{,}575 \text{ psia}$$

$$(p_{tf})_r = \frac{14{,}575}{934} = 15.6$$

From Table A.38(d), for $B = 15.0$, $T_{pr} = 1.45$, and $p_{pr} = 15.6$,

$$\int_{0.2}^{15.6} I(p_r)dp_r = 1.5216$$

Using Eq. 8.46,

$$\int_{0.2}^{(p_{wf})_r} I(p_r)dp_r = 1.5216 + 0.4089 = 1.9305$$

Using this value of the integral, from Table A.38(d), for $B = 15.0$, $T_{pr} = 1.45$,

$$(p_{wf})_r = 20.43$$

Therefore,

$$p_{wf} = (20.43)(934) = 19{,}082 \text{ psia}$$

8.3.3 The Cullender and Smith Method

In the Cullender and Smith method, the following assumptions are made:

1. Steady-state flow.
2. Single-phase gas stream.
3. Change in kinetic energy is small and may be neglected.

The general equation for the flow of gas in inclined pipes may be expressed as follows:

$$\frac{1000\gamma_g Z}{53.34} = \int_{p_{tf}}^{p_{wf}} \frac{\frac{p}{Tz}\,dp}{\frac{2.6665(f/4)q^2}{D^5} + \frac{1}{1000}\frac{Z}{L}\left(\frac{p}{Tz}\right)^2} \tag{8.47}$$

where D is in inches. If

$$F^2 = \frac{2.6665(f/4)q^2}{D^5} \tag{8.48}$$

then

$$\frac{1000\gamma_g Z}{53.34} = \int_{p_{tf}}^{p_{wf}} \frac{\frac{p}{Tz}\,dp}{F^2 + \frac{1}{1000}\frac{Z}{L}\left(\frac{p}{Tz}\right)^2} \tag{8.49}$$

Without making certain assumptions with respect to T and z, Eq. 8.42 does not lend itself to mathematical integration. However, an evaluation of the integral over definite limits can be accomplished by numerical means. For a two-step calculation,

$$\frac{1000\gamma_g Z}{53.34} = \int_{p_{tf}}^{p_{wf}} I\,dp = \frac{(p_{mf} - p_{tf})(I_{mf} + I_{tf})}{2} + \frac{(p_{wf} - p_{mf})(I_{wf} + I_{mf})}{2} \tag{8.50}$$

Thus

$$37.5\gamma_g Z = (p_{mf} - p_{tf})(I_{mf} + I_{tf}) + (p_{wf} - p_{mf})(I_{wf} + I_{mf}) \tag{8.51}$$

where

$$I = \frac{\left(\frac{p}{Tz}\right)}{F^2 + \frac{1}{1000}\frac{Z}{L}\left(\frac{p}{Tz}\right)^2} \tag{8.52}$$

Equation 8.50 may be separated into two expressions, one for each half of the flow string. For the upper half,

$$37.5\gamma_g \frac{Z}{2} = (p_{mf} - p_{tf})(I_{mf} + I_{tf}) \tag{8.53}$$

For the lower half,

$$37.5\gamma_g \frac{Z}{2} = (p_{wf} - p_{mf})(I_{wf} + I_{mf}) \tag{8.54}$$

Again, Simpson's rule gives

$$37.5\gamma_g Z = \frac{p_{wf} - p_{tf}}{3}(I_{tf} + 4I_{mf} + I_{wf}) \tag{8.55}$$

Equation 8.48 may be simplified by using Nikuradse's friction factor equation for fully developed turbulent flow, based on an absolute roughness of 0.000,60 in., to give

$$F_r q = F = \frac{0.107{,}96q}{D^{2.612}} \qquad D < 4.277 \text{ in.} \tag{8.56}$$

and

$$F_r q = F = \frac{0.103{,}37q}{D^{2.582}} \qquad D > 4.277 \text{ in.} \tag{8.57}$$

Values of F_r for various tubing and casing sizes are presented in Table 8.3.

The following procedure is recommended for solving Eq. 8.55 (from *Theory and Practice of the Testing of Gas Wells*):

1. Calculate the left-hand side of Eq. 8.53 for the upper half of the flow string.
2. Calculate F^2 from Eq. 8.56 or Eq. 8.57, or from Table 8.3.
3. Calculate I_{tf} from Eq. 8.52 and wellhead conditions.
4. Assume I_{mf} = I_{tf} for the conditions at the average well depth or at the midpoint of the flow string.
5. Calculate p_{mf} from Eq. 8.53.

TABLE 8.3

Values of F_r for Various Tubing and Casing Sizes

Use Only for ID Less Than 4.277 in.

$$F_r = \frac{0.107,97}{d^{2.612}}$$

Nominal Size in.	OD in.	lb/ft	ID in.	F_r
1	1.315	1.80	1.049	0.095,288
1¼	1.660	2.40	1.380	0.046,552
1½	1.990	2.75	1.610	0.031,122
2	2.375	4.70	1.995	0.017,777
2½	2.875	6.50	2.441	0.010,495
3	3.500	9.30	2.992	0.006,167
3½	4.000	11.00	3.476	0.004,169
4	4.500	12.70	3.958	0.002,970
4½	4.750	16.25	4.082	0.002,740
	4.750	18.00	4.000	0.002,889
4¾	5.000	18.00	4.276	0.002,427
	5.000	21.00	4.154	0.002,617
	Use Only for ID Greater than 4.277 in. $F_r = \frac{0.103,37}{d^{2.582}}$			
4¾	5.000	13.00	4.494	0.002,134,5
	5.000	15.00	4.408	0.002,243,7
5³⁄₁₆	5.500	14.00	5.012	0.001,610,5
	5.500	15.00	4.976	0.001,640,8
	5.500	17.00	4.892	0.001,714,5
	5.500	20.00	4.778	0.001,822,1
	5.500	23.00	4.670	0.001,932,9
	5.500	25.00	4.580	0.002,032,5
5⅝	6.000	15.00	5.524	0.001,252,8
	6.000	17.00	5.450	0.001,297,2
	6.000	20.00	5.352	0.001,359,5
	6.000	23.00	5.240	0.001,435,8
	6.000	26.00	5.140	0.001,509,0
6¼	6.625	20.00	6.049	0.000,991,0
	6.625	22.00	5.989	0.001,016,9
	6.625	24.00	5.921	0.001,047,3
	6.625	26.00	5.855	0.001,078,1
	6.625	28.00	5.791	0.001,109,1
	6.625	31.80	5.675	0.001,168,6
	6.625	34.00	5.595	0.001,212,2
6⅝	7.000	20.00	6.456	0.000,887,6
	7.000	22.00	6.398	0.000,857,4
	7.000	24.00	6.336	0.000,879,2
	7.000	26.00	6.276	0.000,901,1
	7.000	28.00	6.214	0.000,924,5

TABLE 8.3 (Continued)

Values of F_r for Various Tubing and Casing Sizes

Use Only for ID Greater than 4.277 in.

$$F_r = \frac{0.103,37}{d^{2.582}}$$

Nominal Size in.	OD in.	lb/ft	ID in.	F_r
	7.000	30.00	6.154	0.000,947,9
	7.000	40.00	5.836	0.001,087,1
7¼	7.625	26.40	6.969	0.000,687,5
	7.625	29.70	6.875	0.000,712,1
	7.625	33.70	6.765	0.000,742,4
	7.625	38.70	6.625	0.000,783,6
	7.625	45.00	6.445	0.000,841,3
	8.000	26.00	7.386	0.000,591,7
7⅝	8.125	28.00	7.485	0.000,571,7
	8.125	32.00	7.385	0.000,591,9
	8.125	35.50	7.285	0.000,613,2
	8.125	39.50	7.185	0.000,635,4
8¼	8.625	17.50	8.249	0.000,444,8
	8.625	20.00	8.191	0.000,453,0
	8.625	24.00	8.097	0.000,466,7
	8.625	28.00	8.003	0.000,481,0
	8.625	32.00	7.907	0.000,496,2
	8.625	36.00	7.825	0.000,509,8
	8.625	38.00	7.775	0.000,518,3
	8.625	43.00	7.651	0.000,540,3
8⅝	9.000	34.00	8.290	0.000,439,2
	9.000	38.00	8.196	0.000,452,3
	9.000	40.00	8.150	0.000,458,9
	9.000	45.00	8.032	0.000,476,5
9	9.625	36.00	8.921	0.000,363,4
	9.625	40.00	8.835	0.000,372,6
	9.625	43.50	8.775	0.000,381,4
	9.625	47.00	8.681	0.000,389,9
	9.625	53.50	8.535	0.000,407,4
	9.625	58.00	8.435	0.000,420,0
9⅝	10.000	33.00	9.384	0.000,416,7
	10.000	55.50	8.908	0.000,364,8
	10.000	61.20	8.790	0.000,377,5
10	10.750	32.75	10.192	0.000,257,6
	10.750	35.75	10.136	0.000,261,3
	10.750	40.00	10.050	0.000,267,1
	10.750	45.50	9.950	0.000,274,1
	10.750	48.00	9.902	0.000,277,6
	10.750	54.00	9.784	0.000,286,3

Source: After Cullender and Smith.

6. Using the value of p_{mf} calculate in step 5 and the arithmetic average temperature $T_{mf'}$ determine the value of I_{mf} from Eq. 8.52.
7. Recalculate p_{mf} from Eq. 8.53. If this recalculated value is not within 1 psi of the p_{mf} calculated in step 5, repeat steps 6 and 7 until the above criteria are satisfied.
8. Assume $I_{wf} = I_{mf}$ for the conditions at the bottom of the flow string.
9. Repeat steps 5 to 7, using Eq. 8.54 for the lower half of the flow string, and obtain a value of the bottom-hole pressure, p_{wf}.
10. Apply Simpson's rule as expressed by Eq. 8.55 to obtain a more accurate value of the flowing bottom-hole pressure.

The following example illustrates the use of the Cullender and Smith method to calculate flowing bottom-hole pressure from wellhead measurements.

Example 8.8. Calculate the flowing bottom-hole pressure for the gas well by Example 8.6, using Cullender and Smith method.

Solution

$$T_{mf} = \frac{T_{tf} + T_{wf}}{2} = \frac{543 + 620}{2} = 581.5°\text{R}$$

$$\text{Wellhead:} \quad T_{pr} = \frac{T_{tf}}{T_{pc}} = \frac{543}{358} = 1.52$$

$$\text{Midpoint:} \quad T_{pr} = \frac{T_{mf}}{T_{pc}} = \frac{581.5}{358} = 1.62$$

$$\text{Bottom:} \quad T_{pr} = \frac{T_{wf}}{T_{pc}} = \frac{620}{358} = 1.73$$

$$\text{Wellhead:} \quad p_{pr} = \frac{p_{tf}}{p_{pr}} = \frac{2122}{672} = 3.16$$

$$D = 1.9956 \text{ in.} < 4.277 \text{ in.}$$

From Eq. 8.56,

$$F = \frac{(0.107{,}97)(5.153)}{(1.995{,}6)^{2.612}} = 0.091{,}53$$

$$F^2 = 0.008{,}38$$

Left-hand side of Eqs. 8.53 and 8.54:

$$37.5\gamma_g \frac{Z}{2} = (37.5)(0.60)(5700/2) = 64{,}125$$

Calculate I_{tf}

From the z-chart (Fig. 2.4), at $T_{pr} = 152$ and $p_{pr} = 3.16$,

$$z_{tf} = 0.785$$
$$(p_{tf})/(T_{tf} z_{tf}) = (2122)/(543 \times 0.785) = 4.978$$
$$I_{tf} = \frac{p_{tf}/T_{tf} z_{tf}}{F^2 + \dfrac{(p_{tf}/T_{tf} z_{tf})^2}{1000}} = \frac{4.978}{0.008{,}38 + \dfrac{(4.978)^2}{1000}} = 150.11$$

Step 1 (Upper Half of Flow String)

In the first trial assume that

$$I_{mf} = I_{tf} = 150.11$$

Solving Eq. 8.53 for p_{mf},

$$64{,}125 = (p_{mf} - 2122)(150.11 + 150.11)$$
$$p_{mf} = 2336 \text{ psia}$$

In the second trial,

$$p_{pr} = \frac{p_{mf}}{p_{pc}} = \frac{2336}{672} = 3.48$$

From the z-chart,

$$z_{mf} = 0.825 \qquad \text{at } T_{pr} = 1.62$$
$$(p_{mf}/T_{mf} z_{mf}) = (2336)/(581.5 \times 0.825) = 4.869$$
$$I_{mf} = \frac{4.869}{0.008{,}38 + \dfrac{(4.869)^2}{1000}} = 151.74$$

Solving Eq. 8.53 for p_{mf},

$$64{,}125 = (p_{mf} - 2122)(151.74 + 150.11)$$
$$p_{mf} = 2334 \text{ psia}$$

In the third trial,

$$p_{pr} = \frac{2334}{672} = 3.47$$
$$z_{mf} = 0.825 \qquad \text{at } T_{pr} = 1.62$$
$$(p_{mf}/T_{mf} z_{mf}) = (2334)/(581.5 \times 0.825) = 4.865$$
$$I_{mf} = \frac{4.865}{0.008{,}38 + \dfrac{(4.865)^2}{1000}} = 151.80$$

Once more, solving Eq. 8.53 for p_{mf},

$$\begin{aligned} 64{,}125 &= (p_{mf} - 2122)(151.80 + 150.11) \\ p_{mf} &= 2334 \text{ psia} \end{aligned}$$

which is within 1 psi of the old value of p_{mf}.

Step 2 (Lower Half of Flow String)

In the first trial assume that

$$I_{wf} = I_{mf} = 151.80$$

Solving Eq. 8.54 for p_{wf},

$$\begin{aligned} 64{,}125 &= (p_{wf} = 2334)(151.80 + 151.80) \\ p_{wf} &= 2545 \text{ psia} \end{aligned}$$

In the second trial,

$$\begin{aligned} p_{pr} &= \frac{p_{wf}}{p_{pc}} = \frac{2545}{672} = 3.79 \\ z_{wf} &= 0.865 \qquad \text{at } T_{pr} = 1.73 \\ (p_{wf}/T_{wf}z_{wf}) &= (2545)/(620 \times 0.865) = 4.745 \\ I_{wf} &= \frac{4.745}{0.008{,}38 + \dfrac{(4.745)^2}{1000}} = 153.58 \end{aligned}$$

Solving Eq. 8.54 for p_{wf},

$$\begin{aligned} 64{,}125 &= (p_{wf} - 2334)(153.58 + 151.80) \\ p_{wf} &= 2544 \text{ psia} \end{aligned}$$

Parabolic Interpolation

From Eq. 8.55,

$$\begin{aligned} (64{,}125 \times 2) &= \frac{p_{wf} - p_{tf}}{3}[150.00 + 4(151.80) + 153.58] \\ p_{wf} - p_{tf} &= 422 \\ p_{wf} &= 2122 + 422 = 2544 \text{ psia} \end{aligned}$$

The Cullender and Smith method is easily adapted to digital computer calculations. This method is applicable to sour gas wells if approximate corrections are included in the determination of gas deviation factors, as shown in Chapter 2.

8.4 ACCURACY OF CALCULATED BOTTOM-HOLE PRESSURES

For gas and gas condensate wells with normal temperature gradients and with depths of less than about 12,000 ft, all methods presented provide satisfactory accuracy. Because of its simplicity, accuracy, and consistency, the Sukkar and Cornell method is preferred for general use where bottom-hole pressures are less than 10,000 psia. It may be used for pressures higher than 10,000 psia with Messer et al.'s tables. The Cullender and Smith method is the most accurate and is recommended for large temperature gradients and for bottom-hole pressures in excess of 10,000 psia.

In a discussion of the effects of assumptions used to calculate bottom-hole pressures in gas wells Young concludes:

1. An integration interval of 1000 ft should be used to ensure accurate trapezoidal integration of Eq. 8.47 or Eq. 8.58, which considers change in kinetic energy:

$$1000Z = \int_{p_{tf}}^{p_{wf}} \frac{\dfrac{53.34}{\gamma_g}\dfrac{p}{Tz} + \dfrac{111.1q^2}{D^4p}}{\dfrac{2.6665(f/4)q^2}{D^5} + \dfrac{1}{1000}\dfrac{Z}{L}\left(\dfrac{p}{Tz}\right)^2} dp \tag{8.58}$$

   where

   $111.1q^2/D^4p$ = kinetic energy term

2. Simpson's rule should not be applied in an effort to correct for large trapezoidal integration intervals.
3. If the flowing pressure at total depth is the desired quantity, a change in kinetic energy may be ignored when the depth is greater than 4000 ft or the wellhead flowing pressure is above 100 psia. If an accurate pressure traverse is desired, a change in kinetic energy should be considered when the wellhead flowing pressure is below 500 psia.
4. A discontinuity can develop when numerically integrating Eq. 8.47 or 8.58 for the injection case. When a discontinuity occurs, a pressure change in that interval should be set equal to zero. Note, also, that Simpson's rule cannot be applied in this situation.
5. Temperature and gas deviation factors can be assumed constant at their average values for depths up to 8000 ft. The average temperature and gas deviation factor method, however, should not be applied unless a change in kinetic energy is insignificant.

For all normal field situations, Eq. 8.39 should be used to calculate bottom-hole pressure in gas wells. When depth exceeds 8000 ft, the calculation may be broken into two or more intervals. When unusual conditions, such as significant kinetic energy change, prohibit using Eq. 8.39, then Eq. 8.58 should be used.

While the methods with the noted limitations are sufficiently accurate for most engineering purposes, calculated bottom-hole pressures are still subject to serious uncertainties because of:

1. Unknown amounts of liquid hydrocarbons or water in the wellbore and tubing.
2. Departure of actual temperature distribution from that assumed in the method of calculation.

3. Inaccuracies in the determination of gas deviation factor from correlations. Since no correlation charts are available to calculate gas deviation factor or pressures exceeding 10,000 psia (pseudoreduced pressure > 15), it is desirable to obtain z-factors for higher pressure cases from actual laboratory measurements on recombined fluid samples.
4. In the case of the condensate systems, unknown changes of fluid composition with depth.
5. Difficult in the selection of proper friction factor.
6. Inaccuracies in the measurement of specific gravity of the well effluent and in the measurement of flow rate.

8.5 ANNULAR FLOW

In most instances, gas wells are produced through tubing (usually 2 to 3.5 in ID). Occasionally, however, a well may be produced through the casing-tubing annulus. The tubing flow equations may be used for annular flow, provided proper account is taken of the flow diameter variable.

The commonly used method of handling flow in a noncircular cross section is through an "effective diameter." This is defined as

$$D_{\text{eff}} = \frac{4 \times \text{cross-sectional area of flow}}{\text{wetted perimeter}} \tag{8.59}$$

which, for the annulus, reduces to

$$D_{\text{eff}} = \frac{4\,\frac{\pi}{4}\,(D_2^2 - D_1^2)}{(D_2 + D_1)\pi} = (D_2 - D_1) \tag{8.60}$$

where

D_2 = inside diameter of the casing
D_1 = outside diameter of the tubing

In the flow equations derived, the diameter is raised to the fifth power D^5. It is, however, incorrect to use $(D_2 - D_1)^5$. The correct forms are as follows.

The frictional losses for circular tubes are given by

$$lw = \frac{fu^2\,dL}{2g_cD}$$

$$u(\text{ft/sec}) = \frac{\text{ft}^3}{\text{sec}} \times \frac{1}{\text{ft}^2} \tag{8.61}$$

Thus

$$u(\text{ft/sec}) = \frac{q(10^6)}{(24)(3600)} \frac{p_b}{p} \frac{T}{T_b} \frac{z}{1} \times \frac{1}{\frac{\pi}{4}\left[\left(\frac{D_2}{12}\right)^2 - \left(\frac{D_1}{12}\right)^2\right]}$$

or

$$u(\text{ft/sec}) = 59.785 \frac{Tzq}{p(D_2^2 - D_1^2)} \tag{8.62}$$

where

q = gas flow rate in MMcfd measured at 14.65 psia and 60°F

Therefore

$$lw = \frac{f(59.785)^2 T^2 z^2 q^2 \, dL}{(2)(32.17)p^2(D_2^2 - D_1^2)^2 \left(\frac{D_2}{12} - \frac{D_1}{12}\right)}$$

or

$$lw = 667 \left(\frac{Tz}{p}\right)^2 \frac{f}{D_2 - D_1}\left(\frac{q}{D_2^2 - D_1^2}\right)^2 dL \tag{8.63}$$

Note that

$$(D_2 - D_1)(D_2^2 - D_1^2)^2 = (D_2 - D_1)^3(D_2 + D_1)^2 \neq (D_2 - D_1)^5$$

Thus, in the flow equations, D^5 for flow in circular tubes will be replaced by $(D_2 - D_1)^3(D_2 + D_1)^2$ for annual flow. For example, the energy equation (Eq. 8.25) becomes

$$\frac{53.34}{\gamma_g} \frac{Tz}{p} dp + dZ + \frac{0.002,68 \, f \, q^2}{(D_2 - D_1)^3(D_2 + D_1)^2}\left(\frac{Tz}{p}\right)^2 dL = 0 \tag{8.64}$$

The Reynold's number is another term that does not allow direct substitution of $(D_2 - D_1)$ for D. To evaluate the Moody friction factor for annual flow, the Reynold's number becomes

$$(N_{\text{Re}})_{\text{annulus}} \simeq \frac{20,000 q \gamma_g}{\mu(D_2 + D_1)} \tag{8.65}$$

where

q = gas flow rate, MMcfd
γ_g = gas gravity (air = 1)
μ = gas viscosity, cp
D_2, D_1 = casing/tubing diameter, in.

8.6 GAS-LIQUID FLOW IN WELLS

The equations presented in the previous sections are strictly correct for single-phase flow of dry gas in wells. Many gas wells produce condensates and water. In these cases some liquid will be traveling in the well with the gas. Two-phase flow problems in flowing wells can be handled by either gravity adjustment or by applying a two-phase flow correlation.

The best correlations for two-phase vertical flow are those of Hagedorn and Brown, Duns and Ros, Orkiszewski, Beggs and Brill, and Govier and Aziz. Only the Hagedorn and Brown, and Orkiszewski methods will be discussed in detail for vertical flow. The Beggs and Brill method was discussed under pipeline flow. It can also be used for well performance but has been found to sometimes overpredict pressure drop in vertical flow. Discussions and applications of the other correlations are available from Brown's book, *The Technology of Artificial Lift Methods*, Volume 1, or from the original publications of the investigators.

8.6.1 Flow Regimes

Flow patterns or regimes encountered in vertical two-phase flow are shown in Fig. 8.6, after Orkiszewski. Boundaries for the various flow regimes are shown in Fig. 8.7. In *bubble flow*, the pipe is almost completely filled with liquid and the free gas

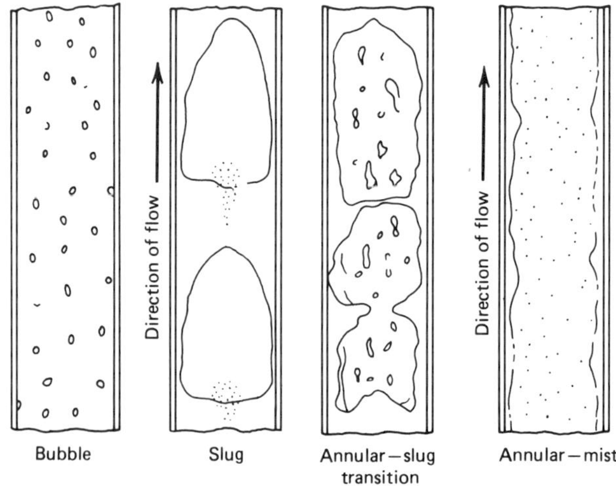

Fig. 8.6 Vertical flow patterns. (After Orkiszewski.)

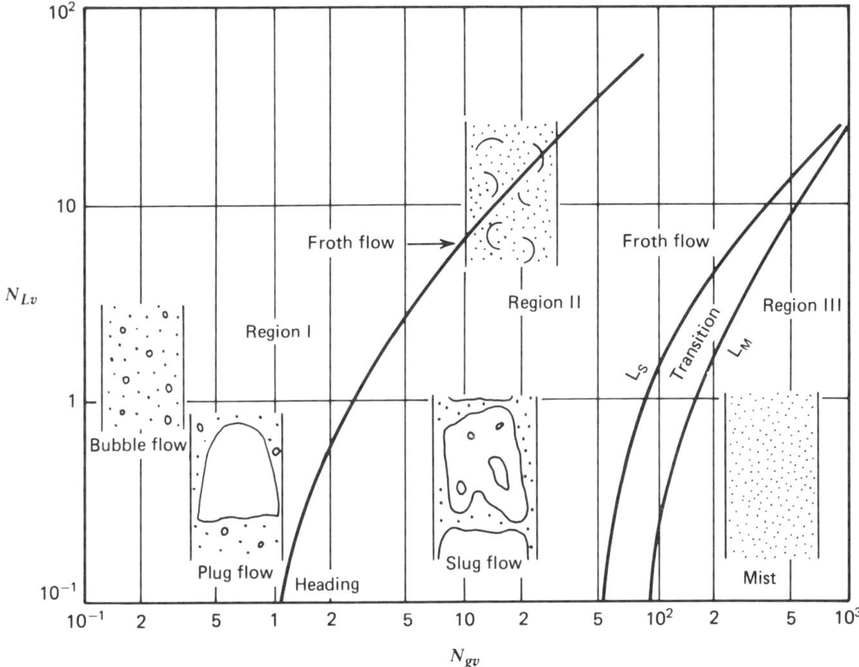

Fig. 8.7 Flow regime map. (After Orkiszewski.)

phase is present in small bubbles. The gas of the pipe is always contacted by the liquid phase and the bubbles have little effect on the pressure gradient. In *slug flow* both gas and liquid have significant effects on the pressure gradient. The gas bubbles coalesce and form slugs; however, the liquid phase is still continuous. In *transition flow*, there is a change from continuous liquid phase to continuous gas phase. The liquid effects are still significant, but the gas effects are predominant. In *mist flow* the liquid is entrained as droplets in the gas phase. The gas phase controls the pressure gradient. The pipe wall is coated with a liquid film.

8.6.2 Gravity Adjustment

The procedures presented for calculating bottom-hole pressures or wellhead tubing pressures for given flow rates may be used for wells producing gas and liquid if the flow rate, gas deviation factor, and specific gravity are adjusted to account for the presence of liquids. The total volumetric gas flow rate, q_{wg}, can be determined using methods presented in Section 8.12. The two-phase gas deviation factor, z_{wg}, can be obtained from Fig. 2.14.

The mixture specific gravity may be calculated using

$$\gamma_{wg} = \frac{\gamma_g + 4591\gamma_L/R}{1 + 1123/R} \tag{8.66}$$

where

γ_{wg} = adjusted fluid gravity (air = 1)
γ_g = dry gas gravity (air = 1)
γ_L = liquid specific gravity (water = 1)
R = producing gas-liquid ratio, scf/STB

The gravity adjustment method may be used for wells producing at high gas-liquid ratios: R greater than 10,000 scf/STB or liquid loading less than 100 bbl/MMscf.

8.6.3 The Hagedorn and Brown Method

The mechanical energy equation, ignoring the acceleration term, may be written as

$$\frac{dP}{dZ} = \frac{g}{g_c} \rho_m \sin\theta + \frac{f \rho_f u_m^2}{2 g_c D} \tag{8.67}$$

where

ρ_m = mixture density = $\rho_L H_L + \rho_g(1 - H_L)$
ρ_L = liquid density
ρ_g = gas density
H_L = liquid holdup (fraction of pipe occupied by liquid)
θ = angle of well from horizontal
$u_m = u_{sL} + u_{sg}$
u_{sL} = superficial liquid velocity = q_L/A_p
u_{sg} = superficial gas velocity = q_g/A_p
A_p = area of flow string = $\pi D^2/4$
D = flow string ID
$\rho_f = \rho_n^2/\rho_m$
$\rho_n = \rho_L\lambda + \rho_g(1 - \lambda)$
$\lambda = u_{sL}/u_m$
f = friction factor

The friction factor may be calculated from the Jain equation or Moody diagram using pipe relative roughness and a Reynolds number given by

$$N_{\text{Re}\,m} = \frac{D u_m \rho_n}{\mu} \tag{8.68}$$

where

$\mu_m = \mu_L^{H_L} \mu_g^{(1-H_L)}$
μ_L = liquid viscosity
μ_g = gas viscosity

The liquid holdup, H_L, used in the Hagedorn and Brown correlation was not measured. It is not a true measure of the portion of the pipe occupied by liquid, but merely a correlating parameter. In order to determine H_L, the following dimensionless numbers are evaluated from known data:

$$N_{Lv} = 1.938 u_{sL} (\rho_L/\sigma)^{0.25} = \text{Liquid velocity number} \tag{8.69}$$

$$N_{gv} = 1.938 u_{sg}(\rho_L/\sigma)^{0.25} = \text{Gas velocity number} \tag{8.70}$$

$$N_d = 120.872 D (\rho_L/\sigma)^{0.5} = \text{Pipe diameter number} \tag{8.71}$$

$$N_L = 0.157{,}26 \mu_L (1.0/\rho_L \sigma^3)^{0.25} = \text{Liquid viscosity number} \tag{8.72}$$

where the following field units are used

u_{sL} = ft/sec
u_{sg} = ft/sec
ρ_m = lbm/cu ft
σ_L = gas-liquid surface tension, dynes/cm
μ_L = cp
D = ft

The following procedure is used to find H_L:

1. Calculate N_L.

2. Determine CN_L from Fig. 8.8.

3. Calculate the holdup factor

$$X_H = \frac{N_{Lv}(CN_L)p^{0.1}}{N_{gv}^{0.575}\, p_b^{0.1}\, N_d} \tag{8.73}$$

where p_b = base pressure (14.7 psia)

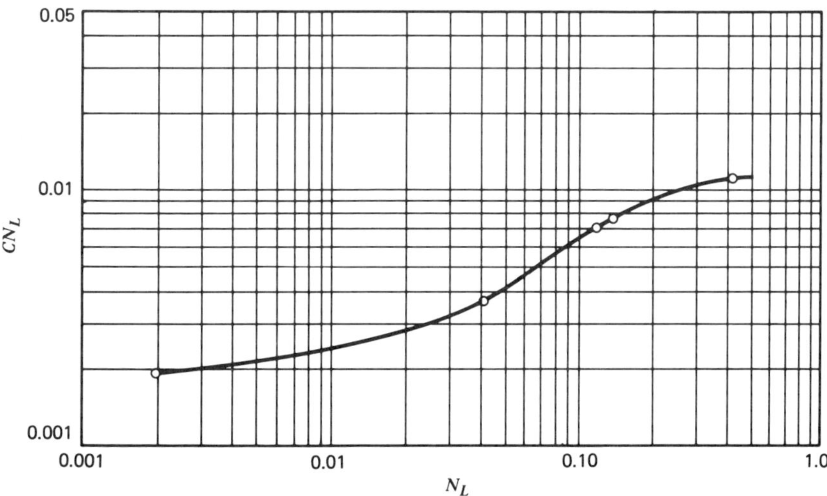

Fig. 8.8 Correlation for viscosity number coefficient C. (After Hagedorn and Brown.)

4. Find H_L/ψ from Fig. 8.9.
5. Calculate

$$X\psi = \frac{N_{gv}N_L^{0.38}}{N_d^{2.14}} \tag{8.74}$$

6. Find ψ from Fig. 8.10.

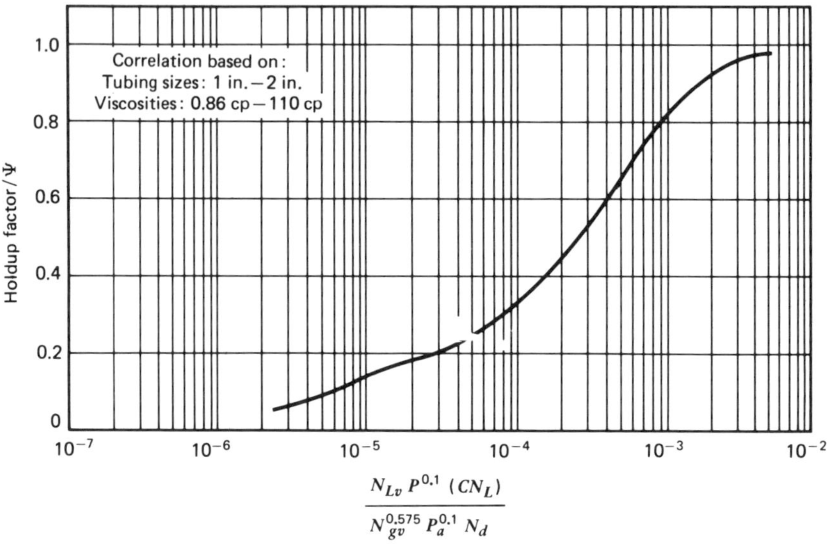

Fig. 8.9 Holdup-factor correlation. (After Hagedorn and Brown.)

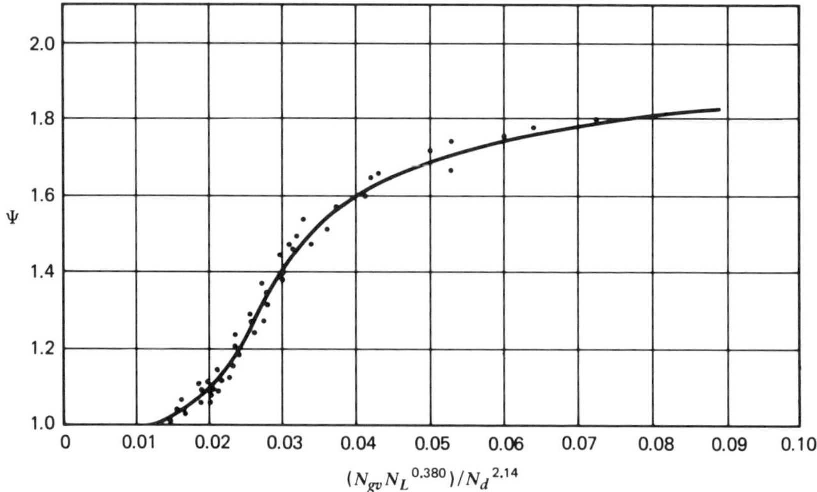

Fig. 8.10 Correlation for secondary correction factor. (After Hagedorn and Brown.)

7. Calculate $H_L = \psi(H_L/\psi)$.

A constraint on liquid holdup is $H_L \geqslant \lambda$.

Knowing H_L, $N_{\text{Re}\ m}$ and hence f can be calculated, and thus the pressure gradient for the interval can be determined.

8.6.4 The Orkiszewski Method

Orkiszewski tested several existing correlations on field data and combined the best of these with his own correlation for slug flow. The slug flow correlation was developed using the data of Hagedorn and Brown. He selected the Griffith and Wallis method for bubble flow and the Duns and Ros method for mist flow. The Orkiszewski correlation is one of the most widely accepted in the petroleum industry. The Orkiszewski equation is

$$\Delta p_k = \frac{1}{144}\left[\frac{\bar{\rho} + \tau_f}{1 - w_t q_g/4637A_p^2\bar{p}}\right]_k \Delta Z_k \tag{8.75}$$

where

$\bar{\rho}$ = average fluid density, lbm/cu ft
Δp = pressure drop, psi
$\bar{p}$ = average pressure, psia
w_t = total mass flow rate, lbm/sec
τ_f = friction loss gradient, psi/ft
q_g = gas volumetric flow rate, cfs
ΔZ = depth change, ft

The flow regime is determined by the limits below. Griffith and Wallis defined the boundary between bubble flow and slug flow. Duns and Ros defined the boundaries for the remaining three regimes.

Flow Regime	Limits	
Bubble flow	$q_g/q_t < L_B$	
Slug flow	$q_g/q_t > L_B$	$\overline{\overline{u}}_g < L_s$
Transition flow	$L_M > \overline{\overline{u}}_g > L_s$	
Mist flow	$\overline{\overline{u}}_g > L_M$	

$$\overline{\overline{u}}_g = \frac{q_g}{A_p}\left(\frac{\rho_L}{g\sigma}\right)^{1/4} \tag{8.76}$$

$$L_B = 1.071 - (0.2218u_t^2/D_h) \tag{8.77}$$

with limit $L_B \geqslant 0.13$.

$$L_s = 50 + 36\overline{\overline{u}}_g q_L/q_g \tag{8.78}$$

$$L_M = 75 + 84\ (\overline{\overline{u}}_g q_L/q_g)^{0.75} \tag{8.79}$$

where

$\bar{\bar{u}}$ = dimensionless gas velocity
u_t = total fluid velocity, q_g/A_p, ft/sec
D_h = hydraulic diameter, ft
L_B, L_s, L_M = bubble-slug, slug-transition, and transition-mist boundaries, respectively, dimensionless

The average density and friction loss gradient are calculated for different flow regimes as follows:

Bubble Flow

The void fraction of gas in bubble flow can be expressed as

$$H_g = \frac{1}{2}\left[1 + \frac{q_t}{u_s A_p} - \sqrt{(1 + q_t/u_s A_p)^2 - \frac{4q_g}{u_s A_p}}\right] \tag{8.80}$$

where

u_s = slip velocity (bubble rise velocity), ft/sec
= 0.8 ft/sec as suggested by Griffith

The average density is computed as

$$\bar{\rho} = (1 - H_g)\,\rho_L + H_g \rho_g \tag{8.81}$$

The friction gradient is

$$\tau_f = f \rho_L u_L^2 / 2 g_c D_h \tag{8.82}$$

where

$$u_L = q_L / [A_p(1 - H_g)] \tag{8.83}$$

The friction factor f is obtained from the Jain equation or Moody diagram using a Reynolds number calculated as

$$N_{\mathrm{Re}} = 1488 D_h u_L \rho_L / \mu_L \tag{8.84}$$

Slug Flow

The average density is given by

$$\bar{\rho} = \frac{w_t + \rho_L u_s A_p}{q_t + u_s A_p} + \delta \rho_L \tag{8.85}$$

where

$$u_s = C_1 C_2 \sqrt{g D_h} \tag{8.86}$$

δ = liquid distribution coefficient

Values of C_1 and C_2 are obtained from Figs. 8.11 and 8.12. The value of δ is dependent on the continuous liquid phase and whether u_s is less than or greater

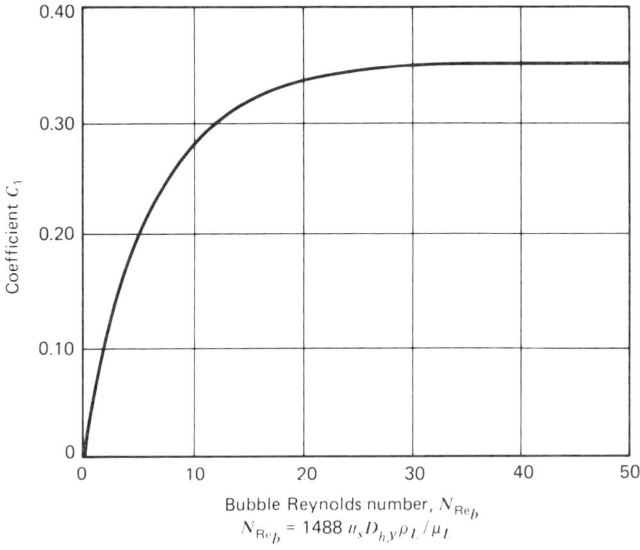

Fig. 8.11 Griffith and Wallis C_1 vs. Bubble Reynolds number.

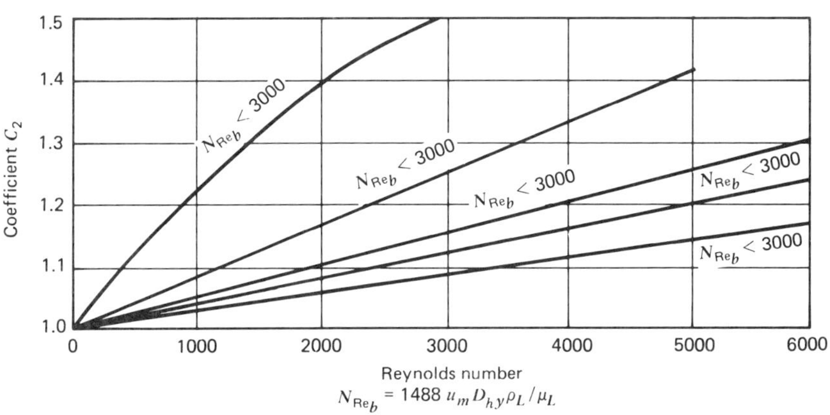

Fig. 8.12 Griffith and Wallis C_2 vs. Bubble Asset Reynolds number.

than 10. σ is calculated from one of the following equations. If $u_t < 10$, for a continuous liquid phase of water,

$$\delta = (0.013 \log \mu_L)/D_h^{1.38} - 0.681 + 0.232 \log u_t - 0.428 \log D_h \quad (8.87)$$

For a continuous liquid phase of oil,

$$\delta = [0.0127 \log (\mu_L + 1)]/D_h^{1.415} - 0.284 + 0.167 \log u_t + 0.113 \log D_h \quad (8.88)$$

If $u_t > 10$, for continuous liquid phase of water,

$$\delta = (0.045 \log \mu_L)/D_h^{0.799} - 0.709 - 0.162 \log\ u_t - 0.888 \log D_h \quad (8.89)$$

For a continuous liquid phase of oil,

$$\begin{aligned}\delta = {} & [0.0274 \log (\mu_L + 1)]/D_h^{1.371} + 0.161 + 0.569 \log D_h \\ & - \log u_t \{[0.01 \log (\mu_L + 1)]/D_h^{1.571} \\ & + 0.397 + 0.63 \log D_h\}\end{aligned} \quad (8.90)$$

The value of δ has the following constraints. If $u_t < 10$,

$$\delta \geqslant -0.065 u_t$$

If $u_t > 10$,

$$\delta \geqslant \frac{-u_s A_p}{q_t + u_s A_p}\left(1 - \frac{\bar{\rho}}{\rho_L}\right)$$

The wall friction term in the slug flow regime is given by

$$\tau_f = \frac{f \rho_L u_t^2}{2 g_c D_h}\left[\frac{q_L + u_s A_p}{q_t + u_s A_p} + \delta\right] \quad (8.91)$$

Transition Flow

The Duns and Ros method of determining the average density and friction gradient for the transition flow regime is to use weighted averages of values for the slug flow and mist flow regimes. Thus

$$\bar{\rho} = \frac{L_M - \overline{\overline{u}}_g}{L_M - L_s}\bar{\rho}_{\text{slug}} + \frac{\overline{\overline{u}}_g - L_s}{L_M - L_s}\bar{\rho}_{\text{mist}} \quad (8.92)$$

$$\tau_f = \frac{L_M - \overline{\overline{u}}_g}{L_M - L_s}(\tau_f)_{\text{slug}} + \frac{\overline{\overline{u}}_g - L_s}{L_M - L_s}(\tau_f)_{\text{mist}} \quad (8.93)$$

Mist Flow

Since there is virtually no slip between the liquid and gas phases in the mist flow regimc, the gas void fraction can be calculated using

$$H_g = \frac{1}{(1 + q_L/q_g)} \quad (8.94)$$

The average density is calculated by

$$\bar{\rho} = (1 - H_g)\rho_L + H_g \rho_g \quad (8.95)$$

The friction loss gradient is given by Dons and Ros as

$$\tau_f = \frac{f \rho_g u_{sg}}{2 g_c D_h} \quad (8.96)$$

The friction factor, f, is obtained from the Moody diagram using the gas Reynolds number

$$N_{RE\ g} = \frac{1488 D_h u_{sg} \rho_g}{\mu_g} \tag{8.97}$$

8.7 PREPARED PRESSURE TRAVERSE CURVES

Pressure traverse curves for vertical single-phase gas flow and gas-liquid flow have been prepared using some of the correlations discussed in Section 8.6. These curves are available from gas-lift valve suppliers and also from Brown's books, *The Technology of Artificial Lift Methods*, Volumes 3a and 3b.

Figure 8.13 is for dry gas flow and is based on the Cullender and Smith method of Section 8.33. Figure 8.14 is for gas-liquid flow and is developed using the Hagedorn and Brown method of Section 8.63.

The procedure for using these working curves is as follows:

1. Select curve for correct tubing size, producing rate and gas-liquid ratio.
2. Locate known pressure on pressure axis and depth corresponding to this pressure on depth axis.
3. Correct this depth for actual well depth.
   (a) If known pressure is surface pressure, add the well depth to the value found in step 2.
   (b) If known pressure is bottom-hole pressure, subtract the well depth from the value found in step 2.
4. The pressure corresponding to the corrected depth is the unknown pressure.

Example 8.9

Given

Tubing size = 1.995 in. ID

Oil rate = 1000 bbl/day

Water rate = 0

Oil gravity = 35 °API

Water specific gravity = 1.07

Gas specific gravity = 0.65

Average flowing temperature = 150°F

Gas producing rate = 500 Mscfd

Tubing length = 8000 ft

Surface flowing pressure = 400 psig

Find the flowing bottom-hole pressure.

Solution

(a) Using Fig. 8.14, locate 400 psig on the pressure axis.
(b) Drop a vertical line to intersect the 500 scf/bbl gas/liquid ratio curve.

(c) From this point of intersection draw a horizontal line to intersect the depth axis at 2500 ft. This is the corresponding depth to the known pressure.
(d) Add the well depth to this depth to obtain 10,500 ft.
(e) From this depth draw a horizontal line to intersect the 500 scf/bbl gas/liquid ratio curve.
(f) From this intersection draw a vertical line to intersect the pressure axis at 2400 psig.
(g) This is the flowing bottom-hole pressure.

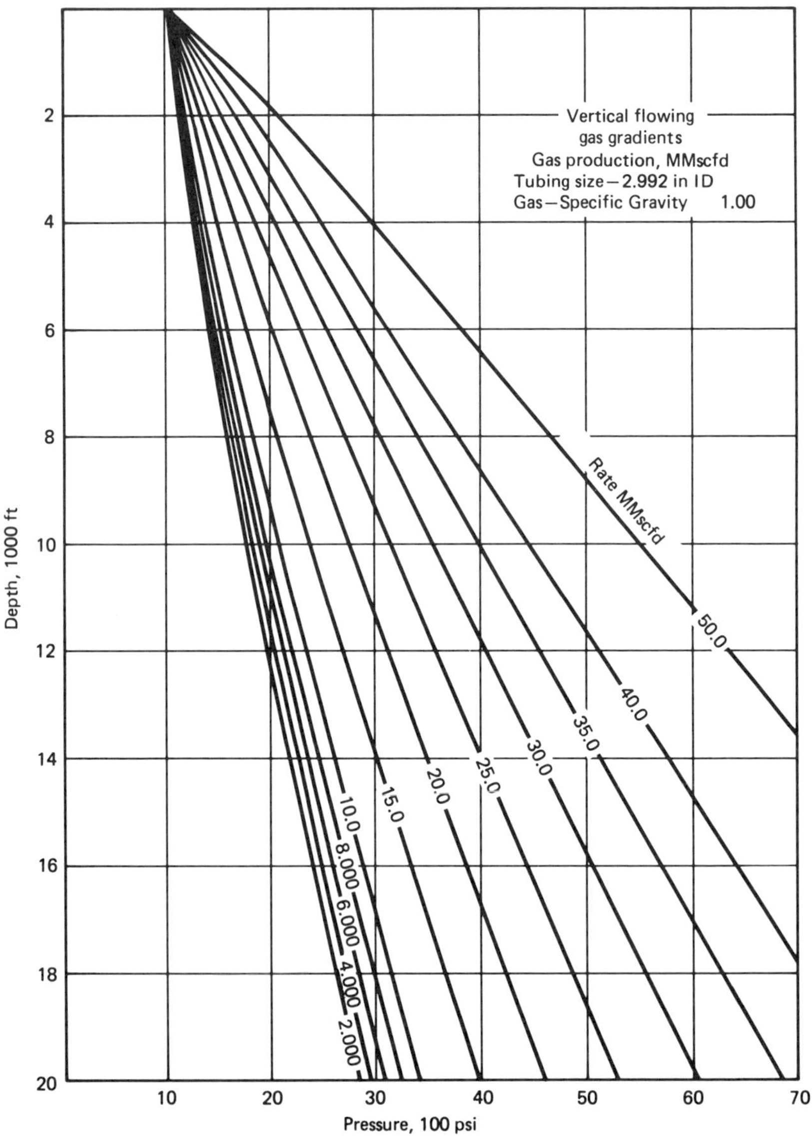

Fig. 8.13 Vertical flowing gas gradients. (After Brown.)

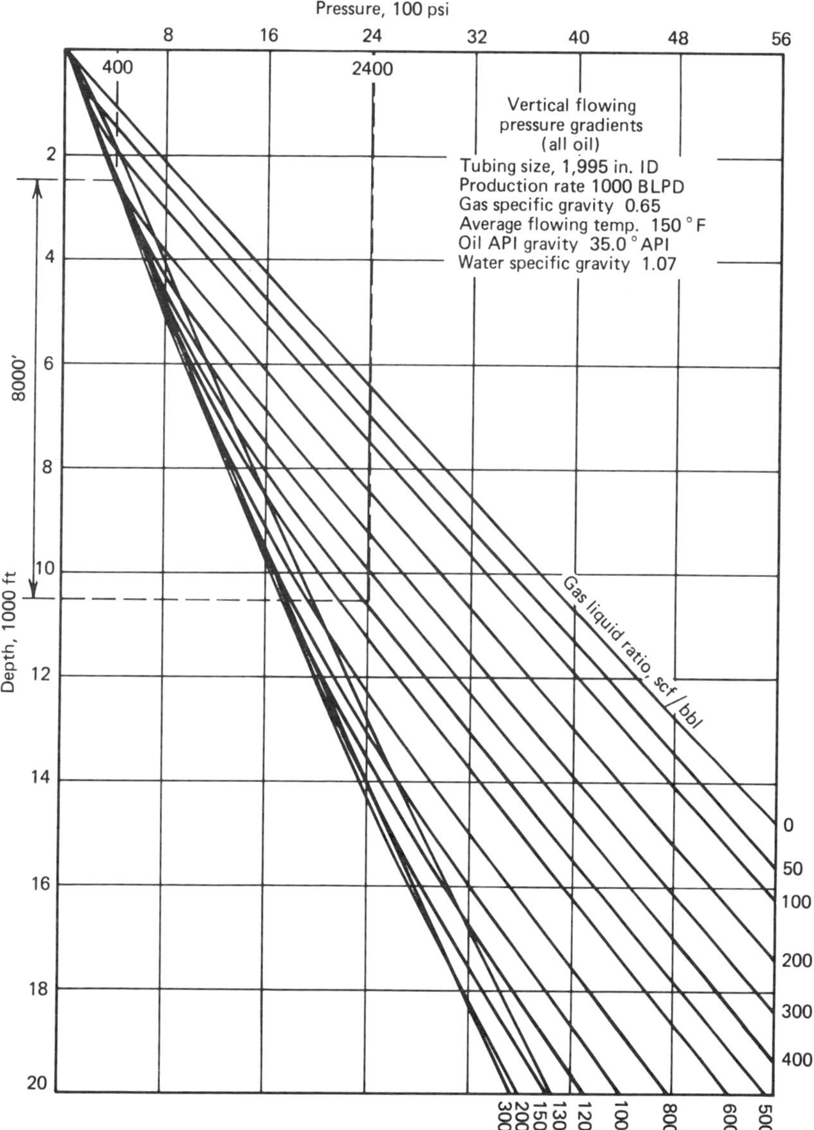

Fig. 8.14 Vertical flowing pressure gradients. (After Brown.)

8.8 GAS FLOW THROUGH RESTRICTIONS

There are several locations in the gas production system where the gas must pass through relatively short restrictions. Surface chokes and subsurface control equipment will be discussed in this section.

The flow through restrictions may be either subcritical or critical. In subcritical flow, the velocity of gas through the restriction is below the speed of sound in the gas. The flow rate will depend on both the upstream and downstream pressures. Subsurface control equipment is sized so that flow is subcritical.

In critical flow, the velocity of gas through the restriction is equal to the speed of sound in the gas. Since pressure disturbances travel at the speed of sound, a disturbance downstream of the restriction in critical flow will not affect the upstream pressure or flow rate. The flow rate depends only on the upstream pressure in critical flow. Surface chokes are usually sized for critical flow.

Practically all flowing wells utilize some surface restriction in order to regulate the flow rate. Only very few produce with absolutely no restrictions in order to obtain maximum production rate. Use of a surface choke to control flow may be necessary because of the following reasons:

1. Maintaining well allowable production rate.
2. Sand control by maintaining sufficient back pressure.
3. Protection of surface equipment.
4. Prevention of water coning.
5. Reservoir management.

Subsurface control equipment comes in a variety of types and sizes. They can be grouped under three main functions:

1. Subsurface tubing safety valves, which shut in the well downhole in the event surface control equipment becomes damaged or is completely removed.
2. Bottom-hole chokes and regulators, which reduce the wellhead flowing pressure and prevent freezing of surface controls and lines by taking pressure drop downhole.
3. Check valves, which prevent back flow of injection wells.

Gas flow through a restriction can be calculated for both critical (sonic) and subcritical (subsonic) flow using the following equation given by Nind:

$$q = 155.5 C_d A p_1 \left\{ \frac{2g}{\gamma_g T}\left(\frac{k}{k-1}\right)\left[\left(\frac{p_2}{p_1}\right)^{2/k} - \left(\frac{p_2}{p_1}\right)^{(k+1)/k}\right]\right\}^{1/2} \tag{8.98}$$

where

q = gas flow rate, Mcfd at standard conditions
C_d = coefficient of discharge $\simeq$ 0.86
A = area of choke, sq. in.
p_1 = upstream pressure, psia
p_2 = downstream pressure, psia
g = gravitational acceleration = 32.17 ft/sec^2
k = specific heat ratio = c_p/c_f (approx. 1.31 for natural gases)
γ_g = specific gravity of gas (air = 1.0)
T = inlet or upstream temperature, °R

Let's call the function of the pressure ratio, $F(p_2/p_1)$, where

$$F\left(\frac{p_2}{p_1}\right) = \left[\left(\frac{p_2}{p_1}\right)^{2/k} - \left(\frac{p_2}{p_1}\right)^{(k+1)/k}\right]^{1/2} \tag{8.99}$$

The function $F(p_2/p_1)$ is plotted in Fig. 8.15 for various values of p_2/p_1 and for a particular gas and upstream temperature.

The maximum value of $F(p_2/p_1)$ in Fig. 8.15 corresponds to the critical flow condition. The critical-flow range, over which the value of the flow rate through the restriction is proportional to the upstream pressure, occurs for values of the ratio of the downstream to upstream pressure less than the value corresponding to the maximum $F(p_2/p_1)$. For this reason the left-hand arch, the curve in Fig. 8.15, has been shown as a broken line, and a full horizontal line has been drawn through the maximum value.

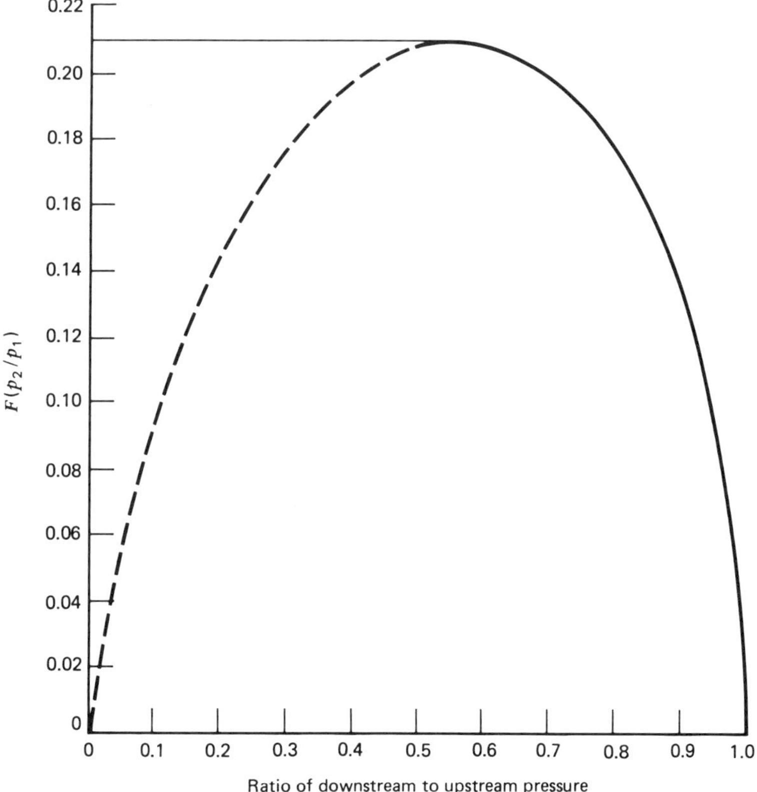

Fig. 8.15 Gas flow through a constriction. (After Nind.)

The pressure ratio at which flow becomes critical depends on the k value for the flowing gas, and is given by

$$R_{pc} = \left(\frac{2}{k + 1}\right)^{k/(k-1)} \tag{8.100}$$

The critical flow pressure ratio for natural gas will generally occur where

$$R_{pc} = 0.55$$

Gas throughput for different orifice sizes may be calculated from Eq. 8.98 or read directly from Fig. 8.16. Values obtained from this chart require correction for operating temperature and gas gravity. The correction factors to be used as multipliers are given in Fig. 8.17.

8.9 GAS WELL PERFORMANCE CURVES

Several well surveillance curves can be used to analyze gas well performance. Inflow performance curves, outflow performance curves, and tubing performance curves are defined. Their uses in analyzing and predicting the performance of gas wells are discussed.

8.9.1 Inflow Performance Relationship

Rawlins and Schellhardt presented an empirical equation in 1935, which is usually called the back-pressure equation:

$$q = C(\bar{p}_R{}^2 - p_{w_f}{}^2)^n \tag{8.101}$$

where

q = gas flow rate, Mcfd
C = a numerical constant
$\bar{p}_R$ = shut-in reservoir pressure, psia
p_{w_f} = flowing bottom-hole pressure, psia
n = a numerical exponent

Equation 8.101 is one of the most widely used methods for describing the downhole performance of a gas well. This equation represents a straight line on log-log graph paper as shown in Fig. 8.18. The constant C represents the horizontal displacement of the performance curve, and the exponent n represents the reciprocal of the curve's slope.

Data for preparing this performance curve are obtained from a four-point test or a back-pressure test. Some important information obtained from this well test is the *absolute open flow* (AOF) potential of the well. This is the projected flow rate of the well at a flowing bottom-hole pressure of zero. The calculated AOF is used for comparing wells from different areas. It is also used by the regulatory agency for establishing the allowable gas production assigned to each well.

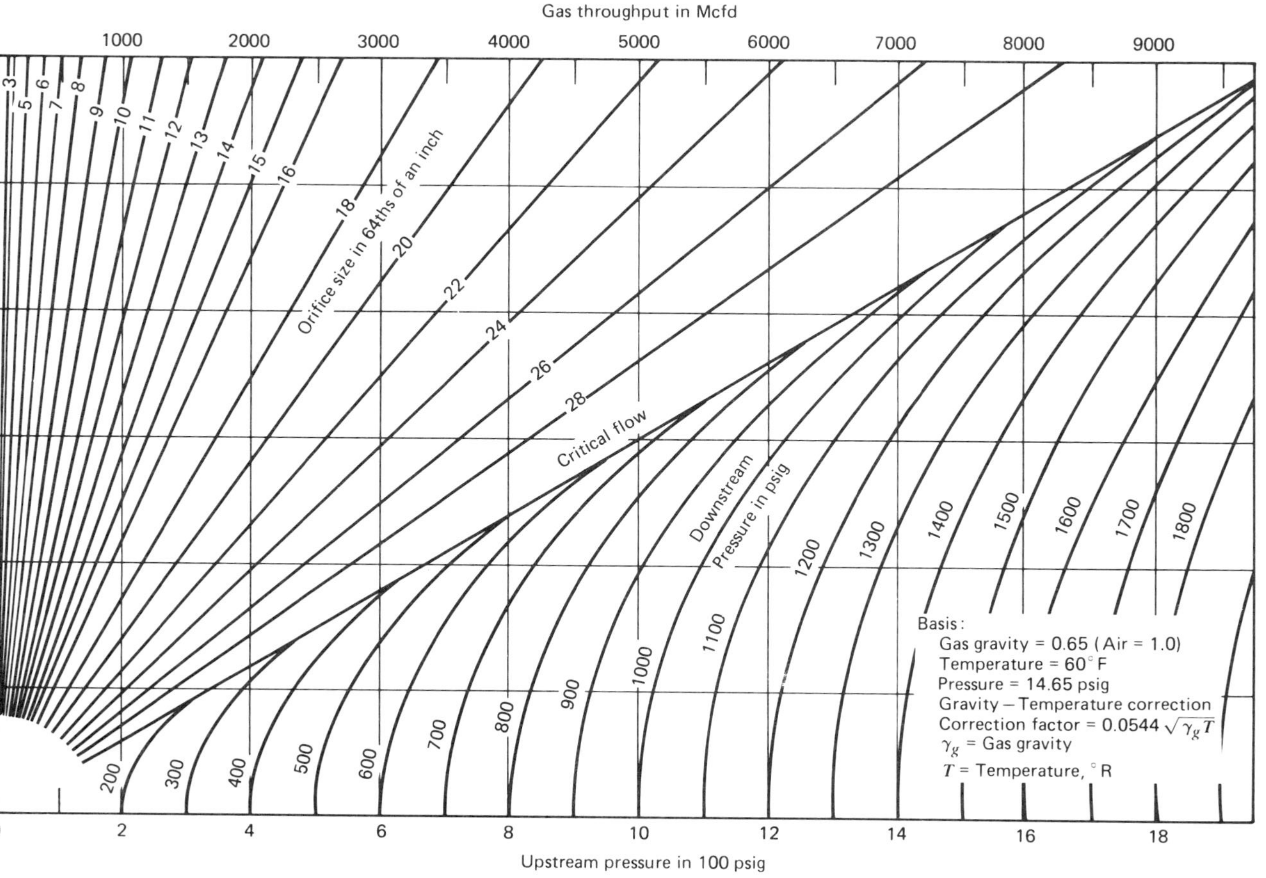

Fig. 8.16 Gas passage chart for various orifice sizes. (After Brown.)

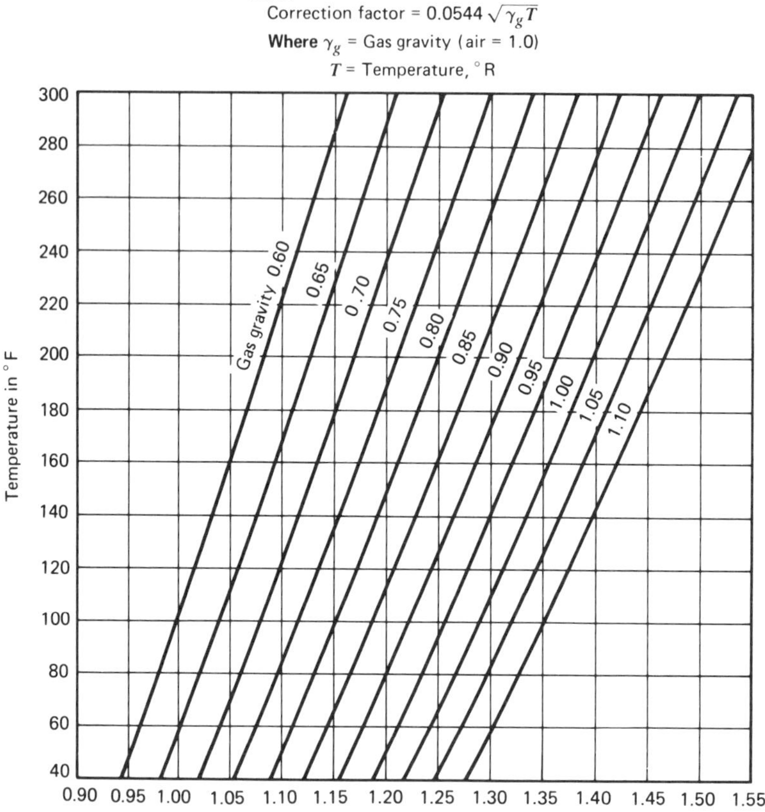

Fig. 8.17 Correction factor chart for gas passage charts.

The curve on Fig. 8.18 can be replotted on Cartesian coordinate paper as shown in Fig. 8.19. This is the inflow performance relationship (IPR) for the gas well. Russell, Goodrick, Perry, and Bruskotter found that the back-pressure equation (Eq. 8.101) predicted producing rates that were too low. They developed a more accurate equation:

$$q = \frac{T_b kh\,(\bar{p}_R^2 - p_{w_f}^2)}{50.304 p_b T\,\bar{\mu}\,\bar{z}\left(\ln \frac{r_e}{r_w} - \frac{3}{4} + s\right)} \tag{8.102}$$

where

q = gas flow rate, Mcfd
T_b = base temperature (520 R), °R
kh = permeability-thickness product, md-ft
$\bar{p}_R$ = shut-in reservoir pressure, psia

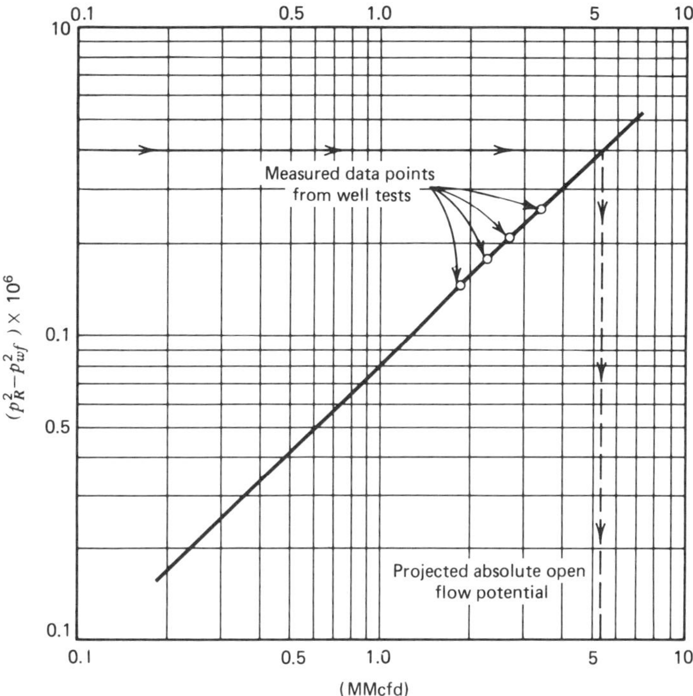

Fig. 8.18 Gaswell inflow performance curve back-pressure or "4-point test" method. (After Greene.)

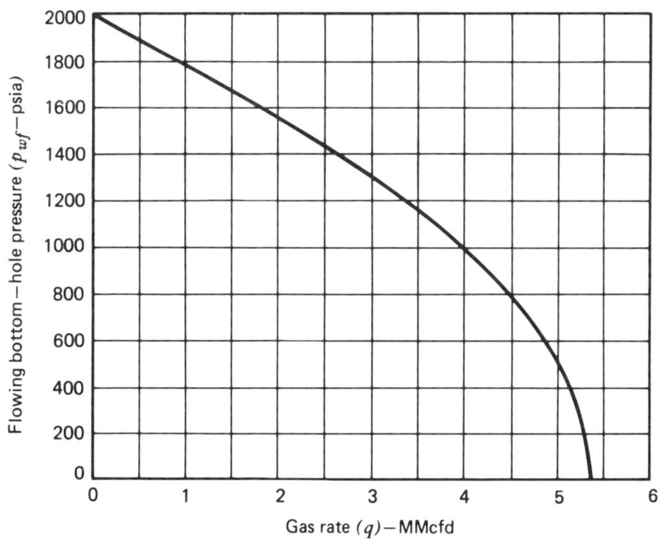

Fig. 8.19 Inflow performance curve log-log plot from Fig. 8.18 replotted on coordinate paper. (After Greene.)

p_{w_f} = flowing bottom-hole pressure, psia
p_b = base pressure (14.7 psia), psia
T = reservoir temperature, °R
$\bar{\mu}$ = gas viscosity at average pressure, $\bar{p}$, cp
$\bar{z}$ = gas deviation factor at average pressure, $\bar{p}$
$\bar{p} = (\bar{p}_R + p_{w_f})/2$ = average pressure, psia
r_e = radius of external boundary, ft
r_w = wellbore radius, ft
s = skin factor

This equation can be written as

$$q = \frac{C'(\bar{p}_R^2 - p_{w_f}^2)}{\bar{\mu}\bar{z}} \tag{8.103}$$

where C' is a constant that can be determined from a single-flowing well test if $\bar{p}_R$ is known. This value of C' should remain constant for the well as flow rate is changed and may be used to construct an inflow performance relationship for the gas well.

8.9.2 Outflow Performance Curves

A well's inflow performance was defined as a pressure versus rate representation of its downhole behavior. A well's *outflow performance curve* is a pressure versus rate representation of its behavior at the surface. The outflow performance curve may be calculated from the inflow performance curve using flowing bottom-hole pressure equations or vertical gradient curves.

A typical outflow curve is shown in Fig. 8.20. The apex of this curve is

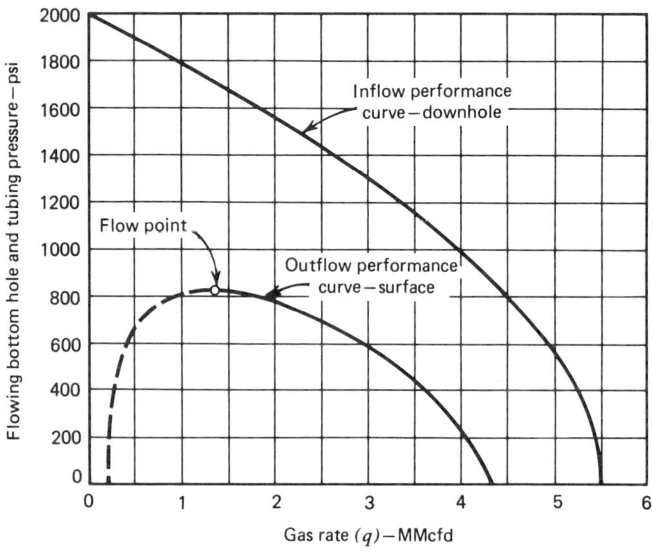

Fig. 8.20 Outflow performance curves. (After Greene.)

designated the *flowpoint*. This marks the *minimum sustainable flow rate* possible from this well. It also marks the *maximum flowing* tubing pressure possible. The solid portion of the outflow performance curve to the right of the flowpoint represents the operating range of the well. The dashed portion of the curve to the left of the flowpoint represents an unstable transition area of flow through which the well must pass as it is being opened up or as it is ceasing to flow. The flowpoint exists, however, only if the gas well produces liquid. A dry gas well that does not produce liquid does not have a flowpoint and will sustain flow at any rate no matter how small.

The effects of changing tubing sizes can be predicted by plotting an outflow performance curve for each tubing size. This has been prepared in Fig. 8.21 for four tubing sizes. It can be seen that at low gas rates the smaller strings have better flow efficiencies and at high rates the larger strings have better efficiencies.

It should be noted that the flow points correspond to the critical flow rates for liquid removal from gas wells. This topic will be discussed later.

8.9.3 Tubing Performance Curves

The *tubing performance curve* is a plot of the flowing bottom-hole pressure required to produce various gas rates through a given size tubing string at some constant-flowing wellhead pressure. Figure 8.22 shows an example of a tubing performance curve. If the wellhead pressure is kept constant, the reservoir abandonment pressure can be determined by declining the inflow performance curve parallel to its present form until it becomes tangent to the tubing preformance curve (dashed line on graph). The value of reservoir pressure at this point will be the abandonment pressure.

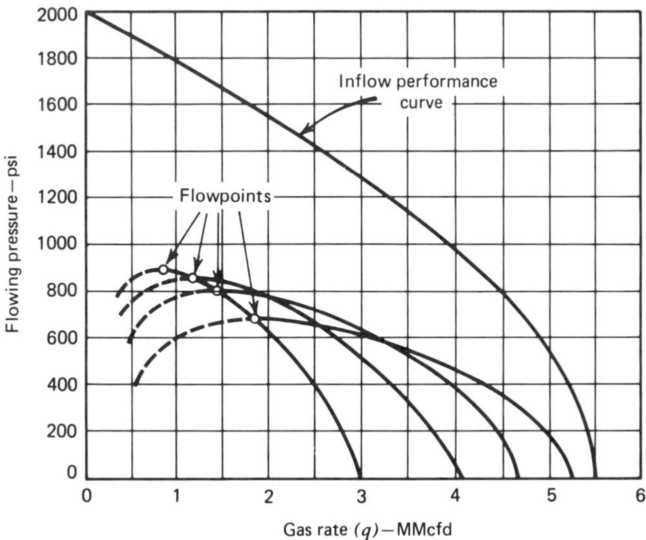

Fig. 8.21 Outflow performance curves for various tubing sizes. (After Greene.)

The point of intersection P of the tubing performance curve and inflow performance curve gives the producing rate and flowing bottom-hole pressure for the constant-flowing tubing pressure. It should be noted that the outflow performance curve provides this information plus the well's producing rate at any other flowing wellhead pressure.

Tubing performance curves for four tubing sizes have been plotted in Fig. 8.23. Each size tubing has an optimum operating range. To the left of these

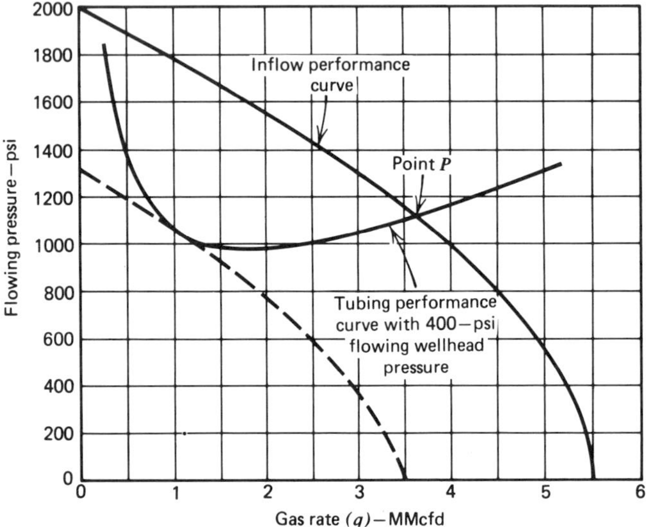

Fig. 8.22 Tubing performance curves. (After Greene.)

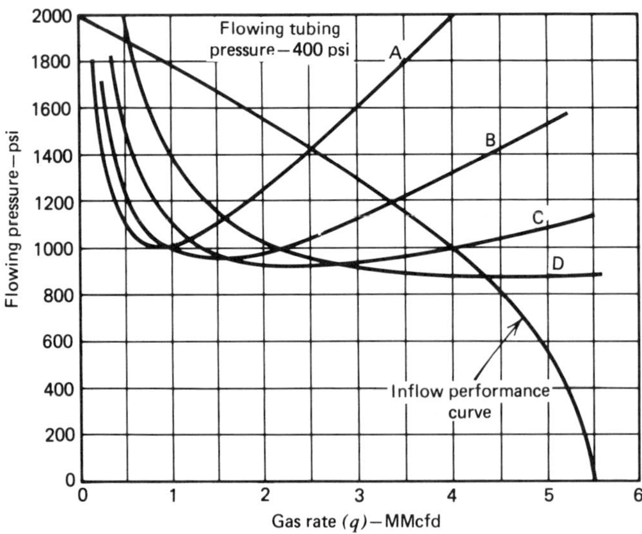

Fig. 8.23 Tubing performance curves—for various tubing sizes. (After Greene.)

optimum flow areas, liquid "fallback" increases and downhole pressure required increases. To the right of these optimum flow areas, friction predominates and downhole pressure required is greater as velocity increases.

8.10 LIQUID LOADING IN GAS WELLS

The presence of a liquid phase in gas reservoirs can affect the flowing characteristics of the gas well. Liquids can come from condensation of hydrocarbon gas (condensate) or from interstitial water in the reservoir matrix. The liquid phase must be transorted as droplets to the surface by the gas. The flow will usually be in the mist flow regime where gas is the continuous phase and liquid the discontinuous phase. Liquid loading or accumulation in gas wells occurs when the gas phase does not provide adequate energy for the continuous removal of liquids from the wellbore. The accumulation of liquid will impose an additional back pressure on the formation, which can restrict well productivity. In low-pressure wells the liquid may completely kill the well; and in high-pressure wells slugging will occur, which can affect well test results.

8.10.1 Flow Rate for Continuous Liquid Removal

Turner, Hubbard, and Dukler analyzed two physical models for the removal of gas well liquids: (1) liquid film movement along the walls of the pipe, and (2) liquid droplets entrained in a high-velocity gas core. The two models probably will exist in actual cases and there will be a continuous exchange of liquid between the gas core and the film. Calculations for the wall film model are more complex, requiring numerical integration. Also the film moving down will eventually break into droplets. Turner et al. tested the models against field data and found the drop model superior. The minimum condition required to unload a gas well is that which will move the largest liquid drops that can exist in a gas stream.

The limiting gas flow velocity for upward drop movement is the terminal free settling velocity of the drop:

$$u_t = \sqrt{\frac{2gm_p\,(\rho_p - \rho)}{\rho_p\,\rho\,A_p\,C_D}} \tag{8.104}$$

where

u_t = terminal velocity of free-falling particle, ft/sec
g = local gravitational acceleration, ft/sec^2
m_p = mass of falling particle, lbm
ρ_p = density of particle, lbm/cu ft
ρ = density of fluid, lbm/cu ft
A_p = projected area of particle, sq ft
C_D = drag coefficient

The general free settling velocity equation (Eq. 8.104) can be written in terms of drop diameter:

$$u_t = 6.55\sqrt{\frac{D(\rho_L - \rho_g)}{\rho_g C_D}} \tag{8.105}$$

where

ρ_L = liquid density, lbm/cu ft
ρ_g = gas density, lbm/cu ft
D = diameter of drop, ft

Equation 8.105 shows that the larger the drop, the higher the gas flow rate necessary to remove it. If the diameter of the largest drop can be determined, the gas flow rate that will ensure upward movement of all drops can be calculated.

The drop is subjected to forces trying to shatter it and surface tension of liquid trying to hold it together. These forces are combined in the Weber number:

$$N_{\mathrm{We}} = \frac{u^2\rho_g/g_c}{\sigma/D} = \frac{\text{velocity pressure}}{\text{surface tension pressure}}$$

or

$$N_{\mathrm{We}} = \frac{u^2\rho_g D}{\sigma g_c} \tag{8.106}$$

The drop will shatter if the Weber number exceeds a critical value of 20 to 30. The maximum drop diameter is

$$D_m = \frac{30\sigma g_c}{\rho_g u_t} \tag{8.107}$$

and the terminal velocity becomes

$$u_t = 17.6\,\frac{\sigma^{1/4}(\rho_L - \rho_g)^{1/4}}{\rho_g^{1/2}} \tag{8.108}$$

where σ is the surface tension in dynes per centimeter and a value of $C_D = 0.44$ has been assumed for field conditions. Comparison with field data indicates that Eq. 8.108 should be adjusted upward by approximately 20% to remove all drop.

$$u_t = 20.4\,\frac{\sigma^{1/4}(\rho_L - \rho_g)^{1/4}}{\rho_g^{1/2}} \tag{8.109}$$

Written in terms of gas volumetric flow rates,

$$q_g(\text{MMcfd}) = \frac{3.06 p u_g A}{Tz} \tag{8.110}$$

where

p = pressure, psia
$u_g = u_t$ = gas velocity, ft/sec
A = flow area of conduit, sq ft
T = flowing temperature, °R
z = gas deviation factor p and T

For field application, specific equations can be derived for water and condensate. For water, using the following values, σ = 60 dynes/cm, ρ_L = 67 lbm/cu ft:

$$u_g(\text{water}) = \frac{5.62(67 - 0.0031p)^{1/4}}{(0.0031p)^{1/2}} \tag{8.111}$$

For condensate, using the following values, σ = 20 dynes/cm, ρ_L = 45 lbm/cu ft:

$$u_g(\text{condensate}) = \frac{4.02(45 - 0.0031p)^{1/4}}{(0.0031p)^{1/2}} \tag{8.112}$$

The equation for water is used when both water and condensate are present.

A nomograph (Fig. 8.24) was developed by Turner et al. for direct solution of these equations. The nomograph allows consideration of all parameters except the gas deviation factor z, which should be known to determine the value of q.

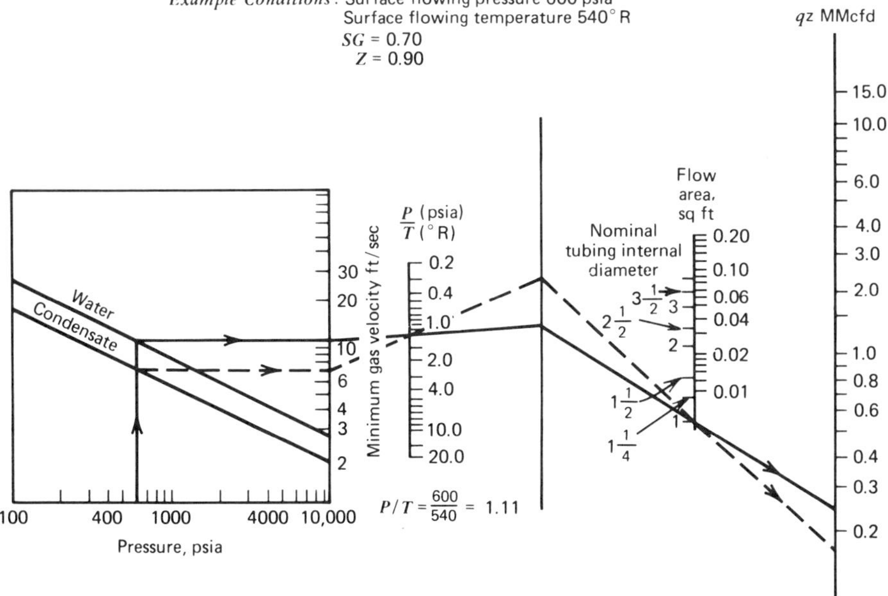

Fig. 8.24 Nomograph for calculating gas rate required to lift liquids through tubing of various sizes. (After Turner et al.)

8.10.2 Methods for Unloading Liquid

Many gas wells do not provide adequate amounts of transport energy for continuous liquid removal from the wellbore. Liquid accumulation or loading will therefore occur in these gas wells. Fluctuations of daily gas rates and casing pressures are characteristics of liquid accumulation in gas wells. There will usually be a buildup of casing pressure which may be sufficient to blow the liquid accumulation in the casing/tubing annulus into the tubing and then to the surface. If enough casing pressure cannot be achieved, the well will die and swabbing becomes necessary to revive it.

Another symptom of liquid loading is abnormally high casing pressure. A casing pressure of more than 200 psi higher than the flowing tubing pressure generally is an indication of excessive liquid accumulation. Flowing bottom-hole pressure surveys and tubing pressure traverse curves can also be used in determining liquid-loading tendencies.

Various methods have been employed to remove liquids from gas wells. We will discuss five basic methods: beam pumping units, plunger lifts, small tubing strings, flow controllers, and soap injection.

Beam Pumping Units

When beam pumps are used to unload liquids from gas wells, the liquids are pumped up the tubing and the gas is produced out of the annulus. It is desirable for the tubing to be set as close to the bottom perforation as possible, and preferably below it. This prevents gas interference problems in the pump by having a liquid cushion. If a diverter is used, no liquid cushion is required.

Beam pumps do not depend on gas velocity for lift. They are quite cheap for shallow wells but become very expensive for deeper and more highly pressured wells. They are best for low gas rates and liquid producing rates of greater than 10 bbl liquid/day.

Plunger Lift

A successful method of liquid removal in gas wells is by plunger lift (Fig., 8.25). A steel plunger with a valve is located in the tubing string. At the bottom of the tubing is an opening through which gas and liquid can pass into the tubing. When the plunger is at the bottom of the tubing, the tubing is closed and all production goes into the annulus. The casing pressure builds up and energy is stored in the annulus for moving the plunger and liquid above the plunger to the surface. A motor valve, operated by a time clock on the flowline, may be used to control the cycle rate of the plunger. Following plunger arrival at the surface a bumper opens the valve in the plunger and the well is allowed to flow for a preset period of time. Plunger lift is one of the most successful methods of unloading gas wells, especially when used with a surface controller.

Small Tubing String

The objective of using small tubing strings is to reduce the flow area and therefore increase the gas velocity so that liquid will be carried to the surface. This is a very

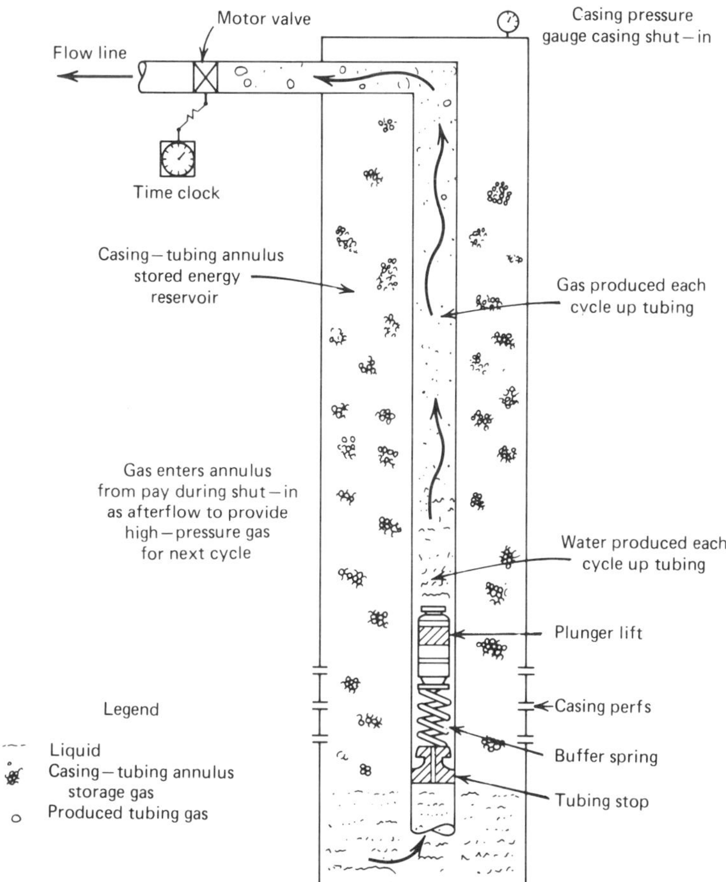

Fig. 8.25 Plunger Lift operations. (After Lisbon and Henry.)

successful method of preventing liquid loading in gas wells. Field experience reported by Libson and Henry indicates that a minimum velocity of 1000 fpm is necessary for continuous liquid removal in drop flow.

Flow Controllers

These are essentially gas-lift valves with time control. They operate on the principles of intermittent gas lift. These flow controllers may not provide consistent production rates because of gas slippage and liquid fallback.

Surface flow controllers in conjunction with plunger lifts have been found to be superior to time controllers. A surface flow controller operates on the principle of permitting the well to flow until a less-than-critical gas rate occurred; it would then shut in the well. Wells having sufficient deliverability to maintain rates above the critical rate for short periods are good candidates for rate or critical velocity control.

Soap Injection

Injection of foaming agents on noncondensate producing wells has become a major method of unloading gas wells. Schematics of soap injection systems are shown in Figs. 8.26 and 8.27. Foaming agents are injected into the casing/tubing annulus with a chemical pump and time clock and water is unloaded continuously in a foamed slug state. Foam injection has not been as successful in condensate producing wells.

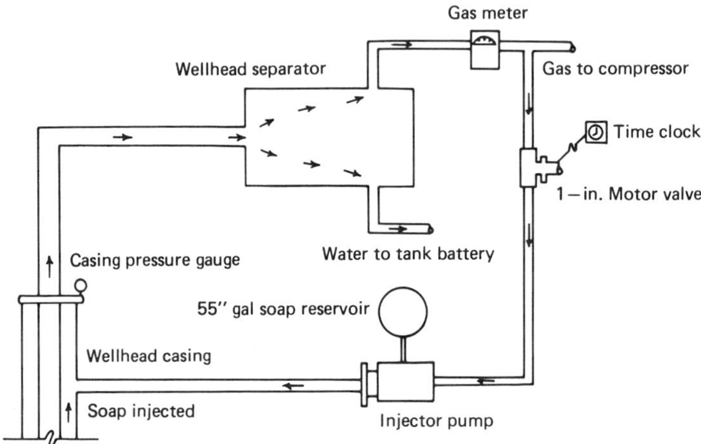

Fig. 8.26 Gas well soap injection system. (After Lisbon and Henry.)

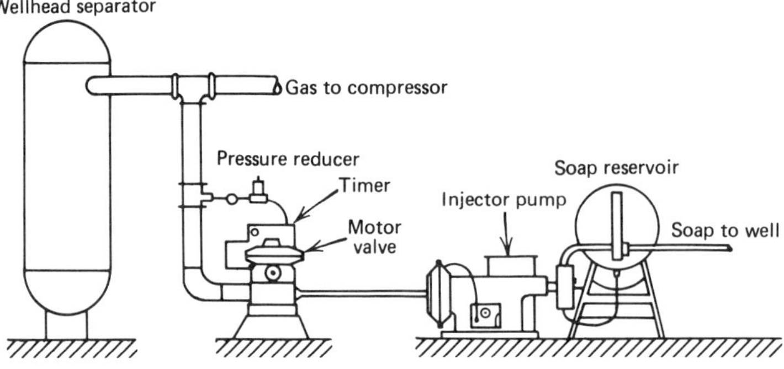

Fig. 8.27 Ground-level view of soap injection system. (After Lisbon and Henry.)

REFERENCES

Aziz, K. "Ways to Calculate Gas Flow and Static Head." *Petroleum Engineer.* Series from November 1960 to September 1963.

Aziz, K., G. W. Govier, M. Fogarasi. "Pressure Drop in Wells Producing Oil and Gas," *Journal of Canadian Petroleum Technology*, p. 38, July–September, 1972.

Beggs, H. D., and J. P. Brill. "A Study of Two-Phase Flow in Inclined Pipes." *Journal of Petroleum Technology*, p. 607, May 1973.

Brown, K. E. *The Technology of Artificial Lift Methods*, Volume 1. Tulsa: PennWell Publishing Co., 1977.

Brown, K. E. *The Technology of Artificial Lift Methods*, Volumes 2a, 3a, and 3b. Tulsa: PennWell Publishing Co., 1980.

Crawford, P. B., and G. H. Fancher. "Calculation of Flowing and Static Bottomhole Pressures of Natural Gas Wells from Surface Measurements." Bulletin 72. Austin: Texas Petroleum Research Committee, 1959.

Cullender, M. H., and R. V. Smith. "Practical Solution of Gas-flow Equations for Wells and Pipelines with Large Temperature Gradients." *Transactions AIME* **207**, pp. 281–87, 1956.

Duns, H., and N. C. J. Ros. "Vertical Flow of Gas and Liquid Mixtures in Wells." 6th World Petroleum Congress, Frankfurt, Germany.

Greene, W. R. "Analyzing the Performance of Gas Wells." *Journal of Petroleum Technology*, p. 1378, July 1983.

Griffith, P., and G. B. Wallis. "Two-Phase Slug Flow." *ASME Journal of Heat Transfer*, p. 307, August 1961.

Hagedorn, A. R., and K. E. Brown. "Experimental Study of Pressure Gradients Occurring During Continuous Two-Phase Flow in Small Diameter Vertical Conduits." *Journal of Petroleum Technology*, p. 475, April 1965.

Hutlas, E. J., and W. R. Granberry. "A Practical Approach to Removing Gas Well Liquids." *Journal of Petroleum Technology*, p. 916, August 1972.

Katz, D. L., et al. *Handbook of Natural Gas Engineering*. New York: McGraw-Hill, 1959.

Libson, T. N., and J. R. Henry. "Case Histories: Identification of and Remedial Action for Liquid Loading in Gas Wells—Intermediate Shelf Gas Play." Journal of Petroleum Technology, p. 685, April 1980.

Messer, P. H., R. Raghavan, and H. J. Ramsy, Jr. "Calculation of Bottom-hole Pressures for Deep, Hot, Sour Gas Wells. *Journal of Petroleum Technology*, pp. 85–92, January 1974.

Nind, T. E. W. *Principles of Oil Well Production*, 2nd Edition. New York: McGraw-Hill, 1981.

Orkiszewski, J. "Predicting Two-Phase Pressure Drops in Vertical Pipe." *Journal Petroleum Technology*, June 1967.

Poettman, F. H. "The Calculation of Pressure Drop in the Flow of Natural Gas Through Pipe." *Trans. AIME* **192**, pp. 317–326. 1951.

Rawlins, E. L., and M. A. Schellhardt. "Back Pressure Data on Natural Gas Wells and Their Application to Production Practices." Bureau of Mines Monograph 7, 1935.

———, "Back Pressure Test for Natural Gas Wells." Railroad Commission of Texas, Oil and Gas Division, Engineering Department, April 1972 (Reprint).

Russell, D. G., J. H. Goodrich, G. E. Perry, and J. F. Bruskotter. "Methods for Predicting Gas Well Performance." *Journal of Petroleum Technology*, p. 99, January 1966.

Rzasa, J. J., and D. L. Datz. "Calculation of Static Pressure Gradients in Gas Wells." *Trans. AIME* **160**, pp. 100–113, 1945.

Standing, M. B., and D. L. Katz. "Density of Natural Gases." *Trans AIME* **146**, pp. 140–149, 1942.

Sukkar, Y. K., and D. Cornell. "Direct Calculation of Bottom-hole Pressures in Natural Gas Wells." *Trans. AIME* **204**, pp. 43–8, 1955.

Theory and Practice of the Testing of Gas Wells, 3rd ed. Energy Resources Conservation Board. 1975.

Turner, R. G., M. G. Hubbard, and A. E. Dukler. "Analysis and Prediction of Minimum Flow Rate for the Continuous Removal of Liquids from Gas Wells." *Journal of Petroleum Technology*, p. 1475, November 1969.

Vitter, A. L., Jr. "Back-pressure Tests of Gas-condensate Wells." *API Drilling and Production Prac.* p. 79, 1942.

Wichert, E., and K. Aziz. "Calculation of Z's for Sour Gases." *Hydrocarbon Processing* **51**, pp. 119–122, 1972.

Young, K. L. "Effect of Assumptions Used to Calculate Bottom-hole Pressure in Gas Wells." *Journal of Petroleum Technology*. pp. 547–550, April 1967.

PROBLEMS

8.1 Calculate the static bottom-hole pressure for a dry, sweet gas well with the following well data:

Gas gravity	= 0.75
Well depth	= 10,000 ft
Wellhead temperature	= 35°F
Formation temperature	= 245°F
Shut-in wellhead pressure	= 2500 psia
T_{pc} = 408R, P_{pc}	= 667 psia

Using:

(a) Average temperature and gas deviation factor method.

(b) Sukkar and Cornell method.

(c) Cullender and Smith method.

8.2 Calculate the flowing bottom-hole pressure in a gas well with the following well data:

Gas gravity	= 0.75
Well depth	= 10,000 ft
Wellhead temperature	= 110°F
Flowing wellhead pressure	= 2000 psia
Formation temperature	= 245°F
Roughness of tubing	= 0.006 in.
Tubing diameter	= 2.441 in.
Gas flow rate	= 4.915 MMscfd
T_{pc} = 408°R, P_{pc}	= 667 psia

Using:

(a) Average temperature and gas deviation factor method.
(b) Sukkar and Cornell method.
(c) Cullender and Smith method.

Appendix

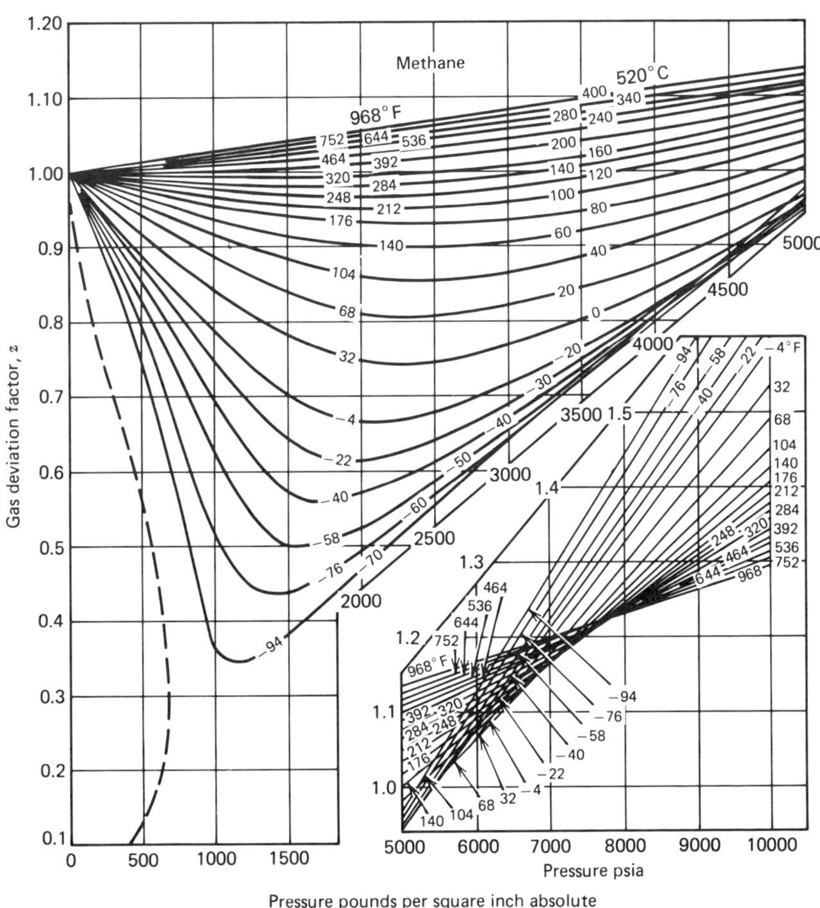

Fig. A.1 Gas deviation factor for methane

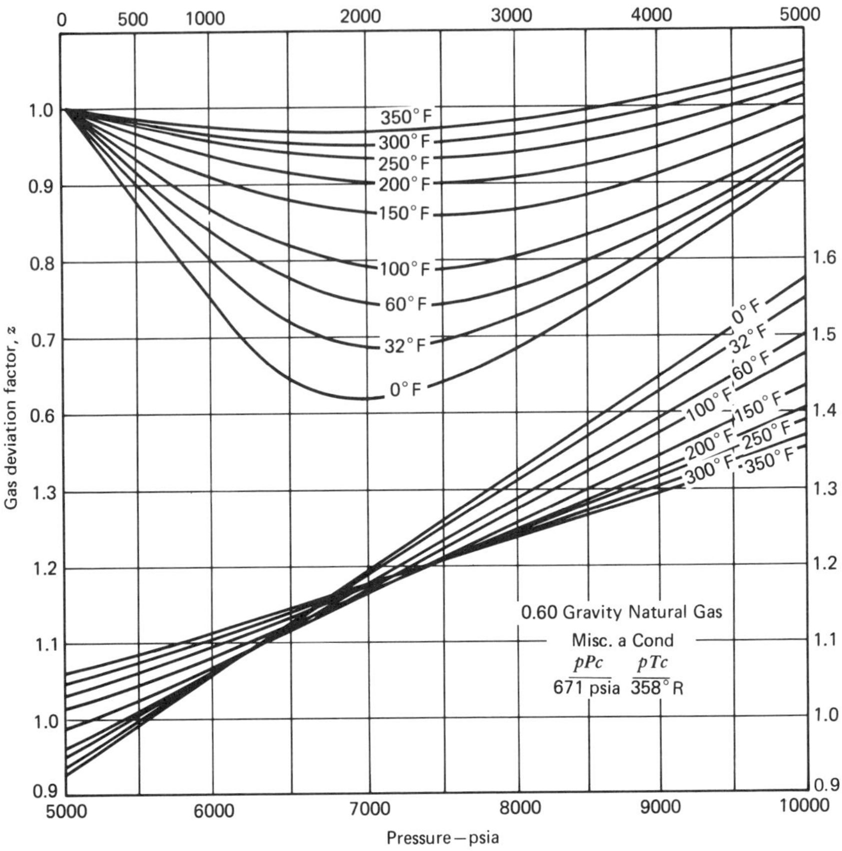

Fig. A.2 0.60 gravity natural gas

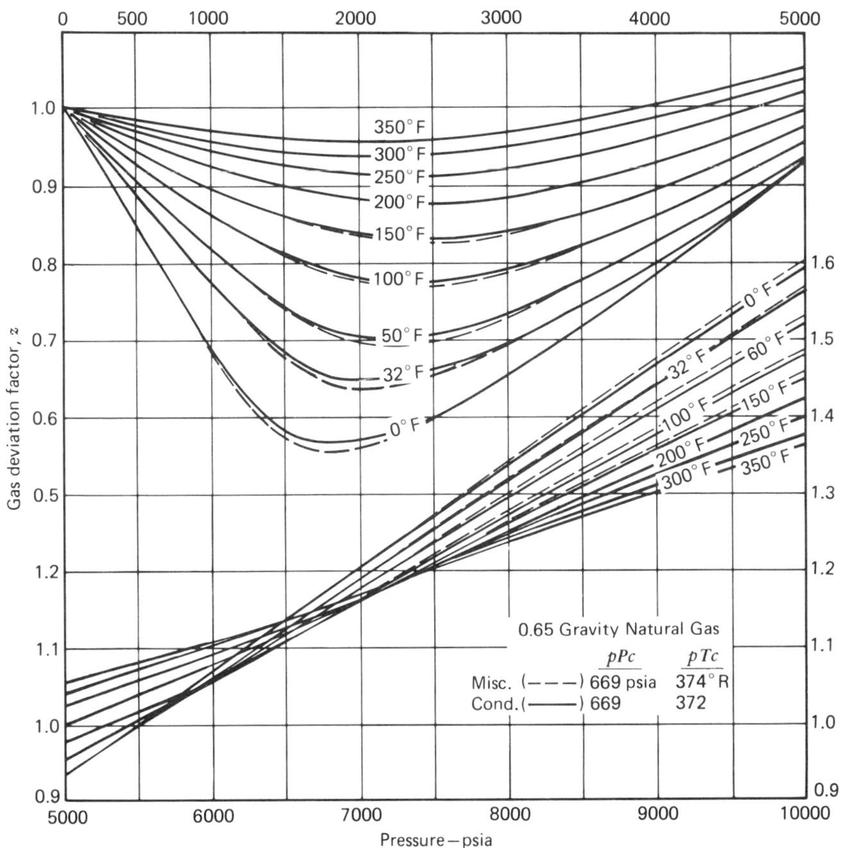

Fig. A.3 0.65 gravity natural gas

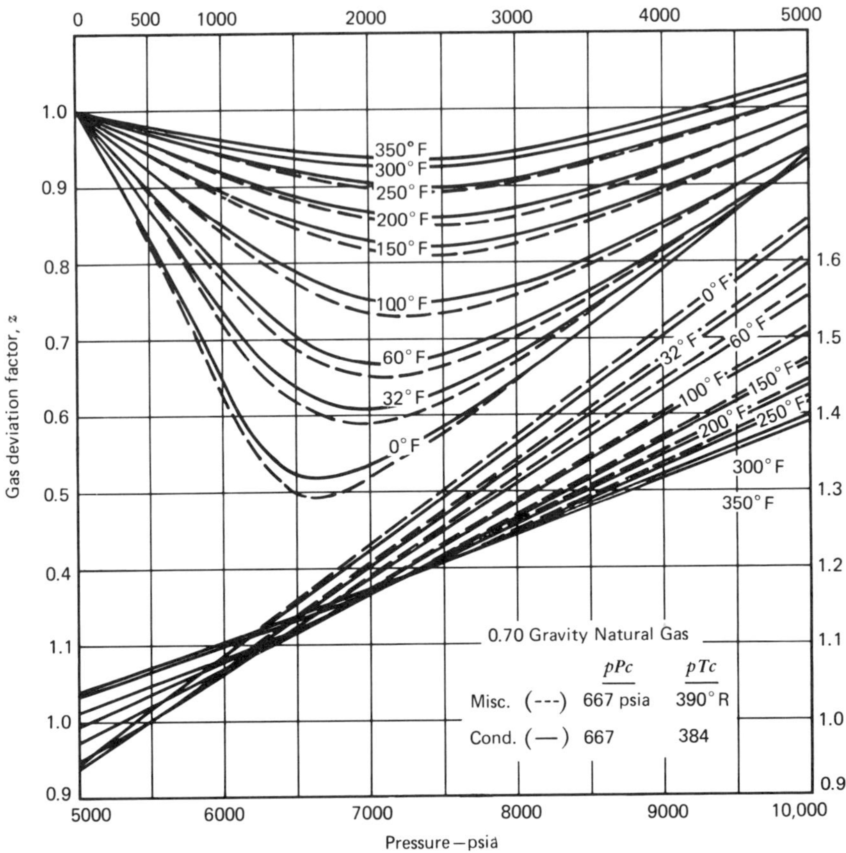

Fig. A.4 0.70 gravity natural gas

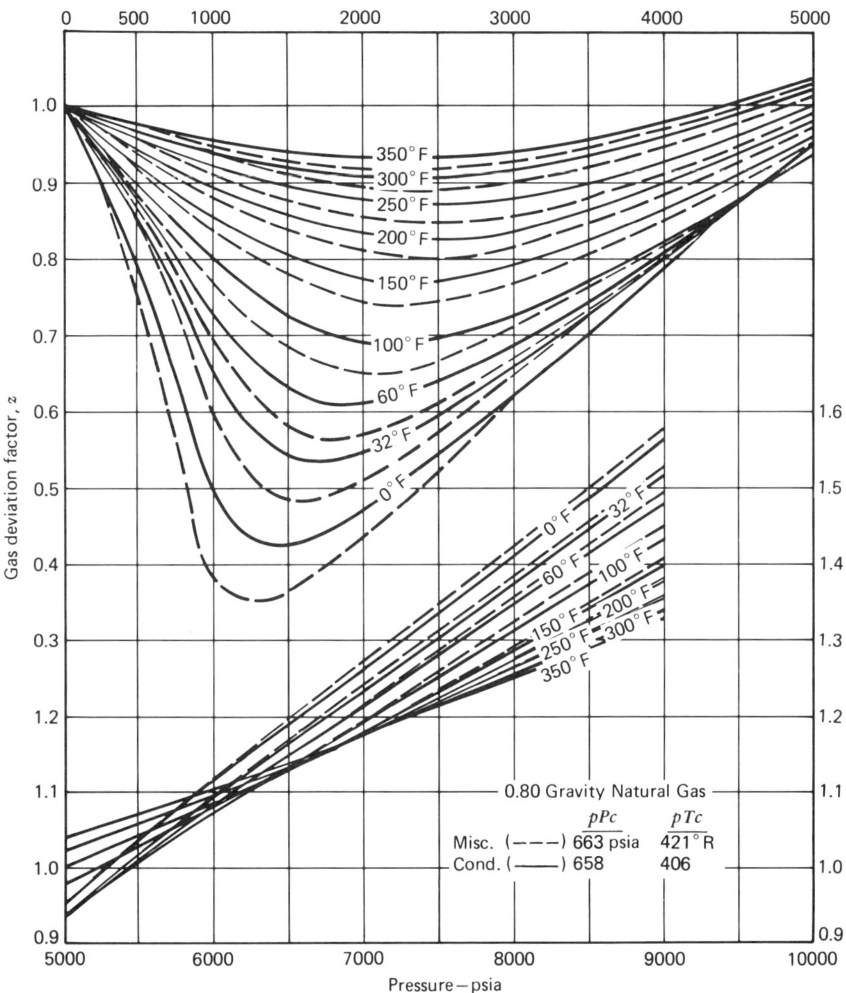

Fig. A.5 0.80 gravity natural gas

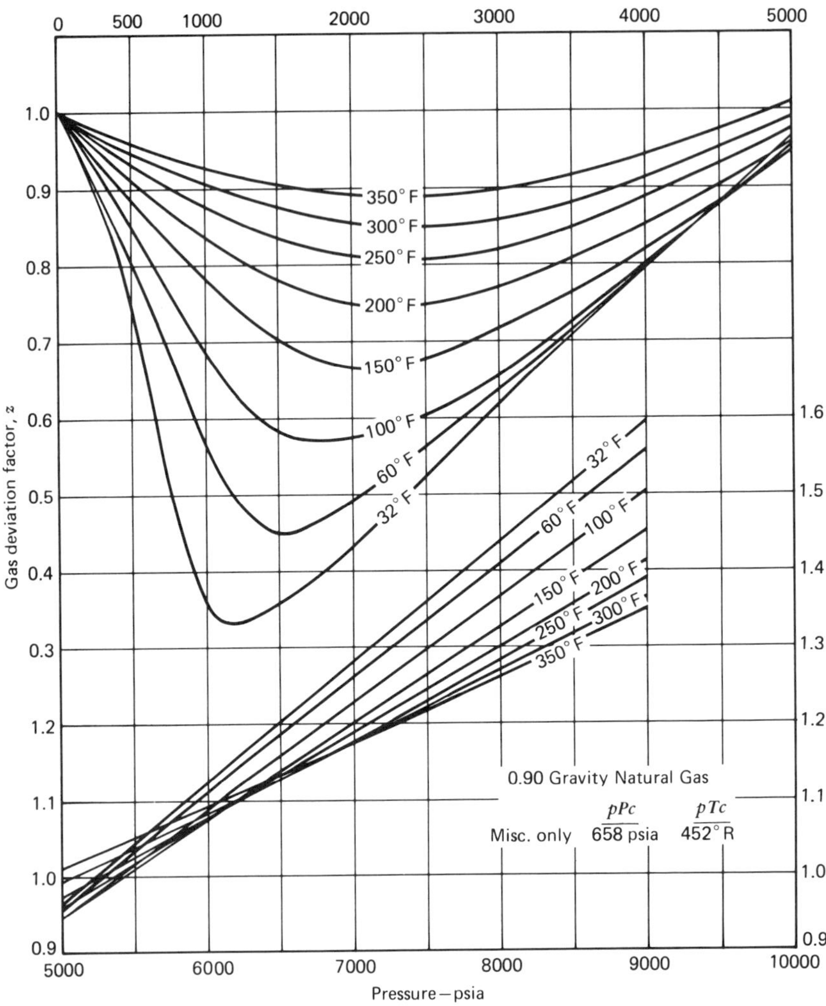

Fig. A.6 0.90 gravity natural gas, miscellaneous only

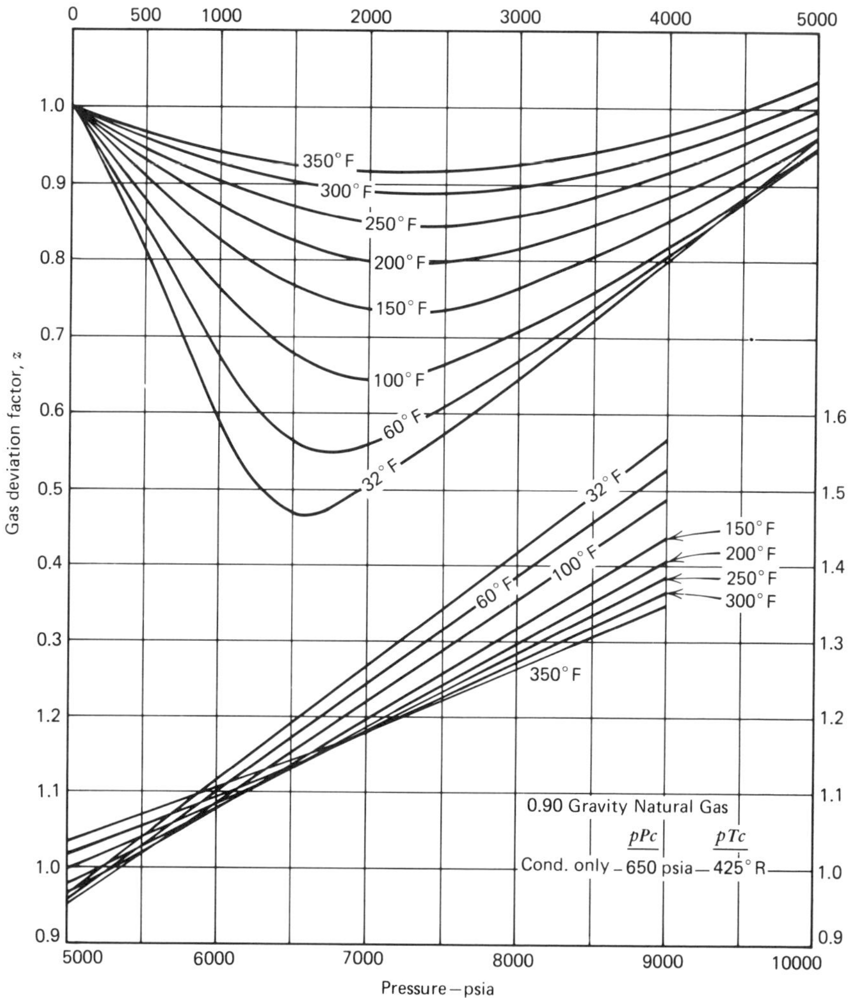

Fig. A.7 0.90 gravity natural gas, condensate only

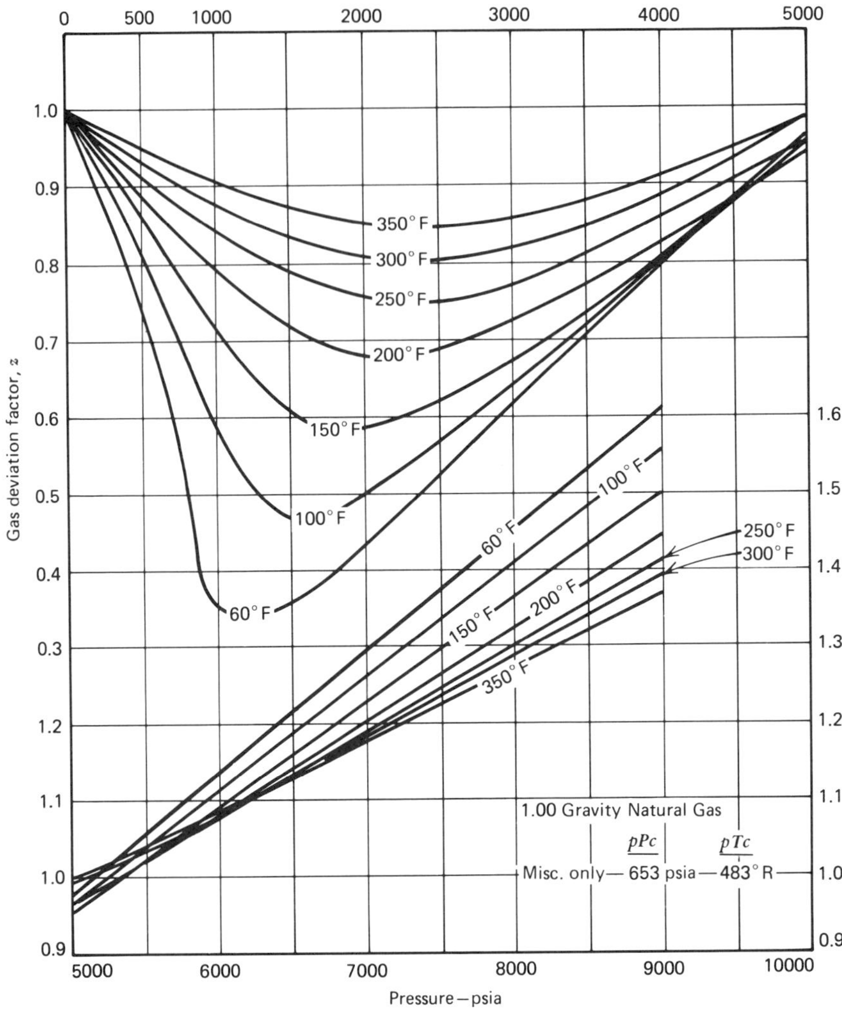

Fig. A.8 1.00 gravity natural gas, miscellaneous only

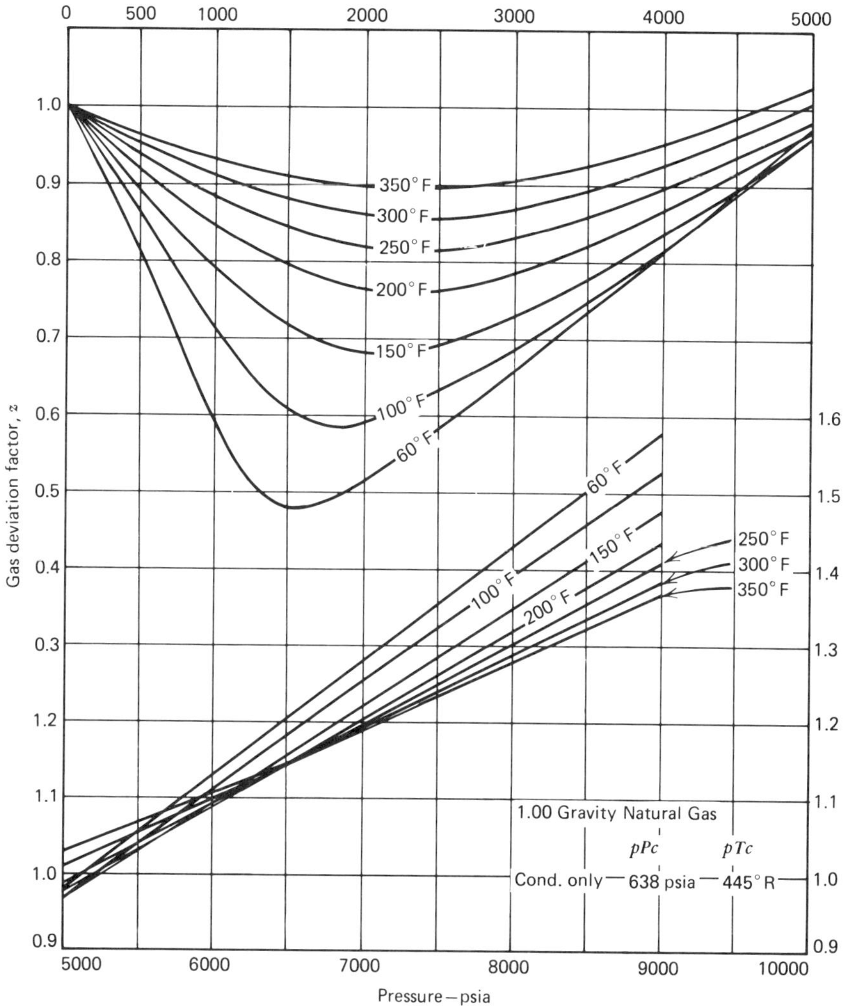

Fig. A.9 1.00 gravity natural gas, condensate only

A.1 PHYSICAL PROPERTIES

TABLE A.1
Physical Constants of Hydrocarbons

No.	Compound	Formula	Molecular weight	Boiling point °F, 14.696 psia	Vapor pressure, 100°F, psia	Freezing point, °F, 14.696 psia	Critical Constants		
							Pressure, psia	Temperature, °F	Volume, cu ft/lb
1	Methane	CH_4	16.043	−258.69	(5000)	−296.46	667.8	−116.63	0.0991
2	Ethane	C_2H_6	30.070	−127.48	(800)	−297.89	707.8	90.09	0.0788
3	Propane	C_3H_8	44.097	−43.67	190.0	−305.84	616.3	206.01	0.0737
4	*n*-Butane	C_4H_{10}	58.124	31.10	51.6	−217.05	550.7	305.65	0.0702
5	Isobutane	C_4H_{10}	58.124	10.90	72.2	−255.29	529.1	274.98	0.0724
6	*n*-Pentane	C_5H_{12}	72.151	96.92	15.570	−201.51	488.6	385.7	0.0675
7	Isopentane	C_5H_{12}	72.151	82.12	20.44	−255.83	490.4	369.10	0.0679
8	Neopentane	C_5H_{12}	72.151	49.10	35.9	2.17	464.0	321.13	0.0674
9	*n*-Hexane	C_6H_{14}	86.178	155.72	4.956	−139.58	436.9	453.7	0.0688
10	2-Methylpentane	C_6H_{14}	86.178	140.47	6.767	−244.63	436.6	435.83	0.0681
11	3-Methylpentane	C_6H_{14}	86.178	145.89	6.098	—	453.1	448.3	0.0681
12	Neohexane	C_6H_{14}	86.178	121.52	9.856	−147.72	446.8	420.13	0.0667
13	2,3-Dimethylbutane	C_6H_{14}	86.178	136.36	7.404	−199.38	453.5	440.29	0.0665
14	*n*-Heptane	C_7H_{16}	100.205	209,17	1.620	−131.05	396.8	512.8	0.0691
15	2-Methylhexane	C_7H_{16}	100.205	194.09	2.271	−180.89	396.5	495.00	0.0673
16	3-Methylhexane	C_7H_{16}	100.205	197.32	2.130	—	408.1	503.78	0.0646
17	3-Ethylpentane	C_7H_{16}	100.205	200.25	2.012	−181.48	419.3	513.48	0.0665
18	2,2-Dimethylpentane	C_7H_{16}	100.205	174.54	3.492	−190.86	402.2	477.23	0.0665
19	2,4-Dimethylpentane	C_7H_{16}	100.205	176.89	3.292	−182.63	396.9	475.95	0.0668
20	3,3-Dimethylpentane	C_yH_{16}	100.205	186.91	2.773	−210.01	427.2	505.85	0.0662
21	Triptane	C_7H_{16}	100.205	177.58	3.374	−12.82	428.4	496.44	0.0636
22	*n*-Octane	C_8H_{18}	114.232	258.22	0.537	−70.18	360.6	564.22	0.0690
23	Diisobutyl	C_8H_{18}	114.232	228.39	1.101	−132.07	360.6	530.44	0.0676
24	Isooctane	C_8H_{18}	114.232	210.63	1.708	−161.27	372.4	519.46	0.0656
25	*n*-Nonane	C_9H_{20}	128.259	303.47	0.179	−64.28	332.0	610.68	0.0684
26	*n*-Decane	$C_{10}H_{22}$	142.286	345.48	0.0597	−21.36	304.0	652.1	0.0679
27	Cyclopentane	C_5H_{10}	70.135	120.65	9.914	−136.91	653.8	461.5	0.059
28	Methylcyclopentane	C_6H_{12}	84.162	161.25	4.503	−224.44	548.9	499.35	0.0607
29	Cyclohexane	C_6H_{12}	84.162	177.29	3.264	43.77	591.0	536.7	0.0586
30	Methylcyclohexane	C_7H_{14}	98.189	213.68	1.609	−195.87	503.5	570.27	0.0600
31	Ethylene	C_2H_4	28.054	−154.62	—	−272.45	729.8	48.58	0.0737
32	Propene	C_3H_6	42.081	−53.90	226.4	−301.45	669.0	196.9	0.0689
33	1-Butene	C_4H_8	56.108	20.75	63.05	−301.63	583.0	295.6	0.0685

TABLE A.1 (Continued)

No.	Density of liquid, 60°F, 14.696 psia: Specific gravity 60°F/60°F	lb/gal (Wt in vacuum)	lb/gal (Wt in air)	Gal/lb-mole	Temperature Coefficient of Density	Pitzer acentric factor	Compressibility factor of real gas, Z 14.696 psia, 60°F	Gas density, 60°F, 14.696 psia Ideal gas: Specific gravity Air = 1°	cu ft gas /lb	cu ft gas/gal liquid	Specific heat 60°F, 14.696 psia Cp. Btu/lb °F: Ideal gas	Liquid
1	0.3	2.5	2.5	6.4	—	0.0104	0.9881	0.5539	23.65	59.	0.5266	—
2	0.3564	2.971	2.962	10.12	—	0.0986	0.9916	1.0382	12.62	37.5	0.4097	0.9256
3	0.5077	4.233	4.223	10.42	0.00152	0.1524	0.9820	1.5225	8.606	36.43	0.3881	0.5920
4	0.5844	4.872	4.865	11.93	0.00117	0.2010	0.9667	2.0068	6.529	31.81	0.3867	0.5636
5	0.5631	4.695	4.686	12.38	0.00119	0.1848	0.9696	2.0068	6.529	30.65	0.3872	0.5695
6	0.6310	2.261	5.251	13.71	0.00087	0.2539	0.9549	2.4911	5.260	27.67	0.3883	0.5441
7	0.6247	5.208	5.199	13.85	0.00090	0.2223	0.9544	2.4911	5.260	27.39	0.3827	0.5353
8	0.5967	4.975	4.965	14.50	0.00104	0.1969	0.9510	2.4911	5.260	26.17	(0.3866)	0.554
9	0.6640	5.536	5.526	15.57	0.00075	0.3007	—	2.9753	4.404	24.38	0.3864	0.5332
10	0.6579	5.485	5.475	15.71	0.00073	0.2825	—	2.9753	4.404	24.15	0.3872	0.5264
11	0.6689	5.577	5.568	15.45	0.00075	0.2741	—	2.9753	4.404	24.56	0.3815	0.507
12	0.6540	5.453	5.443	15.81	0.00078	0.2369	—	2.9753	4.404	24.01	0.3809	0.5165
13	0.6664	5.556	5.546	15.51	0.00075	0.2495	—	2.9753	4.404	24.47	0.378	0.5127
14	0.6882	5.738	5.728	17.46	0.00069	0.3498	—	3.4596	3.787	21.73	0.3875	0.5283
15	0.6830	5.694	5.685	17.60	0.00068	0.3336	—	3.4596	3.787	21.57	(0.390)	0.5223
16	0.6917	5.767	5.757	17.38	0.00069	0.3257	—	3.4596	3.787	21.84	(0.390)	0.511
17	0.7028	5.859	5.850	17.10	0.00070	0.3095	—	3.4596	3.787	22.19	(0.390)	0.5145
18	0.6782	5.654	5.645	17.72	0.00072	0.2998	—	3.4596	3.787	21.41	(0.395)	0.5171
19	0.6773	5.647	5.637	17.75	0.00072	0.3048	—	3.4596	3.787	21.39	0.3906	0.5247
20	0.6976	5.816	5.807	17.23	0.00065	0.2840	—	3.4596	3.787	22.03	(0.395)	0.502
21	0.6946	5.791	5.782	17.30	0.00069	0.2568	—	3.4596	3.787	21.93	0.3812	0.4995
22	0.7068	5.893	5.883	19.39	0.00062	0.4018	—	3.9439	3.322	19.58	(0.3876)	0.5239
23	0.6979	5.819	5.810	19.63	0.00065	0.3596	—	3.9439	3.322	19.33	(0.373)	0.5114
24	0.6962	5.804	5.795	19.68	0.00065	0.3041	—	3.9439	3.322	19.28	0.3758	0.4892
25	0.7217	6.017	6.008	21.32	0.00063	0.4455	—	4.4282	2.959	17.80	0.3840	0.5228
26	0.7342	6.121	6.112	23.24	0.00055	0.4885	—	4.9125	2.667	16.33	0.3835	0.5208
27	0.7504	6.256	6.247	11.21	0.00070	0.1955	0.9657	2.4215	5.411	33.85	0.2712	0.4216
28	0.7536	6.283	6.274	13.40	0.00071	0.2306	—	2.9057	4.509	28.33	0.3010	0.4407
29	0.7834	6.531	6.522	12.89	0.00068	0.2133	—	2.9057	4.509	29.45	0.2900	0.4322
30	0.7740	6.453	6.444	15.22	0.00063	0.2567	—	3.3900	3.865	24.94	0.3170	0.4397
31	—	—	—	—	—	0.0868	0.9938	0.9686	13.53	—	0.3622	—
32	0.5220	4.352	4.343	9.67	0.00189	0.1405	0.9844	1.4529	49.018	39.25	0.3541	0.585
33	0.6013	5.013	5.004	11.19	0.00116	0.1906	0.9704	1.9372	6.764	33.91	0.3548	0.535

TABLE A.1 (Continued)

No.	Compound	Formula	Molecular weight	Boiling point °F, 14.696 psia	Vapor pressure, 100°F, psia	Freezing point, °F, 14.696 psia	Critical Constants		
							Pressure, psia	Temperature, °F	Volume, cu ft/lb
34	*Cis*-2-Butene	C_4H_8	56.108	38.69	45.54	−218.06	610.0	324.37	0.0668
35	*Trans*-2-Butene	C_4H_8	56.108	33.58	49.80	−157.96	595.0	311.86	0.0680
36	Isobutene	C_4H_8	56.108	19.59	63.40	−220.61	580.0	292.55	0.0682
37	1-Pentene	C_5H_{10}	70.135	85.93	19.115	−265.39	590.0	376.93	0.0697
38	1,2-Butodiene	C_4H_6	54.092	51.53	(20.)	−213.16	(653.)	(339.)	(0.0649)
39	1,3-Butodiene	C_4H_6	54.092	24.06	(60.)	−164.02	628.0	306.0	0.0654
40	Isoprene	C_5H_9	68.119	93.30	16.672	−230.74	(558.4)	(412.)	(0.0650)
41	Acetylene	C_2H_2	26.038	−119	—	−114.	890.4	95.31	0.0695
42	Benzene	C_6H_6	78.114	176.17	3.224	41.96	710.4	552.22	0.0531
43	Toluene	C_7H_8	92.141	231.13	1.032	−138.94	595.9	605.55	0.0549
44	Ethylbenzene	C_8H_{10}	106.168	227.16	0.371	−138.91	523.5	651.24	0.0564
45	*o*-Xylene	C_8H_{10}	106.168	291.97	0.264	−13.30	541.4	675.0	0.0557
46	*m*-Xylene	C_8H_{10}	106.168	282.41	0.326	−54.12	513.6	651.02	0.0567
47	*p*-Xylene	C_8H_{10}	106.168	281.05	0.347	55.86	509.2	649.6	0.0572
48	Styrene	C_8H_8	104.152	293.29	(0.24)	−23.10	580.0	706.0	0.0541
49	Isopropylbenzene	C_9H_{12}	120.195	306.34	0.188	−140.82	465.4	676.4	0.0570
50	Methyl alcohol	CH_4O	32.042	148.1	4.63	−143.82	1174.2	462.97	0.0589
51	Ethyl alcohol	C_2H_6O	46.069	172.92	2.3	−173.4	925.3	469.58	0.0580
52	Carbon monoxide	CO	28.010	−313.6	—	−340.6	507.	−220.	0.0532
53	Carbon dioxide	CO_2	44.010	−109.3	—	—	1071.	87.9	0.0342
54	Hydrogen sulfide	H_2S	34.076	−76.6	394.0	−117.2	1306.	212.7	0.0459
55	Sulfur dioxide	SO_2	64.059	14.0	88.	−103.9	1145.	315.5	0.0306
56	Ammonia	NH_3	17.031	−28.2	212.	−107.9	1636.	270.3	0.0681
57	Air	N_2O_2	28.964	−317.6	—	—	547.	−221.3	0.0517
58	Hydrogen	H_2	2.016	−423.0	—	−434.8	188.1	−399.8	0.5167
59	Oxygen	O_2	31.999	−297.4	—	−361.8	736.9	−181.1	0.0382
60	Nitrogen	H_2	28.013	−320.4	—	−346.0	493.0	−232.4	0.0514
61	Chlorine	Cl_2	70.906	−29.3	158.	−149.8	1118.4	291.	0.0281
62	Water	H_2O	18.015	212.0	0.9492	32.0	3208.	705.6	0.0500
63	Helium	He	4.003	—	—	—	—	—	—
64	Hydrogen chloride	HCl	36.461	−121.	925.	−173.6	1198.	124.5	0.0208

Source: Courtesy of GPSA.

TABLE A.1 (Continued)

No.	Density of liquid, 60°F, 14.696 psia: Specific gravity 60°F/60°F	Density of liquid, 60°F, 14.696 psia: lb/gal (Wt in vacuum)	Density of liquid, 60°F, 14.696 psia: lb/gal (Wt in air)	Density of liquid, 60°F, 14.696 psia: Gal/lb-mole	Temperature Coefficient of Density	Pitzer acentric factor	Compressibility factor of real gas, Z 14.696 psia, 60°F	Gas density, 60°F, 14.696 psia, Ideal gas: Specific gravity Air = 1°	Gas density, 60°F, 14.696 psia, Ideal gas: cu ft gas /lb	Gas density, 60°F, 14.696 psia, Ideal gas: cu ft gas/gal liquid	Specific heat 60°F, 14.696 psia, Cp. Btu/lb °F: Ideal gas	Specific heat 60°F, 14.696 psia, Cp. Btu/lb °F: Liquid
34	0.6271	5.228	5.219	10.73	0.00098	0.1953	0.9661	1.9372	6.764	35.36	0.3269	0.5271
35	0.6100	5.086	5.076	11.03	0.00107	0.2220	0.9662	1.9372	6.764	34.40	0.3654	0.5351
36	0.6004	5.006	4.996	11.21	0.00120	0.1951	0.9689	1.9372	6.764	33.86	0.3701	0.549
37	0.6457	5.383	5.374	13.03	0.00089	0.2925	0.9550	2.4215	5.411	29.13	0.3635	0.5196
38	0.658	5.486	5.470	9.86	0.00098	0.2485	(0.969)	1.8676	7.016	38.49	0.3458	0.5408
39	0.6272	5.229	5.220	10.34	0.00113	0.1955	(0.965)	1.8676	7.016	36.69	0.3412	0.5079
40	0.6861	5.720	5.711	11.91	0.00086	0.2323	(0.962)	2.3519	5.571	31.87	0.357	0.5192
41	0.615	—	—	—	—	0.1803	0.9925	0.8990	14.57	—	0.3966	—
42	0.8844	7.373	7.365	10.59	0.00066	0.2125	0.929	2.6969	4.858	35.82	0.2429	0.4098
43	0.8718	7.268	7.260	12.68	0.00060	0.2596	0.903	3.1812	4.119	29.94	0.2598	0.4012
44	0.8718	7.268	7.259	14.61	0.00054	0.3169	—	3.6655	3.574	25.98	0.2795	0.4114
45	0.8848	7.377	7.367	14.39	0.00055	0.3023	—	3.6655	3.574	26.37	0.2914	0.4418
46	0.8687	7.243	7.234	14.66	0.00054	0.3278	—	3.6655	3.574	25.89	0.2782	0.4045
47	0.8657	7.218	7.209	14.71	0.00054	0.3138	—	3.6655	3.574	25.80	0.2769	0.4083
48	0.9110	7.595	7.586	13.71	0.00057	—	—	3.5959	3.644	27.67	0.2711	0.4122
49	0.8663	7.223	7.214	16.64	0.00054	0.2862	—	4.1498	3.157	22.80	0.2917	(0.414)
50	0.796	6.64	6.63	4.83	—	—	—	1.1063	11.84	78.6	0.3231	0.594
51	0.794	6.62	6.61	6.96	—	—	—	1.5906	8.237	54.5	0.3323	0.562
52	0.801	6.68	6.67	4.19	—	0.041	0.9995	0.9671	13.55	—	0.2484	—
53	0.827	6.89	6.88	6.38	—	0.225	0.9943	1.5195	8.623	59.5	0.1991	—
54	0.79	6.59	6.58	5.17	—	0.100	0.9903	1.1765	11.14	73.3	0.238	—
55	1.397	11.65	11.64	5.50	—	0.246	—	2.2117	5.924	69.0	0.145	0.325
56	0.6173	5.15	5.14	3.31	—	0.255	—	0.5880	22.28	114.7	0.5002	1.114
57	0.856	7.14	7.13	4.06	—	—	0.9996	1.0000	13.10	—	0.2400	—
58	0.07	—	—	—	—	0.000	1.0006	0.0696	188.2	—	3.408	—
59	1.14	9.50	9.49	3.37	—	0.0213	—	1.1048	11.86	—	0.2188	—
60	0.808	6.74	6.73	4.16	—	0.040	0.9997	0.9672	13.55	—	0.2482	—
61	1.414	11.79	11.78	6.01	—	—	—	2.4481	5.352	63.1	0.119	—
62	1.000	8.337	8.328	2.16	—	0.348	—	0.6220	21.06	175.6	0.4446	1.0009
63	—	—	—	—	—	—	—	—	—	—	—	—
64	0.8558	7.135	7.126	5.11	0.00335	—	—	1.2588	10.41	74.3	0.190	—

Source: Courtesy of GPSA.

TABLE A.1 (Continued)

No.	Compound	Calorific value, 60°F — Net: Btu/cu ft, Ideal gas, 14.696 psia (20)	Calorific value, 60°F — Gross: Btu/cu ft, Ideal gas, 14.696 psia (20)	Gross: Btu/lb liquid	Gross: (wt in vacuum) Btu/gal liquid	Heat of vaporization, 14.696 psia at boiling point, Btu/lb	Refractive index, nD 68°F	Air required for combustion ideal gas cu ft/cu ft	Flammability Limits, Vol % in Air Mixture — Lower	Flammability Limits — Higher	ASTM Octane Number — Motor method D-357	ASTM Octane Number — Research method D-908
1	Methane	909.1	1009.7	—	—	219.22	—	9.54	5.0	15.0	—	—
2	Ethane	1617.8	1768.8	—	—	210.41	—	16.70	2.9	13.0	+.05	+1.6
3	Propane	2316.1	2517.4	21,513	91,065	183.05	—	23.86	2.1	9.5	97.1	+1.8
4	*n*-Butane	3010.4	3262.1	21,139	102,989	165.65	1.3326	31.02	1.8	8.4	89.6	93.8
5	Isobutane	3001.1	3252.7	21,091	99,022	157.53	—	31.02	1.8	8.4	97.6	+.10
6	*n*-Pentane	3707.5	4009.5	20,928	110,102	153.59	1.357,48	38.18	1.4	8.3	62.6	61.7
7	Isopentane	3698.3	4000.3	20,889	108,790	147.13	1.353,73	38.18	1.4	(8.3)	90.3	92.3
8	Neopentane	3682.6	3984.6	20,824	103,599	135.58	1.342	38.18	1.4	(8.3)	80.2	85.5
9	*n*-Hexane	4403.7	4756.1	20,784	115,060	143.95	1.374,86	45.34	1.2	7.7	26.0	24.8
10	2-Methylpentane	4395.8	4748.1	20,757	113,852	138.67	1.371,45	45.34	1.2	(7.7)	73.5	73.4
11	3-Methylpentane	4398.7	4751.0	20,768	115,823	140.09	1.376,52	45.34	(1.2)	(7.7)	74.3	74.5
12	Neohexane	4382.6	4735.0	20,710	112,932	131.24	1.368,76	45.34	1.2	(7.7)	93.4	91.8
13	2,3-Dimethylbutane	4391.7	4744.0	20,742	115,243	136.08	1.374,95	45.34	(1.2)	(7.7)	94.3	+0.3
14	*n*-Heptane	5100.2	5502.9	20,681	118,668	136.01	1.387,64	52.50	1.0	7.0	0.0	0.0
15	2-Methylhexane	5092.1	5494.8	20,658	117,627	131.59	1.384,85	52.50	(1.0)	(7.0)	46.4	42.4
16	3-Methylhexane	5095.2	5497.8	20,668	119,192	131.11	1.388,64	52.50	(1.0)	(7.0)	55.8	52.0
17	3-Ethylpentane	5098.2	5500.9	20,679	121,158	132.83	1.393,39	52.50	(1.0)	(7.0)	69.3	65.0

18	2,2-Dimethylpentane	5079.4	5482.1	20,620	116,585	125.13	1.382,15	52.50	(1.0)	(7.0)	95.6	92.8
19	2,4-Dimethylpentane	5084.3	5487.0	20,636	116,531	126.58	1.381,45	52.50	(1.0)	(7.0)	83.8	83.1
20	3,3-Dimethylpentane	5085.0	5487.6	20,638	120,031	127.21	1.390,92	52.50	(1.0)	(7.0)	86.6	80.8
21	Triptane	5081.0	5483.6	20,627	119,451	124.21	1.389,44	52.50	(1.0)	(7.0)	+0.1	+1.8
22	*n*-Octane	5796.7	6249.7	20,604	121,419	129.53	1.397,43	59.65	0.96	—	—	—
23	Diisobutyl	5781.3	6234.3	20,564	119,662	122.8	1.392,46	59.65	(0.98)	—	55.7	55.2
24	Isooctane	5779.8	6232.8	20,570	119,388	116.71	1.391,45	59.65	1.0	—	100.	100.
25	*n*-Nonane	6493.3	6996.6	20,544	123,613	123.76	1.405,42	66.81	0.87	2.9	—	—
26	*n*-Decane	7188.6	7742.3	20,494	125,444	118.68	1.411,89	73.97	0.78	2.6	—	—
27	Cyclopentane	3512.0	3763.7	20,188	126,296	167.34	1.406,45	35.79	(1.4)	—	84.9	+0.1
28	Methylcyclopentane	4198.4	4500.4	20,130	126,477	147.83	1.409,70	42.95	(1.2)	8.35	80.0	91.3
29	Cyclohexane	4178.8	4480.8	20,035	130,849	153.0	1.426,23	42.95	1.3	7.8	77.2	83.0
30	Methylcyclohexane	4862.8	5215.2	20,001	129,066	136.3	1.423,12	50.11	1.2	—	71.1	74.8
31	Ethylene	1499.0	1599.7	—	—	207.57	—	14.32	2.7	34.0	75.6	+.03
32	Propene	2182.7	2333.7	—	—	188.18	—	21.48	2.0	10.0	84.9	+0.2
33	1-Butene	2879.4	3080.7	20,678	103,659	167.94	—	28.63	1.6	9.3	80.8	97.4
34	*Cis*-2-Butene	2871.7	3073.1	20,611	107,754	178.91	—	28.63	(1.6)	—	83.5	100.
35	*Trans*-2-Butene	2866.8	3068.2	20,584	104,690	174.39	—	28.63	(1.6)	—	—	—
36	Isobutene	2860.4	2061.8	20,548	102,863	169.48	—	28.63	(1.6)	—	—	—
37	1-Pentene	3575.2	3826.9	20,548	110,610	154.46	1.371,48	35.79	1.4	8.7	77.1	90.9
38	1,2-Butadiene	2789.0	2940.0	20,447	112,172	(181.)	—	26.25	(2.0)	(12.)	—	—
39	1,3-Butadiene	2730.0	2881.0	20,047	104,826	(174.)	—	26.25	2.0	11.5	—	—
40	Isoprene	3410.8	3612.1	19,964	114,194	(153.)	1.421,94	33.41	(1.5)	—	81.0	99.1
41	Acetylene	1422.4	1472.8	—	—	—	—	11.93	2.5	80.	—	—
42	Benzene	3590.7	3741.7	17,992	132,655	169.31	1.501,12	35.79	1.3	7.9	+2.8	—
43	Toluene	4273.3	4474.7	18,252	132,656	154.84	1.496,93	42.95	1.2	7.1	+0.3	+5.8
44	Ethylbenzene	4970.0	5221.7	18,494	134,414	144.0	1.495,88	50.11	0.99	6.7	97.9	+0.8
45	*o*-Xylene	4958.3	5210.0	18,445	136,069	149.	1.505,45	50.11	1.1	6.4	100.	—
46	*m*-Xylene	4956.8	5208.5	18,441	133,568	147.2	1.497,22	50.11	1.1	6.4	+2.8	+4.0
47	*p*-Xylene	4956.9	5208.5	18,445	133,136	144.52	1.495,82	50.11	1.1	6.6	+1.2	+3.4
48	Styrene	4828.7	5030.0	18,150	137,849	(151.)	1.546,82	47.72	1.1	6.1	+0.2	>+3.
49	Isopropylbenzene	5661.4	5963.4	18,665	134,817	134.3	1.491,45	57.27	0.88	6.5	99.3	+2.1

(Continued)

TABLE A.1 (Continued)

No.	Compound	Calorific value, 60°F: Net, Btu/cu ft, Ideal gas, 14.696 psia (20)	Calorific value, 60°F: Gross, Btu/cu ft, Ideal gas, 14.696 psia (20)	Gross, Btu/lb liquid (wt in vacuum)	Gross, Btu/gal liquid	Heat of vaporization, 14.696 psia at boiling point, Btu/lb	Refractive index, nD, 68°F	Air required for combustion ideal gas, cu ft/cu ft	Flammability Limits, Vol % in Air Mixture: Lower	Flammability Limits: Higher	ASTM Octane Number: Motor method D-357	ASTM Octane Number: Research method D-908
50	Methyl alcohol	—	—	9,760	64,771	473.	1.3288	7.16	6.72	36.50	—	—
51	Ethyl alcohol	—	—	12,780	84,600	367.	1.3614	14.32	3.28	18.95	—	—
52	Carbon monoxide	—	321.	—	—	92.7	—	2.39	12.50	74.20	—	—
53	Carbon dioxide	—	—	—	—	238.2	—	—	—	—	—	—
54	Hydrogen sulfide	588.	637.	—	—	235.6	—	7.16	4.30	45.50	—	—
55	Sulfur dioxide	—	—	—	—	166.7	—	—	—	—	—	—
56	Ammonia	359.	434.	—	—	587.2	—	3.58	15.50	27.00	—	—
57	Air	—	—	—	—	92.	—	—	—	—	—	—
58	Hydrogen	274.	324.	—	—	193.9	—	2.39	4.00	74.20	—	—
59	Oxygen	—	—	—	—	91.6	—	—	—	—	—	—
60	Nitrogen	—	—	—	—	87.8	—	—	—	—	—	—
61	Chlorine	—	—	—	—	123.8	—	—	—	—	—	—
62	Water	—	—	—	—	970.3	1.3330	—	—	—	—	—
63	Helium	—	—	—	—	—	—	—	—	—	—	—
64	Hydrogen Chloride	—	—	—	—	185.5	—	—	—	—	—	—

Source: Courtesy of GPSA.

A.2 SEPARATOR SIZING TABLES FROM SIVALLS

A.2.1 Tables Applying to Standard Vertical High- or Low-Pressure Separators

Table A.2 Specifications of Standard Vertical Low-Pressure Separators

Table A.3 Settling Volumes of Standard Vertical Low-Pressure Separators, 125 psi WP

Table A.4 Specifications of Standard Vertical High-Pressure Separators

Table A.5 Settling Volumes of Standard Vertical High-Pressure Separators, 230 to 2000 psi WP

Table A.6 Specifications of Standard Horizontal Low-Pressure Separators

Table A.7 Settling Volumes of Standard Horizontal Low-Pressure Separators, 125 psi WP

Table A.8 Specifications of Standard Horizontal High-Pressure Separators

Table A.9 Settling Volumes of Standard Horizontal High-Pressure Separators, 230 to 2000 psi WP

Table A.10 Specifications of Standard Spherical Separators

Table A.11 Settling Volumes of Standard Spherical Low-Pressure Separators, 125 psi WP

Table A.12 Settling Volumes of Standard Spherical High-Pressure Separators, 230 to 3000 psi WP

TABLE A.2

Specifications of Standard Vertical Low-pressure Separators

Size, Dia × Ht	Working Pressure, psi	Inlet and Gas Outlet Conn.	Oil Outlet Conn.	Standard Valves		Shipping Weight, lb
				Oil or Oil and Water	Gas	
24′ × 5′	125	2″ Thd	2″ Thd	2″	2″	950
24″ × 7½′	125	2″ Thd	2″ Thd	2″	2″	1150
30″ × 10′	125	3″ Thd	3″ Thd	2″	2″	2000
36″ × 5′	125	4″ Thd	2″ Thd	2″	2″	2000
36″ × 7½′	125	4″ Thd	3″ Thd	2″	2″	2350
36″ × 10′	125	4″ Thd	4″ Thd	2″	2″	2700
48″ × 10′	125	6″ Flg	4″ Thd	2″	2″	3400
48″ × 15′	125	6″ Flg	4″ Thd	2″	2″	4500
60″ × 10′	125	6″ Flg	4″ Thd	3″	3″	5200
60″ × 15′	125	6″ Flg	4″ Thd	3″	3″	6400
60″ × 20′	125	6″ Flg	4″ Thd	3″	3″	7600

TABLE A.3
Settling Volumes of Standard Vertical Low-pressure Separators, 125 psi WP

Size, Dia × Ht	Settling Volume, bbl	
	Oil-Gas Separators	Oil-Gas-Water Separators[a]
24″ × 5′	0.65	1.10
24″ × 7½′	1.01	1.82
30″ × 10′	2.06	3.75
36″ × 5′	1.61	2.63
36″ × 7½′	2.43	4.26
36″ × 10′	3.04	5.48
48″ × 10′	5.67	10.06
48″ × 15′	7.86	14.44
60″ × 10′	9.23	16.08
60″ × 15′	12.65	12.93
60″ × 20′	15.51	18.64

[a]Total settling volume is usually split even between oil and water.

TABLE A.4
Specifications of Standard Vertical High Pressure Separators

Size, Dia × Ht	Working Pressure, psi[a]	Inlet and Gas Outlet Connection	Standard Liquid Valve	Shipping Weight, lb
16″ × 5′	1000	2″ Thd	1″	1,100
16″ × 7½′		2″ Thd	1″	1,200
16″ × 10′		2″ Thd	1″	1,500
20″ × 5′	1000	3″ Flg	1″	1,600
20″ × 7½′		3″ Flg	1″	1,900
20″ × 10′		3″ Flg	1″	2,200
24″ × 5′	1000	3″ Flg	1″	2,500
24″ × 7½′		3″ Flg	1″	2,850
24″ × 10′		3″ Flg	1″	3,300
30″ × 5′	1000	4″ Flg	1″	3,200
30″ × 7½′		4″ Flg	1″	3,650
30″ × 10′		4″ Flg	1″	4,200
36″ × 7½′	1000	4″ Flg	1″	5,400
36″ × 10′		4″ Flg	1″	6,400
36″ × 15′		4″ Flg	1″	8,700
42″ × 7½′	1000	6″ Flg	2″	7,700
42″ × 10′		6″ Flg	2″	9,100
42″ × 15′		6″ Flg	2″	12,000
48″ × 7½′	1000	6″ Flg	2″	10,400
48″ × 10′		6″ Flg	2″	12,400
48″ × 15′		6″ Flg	2″	16,400
54″ × 7½′	1000	6″ Flg	2″	12,300
54″ × 10′		6″ Flg	2″	14,900
54″ × 15′		6″ Flg	2″	20,400
60″ × 7½′	1000	6″ Flg	2″	17,500
60″ × 10′		6″ Flg	2″	20,500
60″ × 15′		6″ Flg	2″	26,500
60″ × 20′		6″ Flg	2″	32,500

[a]Other standard working pressures available are 230, 500, 600, 1200, 1440, 1500, and 2000 psi.

TABLE A.5
Settling Volumes of Standard Vertical High-pressure Separators, 230 to 2000 psi WP[c]

Size, Dia × Ht	Settling Volume, bbl[a]	
	Oil-Gas Separators	Oil-Gas-Water Separators[b]
16″ × 5′	0.27	0.44
16″ × 7½′	0.41	0.72
16″ × 10′	0.51	0.94
20″ × 5′	0.44	0.71
20″ × 7½′	0.65	1.15
20″ × 10′	0.82	1.48
24″ × 5′	0.66	1.05
24″ × 7½′	0.97	1.68
24″ × 10′	1.21	2.15
30″ × 5′	1.13	1.76
30″ × 7½′	1.64	2.78
30″ × 10′	2.02	3.54
36″ × 7½′	2.47	4.13
36″ × 10′	3.02	5.24
36″ × 15′	4.13	7.45
42″ × 7½′	3.53	5.80
42″ × 10′	4.29	7.32
42″ × 15′	5.80	10.36
48″ × 7½′	4.81	7.79
48″ × 10′	5.80	9.78
48″ × 15′	7.79	13.76
54″ × 7½′	6.33	10.12
54″ × 10′	7.60	12.65
54″ × 15′	10.12	17.70
60″ × 7½′	8.08	12.73
60″ × 10′	9.63	15.83
60″ × 15′	12.73	22.03
60″ × 20′	15.31	27.20

[a]Based on 1000 psi WP separators.

[b]Total settling volume is usually split even between oil and water.

[c]Standard working pressures available are 230, 500, 1000, 1200, 1440, 1500, and 2000 psi.

TABLE A.6
Specifications of Standard Horizontal Low-pressure Separators

Size, Dia × Ht	Working Pressure, psi	Inlet and Gas Outlet Connection	Oil Outlet Connection	Standard Valves: Oil or Oil and Water	Standard Valves: Gas	Shipping Weight, lb
24″ × 5′	125	2″ Thd	2″ Thd	2″	2″	1,000
24″ × 7½′	125	2″ Thd	2″ Thd	2″	2″	1,200
24″ × 10′	125	3″ Thd	2″ Thd	2″	2″	1,600
30″ × 5′	125	3″ Thd	3″ Thd	2″	2″	1,200
30″ × 7½′	125	3″ Thd	3″ Thd	2″	2″	1,600
30″ × 10′	125	4″ Thd	4″ Thd	2″	2″	2,100
36″ × 10′	125	4″ Thd	4″ Thd	2″	2″	2,900
36″ × 15′	125	4″ Thd	4″ Thd	2″	2″	3,800
48″ × 10′	125	6″ Flg	4″ Thd	2″	2″	3,500
48″ × 15′	125	6″ Flg	4″ Thd	3″	3″	4,600
60″ × 10′	125	6″ Flg	4″ Thd	3″	3″	6,200
60″ × 15′	125	6″ Flg	4″ Thd	3″	3″	8,100
60″ × 20′	125	6″ Flg	4″ Thd	4″	4″	10,000

TABLE A.7
Settling Volumes of Standard Horizontal Low-pressure Separators, 125 psi WP

Size, Dia × Length	Settling Volume, bbl: 1/2 Full	1/3 Full	1/4 Full
24″ × 5′	1.55	0.89	0.59
24″ × 7½′	2.22	1.28	0.86
24″ × 10′	2.89	1.67	1.12
30″ × 5′	2.48	1.43	0.94
30″ × 7½′	3.54	2.04	1.36
30″ × 10′	4.59	2.66	1.77
36″ × 10′	6.71	3.88	2.59
36″ × 15′	9.76	5.66	3.79
48″ × 10′	12.24	7.07	4.71
48″ × 15′	17.72	10.26	6.85
60″ × 10′	19.50	11.24	7.47
60″ × 15′	28.06	16.23	10.82
60″ × 20′	36.63	21.21	14.16

TABLE A.8
Specifications of Standard Horizontal High-pressure Separators

Size, Dia × Ht	Working Pressure, psi[a]	Inlet and Gas Outlet Connection	Standard Liquid Valve	Shipping Weight, lb
12¾″ × 5′	1000	2″ Thd	1″	1,100
12¾″ × 7½′		2″ Thd	1″	1,200
12¾″ × 10′		2″ Thd	1″	1,300
16″ × 5′	1000	2″ Thd	1″	1,400
16″ × 7½′		2″ Thd	1″	1,750
16″ × 10′		2″ Thd	1″	2,100
20″ × 5′	1000	3″ Flg	1″	1,800
20″ × 7½′		3″ Flg	1″	2,300
20″ × 10′		3″ Flg	1″	2,900
24″ × 5′	1000	4″ Flg	1″	2,200
24″ × 7½′		4″ Flg	1″	3,000
24″ × 10′		4″ Flg	1″	3,800
24″ × 15′		4″ Flg	1″	5,400
30″ × 5′	1000	4″ Flg	1″	3,200
30″ × 7½′		4″ Flg	1″	4,300
30″ × 10′		4″ Flg	1″	5,500
30″ × 15′		4″ Flg	2″	7,800
36″ × 7½′	1000	6″ Flg	2″	6,100
36″ × 10′		6″ Flg	2″	7,500
36″ × 15′		6″ Flg	2″	10,200
36″ × 20′		6″ Flg	2″	12,000
42″ × 7½′	1000	6″ Flg	2″	8,200
42″ × 10′		6″ Flg	2″	9,900
42″ × 15′		6″ Flg	2″	13,400
42″ × 20′		6″ Flg	2″	16,900
48″ × 7½′	1000	8″ Flg	2″	10,900
48″ × 10′		8″ Flg	2″	12,700
48″ × 15′		8″ Flg	2″	17,500
48″ × 20′		8″ Flg	2″	22,100
54″ × 7½′	1000	8″ Flg	2″	13,400
54″ × 10′		8″ Flg	2″	16,000
54″ × 15′		8″ Flg	2″	21,200
54″ × 20′		8″ Flg	2″	26,400
60″ × 7½′	1000	8″ Flg	2″	16,700
60″ × 10′		8″ Flg	2″	19,900
60″ × 15′		8″ Flg	2″	26,400
60″ × 20′		8″ Flg	2″	32,900

[a]Other standard working pressures available are 230, 500, 600, 1200, 1440, 1500, and 2000 psi.

TABLE A.9
Settling Volumes of Standard Horizontal High-pressure Separators, 230 to 2000 psi WP[b]

Size Dia × Len	Settling Volume, bbl[a]		
	1/2 Full	1/3 Full	1/4 Full
12¾″ × 5′	0.38	0.22	0.15
12¾″ × 7½′	0.55	0.32	0.21
12¾″ × 10′	0.72	0.42	0.28
16″ × 5′	0.61	0.35	0.24
16″ × 7½′	0.88	0.50	0.34
16″ × 10′	1.14	0.66	0.44
20″ × 5′	0.98	0.55	0.38
20″ × 7½′	1.39	0.79	0.54
20″ × 10′	1.80	1.03	0.70
24″ × 5′	1.45	0.83	0.55
24″ × 7½′	2.04	1.18	0.78
24″ × 10′	2.63	1.52	0.01
24″ × 15′	3.81	2.21	1.47
30″ × 5′	2.43	1.39	0.91
30″ × 7½′	3.40	1.96	1.29
30″ × 10′	4.37	2.52	1.67
30″ × 15′	6.30	3.65	2.42
36″ × 7½′	4.99	2.87	1.90
36″ × 10′	6.38	3.68	2.45
36″ × 15′	9.17	5.30	3.54
36″ × 20′	11.96	6.92	4.63
42″ × 7½′	6.93	3.98	2.61
42″ × 10′	8.83	5.09	3.35
42″ × 15′	12.62	7.30	4.83
42″ × 20′	16.41	9.51	6.32
48″ × 7½′	9.28	5.32	3.51
48″ × 10′	11.77	6.77	4.49
48″ × 15′	16.74	9.67	6.43
48″ × 20′	21.71	12.57	8.38
54″ × 7½′	12.02	6.87	4.49
54″ × 10′	15.17	8.71	5.73
54″ × 15′	12.49	12.40	8.20
54″ × 20′	27.81	16.08	10.68
60″ × 7½′	15.05	8.60	5.66
60″ × 10′	18.93	10.86	7.17
60″ × 15′	26.68	15.38	10.21
60″ × 20′	34.44	19.90	13.24

[a]Based on 1000 psi WP separator.

[b]Standard working pressures available are 230, 500, 600, 1000, 1200, 1440, 1500 and 2000 psi.

TABLE A.10
Specifications of Standard Spherical Separators

Diameter	Working Pressure, psi	Inlet and Gas Outlet Connection	Standard Liquid Valve	Shipping Weight, lb
41″	125	4″ Thd	2″	1000
46″		4″ Thd	3″	1300
54″		4″ Thd	4″	1700
42″	250	3″ Flg	2″	1100
48″		4″ Flg	2″	1400
60″		6″ Flg	2″	3400
24″	1000[a]	2″ Flg	1″	1300
30″		2″ Flg	1″	1400
36″		3″ Flg	1″	1800
42″		4″ Flg	2″	2800
48″		4″ Flg	2″	3700
60″		6″ Flg	2″	4300

[a]Other standard working pressures available are 500, 600, 1200, 1440, 2000, and 3000 psi.

TABLE A.11
Settling Volumes of Standard Spherical Low-pressure Separators, 125 psi WP

Size, OD	Settling Volume, bbl
41″	0.77
46″	1.02
54″	1.60

TABLE A.12
Settling Volumes of Standard Spherical High-pressure Separators, 230 to 3000 psi WP[a]

Size, OD	Settling Volume, bbl[b]
24″	0.15
30″	0.30
36″	0.54
42″	0.88
48″	1.33
60″	2.20

[a]Standard working pressures available are 230, 500, 600, 1000, 1200, 1440, 1500, 2000, and 3000 psi.
[b]Based on 1000 psi WP separator.

A.3 GLYCOL DEHYDRATOR DESIGN TABLES FROM SIVALLS

TABLE A.13
Physical and Chemical Properties of Glycols

	Ethylene Glycol	Diethylene Glycol	Triethylene Glycol
Molecular weight	62.07	106.12	150.17
Specific gravity at 68°F	1.1155	1.1184	1.1255
Specific weight, lb/gal	9.292	9.316	9.375
Boil point at 760 MMHg, °F	387.7	474.4	550.4
Freezing point, °F	9.1	18.0	24.3
Surface tension at 77°F, dynes/cm	47.0	44.8	45.2
Heat of vaporization at 760 MMHg, Btu/lb	364	232	174

(Continued)

TABLE A.13 (Continued)

100% Diethylene Glycol

Temp., °F	Sp Gr	Viscosity, cps	Sp Heat, Btu/lb °F	Thermal Conductivity, Btuh/sq ft °F ft
50	1.127	72	0.53	0.146
75	1.117	45	0.54	0.14
100	1.107	18	0.56	0.135
125	1.098	12.7	0.57	0.13
150	1.089	7.3	0.58	0.125
175	1.076	5.5	0.59	0.12
200	1.064	3.6	0.60	0.115
225	1.054	2.8	0.61	0.11
250	1.043	1.9	0.63	0.105
275	1.032	1.6	0.62	
300	1.021	1.3	0.66	

100% Triethylene Glycol

Temp., °F	Sp Gr	Viscosity	Sp Heat, Btu/lb °F	Thermal Conductivity, Btuh/sq ft °F ft
50	1.134	88	0.485	0.14
75	1.123	56	6.50	0.138
100	1.111	23	0.52	0.132
125	1.101	15.5	0.535	0.130
150	1.091	8.1	0.55	0.125
175	1.080	6.1	0.57	0.121
200	1.068	4.0	0.585	0.118
225	1.057	3.1	0.60	0.113
250	1.034	1.9	0.635	
300	1.022	1.5	0.65	

TABLE A.14
Vertical Inlet Scrubbers—Specifications

Nominal WP psig	Size, OD	Nominal Gas Capacity, MMscfd[a]	Inlet and Gas Outlet Connection	Std Oil Valve	Shipping Weight, lb
230	16″	1.8	2″	1″	900
	20″	2.9	3″	1″	1,000
	24″	4.1	3″	1″	1,200
	30″	6.5	4″	1″	1,400
	36″	9.4	4″	1″	1,900
	42″	12.7	6″	2″	2,600
	48″	16.7	6″	2″	3,000
	54″	21.1	6″	2″	3,500
	60″	26.1	6″	2″	4,500
500	16″	2.7	2″	1″	1,000
	20″	4.3	3″	1″	1,300
	24″	6.1	3″	1″	2,100
	30″	9.3	4″	1″	2,700
	36″	13.3	4″	1″	3,800
	42″	18.4	6″	2″	4,200
	48″	24.3	6″	2″	5,000
	54″	30.6	6″	2″	5,400
	60″	38.1	6″	2″	7,500
600	16″	3.0	2″	1″	1,100
	20″	4.6	3″	1″	1,400
	24″	6.3	3″	1″	2,200
	30″	9.8	4″	1″	2,800
	36″	14.7	4″	1″	3,900
	42″	20.4	6″	2″	4,500
	48″	27.1	6″	2″	5,100
	54″	34.0	6″	2″	6,000
	60″	42.3	6″	2″	8,100
1000	16″	3.9	2″	1″	1,100
	20″	6.1	3″	1″	1,600
	24″	8.8	3″	1″	2,500
	30″	13.6	4″	1″	3,200
	36″	20.7	4″	1″	4,400
	42″	27.5	6″	2″	6,300
	48″	36.9	6″	2″	8,400
	54″	46.1	6″	2″	9,700
	60″	57.7	6″	2″	14,500
1200	16″	4.2	2″	1″	1,150
	20″	6.5	3″	1″	1,800

TABLE A.14 (Continued)

Nominal WP psig	Size, OD	Nominal Gas Capacity, MMscfd[a]	Inlet and Gas Outlet Connection	Std Oil Valve	Shipping Weight, lb
	24″	6.5	3″	1″	2,600
	30″	15.3	4″	1″	3,400
	36″	23.1	4″	1″	4,700
	42″	31.0	6″	2″	6,700
	48″	40.5	6″	2″	8,500
	54″	51.4	6″	2″	11,300
	60″	62.3	6″	2″	14,500
1440	16″	4.8	2″	1″	1,500
	20″	6.7	3″	1″	2,100
	24″	11.2	3″	1″	2,800
	30″	17.7	4″	1″	3,900
	36″	25.5	4″	1″	5,400
	42″	34.7	6″	2″	7,800
	48″	45.3	6″	2″	9,200
	54″	56.1	6″	2″	12,900
	60″	69.6	6″	2″	16,000

[a]Gas capacity based on 100°F, 0.7 sp gr, and vessel working pressure.

TABLE A.15
Settling Volumes and Liquid Capacities of Scrubbers

	Two-phase Scrubber			Three-phase Scrubber			
						Liquid Capacity, bbl/day[c]	
Size OD	Shell Height	Settling Volume, bbl[a]	Liquid Capacity, bbl/day[b]	Shell Height	Settling Volume, bbl[a]	Oil	Water
16″	5′	0.27	340	7½′	0.72	100	100
20″	5′	0.44	530	7½′	1.15	160	160
24″	5′	0.66	760	7½′	1.68	240	240
30″	5′	1.13	1180	7½′	2.78	400	400
36″	5′	1.73	2000	7½′	4.13	590	590
42″	5′	2.52	3000	7½′	5.80	830	830
48″	5′	3.48	4000	7½′	7.79	1120	1120
54″	5′	4.65	5000	7½′	10.12	1450	1450
60″	5′	6.01	6000	7½′	12.73	1830	1830

[a]Based on nominal 1000 psig WP scrubber.
[b]Based on 1.0 minute retention time.
[c]Based on 5.0 minute retention time.

TABLE A.16
Tray-type Glycol/gas Contactors—Specifications

Nominal WP psig	Size OD	Nominal Gas Capacity, MMscfd[a]	Gas Inlet and Outlet Size	Glycol Inlet and Outlet Size	Glycol Cooler Size	Shipping Weight, lb
250	12¾″	1.5	2″	½″	2″ × 4″	800
	16″	2.4	2″	¾″	2″ × 4″	900
	18″	3.2	3″	¾″	3″ × 5″	1,100
	20″	4.0	3″	1″	3″ × 5″	1,400
	24″	6.1	3″	1″	3″ × 5″	2,000
	30″	9.9	4″	1″	4″ × 6″	2,400
	36″	14.7	4″	1½″	4″ × 6″	3,200
	42″	19.7	4″	1½″	4″ × 6″	4,400
	48″	26.3	6″	2″	6″ × 8″	6,300
	54″	32.7	6″	2″	6″ × 8″	7,700
	60″	40.6	6″	2″	6″ × 8″	9,600
500	12¾″	2.0	2″	½″	2″ × 4″	1,000
	16″	3.2	2″	¾″	2″ × 4″	1,200
	18″	4.3	3″	¾″	3″ × 5″	1,500
	20″	5.3	3″	1″	3″ × 5″	1,700
	24″	8.3	3″	1″	3″ × 5″	2,900
	30″	13.1	3″	1″	3″ × 5″	3,900
	36″	19.2	4″	1½″	4″ × 6″	6,000
	42″	27.4	4″	1½″	4″ × 6″	7,700
	48″	35.1	6″	2″	6″ × 8″	10,000
	54″	44.5	6″	2″	6″ × 8″	12,000
	60″	55.2	6″	2″	6″ × 8″	15,300
600	12¾″	2.2	2″	½″	2″ × 4″	1,100
	16″	3.4	2″	¾″	2″ × 4″	1,300
	18″	4.5	3″	¾″	3″ × 5″	1,600
	20″	5.5	3″	1″	3″ × 5″	1,800
	24″	8.5	3″	1″	3″ × 5″	3,000
	30″	14.3	3″	1″	3″ × 5″	4,000
	36″	21.2	4″	1½″	4″ × 6″	6,300
	42″	29.4	4″	1½″	4″ × 6″	8,400
	48″	39.2	6″	2″	6″ × 8″	11,300
	54″	49.3	6″	2″	6″ × 8″	13,400
	60″	61.3	6″	2″	6″ × 8″	16,500
1000	12¾″	2.7	2″	½″	2″ × 4″	1,300
	16″	4.3	2″	¾″	2″ × 4″	1,600
	18″	5.5	3″	¾″	3″ × 5″	2,100
	20″	7.3	3″	1″	3″ × 5″	2,600
	24″	11.3	3″	1″	3″ × 5″	4,200

(Continued)

TABLE A.16 (Continued)

Nominal WP psig	Size OD	Nominal Gas Capacity, MMscfd[a]	Gas Inlet and Outlet Size	Glycol Inlet and Outlet Size	Glycol Cooler Size	Shipping Weight, lb
	30″	18.4	3″	1″	3″ × 5″	5,500
	36″	27.5	4″	1½″	4″ × 6″	8,500
	42″	37.1	4″	1½″	4″ × 6″	11,800
	48″	49.6	6″	2″	6″ × 8″	16,200
	54″	62.0	6″	2″	6″ × 8″	20,200
	60″	77.5	6″	2″	6″ × 8″	26,300
1200	12¾″	3.0	2″	½″	2″ × 4″	1,500
	16″	4.7	2″	¾″	2″ × 4″	1,900
	18″	6.0	3″	¾″	3″ × 5″	2,300
	20″	7.8	3″	1″	3″ × 5″	3,000
	24″	12.0	3″	1″	3″ × 5″	4,900
	30″	20.1	3″	1″	3″ × 5″	6,400
	36″	29.8	4″	1½″	4″ × 6″	10,000
	42″	41.4	4″	1½″	4″ × 6″	13,100
	48″	54.1	6″	2″	6″ × 8″	18,400
	54″	68.4	6″	2″	6″ × 8″	23,500
	60″	85.0	6″	2″	6″ × 8″	29,000
1440	12¾″	3.1	2″	½″	2″ × 4″	1,800
	16″	4.9	2″	¾″	2″ × 4″	2,200
	18″	6.5	3″	¾″	3″ × 5″	2,800
	20″	8.3	3″	1″	3″ × 5″	3,500
	24″	13.3	3″	1″	3″ × 5″	5,800
	30″	22.3	3″	1″	3″ × 5″	7,500
	36″	32.8	4″	1½″	4″ × 6″	11,700
	42″	44.3	4″	1½″	4″ × 6″	14,400
	48″	58.3	6″	2″	6″ × 8″	20,000
	54″	74.0	6″	2″	6″ × 8″	25,800
	60″	91.1	6″	2″	6″ × 8″	32,000

[a]Gas capacity based on 100°F, 0.7 sp gr and contactor working pressure.

TABLE A.17
Tray-type Glycol/Gas Contactors—Specifications

Size OD	Standard Shell Height[a]	Standard Glycol Cooler Height[a]	Add to Height For Add. Tray, ea.	Glycol Charge, gal	
				Standard[a]	For Each Add. Tray
12¾″	13′	9′	2′	10	1.5
16″	13′	9′	2′	13	2.2
18″	13′	9′	2′	16	2.8
20″	13′	9′	2′	19	3.6
24″	13′	9′	2′	25	5.0
30″	13′	9′	2′	38	8.2
36″	13′	9′	2′	53	11.8
42″	13′	9′	2′	73	16.8
48″	13′	9′	2′	90	20.9
54″	13′	9′	2′	112	26.6
60″	13′	9′	2′	137	32.6

[a]For standard four-tray contactor.

TABLE A.18
Packed Column Glycol/Gas Contactors—Specifications

Nominal WP psig	Size OD	Nominal Gas Capacity, MMscfd[a]	Gas Inlet and Outlet Size	Glycol Inlet and Outlet Size	Glycol Cooler Size	Shipping Weight, lb
250	10¾″	1.1	2″	½″	2″ × 4″	500
	12¾″	1.6	2″	½″	2″ × 4″	600
	14″	1.9	2″	½″	2″ × 4″	650
	16″	2.5	2″	½″	2″ × 4″	800
	18″	3.4	3″	¾″	3″ × 5″	900
	20″	4.0	3″	¾″	3″ × 5″	1100
	24″	5.5	3″	1″	3″ × 5″	1800
500	10¾″	1.5	2″	½″	2″ × 4″	600
	12¾″	2.2	2″	½″	2″ × 4″	700
	14″	2.6	2″	½″	2″ × 4″	750
	16″	3.4	2″	½″	2″ × 4″	900
	18″	4.4	3″	¾″	3″ × 5″	1000
	20″	5.5	3″	¾″	3″ × 5″	1500
	24″	7.5	3″	1″	3″ × 5″	2500
600	10¾″	1.7	2″	½″	2″ × 4″	650
	12¾″	2.4	2″	½″	2″ × 4″	750
	14″	2.9	2″	½″	2″ × 4″	800
	16″	3.8	2″	½″	2″ × 4″	950
	18″	4.8	3″	¾″	3″ × 5″	1100
	20″	6.0	3″	¾″	3″ × 5″	1700
	24″	8.1	3″	1″	3″ × 5″	2700

(Continued)

TABLE A.18 (Continued)

Nominal WP psig	Size OD	Nominal Gas Capacity, MMscfd[a]	Gas Inlet and Outlet Size	Glycol Inlet and Outlet Size	Glycol Cooler Size	Shipping Weight, lb
1000	10¾″	2.3	2″	½″	2″ × 4″	900
	12¾″	3.3	2″	½″	2″ × 4″	1000
	14″	4.0	2″	½″	2″ × 4″	1100
	16″	5.2	2″	½″	2″ × 4″	1300
	18″	6.6	3″	¾″	3″ × 5″	1800
	20″	8.2	3″	¾″	3″ × 5″	2300
	24″	11.8	3″	1″	3″ × 5″	3500
1200	10¾″	2.5	2″	½″	2″ × 4″	1200
	12¾″	3.6	2″	½″	2″ × 4″	1300
	14″	4.1	2″	½″	2″ × 4″	1500
	16″	5.4	2″	½″	2″ × 4″	1700
	18″	6.9	3″	¾″	3″ × 5″	2200
	20″	8.5	3″	¾″	3″ × 5″	2800
	24″	12.3	3″	¾″	3″ × 5″	4000
1440	10¾″	2.6	2″	½″	2″ × 4″	1300
	12¾″	3.7	2″	½″	2″ × 4″	1400
	14″	4.5	2″	½″	2″ × 4″	1600
	16″	5.9	2″	½″	2″ × 4″	1900
	18″	7.5	3″	¾″	3″ × 5″	2500
	20″	9.3	3″	¾″	3″ × 5″	3100
	24″	12.7	3″	1″	3″ × 5″	4500

[a]Gas capacity based on 100°F, 0.7 sp gr and contactor working pressure.

TABLE A.19
Packed Column Glycol/Gas Contactors—Specifications

Size, OD	Standard Shell Height	Standard Glycol Cooler Height	Standard Contacting Element[a]	Glycol Charge, gal
10¾″	9′	7′	1″ × 4′	6
12¾″	9′	7′	1″ × 4′	7
14″	9′	7′	1″ × 4′	8
16″	9′	7′	1″ × 4′	10
18″	9′	7′	1″ × 4′	12
20″	9′	7′	1″ × 4′	14
24″	9′	7′	1″ × 4′	18

[a]Standard contacting element is carbon steel metal pall rings of size listed in Table.

TABLE A.20
Glycol Reconcentrators—Specifications

Reboiler Capacity, Btuh	Glycol Capacity, gph[a]	Reboiler Size, Dia. × Len.	Heat Exchanger Surge Tank, Size, Dia. × Len.	Stripping Still Size, Dia. × Ht.	Reflex Condenser Size, Dia. × Ht.
75,000	20	18″ × 3″, 6″	18″ × 3′, 6″	6⅝″ × 4′, 6″	6⅝″ × 2′, 0″
75,000	35	18″ × 3′, 6″	18″ × 3′, 6″	6⅝″ × 4′, 6″	6⅝″ × 2′, 0″
125,000	40	18″ × 5′	18″ × 5′	6⅝″ × 4′, 6″	6⅝″ × 2′, 0″
125,000	70	18″ × 5′	18″ × 5′	6⅝″ × 4′, 6″	6⅝″ × 2′, 0″
175,000	90	24″ × 5′	24″ × 5′	8⅝″ × 4′, 6″	8⅝″ × 2′, 0″
175,000	100	24″ × 5′	24″ × 5′	8⅝″ × 4′, 6″	8⅝″ × 2′, 0″
250,000	150	24″ × 7′	24″ × 7′	8⅝″ × 5′, 0″	8⅝″ × 2′, 0″
350,000	210	24″ × 10′	24″ × 10′	10¾″ × 5′, 0″	10¾″ × 2′, 6″
400,000	250	30″ × 10′	30″ × 10′	10¾″ × 6′, 0″	10¾″ × 2′, 6″
500,000	315	36″ × 10′	36″ × 10′	12¾″ × 7′, 0″	12¾″ × 2′, 6″
750,000	450	36″ × 15′	36″ × 10′	14″ × 8′, 0″	14″ × 3′, 0″
850,000	450	42″ × 15′	36″ × 10′	14″ × 8′, 0″	14″ × 3′, 0″
1,000,000	450	48″ × 16′	36″ × 10′	16″ × 8′, 0″	16″ × 3′, 0″

Flash Separator Size, Dia × Ht.	Heat Exchange Coil		Glycol Pump Model	High-Pressure Glycol Filter Size	Glycol Charge, gal	Shipping Wt, lb
	Size	Coil Area, sq ft				
12″ × 48″	½″	12.9	1715PV	1″	75	2,100
12″ × 48″	½″	12.9	4015PV	1″	75	2,100
16″ × 48″	½″	23.3	4015PV	1″	105	2,200
16″ × 48″	½″	23.3	9015PV	1″	105	2,250
16″ × 48″	½″	31.1	9015PV	1″	190	3,200
16″ × 48″	½″	31.1	21015PV	1½″	190	3,200
16″ × 48″	¾″	44.6	21015PV	1½″	260	3,700
20″ × 48″	¾″	64.8	21015PV	1½″	375	4,000
20″ × 48″	¾″	64.8	45015PV	2″	445	4,500
24″ × 48″	1″	82.1	45015PV	2″	680	6,500
30″ × 48″	1″	102.6	45015PV	2″	990	7,000
30″ × 48″	1″	102.6	45015PV	2″	1175	7,500
30″ × 48″	1″	102.6	45015PV	2″	1425	10,000

[a]Glycol capacity is based on circulating 2.5 gal TEG/lb H_2O and is controlled by the reboiler capacity or pump capacity, whichever is smaller.

TABLE A.21
Glycol Pumps—Standard High-pressure Pumps

Model Number	Circulation Rate—Gallons/Hour [a]Pump Speed—Strokes/Minute Count one stroke for each discharge of pump																
	8	10	12	14	16	18	20	22	24	26	28	30	32	34	36	38	40
1715 PV	8	10	12	14	16	18	20	22	24	26	28	30	32	34	36	38	40
4015 PV			12	14	16	18	20	22	24	26	28	30	32	34	36	38	40
9015 PV			27	31.5	36	40.5	45	49.5	54	48.5	63	67.5	72	76.5	81	85.5	90
21015 PV		66	79	92	105	118	131	144	157	171	184	197	210				
45015 PV		166	200	233	266	300	333	366	400	433	466						

[a]It is not recommended to attempt to run pumps at speeds less or greater than those indicated in the above table.

Gas Consumption

Operating pressure, psig	300	400	500	600	700	800	900	1000	1100	1200	1300	1400	1500
Cu ft/gal at 14.4 and 60°F	1.7	2.3	2.8	3.4	3.7	4.5	5.0	5.6	6.1	6.7	7.2	7.9	8.3

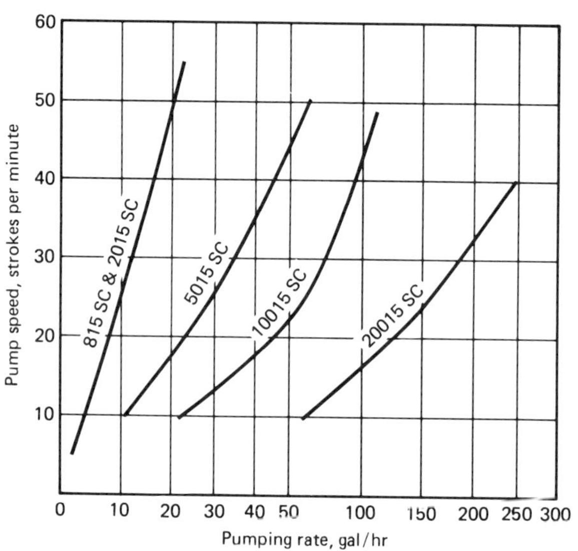

Pump Model	Pump Conn.	Size Strainer	High-pressure Filter Size	High-pressure Filter Elements	Low-pressure Filter Size	Low-pressure Filter Elements
315PV	¼″	½″	1″	1-2¾″ × 9¾″	½″	1-3″ × 18″
1715PV and 815SC	½″	¾″	1″	1-2¾″ × 9¾″	½″	1-3″ × 18″
4015PV and 2015SC	½″	¾″	1″	1-2¾″ × 9¾″	½″	1-3″ × 18″
9015PV and 5015SC	¾″	1″	1″	2-2¾″ × 9¾″	¾″	1-3″ × 36″
21015PV and 10015SC	1″	1″	1½″	4-2¾″ × 9¾″	1″	4-3″ × 18″
45015PV and 20015SC	1½″	1½″	2″	8-2¾″ × 9¾″	1½″	4-3″ × 36″

TABLE A.21 (Continued)

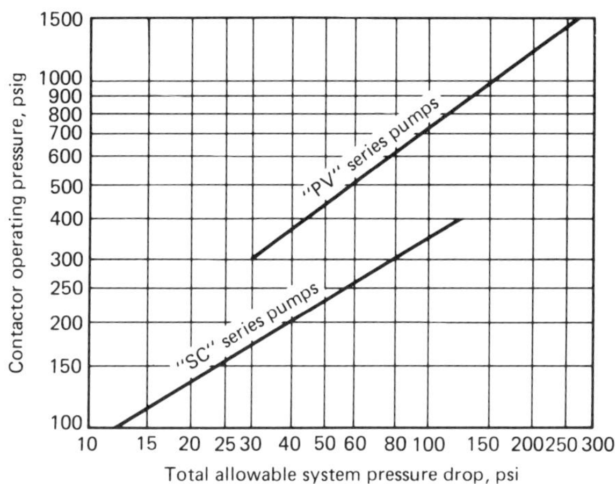

A.4 ORIFICE METER TABLES FOR NATURAL GAS

The tolerances necessary in the use of any orifice meter do not warrant taking the values in these tables to be accurate beyond one in 500. Four figures are given in all cases solely to enable different computers to agree within 1 or 2 in the fourth significant figure regardless of whether it is on the right or left of the decimal.

In some of the tables values of the constants for a few of the smaller orifices are marked with an asterisk, these orifices have diameter ratios lower than the minimum value for which the formulas used were derived and this size of plate should not be used unless it is understood that the accuracy of measurement may be relatively low.

A.4.1 Tables Applying to Orifice Flow Constants for Flange Tap Installations

Table A.22 Basic Orifice Factors

Table A.23 "*b*" Values for Reynolds Number Factor Determination

Table A.24 Expansion Factors, Static Pressure Upstream

Table A.25 Expansion Factors, Static Pressure Downstream

Table A.26 Expansion Factors, Static Pressure Mean of Upstream and Downstream

TABLE A.22

F_b Basic Orifice Factors—Flange Taps

Base temperature = 60°F Flowing temperature = 60°F $\sqrt{h_w p_f} = \infty$

Base pressure = 14.73 psia Specific gravity = 1.0 $h_w/p_f = 0$

Pipe Sizes—Nominal and Published Inside Diameters, Inches

Orifice Diameter, in.	2			3				4	
	1.689	1.939	2.067	2.300	2.626	2.900	3.068	3.152	3.438
0.250	12.695	12.707	12.711	12.714	12.712	12.708	12.705	12.703	12.697
0.375	28.474	28.439	28.428	28.411	28.393	28.382	28.376	28.373	28.364
0.500	50.777	50.587	50.521	50.435	50.356	50.313	50.292	50.284	50.258
0.625	80.090	79.509	79.311	79.052	78.818	78.686	78.625	78.598	78.523
0.750	117.09	115.62	115.14	114.52	113.99	113.70	113.56	113.50	113.33
0.875	162.95	159.56	158.47	157.12	156.00	155.41	155.14	155.03	154.71
1.000	219.77	212.47	210.22	207.44	205.18	204.04	203.54	203.33	202.75
1.125	290.99	276.20	271.70	266.35	262.06	259.95	259.04	258.65	257.63
1.250	385.78	353.58	345.13	335.12	327.39	323.63	322.03	321.37	319.61
1.375		448.57	433.50	415.75	402.18	395.80	393.09	391.97	389.03
1.500			542.26	510.86	487.98	477.36	472.96	471.14	466.39
1.625				623.91	586.82	569.65	562.58	559.72	552.31
1.750					701.27	674.44	663.42	658.96	647.54
1.875					834.88	793.88	777.18	770.44	753.17
2.000						930.65	906.01	896.06	870.59
2.125						1091.2	1052.5	1038.1	1001.4
2.250							1223.2	1199.9	1147.7
2.375									1311.7
2.500									1498.4

Orifice Diameter, in.	4		6				8		
	3.826	4.026	4.897	5.182	5.761	6.065	7.625	7.981	8.071
0.250	12.687	12.683							
0.375	28.353	28.348							
0.500	50.234	50.224	50.197	50.191	50.182	50.178			
0.625	78.450	78.421	78.338	78.321	78.296	78.287			
0.750	113.15	113.08	112.87	112.82	112.75	112.72			
0.875	154.40	154.27	153.88	153.78	153.63	153.56	153.34	153.31	153.31
1.000	202.20	201.99	201.34	201.19	200.96	200.85	200.46	200.39	200.38
1.125	256.69	256.33	255.31	255.08	254.72	254.56	253.99	253.69	253.87
1.250	318.03	317.45	315.83	315.48	314.95	314.72	313.91	313.78	313.74
1.375	386.45	385.51	382.99	382.47	381.70	381.37	380.25	380.06	380.02
1.500	462.27	460.79	456.93	456.16	455.03	454.57	453.02	452.78	452.72
1.625	545.89	543.61	537.77	536.64	535.03	534.38	532.27	531.95	531.87
1.750	637.84	634.39	625.73	624.09	621.79	620.88	618.02	617.60	617.50
1.875	738.75	733.68	721.03	718.69	715.44	714.19	710.32	709.77	709.64
2.000	849.41	842.12	823.99	820.68	816.13	814.41	809.22	808.50	808.34
2.125	970.95	960.48	934.97	930.35	924.07	921.71	914.79	913.86	913.64
2.250	1104.7	1089.9	1054.4	1048.1	1039.5	1036.3	1027.1	1025.9	1025.6
2.375	1252.1	1231.7	1182.9	1174.2	1162.6	1158.3	1146.2	1144.7	1144.3
2.500	1415.0	1387.2	1320.9	1309.3	1293.8	1288.2	1272.3	1270.3	1269.8
2.625	1595.6	1558.2	1469.2	1453.9	1433.5	1426.0	1405.4	1402.9	1402.3
2.750	1797.1	1746.7	1628.9	1608.7	1582.1	1572.3	1545.7	1542.5	1541.8
2.875		1955.5	1801.0	1774.5	1740.0	1727.5	1693.4	1689.3	1688.4
3.000		2194.9	1986.6	1952.4	1907.8	1891.9	1848.6	1843.5	1842.3
3.125			2187.2	2143.4	2086.4	2066.1	2011.6	2005.2	2003.8
3.250			2404.2	2348.8	2276.5	2250.8	2182.6	2174.6	2172.9
3.375			2639.5	2569.8	2479.1	2446.8	2361.8	2352.0	2349.9
3.500			2895.5	2808.1	2695.1	2654.9	2654.9	2537.7	2535.0

(Continued)

TABLE A.22 (Continued)

Orifice Diameter, in.	4			6			8		
	3.826	4.026	4.897	5.182	5.761	6.065	7.625	7.981	8.071
3.625			3180.8	3065.3	2925.7	2876.0	2746.5	2731.8	2728.6
3.750				3345.5	3172.1	3111.2	2952.6	2934.8	2930.8
3.875				3657.7	3435.7	3361.5	3168.3	3146.9	3142.1
4.000					3718.2	3628.2	3394.3	3368.5	3362.9
4.250					4354.8	4216.6	3879.4	3842.3	3834.2
4.500						4900.9	4412.8	4360.5	4349.0
4.750							5000.7	4928.1	4912.2
5.000							5650.0	5551.1	5529.5
5.250							6369.3	6236.4	6207.3
5.500							7170.9	6992.0	6953.6
5.750								7830.0	7777.8
6.000									8706.9

Orifice Diameter, in.	10			12			16		
	9.564	10.020	10.136	11.376	11.938	12.090	14.688	15.000	15.250
1.000	200.20								
1.125	253.55	253.48	253.47						
1.250	313.31	313.20	313.18	312.94	312.85	312.83			
1.375	379.44	379.29	379.26	378.94	378.82	378.79			
1.500	451.95	451.76	451.72	451.30	451.14	451.10	450.53	450.48	
1.625	530.87	530.63	530.57	530.04	529.83	529.78	529.06	528.99	528.94
1.750	616.21	615.90	615.83	615.16	614.90	614.84	613.94	613.85	613.78
1.875	707.99	707.61	707.51	706.68	706.36	706.28	705.18	705.07	704.99
2.000	806.23	805.76	805.65	804.61	804.23	804.13	802.78	802.65	802.55
2.125	910.97	910.38	910.24	908.98	908.51	908.39	906.77	906.61	906.49
2.250	1,022.2	1,021.5	1,021.3	1,019.8	1,019.2	1,019.1	1,017.1	1,017.0	1,016.8
2.375	1,140.1	1,139.2	1,139.0	1,137.1	1,136.4	1,136.2	1,133.9	1,133.7	1,133.5
2.500	1,264.5	1,263.4	1,263.1	1,260.8	1,260.0	1,259.8	1,257.1	1,256.8	1,256.6
2.625	1,395.6	1,394.2	1,393.9	1,391.1	1,390.1	1,389.9	1,386.7	1,386.4	1,386.1
2.750	1,533.4	1,531.7	1,531.3	1,528.0	1,526.8	1,526.5	1,522.7	1,522.4	1,522.1
2.875	1,678.0	1,675.9	1,675.4	1,671.4	1,670.0	1,669.6	1,665.2	1,664.8	1,664.5
3.000	1,829.4	1,826.9	1,826.3	1,821.4	1,819.7	1,819.3	1,814.1	1,813.7	1,813.3
3.125	1,987.8	1,984.7	1,984.0	1,978.1	1,976.1	1,975.6	1,969.6	1,969.0	1,966.6
3.250	2,153.2	2,149.5	2,148.6	2,141.5	2,139.2	2,130.6	2,131.5	2,130.9	2,130.4
3.375	2,325.7	2,321.2	2,320.2	2,311.7	2,308.9	2,308.2	2,299.9	2,299.2	2,293.7
3.500	2,505.6	2,500.1	2,498.9	2,488.7	2,485.4	2,484.6	2,474.9	2,474.1	2,473.5
3.625	2,692.8	2,686.2	2,684.7	2,672.6	2,668.7	2,667.7	2,656.4	2,655.5	2,654.8
3.750	2,887.6	2,879.7	2,877.9	2,863.5	2,858.8	2,857.7	2,844.6	2,843.5	2,842.7
3.875	3,090.1	3,080.7	3,078.5	3,061.4	3,055.9	3,054.6	3,039.4	3,038.1	3,037.2
4.000	3,300.6	3,289.3	3,286.8	3,266.4	3,260.0	3,258.5	3,240.8	3,239.4	3,238.3

(Continued)

TABLE A.22 (Continued)

Orifice Diameter, in.	10			12			16		
	9.564	10.020	10.136	11.376	11.938	12.090	14.688	15.000	15.250
4.250	3,746.1	3,730.2	3,726.7	3,698.4	3,689.6	3,687.5	3,663.8	3,661.9	3,660.5
4.500	4,226.0	4,204.1	4,199.2	4,160.4	4,148.4	4,145.5	4,113.9	4,111.5	4,109.7
4.750	4,742.7	4,712.8	4,706.2	4,653.4	4,637.2	4,633.4	4,591.5	4,508.4	4,586.0
5.000	5,298.6	5,258.5	5,249.6	5,179.0	5,157.4	5,152.3	5,097.2	5,093.1	5,090.1
5.250	5,897.4	5,843.6	5,831.8	5,738.5	5,710.0	5,703.3	5,631.4	5,626.1	5,622.3
5.500	6,543.1	6,471.9	6,456.3	6,333.8	6,296.6	6,287.9	6,194.8	6,180.1	6,183.1
5.750	7,240.0	7,146.9	7,126.5	6,966.9	6,919.0	6,907.8	6,788.1	6,779.5	6,773.3
6.000	7,993.3	7,873.0	7,846.6	7,640.4	7,579.0	7,564.7	7,412.3	7,401.5	7,393.6
6.250	8,808.9	8,654.8	8,621.1	8,357.3	8,278.9	8,260.7	8,060.4	8,054.8	8,044.8
6.500	9,693.3	9,498.1	9,455.3	9,121.0	9,021.7	8,998.7	8,757.3	8,740.3	8,727.9
6.750	10,654	10,409	10,355	9,935.2	9,810.5	9,781.6	9,480.4	9,459.4	9,444.0
7.000	11,711	11,394	11,327	10,804	10,649	10,613	10,239	10,213	10,194
7.250		12,467	12,381	11,732	11,540	11,496	11,035	11,003	10,980
7.500		13,656	13,541	12,725	12,489	12,434	11,869	11,831	11,803
7.750				13,787	13,500	13,433	12,745	12,698	12,664
8.000				14,927	14,578	14,498	13,664	13,607	13,566

TABLE A.22 (Continued)

Orifice Diameter, in.	10			12			16		
	9.564	10.020	10.136	11.376	11.938	12.090	14.688	15.000	15.250
8.250				16,156	15,730	15,633	14,628	14,560	14,511
8.500				17,505	16,962	16,845	15,642	15,560	15,501
8.750					18,296	18,148	16,706	16,609	16,539
9.000						19,565	17,826	17,711	17,628
9.250							19,004	18,868	18,770
9.500							20,245	20,085	19,969
9.750							21,552	21,365	21,230
10.000							22,930	22,712	22,555
10.250							24,385	24,132	23,948
10.500							25,924	25,628	25,416
10.750							27,567	27,210	26,962
11.000							29,331	28,899	28,600
11.250								30.710	30,348

TABLE A.22 (Continued)

F_b Basic Orifice Factors—Flange Taps

Base temperature = 60°F Flowing temperature = 60°F $\sqrt{h_w p_f} = \infty$
Base pressure = 14.73 psia Specific gravity = 1.0 $h_w/p_f = 0$

Pipe Sizes—Nominal and Published Inside Diameters, Inches

Orifice Diameter	20			24			30		
inch.	18.814	19.000	19.250	22.626	23.000	23.250	28.628	29.000	29.250
2.000	801.40	801.35	801.29						
2.125	905.11	905.06	904.98						
2.250	1,015.2	1,015.1	1,015.0						
2.375	1,131.6	1,131.5	1,131.4	1,130.2	1,130.1	1,130.0			
2.500	1,254.4	1,254.3	1,254.2	1,252.8	1,252.6	1,252.6			
2.625	1,383.6	1,383.5	1,383.3	1,381.7	1,381.5	1,381.4			
2.750	1,519.1	1,519.0	1,518.8	1,517.0	1,516.8	1,516.7			
2.875	1,661.0	1,660.9	1,660.7	1,658.6	1,658.4	1,658.3	1,656.0		
3.000	1,809.4	1,809.2	1,809.0	1,806.6	1,806.4	1,806.2	1,803.7	1,803.5	1,803.4
3.125	1,964.1	1,963.9	1,963.7	1,961.0	1,960.7	1,960.6	1,957.7	1,957.5	1,957.4
3.250	2,125.3	2,125.1	2,124.8	2,121.7	2,121.5	2,121.3	2,118.0	2,117.9	2,117.7
3.375	2,292.9	2,292.6	2,292.3	2,280.9	2,288.6	2,288.4	2,284.8	2,284.5	2,284.4
3.500	2,466.9	2,466.6	2,466.3	2,462.4	2,462.1	2,461.8	2,457.8	2,457.6	2,457.5
3.625	2,647.3	2,647.0	2,646.6	2,642.4	2,642.0	2,641.7	2,637.3	2,637.0	2,636.8
3.750	2,834.2	2,833.9	2,833.5	2,828.7	2,828.3	2,828.0	2,823.1	2,822.8	2,822.6
3.875	3,027.5	3,027.3	3,026.8	3,021.5	3,021.0	3,020.7	3,015.2	3,014.9	3,014.7
4.000	3,227.5	3,227.1	3,226.5	3,220.6	3,220.1	3,219.8	3,213.8	3,213.5	3,213.2
4.250	3,646.7	3,646.2	3,645.6	3,638.3	3,637.7	3,637.2	3,630.1	3,629.7	3,629.4
4.500	4,092.1	4,091.5	4,090.6	4,081.8	4,081.0	4,080.5	4,071.9	4,071.4	4,071.1
4.750	4,563.7	4,562.9	4,561.9	4,551.1	4,550.1	4,549.5	4,539.4	4,538.8	4,538.4
5.000	5,061.8	5,060.8	5,050.6	5,046.4	5,045.2	5,044.5	5,032.5	5,031.8	5,031.4

5.250	5,586.6	5,585.4	5,583.8	5,567.7	5,566.4	5,565.5	5,551.3	5,550.5	5,550.0
5.500	6,138.2	6,136.7	6,134.8	6,115.3	6.113.6	6,112.6	6,095.8	6,094.9	6,094.4
5.750	6,717.0	6,715.2	6,712.3	6,689.1	6,687.2	6,685.9	6,666.2	6,665.2	6,664.5
6.000	7,323.4	7,321.1	7,318.2	7,289.4	7,287.1	7,285.6	7,262.5	7,261.3	7,260.5
6.250	7,957.5	7,954.7	7,951.2	7,916.4	7,913.6	7,911.9	7,864.7	7,883.4	7,882.5
6.500	8,620.0	8,616.5	8,612.2	8,570.2	8,566.9	8,564.8	8,533.0	8,531.4	8,530.4
6.750	9,311.1	9,306.9	9,301.6	9,251.1	9,247.2	9,244.7	9,207.4	9,205.6	9,204.4
7.000	10,031	10,026	10,020	9,959.3	9,954.6	9,951.7	9,908.0	9,905.9	9,904.6
7.250	10,782	10,776	10,768	10,695	10,669	10,686	10,635	10,633	10,631
7.500	11,562	11,555	11,546	11,459	11,452	11,448	11,388	11,386	11,384
7.750	12,374	12,365	12,354	12,250	12,243	12,238	12,168	12,165	12,163
8.000	13,218	13,207	13,194	13,071	13,062	13,056	12,975	12,971	12,969
8.250	14,095	14,082	14,066	13,920	13,910	13,903	13,809	13,805	13,802
8.500	15,005	14,990	14,971	14,799	14,787	14,779	14,669	14,665	14,661
8.750	15,950	15,933	15,911	15,708	15,693	15,684	15,557	15,552	15,548
9.000	16,932	16,911	16,885	16,648	16,630	16,620	16,473	16,466	16,462
9.250	17,950	17,926	17,895	17,618	17,596	17,585	17,416	17,409	17,404
9.500	19,007	18,979	18,943	18,620	18,597	18,582	18,387	18,379	18,373
9.750	20,104	20,071	20,030	19,655	19,628	19,611	19,386	19,377	19,371
10.000	21,243	21,205	21,157	20,723	20,692	20,672	20,414	20,403	20,396
10.250	22,426	22,332	22,326	21,825	21,789	21,767	21,471	21,458	21,450
10.500	23,654	23,603	23,538	22,926	22,921	22,895	22,556	22,542	22,533
10.750	24,931	24,672	24,797	24,134	24,087	24,058	23,672	23,656	23,646
11.000	26,257	26,190	26,104	25,344	25,290	25,257	24,817	24,799	24,787
11.250	27,636	27,559	27,460	26,592	26,531	26,492	25,992	25,972	25,959
11.500	29,070	28,982	28,870	27,878	27,809	27,766	27,199	27,176	27,161
11.750	30,562	30,462	30,334	29,205	29,126	29,077	28,437	28,411	28,394
12.000	32,116	32,001	31,856	30,574	30,485	30,429	29,706	29,677	29,659

(Continued)

TABLE A.22 (Continued)

Orifice Diameter inch.	20			24			30		
	18.814	19.000	19.250	22.626	23.000	23.250	28.628	29.000	29.250
12.500	35,417	35,270	35,084	33,444	33,330	33,259	32,343	32,306	32,283
13.000	39,003	38,817	38,581	36,502	36,357	36,267	35,114	35,068	35,039
13.500	42,913	42,673	42,375	39,762	39,581	39,467	38,025	37,968	37,932
14.000	47,244	46,921	46,523	43,241	43,015	42,874	41,082	41,012	40,968
14.500				46,958	46,679	46,505	44,291	44,206	44,151
15.000				50,934	50,591	50,378	47,622	47,557	47,490
15.500				55,192	54,774	54,513	51,202	51,075	50,993
16.000				59,759	59,251	58,935	54,923	54,769	54,671
16.500				64,701	64,060	63,670	58,835	58,649	58,531
17.000					69,288	68,792	62,950	62,728	62,586
17.500							67,282	67,017	66,848
18.000							71,844	71,530	71,330
18.500							76,653	76,282	76,046
19.000							81,725	81,289	81,012
19.500							87,079	86,568	86,244
20.000							92,734	92,140	91,761
20.500							98,728	98,025	97,564
21.000							105,130	104,280	103,750
21.500								110,980	110,340

TABLE A.23

"*b*" Values for Reynolds Number Factor F_r Determination—Flange Taps

$$F_r = 1 + \frac{b}{\sqrt{h_w p_f}}$$

Pipe Sizes—Nominal and Published Inside Diameters, Inches

Orifice Diameter, In.	2				3			4	
	1.689	1.939	2.067	2.300	2.626	2.900	3.068	3.152	3.438
0.250	0.0879	0.0911	0.0926	0.0950	0.0979	0.0999	0.1010	0.1014	0.1030
0.375	0.0677	0.0709	0.0726	0.0755	0.0792	0.0820	0.0836	0.0844	0.0867
0.500	0.0562	0.0576	0.0588	0.0612	0.0648	0.0677	0.0695	0.0703	0.0728
0.625	0.0520	0.0505	0.0506	0.0516	0.0541	0.0566	0.0583	0.0591	0.0618
0.750	0.0536	0.0485	0.0471	0.0462	0.0470	0.0486	0.0498	0.0504	0.0528
0.875	0.0595	0.0506	0.0478	0.0445	0.0429	0.0433	0.0438	0.0442	0.0460
1.000	0.0677	0.0559	0.0515	0.0458	0.0416	0.0403	0.0402	0.0403	0.0411
1.125	0.0762	0.0630	0.0574	0.0495	0.0427	0.0396	0.0386	0.0383	0.0380
1.250	0.0824	0.0707	0.0646	0.0550	0.0456	0.0408	0.0388	0.0381	0.0365
1.375		0.0772	0.0715	0.0614	0.0501	0.0435	0.0406	0.0394	0.0365
1.500			0.0773	0.0679	0.0554	0.0474	0.0436	0.0420	0.0378
1.625				0.0735	0.0613	0.0522	0.0477	0.0457	0.0402
1.750					0.0669	0.0575	0.0524	0.0500	0.0434
1.875					0.0717	0.0628	0.0574	0.0549	0.0473
2.000						0.0676	0.0624	0.0598	0.0517
2.125						0.0715	0.0669	0.0642	0.0563
2.250							0.0706	0.0685	0.0607
2.375									0.0648
2.500									0.0683

Orifice Diameter in.	4			6				8	
	3.826	4.026	4.897	5.189	5.761	6.065	7.625	7.981	8.071
0.250	0.1047	0.1054							
0.375	0.0894	0.0907							
0.500	0.0763	0.0779	0.0836	0.0852	0.0880	0.0892			
0.625	0.0653	0.0670	0.0734	0.0753	0.0785	0.0801			
0.750	0.0561	0.0578	0.0645	0.0665	0.0701	0.0718			
0.875	0.0487	0.0502	0.0567	0.0587	0.0625	0.0643	0.0723	0.0738	0.0742
1.000	0.0430	0.0442	0.0500	0.0520	0.0557	0.0576	0.0660	0.0676	0.0680
1.125	0.0388	0.0396	0.0444	0.0462	0.0498	0.0517	0.0602	0.0619	0.0623
1.250	0.0361	0.0364	0.0399	0.0414	0.0447	0.0464	0.0549	0.0566	0.0571
1.375	0.0347	0.0344	0.0363	0.0375	0.0403	0.0419	0.0501	0.0518	0.0523

(Continued)

TABLE A.23 (Continued)

Orifice Diameter in.	4		6				8		
	3.826	4.026	4.897	5.189	5.761	6.065	7.625	7.981	8.071
1.500	0.0345	0.0336	0.0336	0.0344	0.0367	0.0381	0.0457	0.0474	0.0479
1.625	0.0354	0.0338	0.0318	0.0322	0.0337	0.0348	0.0418	0.0435	0.0439
1.750	0.0372	0.0350	0.0307	0.0306	0.0314	0.0322	0.0383	0.0399	0.0403
1.875	0.0398	0.0370	0.0305	0.0298	0.0298	0.0303	0.0353	0.0366	0.0371
2.000	0.0430	0.0395	0.0308	0.0296	0.0287	0.0288	0.0327	0.0340	0.0343
2.125	0.0467	0.0427	0.0318	0.0300	0.0281	0.0278	0.0304	0.0315	0.0318
2.250	0.0507	0.0462	0.0334	0.0310	0.0281	0.0274	0.0286	0.0295	0.0297
2.375	0.0548	0.0501	0.0354	0.0324	0.0286	0.0274	0.0271	0.0278	0.0280
2.500	0.0589	0.0540	0.0378	0.0342	0.0295	0.0279	0.0259	0.0264	0.0265
2.625	0.0626	0.0579	0.0406	0.0365	0.0308	0.0287	0.0251	0.0253	0.0254
2.750	0.0659	0.0615	0.0436	0.0391	0.0324	0.0300	0.0246	0.0245	0.0245
2.875		0.0647	0.0468	0.0418	0.0343	0.0314	0.0244	0.0240	0.0240
3.000		0.0673	0.0500	0.0448	0.0366	0.0332	0.0245	0.0238	0.0237
3.125			0.0533	0.0479	0.0389	0.0353	0.0248	0.0239	0.0237
3.250			0.0564	0.0510	0.0416	0.0375	0.0254	0.0242	0.0240
3.375			0.0594	0.0541	0.0443	0.0400	0.0263	0.0248	0.0244
3.500			0.0620	0.0569	0.0472	0.0426	0.0273	0.0255	0.0251
3.625			0.0643	0.0597	0.0500	0.0452	0.0286	0.0265	0.0260
3.750				0.0621	0.0527	0.0479	0.0300	0.0274	0.0271
3.875				0.0640	0.0553	0.0505	0.0316	0.0289	0.0283
4.000					0.0578	0.0531	0.0334	0.0304	0.0297
4.250					0.0620	0.0579	0.0372	0.0338	0.0330
4.500						0.0618	0.0414	0.0386	0.0366
4.750							0.0457	0.0416	0.0405
5.000							0.0500	0.0457	0.0446
5.250							0.0539	0.0497	0.0487
5.500							0.0574	0.0535	0.0524
5.750								0.0569	0.0559
6.000									0.0588

TABLE A.23 (Continued)

Orifice Diameter, in.	10			12			16		
	9.564	10.020	10.136	11.376	11.938	12.090	14.688	15.000	15.250
1.000	0.0738								
1.125	0.0685	0.0701	0.0705						
1.250	0.0635	0.0652	0.0656	0.0698	0.0714	0.0718			
1.375	0.0588	0.0606	0.0610	0.0654	0.0671	0.0676			
1.500	0.0545	0.0563	0.0568	0.0612	0.0631	0.0635	0.0706	0.0713	
1.625	0.0504	0.0523	0.0527	0.0573	0.0592	0.0597	0.0670	0.0678	0.0684
1.750	0.0467	0.0485	0.0490	0.0536	0.0555	0.0560	0.0636	0.0644	0.0650
1.875	0.0433	0.0451	0.0455	0.0501	0.0521	0.0526	0.0604	0.0612	0.0618
2.000	0.0401	0.0419	0.0414	0.0469	0.0488	0.0492	0.0572	0.0581	0.0587
2.125	0.0372	0.0389	0.0383	0.0438	0.0458	0.0463	0.0542	0.0551	0.0558
2.250	0.0346	0.0362	0.0356	0.0410	0.0429	0.0434	0.0514	0.0523	0.0529
2.375	0.0322	0.0337	0.0330	0.0383	0.0402	0.0407	0.0467	0.0496	0.0502
2.500	0.0302	0.0315	0.0308	0.0359	0.0377	0.0382	0.0461	0.0470	0.0476
2.625	0.0283	0.0296	0.0287	0.0336	0.0354	0.0358	0.0436	0.0445	0.0452
2.750	0.0267	0.0278	0.0269	0.0316	0.0332	0.0336	0.0413	0.0422	0.0428
2.875	0.0254	0.0263	0.0253	0.0297	0.0312	0.0317	0.0391	0.0399	0.0406
3.000	0.0243	0.0250	0.0252	0.0278	0.0294	0.0298	0.0370	0.0378	0.0385
3.125	0.0234	0.0239	0.0241	0.0264	0.0278	0.0282	0.0350	0.0358	0.0365
3.250	0.0226	0.0230	0.0231	0.0251	0.0263	0.0266	0.0331	0.0339	0.0346
3.375	0.0221	0.0223	0.0224	0.0239	0.0250	0.0253	0.0314	0.0321	0.0328
3.500	0.0219	0.0218	0.0218	0.0229	0.0238	0.0241	0.0298	0.0305	0.0311
3.625	0.0218	0.0214	0.0214	0.0221	0.0226	0.0230	0.0282	0.0290	0.0295
3.750	0.0218	0.0213	0.0212	0.0214	0.0219	0.0221	0.0268	0.0275	0.0281
3.875	0.0221	0.0213	0.0211	0.0208	0.0212	0.0213	0.0255	0.0262	0.0267
4.000	0.0225	0.0214	0.0212	0.0204	0.0206	0.0207	0.0243	0.0249	0.0254
4.250	0.0238	0.0222	0.0219	0.0200	0.0198	0.0198	0.0223	0.0228	0.0232
4.500	0.0256	0.0236	0.0231	0.0201	0.0195	0.0194	0.0206	0.0210	0.0213
4.750	0.0279	0.0254	0.0249	0.0207	0.0196	0.0194	0.0193	0.0196	0.0198
5.000	0.0307	0.0277	0.0270	0.0217	0.0202	0.0199	0.0184	0.0185	0.0187
5.250	0.0337	0.0303	0.0295	0.0231	0.0212	0.0208	0.0178	0.0178	0.0179
5.500	0.0370	0.0332	0.0323	0.0249	0.0226	0.0221	0.0176	0.0174	0.0174
5.750	0.0404	0.0363	0.0354	0.0270	0.0243	0.0237	0.0176	0.0174	0.0172
6.000	0.0438	0.0396	0.0386	0.0294	0.0263	0.0255	0.0180	0.0176	0.0173
6.250	0.0473	0.0437	0.0418	0.0320	0.0285	0.0277	0.0186	0.0160	0.0177
6.500	0.0505	0.0462	0.0451	0.0347	0.0309	0.0300	0.0195	0.0188	0.0183
6.750	0.0536	0.0493	0.0483	0.0376	0.0335	0.0325	0.0206	0.0198	0.0192
7.000	0.0562	0.0523	0.0513	0.0406	0.0362	0.0351	0.0220	0.0210	0.0202

(Continued)

TABLE A.23 (Continued)

Orifice Diameter, in.	10			12			16		
	9.564	10.020	10.136	11.376	11.938	12.090	14.688	15.000	15.250
7.250		0.0550	0.0540	0.0435	0.0390	0.0379	0.0235	0.0224	0.0216
7.500		0.0572	0.0564	0.0463	0.0418	0.0407	0.0252	0.0240	0.0230
7.750				0.0491	0.0446	0.0434	0.0271	0.0257	0.0246
8.000				0.0517	0.0473	0.0461	0.0291	0.0276	0.0264
8.250				0.0540	0.0498	0.0487	0.0312	0.0296	0.0283
8.500				0.0560	0.0522	0.0511	0.0334	0.0317	0.0303
8.750					0.0543	0.0534	0.0357	0.0338	0.0324
9.000						0.0553	0.0380	0.0361	0.0346
9.250							0.0402	0.0383	0.0368
9.500							0.0425	0.0406	0.0390
9.750							0.0447	0.0427	0.0412
10.000							0.0469	0.0449	0.0434
10.250							0.0489	0.0470	0.0455
10.500							0.0508	0.0490	0.0475
10.750							0.0526	0.0509	0.0495
11.000							0.0541	0.0526	0.0513
11.250								0.0541	0.0528

TABLE A.23 (Continued)

Orifice Diameter, in.	20			24			30		
	18.814	19.000	19.250	22.626	23.000	23.250	28.628	29.000	29.250
2.000	0.0667	0.0671	0.0676						
2.125	0.0640	0.0644	0.0649						
2.250	0.0614	0.0618	0.0622						
2.375	0.0588	0.0592	0.0597	0.0659	0.0665	0.0669			
2.500	0.0563	0.0568	0.0573	0.0636	0.0642	0.0646			
2.625	0.0540	0.0544	0.0549	0.0614	0.0620	0.0624			
2.750	0.0517	0.0521	0.0526	0.0592	0.0599	0.0603			
2.875	0.0494	0.0499	0.0504	0.0571	0.0578	0.0582	0.0662		
3.000	0.0473	0.0477	0.0483	0.0551	0.0557	0.0562	0.0644	0.0649	0.0652
3.125	0.0452	0.0457	0.0462	0.0531	0.0538	0.0542	0.0626	0.0631	0.0634
3.250	0.0433	0.0437	0.0442	0.0511	0.0520	0.0523	0.0608	0.0613	0.0616
3.375	0.0414	0.0418	0.0423	0.0493	0.0500	0.0504	0.0590	0.0596	0.0599
3.500	0.0395	0.0399	0.0405	0.0474	0.0481	0.0486	0.0574	0.0579	0.0582
3.625	0.0378	0.0382	0.0387	0.0457	0.0464	0.0468	0.0557	0.0562	0.0566
3.750	0.0361	0.0365	0.0370	0.0440	0.0447	0.0451	0.0541	0.0546	0.0550
3.875	0.0345	0.0349	0.0354	0.0423	0.0430	0.0435	0.0525	0.0530	0.0534
4.000	0.0329	0.0333	0.0339	0.0407	0.0414	0.0419	0.0509	0.0515	0.0518
4.250	0.0301	0.0304	0.0310	0.0376	0.0384	0.0388	0.0479	0.0485	0.0488
4.500	0.0275	0.0279	0.0283	0.0348	0.0355	0.0360	0.0450	0.0456	0.0460
4.750	0.0252	0.0256	0.0260	0.0322	0.0328	0.0333	0.0423	0.0429	0.0433
5.000	0.0232	0.0235	0.0239	0.0297	0.0304	0.0308	0.0397	0.0403	0.0407
5.250	0.0214	0.0217	0.0220	0.0275	0.0281	0.0285	0.0373	0.0378	0.0382
5.500	0.0199	0.0201	0.0204	0.0254	0.0260	0.0264	0.0349	0.0355	0.0359
5.750	0.0186	0.0188	0.0191	0.0236	0.0241	0.0245	0.0327	0.0333	0.0337
6.000	0.0176	0.0177	0.0179	0.0219	0.0224	0.0228	0.0306	0.0312	0.0316
6.250	0.0167	0.0168	0.0170	0.0204	0.0208	0.0212	0.0287	0.0292	0.0296
6.500	0.0161	0.0162	0.0163	0.0191	0.0195	0.0198	0.0269	0.0274	0.0277
6.750	0.0157	0.0157	0.0157	0.0179	0.0183	0.0185	0.0252	0.0257	0.0260
7.000	0.0155	0.0155	0.0154	0.0169	0.0172	0.0174	0.0236	0.0240	0.0244
7.250	0.0155	0.0154	0.0153	0.0161	0.0163	0.0165	0.0221	0.0226	0.0229
7.500	0.0157	0.0155	0.0154	0.0154	0.0156	0.0157	0.0208	0.0212	0.0215
7.750	0.0160	0.0158	0.0156	0.0148	0.0150	0.0151	0.0195	0.0199	0.0202
8.000	0.0166	0.0163	0.0160	0.0144	0.0145	0.0146	0.0184	0.0187	0.0190
8.250	0.0172	0.0169	0.0165	0.0142	0.0142	0.0142	0.0174	0.0177	0.0179
8.500	0.0180	0.0177	0.0172	0.0141	0.0140	0.0140	0.0164	0.0168	0.0170
8.750	0.0190	0.0186	0.0180	0.0141	0.0140	0.0139	0.0156	0.0159	0.0161
9.000	0.0201	0.0196	0.0190	0.0143	0.0141	0.0140	0.0149	0.0152	0.0153

(Continued)

TABLE A.23 (Continued)

Orifice Diameter, in.	20			24			30		
	18.814	19.000	19.250	22.626	23.000	23.250	28.628	29.000	29.250
9.250	0.0213	0.0208	0.0201	0.0146	0.0143	0.0141	0.0143	0.0145	0.0146
9.500	0.0226	0.0220	0.0213	0.0150	0.0146	0.0144	0.0138	0.0139	0.0141
9.750	0.0240	0.0234	0.0226	0.0155	0.0150	0.0147	0.0133	0.0135	0.0136
10.000	0.0256	0.0249	0.0240	0.0161	0.0155	0.0152	0.0130	0.0131	0.0132
10.250	0.0271	0.0264	0.0255	0.0168	0.0162	0.0158	0.0128	0.0128	0.0128
10.500	0.0288	0.0280	0.0270	0.0176	0.0169	0.0164	0.0126	0.0126	0.0126
10.750	0.0305	0.0297	0.0286	0.0185	0.0176	0.0172	0.0125	0.0125	0.0125
11.000	0.0322	0.0314	0.0303	0.0194	0.0186	0.0181	0.0125	0.0124	0.0124
11.250	0.0340	0.0332	0.0320	0.0205	0.0196	0.0190	0.0126	0.0125	0.0124
11.500	0.0358	0.0349	0.0338	0.0216	0.0207	0.0200	0.0128	0.0126	0.0125
11.750	0.0376	0.0367	0.0355	0.0228	0.0218	0.0211	0.0130	0.0128	0.0127
12.000	0.0394	0.0385	0.0373	0.0241	0.0230	0.0223	0.0134	0.0131	0.0129
12.500	0.0429	0.0420	0.0408	0.0267	0.0255	0.0248	0.0142	0.0138	0.0136
13.000	0.0463	0.0454	0.0442	0.0296	0.0282	0.0274	0.0153	0.0148	0.0145
13.500	0.0494	0.0485	0.0474	0.0326	0.0311	0.0302	0.0166	0.0160	0.0157
14.000	0.0520	0.0512	0.0502	0.0356	0.0341	0.0331	0.0182	0.0175	0.0171
14.500				0.0386	0.0370	0.0360	0.0199	0.0192	0.0187
15.000				0.0415	0.0400	0.0390	0.0218	0.0209	0.0204
15.500				0.0443	0.0426	0.0418	0.0239	0.0230	0.0224
16.000				0.0470	0.0455	0.0446	0.0260	0.0250	0.0244
16.500				0.0494	0.0480	0.0471	0.0283	0.0273	0.0266
17.000					0.0503	0.0494	0.0307	0.0296	0.0288
17.500							0.0331	0.0319	0.0312
18.000							0.0355	0.0343	0.0335
18.500							0.0379	0.0366	0.0358
19.000							0.0402	0.0390	0.0382
19.500							0.0424	0.0412	0.0404
20.000							0.0446	0.0434	0.0426
20.500							0.0466	0.0455	0.0448
21.000							0.0485	0.0475	0.0467
21.500								0.0492	0.0485

TABLE A.24

Y_1 Expansion Factors—Flange Taps

Static Pressure Taken from Upstream Taps

$\frac{h_w}{p_{f1}}$ Ratio	$\beta = \frac{d}{D}$ Ratio													
	0.1	0.2	0.3	0.4	0.45	0.50	0.52	0.54	0.56	0.58	0.60	0.61	0.62	0.63
0.0	1.0000	1.0000	1.0000	1.0000	1.0000	1.0000	1.0000	1.0000	1.0000	1.0000	1.0000	1.0000	1.0000	1.0000
0.1	0.9989	0.9989	0.9989	0.9988	0.9988	0.9988	0.9988	0.9988	0.9988	0.9988	0.9987	0.9987	0.9987	0.9987
0.2	0.9977	0.9977	0.9977	0.9977	0.9976	0.9976	0.9976	0.9976	0.9975	0.9975	0.9975	0.9975	0.9974	0.9974
0.3	0.9966	0.9966	0.9966	0.9965	0.9965	0.9964	0.9964	0.9963	0.9963	0.9963	0.9962	0.9962	0.9962	0.9961
0.4	0.9954	0.9954	0.9954	0.9953	0.9953	0.9952	0.9952	0.9951	0.9951	0.9950	0.9949	0.9949	0.9949	0.9948
0.5	0.9943	0.9943	0.9943	0.9942	0.9941	0.9940	0.9940	0.9939	0.9938	0.9938	0.9937	0.9936	0.9936	0.9935
0.6	0.9932	0.9932	0.9931	0.9930	0.9929	0.9928	0.9927	0.9927	0.9926	0.9925	0.9924	0.9924	0.9923	0.9923
0.7	0.9920	0.9920	0.9920	0.9919	0.9918	0.9916	0.9915	0.9915	0.9914	0.9913	0.9912	0.9911	0.9910	0.9910
0.8	0.9909	0.9909	0.9908	0.9907	0.9906	0.9904	0.9903	0.9902	0.9901	0.9900	0.9899	0.9898	0.9897	0.9897
0.9	0.9898	0.9897	0.9897	0.9895	0.9894	0.9892	0.9891	0.9890	0.9889	0.9888	0.9886	0.9885	0.9885	0.9884
1.0	0.9886	0.9886	0.9885	0.9884	0.9882	0.9880	0.9879	0.9878	0.9877	0.9875	0.9874	0.9873	0.9872	0.9871
1.1	0.9875	0.9875	0.9874	0.9872	0.9870	0.9868	0.9867	0.9866	0.9864	0.9863	0.9861	0.9860	0.9859	0.9858
1.2	0.9863	0.9863	0.9862	0.9860	0.9859	0.9856	0.9855	0.9853	0.9852	0.9850	0.9848	0.9847	0.9846	0.9845
1.3	0.9852	0.9852	0.9851	0.9849	0.9847	0.9844	0.9843	0.9841	0.9840	0.9838	0.9836	0.9835	0.9833	0.9832
1.4	0.9841	0.9840	0.9840	0.9837	0.9835	0.9832	0.9831	0.9829	0.9827	0.9825	0.9823	0.9822	0.9821	0.9819
1.5	0.9829	0.9829	0.9828	0.9826	0.9823	0.9820	0.9819	0.9817	0.9815	0.9813	0.9810	0.9809	0.9808	0.9806
1.6	0.9818	0.9818	0.9817	0.9814	0.9811	0.9808	0.9806	0.9805	0.9803	0.9800	0.9798	0.9796	0.9795	0.9793
1.7	0.9806	0.9806	0.9805	0.9802	0.9800	0.9796	0.9794	0.9792	0.9790	0.9788	0.9785	0.9784	0.9782	0.9780
1.8	0.9795	0.9795	0.9794	0.9791	0.9788	0.9784	0.9782	0.9780	0.9778	0.9775	0.9772	0.9771	0.9769	0.9768
1.9	0.9784	0.9783	0.9782	0.9779	0.9776	0.9772	0.9770	0.9768	0.9766	0.9763	0.9760	0.9758	0.9756	0.9755
2.0	0.9772	0.9772	0.9771	0.9767	0.9764	0.9760	0.9758	0.9756	0.9753	0.9750	0.9747	0.9745	0.9744	0.9742
2.1	0.9761	0.9761	0.9759	0.9756	0.9753	0.9748	0.9746	0.9744	0.9741	0.9738	0.9734	0.9733	0.9731	0.9729
2.2	0.9750	0.9749	0.9748	0.9744	0.9741	0.9736	0.9734	0.9731	0.9729	0.9725	0.9722	0.9720	0.9718	0.9716
2.3	0.9738	0.9738	0.9736	0.9732	0.9729	0.9724	0.9722	0.9719	0.9716	0.9713	0.9709	0.9707	0.9705	0.9703
2.4	0.9727	0.9726	0.9725	0.9721	0.9717	0.9712	0.9710	0.9707	0.9704	0.9700	0.9697	0.9694	0.9692	0.9690
2.5	0.9715	0.9715	0.9713	0.9709	0.9705	0.9700	0.9698	0.9695	0.9692	0.9688	0.9684	0.9682	0.9680	0.9677
2.6	0.9704	0.9704	0.9702	0.9698	0.9694	0.9688	0.9686	0.9683	0.9679	0.9675	0.9671	0.9669	0.9667	0.9664
2.7	0.9693	0.9692	0.9691	0.9686	0.9682	0.9676	0.9673	0.9670	0.9667	0.9663	0.9659	0.9656	0.9654	0.9651
2.8	0.9681	0.9681	0.9679	0.9674	0.9670	0.9664	0.9661	0.9658	0.9654	0.9650	0.9646	0.9644	0.9641	0.9638
2.9	0.9670	0.9669	0.9668	0.9663	0.9658	0.9652	0.9649	0.9646	0.9642	0.9638	0.9633	0.9631	0.9628	0.9625
3.0	0.9658	0.9658	0.9656	0.9651	0.9647	0.9640	0.9637	0.9634	0.9630	0.9626	0.9621	0.9618	0.9615	0.9613
3.1	0.9647	0.9647	0.9645	0.9639	0.9635	0.9628	0.9625	0.9622	0.9617	0.9613	0.9608	0.9605	0.9603	0.9600
3.2	0.9636	0.9635	0.9633	0.9628	0.9623	0.9616	0.9613	0.9609	0.9605	0.9601	0.9595	0.9593	0.9590	0.9587
3.3	0.9624	0.9624	0.9622	0.9616	0.9611	0.9604	0.9601	0.9597	0.9593	0.9588	0.9583	0.9580	0.9577	0.9574
3.4	0.9613	0.9612	0.9610	0.9604	0.9599	0.9592	0.9589	0.9585	0.9580	0.9576	0.9570	0.9567	0.9564	0.9561
3.5	0.9602	0.9601	0.9599	0.9593	0.9588	0.9580	0.9577	0.9573	0.9568	0.9563	0.9558	0.9554	0.9551	0.9548
3.6	0.9590	0.9590	0.9587	0.9581	0.9576	0.9568	0.9565	0.9560	0.9556	0.9551	0.9545	0.9542	0.9538	0.9535
3.7	0.9579	0.9578	0.9576	0.9570	0.9561	0.9556	0.9553	0.9548	0.9543	0.9538	0.9532	0.9529	0.9526	0.9522
3.8	0.9567	0.9567	0.9564	0.9558	0.9552	0.9544	0.9540	0.9536	0.9531	0.9526	0.9520	0.9516	0.9513	0.9509
3.9	0.9556	0.9555	0.9553	0.9546	0.9540	0.9532	0.9528	0.9524	0.9519	0.9513	0.9507	0.9504	0.9500	0.9496
4.0	0.9545	0.9544	0.9542	0.9535	0.9529	0.9520	0.9516	0.9512	0.9506	0.9501	0.9494	0.9491	0.9487	0.9483

(Continued)

TABLE A.24 (Continued)

$\frac{h_w}{p_{fl}}$ Ratio	$\beta = \frac{d}{D}$ Ratio											
	0.64	0.65	0.66	0.67	0.68	0.69	0.70	0.71	0.72	0.73	0.74	0.75
0.0	1.0000	1.0000	1.0000	1.0000	1.0000	1.0000	1.0000	1.0000	1.0000	1.0000	1.0000	1.0000
0.1	0.9987	0.9987	0.9987	0.9987	0.9987	0.9986	0.9986	0.9986	0.9986	0.9986	0.9986	0.9986
0.2	0.9974	0.9974	0.9974	0.9973	0.9973	0.9973	0.9973	0.9972	0.9972	0.9972	0.9971	0.9971
0.3	0.9961	0.9961	0.9960	0.9960	0.9960	0.9959	0.9959	0.9958	0.9958	0.9958	0.9957	0.9957
0.4	0.9948	0.9948	0.9947	0.9947	0.9946	0.9946	0.9945	0.9945	0.9944	0.9943	0.9943	0.9942
0.5	0.9935	0.9934	0.9934	0.9933	0.9933	0.9932	0.9931	0.9931	0.9930	0.9929	0.9929	0.9928
0.6	0.9922	0.9921	0.9921	0.9920	0.9919	0.9918	0.9918	0.9917	0.9916	0.9915	0.9914	0.9913
0.7	0.9909	0.9908	0.9907	0.9907	0.9906	0.9905	0.9904	0.9903	0.9902	0.9901	0.9900	0.9899
0.8	0.9896	0.9895	0.9894	0.9893	0.9892	0.9891	0.9890	0.9889	0.9888	0.9887	0.9886	0.9884
0.9	0.9883	0.9882	0.9881	0.9880	0.9879	0.9878	0.9877	0.9875	0.9874	0.9873	0.9871	0.9870
1.0	0.9870	0.9869	0.9868	0.9867	0.9865	0.9864	0.9863	0.9861	0.9860	0.9859	0.9857	0.9855
1.1	0.9857	0.9856	0.9854	0.9853	0.9852	0.9851	0.9849	0.9848	0.9846	0.9844	0.9843	0.9841
1.2	0.9844	0.9843	0.9841	0.9840	0.9838	0.9837	0.9835	0.9834	0.9832	0.9830	0.9828	0.9826
1.3	0.9831	0.9829	0.9828	0.9827	0.9825	0.9823	0.9822	0.9820	0.9818	0.9816	0.9814	0.9812
1.4	0.9818	0.9816	0.9815	0.9813	0.9812	0.9810	0.9808	0.9806	0.9804	0.9802	0.9800	0.9798
1.5	0.9805	0.9803	0.9802	0.9800	0.9798	0.9796	0.9794	0.9792	0.9790	0.9788	0.9786	0.9783
1.6	0.9792	0.9790	0.9788	0.9787	0.9785	0.9783	0.9781	0.9778	0.9776	0.9774	0.9771	0.9769
1.7	0.9779	0.9777	0.9775	0.9773	0.9771	0.9769	0.9767	0.9764	0.9762	0.9760	0.9757	0.9754
1.8	0.9766	0.9764	0.9762	0.9760	0.9758	0.9755	0.9753	0.9751	0.9748	0.9745	0.9743	0.9740
1.9	0.9753	0.9751	0.9749	0.9747	0.9744	0.9742	0.9739	0.9737	0.9734	0.9731	0.9728	0.9725
2.0	0.9740	0.9738	0.9735	0.9733	0.9731	0.9728	0.9726	0.9723	0.9720	0.9717	0.9714	0.9711
2.1	0.9727	0.9725	0.9722	0.9720	0.9717	0.9715	0.9712	0.9709	0.9706	0.9703	0.9700	0.9696
2.2	0.9714	0.9711	0.9709	0.9706	0.9704	0.9701	0.9698	0.9695	0.9692	0.9689	0.9685	0.9682
2.3	0.9701	0.9698	0.9696	0.9693	0.9690	0.9688	0.9685	0.9681	0.9678	0.9675	0.9671	0.9667
2.4	0.9688	0.9685	0.9683	0.9680	0.9677	0.9674	0.9671	0.9668	0.9664	0.9661	0.9657	0.9653
2.5	0.9675	0.9672	0.9669	0.9666	0.9663	0.9660	0.9657	0.9654	0.9650	0.9646	0.9643	0.9639
2.6	0.9662	0.9659	0.9656	0.9653	0.9650	0.9647	0.9643	0.9640	0.9636	0.9632	0.9628	0.9624
2.7	0.9649	0.9646	0.9643	0.9640	0.9637	0.9633	0.9630	0.9626	0.9622	0.9618	0.9614	0.9610
2.8	0.9636	0.9633	0.9630	0.9626	0.9623	0.9620	0.9616	0.9612	0.9608	0.9604	0.9600	0.9595
2.9	0.9623	0.9620	0.9616	0.9613	0.9610	0.9606	0.9602	0.9598	0.9594	0.9590	0.9585	0.9581
3.0	0.9610	0.9606	0.9603	0.9600	0.9596	0.9592	0.9588	0.9584	0.9580	0.9576	0.9571	0.9566
3.1	0.9597	0.9593	0.9590	0.9586	0.9583	0.9579	0.9575	0.9571	0.9566	0.9562	0.9557	0.9552
3.2	0.9584	0.9580	0.9577	0.9573	0.9569	0.9565	0.9561	0.9557	0.9552	0.9547	0.9542	0.9537
3.3	0.9571	0.9567	0.9564	0.9560	0.9556	0.9552	0.9547	0.9543	0.9538	0.9533	0.9528	0.9523
3.4	0.9558	0.9554	0.9550	0.9546	0.9542	0.9538	0.9534	0.9529	0.9524	0.9519	0.9514	0.9508
3.5	0.9545	0.9541	0.9537	0.9533	0.9529	0.9524	0.9520	0.9515	0.9510	0.9505	0.9500	0.9494
3.6	0.9532	0.9528	0.9524	0.9520	0.9515	0.9511	0.9506	0.9501	0.9496	0.9491	0.9485	0.9480
3.7	0.9518	0.9515	0.9511	0.9506	0.9502	0.9497	0.9492	0.9487	0.9482	0.9477	0.9471	0.9465
3.8	0.9505	0.9502	0.9497	0.9493	0.9488	0.9484	0.9479	0.9474	0.9468	0.9463	0.9457	0.9451
3.9	0.9492	0.9488	0.9484	0.9480	0.9475	0.9470	0.9465	0.9460	0.9454	0.9448	0.9442	0.9436
4.0	0.9479	0.9475	0.9471	0.9465	0.9462	0.9457	0.9451	0.9446	0.9440	0.9434	0.9428	0.9422

TABLE A.25
Y_2 Expansion Factors—Flange Taps
Static Pressure Taken from Downstream Taps

$\frac{h_w}{p_{f2}}$ Ratio	$\beta = \frac{d}{D}$ Ratio 0.1	0.2	0.3	0.4	0.45	0.50	0.52	0.54	0.56	0.58	0.60	0.61	0.62	0.63
0.0	1.0000	1.0000	1.0000	1.0000	1.0000	1.0000	1.0000	1.0000	1.0000	1.0000	1.0000	1.0000	1.0000	1.0000
0.1	1.0007	1.0007	1.0006	1.0006	1.0006	1.0006	1.0006	1.0006	1.0006	1.0006	1.0005	1.0005	1.0005	1.0005
0.2	1.0013	1.0013	1.0013	1.0013	1.0012	1.0012	1.0012	1.0012	1.0011	1.0011	1.0011	1.0011	1.0010	1.0010
0.3	1.0020	1.0020	1.0020	1.0019	1.0019	1.0018	1.0018	1.0018	1.0017	1.0017	1.0016	1.0016	1.0016	1.0015
0.4	1.0027	1.0027	1.0026	1.0026	1.0025	1.0024	1.0024	1.0023	1.0023	1.0022	1.0022	1.0021	1.0021	1.0021
0.5	1.0033	1.0033	1.0033	1.0032	1.0031	1.0030	1.0030	1.0029	1.0029	1.0028	1.0027	1.0027	1.0026	1.0026
0.6	1.0040	1.0040	1.0040	1.0039	1.0038	1.0036	1.0036	1.0035	1.0034	1.0034	1.0033	1.0032	1.0032	1.0031
0.7	1.0047	1.0047	1.0046	1.0045	1.0044	1.0043	1.0042	1.0041	1.0040	1.0039	1.0038	1.0038	1.0037	1.0036
0.8	1.0054	1.0053	1.0053	1.0052	1.0050	1.0049	1.0048	1.0047	1.0046	1.0045	1.0044	1.0043	1.0042	1.0042
0.9	1.0060	1.0060	1.0060	1.0058	1.0057	1.0055	1.0054	1.0053	1.0052	1.0050	1.0049	1.0048	1.0048	1.0047
1.0	1.0067	1.0067	1.0066	1.0065	1.0063	1.0061	1.0060	1.0059	1.0058	1.0056	1.0055	1.0054	1.0053	1.0052
1.1	1.0074	1.0074	1.0073	1.0071	1.0069	1.0067	1.0066	1.0065	1.0063	1.0062	1.0060	1.0059	1.0058	1.0057
1.2	1.0080	1.0080	1.0080	1.0078	1.0076	1.0073	1.0072	1.0071	1.0069	1.0068	1.0066	1.0065	1.0064	1.0062
1.3	1.0087	1.0087	1.0086	1.0084	1.0082	1.0080	1.0078	1.0077	1.0075	1.0073	1.0071	1.0070	1.0069	1.0068
1.4	1.0094	1.0094	1.0093	1.0091	1.0089	1.0086	1.0084	1.0083	1.0081	1.0079	1.0077	1.0076	1.0074	1.0073
1.5	1.0101	1.0101	1.0100	1.0097	1.0095	1.0092	1.0090	1.0089	1.0087	1.0085	1.0082	1.0081	1.0080	1.0078
1.6	1.0108	1.0107	1.0106	1.0104	1.0101	1.0098	1.0096	1.0095	1.0093	1.0090	1.0088	1.0087	1.0085	1.0084
1.7	1.0114	1.0114	1.0113	1.0110	1.0108	1.0104	1.0103	1.0101	1.0099	1.0096	1.0094	1.0092	1.0091	1.0089
1.8	1.0121	1.0121	1.0120	1.0117	1.0114	1.0111	1.0109	1.0107	1.0104	1.0102	1.0099	1.0098	1.0096	1.0094
1.9	1.0128	1.0128	1.0126	1.0123	1.0121	1.0117	1.0115	1.0113	1.0110	1.0108	1.0105	1.0103	1.0102	1.0100
2.0	1.0135	1.0134	1.0133	1.0130	1.0127	1.0123	1.0121	1.0119	1.0116	1.0114	1.0110	1.0109	1.0107	1.0105
2.1	1.0142	1.0141	1.0140	1.0136	1.0134	1.0129	1.0127	1.0125	1.0122	1.0119	1.0116	1.0114	1.0112	1.0111
2.2	1.0148	1.0148	1.0147	1.0143	1.0140	1.0136	1.0133	1.0131	1.0128	1.0125	1.0122	1.0120	1.0118	1.0116
2.3	1.0155	1.0155	1.0154	1.0150	1.0146	1.0142	1.0140	1.0137	1.0134	1.0131	1.0127	1.0126	1.0124	1.0121
2.4	1.0162	1.0162	1.0160	1.0156	1.0153	1.0148	1.0146	1.0143	1.0140	1.0137	1.0133	1.0131	1.0129	1.0127
2.5	1.0169	1.0168	1.0167	1.0163	1.0159	1.0154	1.0152	1.0149	1.0146	1.0142	1.0139	1.0137	1.0134	1.0132
2.6	1.0176	1.0175	1.0174	1.0170	1.0166	1.0161	1.0158	1.0155	1.0152	1.0148	1.0144	1.0142	1.0140	1.0138
2.7	1.0182	1.0182	1.0180	1.0176	1.0172	1.0167	1.0164	1.0161	1.0158	1.0154	1.0150	1.0148	1.0146	1.0143
2.8	1.0189	1.0189	1.0187	1.0183	1.0179	1.0173	1.0170	1.0167	1.0164	1.0160	1.0156	1.0154	1.0151	1.0148
2.9	1.0196	1.0196	1.0194	1.0189	1.0185	1.0180	1.0177	1.0173	1.0170	1.0166	1.0162	1.0159	1.0157	1.0154
3.0	1.0203	1.0203	1.0201	1.0196	1.0192	1.0186	1.0183	1.0180	1.0176	1.0172	1.0167	1.0165	1.0162	1.0160
3.1	1.0210	1.0210	1.0208	1.0203	1.0198	1.0192	1.0189	1.0186	1.0182	1.0178	1.0173	1.0170	1.0168	1.0165
3.2	1.0217	1.0216	1.0214	1.0209	1.0205	1.0198	1.0195	1.0192	1.0188	1.0184	1.0179	1.0176	1.0173	1.0170
3.3	1.0224	1.0223	1.0221	1.0216	1.0211	1.0205	1.0202	1.0198	1.0194	1.0189	1.0184	1.0182	1.0179	1.0176
3.4	1.0230	1.0230	1.0228	1.0223	1.0218	1.0211	1.0208	1.0204	1.0200	1.0195	1.0190	1.0187	1.0184	1.0181
3.5	1.0237	1.0237	1.0235	1.0229	1.0224	1.0217	1.0214	1.0210	1.0206	1.0201	1.0196	1.0193	1.0190	1.0187
3.6	1.0244	1.0244	1.0242	1.0236	1.0231	1.0224	1.0220	1.0216	1.0212	1.0207	1.0202	1.0199	1.0196	1.0192
3.7	1.0251	1.0251	1.0248	1.0243	1.0237	1.0230	1.0226	1.0222	1.0218	1.0213	1.0207	1.0204	1.0201	1.0198
3.8	1.0258	1.0258	1.0255	1.0249	1.0244	1.0236	1.0233	1.0229	1.0224	1.0219	1.0213	1.0210	1.0207	1.0204
3.9	1.0265	1.0264	1.0262	1.0256	1.0250	1.0243	1.0239	1.0235	1.0230	1.0225	1.0219	1.0216	1.0213	1.0209
4.0	1.0272	1.0271	1.0269	1.0263	1.0257	1.0249	1.0245	1.0241	1.0236	1.0231	1.0225	1.0222	1.0218	1.0215

(Continued)

TABLE A.25 (Continued)

$\frac{h_w}{p_{f2}}$ Ratio	$\beta = \frac{d}{D}$ Ratio											
	0.64	0.65	0.66	0.67	0.68	0.69	0.70	0.71	0.72	0.73	0.74	0.75
0.0	1.0000	1.0000	1.0000	1.0000	1.0000	1.0000	1.0000	1.0000	1.0000	1.0000	1.0000	1.0000
0.1	1.0005	1.0005	1.0005	1.0005	1.0004	1.0004	1.0004	1.0004	1.0004	1.0004	1.0004	1.0004
0.2	1.0010	1.0010	1.0010	1.0009	1.0009	1.0009	1.0009	1.0008	1.0008	1.0008	1.0008	1.0007
0.3	1.0015	1.0015	1.0014	1.0014	1.0014	1.0013	1.0013	1.0013	1.0012	1.0012	1.0011	1.0011
0.4	1.0020	1.0020	1.0019	1.0019	1.0018	1,0018	1.0017	1.0017	1.0016	1.0016	1.0015	1.0014
0.5	1.0025	1.0025	1.0024	1.0024	1.0023	1.0022	1.0022	1.0021	1.0020	1.0020	1.0019	1.0018
0.6	1.0030	1.0030	1.0029	1.0028	1.0028	1.0027	1.0026	1.0025	1.0025	1.0024	1.0023	1.0022
0.7	1.0036	1.0035	1.0034	1.0033	1.0032	1.0032	1.0031	1.0030	1.0029	1.0028	1.0027	1.0026
0.8	1.0041	1.0040	1.0039	1.0038	1.0037	1.0036	1.0035	1.0034	1.0033	1.0032	1.0030	1.0029
0.9	1.0046	1.0045	1.0044	1.0043	1.0042	1.0041	1.0040	1.0038	1.0037	1.0036	1.0034	1.0033
1.0	1.0051	1.0050	1.0049	1.0048	1.0047	1.0045	1.0044	1.0043	1.0041	1.0040	1.0038	1.0037
1.1	1.0056	1.0055	1.0054	1.0053	1.0051	1.0050	1.0049	1.0047	1.0046	1.0044	1.0042	1.0041
1.2	1.0061	1.0060	1.0059	1.0058	1.0056	1.0055	1.0053	1.0052	1.0050	1.0048	1.0046	1.0044
1.3	1.0066	1.0065	1.0064	1.0062	1.0061	1.0059	1.0058	1.0056	1.0054	1.0052	1.0050	1.0048
1.4	1.0072	1.0070	1.0069	1.0067	1.0066	1.0064	1.0062	1.0060	1.0058	1.0056	1.0054	1.0052
1.5	1.0077	1.0076	1.0074	1.0072	1.0070	1.0069	1.0067	1.0065	1.0063	1.0060	1.0058	1.0056
1.6	1.0082	1.0081	1.0079	1.0077	1.0075	1.0073	1.0071	1.0069	1.0067	1.0065	1.0062	1.0060
1.7	1.0088	1.0086	1.0084	1.0082	1.0080	1.0078	1.0076	1.0074	1.0071	1.0069	1.0066	1.0064
1.8	1.0093	1.0091	1.0089	1.0087	1.0085	1.0083	1.0080	1.0078	1.0076	1.0073	1.0070	1.0068
1.9	1.0098	1.0096	1.0094	1.0092	1.0090	1.0088	1.0085	1.0083	1.0080	1.0077	1.0074	1.0071
2.0	1.0103	1.0101	1.0099	1.0097	1.0095	1.0092	1.0090	1.0087	1.0084	1.0081	1.0078	1.0075
2.1	1.0109	1.0106	1.0104	1.0102	1.0100	1.0097	1.0094	1.0092	1.0089	1.0086	1.0083	1.0079
2.2	1.0114	1.0112	1.0109	1.0107	1.0104	1.0102	1.0099	1.0096	1.0093	1.0090	1.0087	1.0083
2.3	1.0119	1.0117	1.0114	1.0112	1.0109	1.0106	1.0104	1.0101	1.0098	1.0094	1.0091	1.0087
2.4	1.0124	1.0122	1.0120	1.0117	1.0114	1.0111	1.0108	1.0105	1.0102	1.0098	1.0095	1.0091
2.5	1.0130	1.0127	1.0125	1.0122	1.0119	1.0116	1.0113	1.0110	1.0106	1.0103	1.0099	1.0095
2.6	1.0135	1.0133	1.0130	1.0127	1.0124	1.0121	1.0118	1.0114	1.0111	1.0107	1.0103	1.0099
2.7	1.0140	1.0138	1.0135	1.0132	1.0129	1.0126	1.0122	1.0119	1.0115	1.0111	1.0107	1.0103
2.8	1.0146	1.0143	1.0140	1.0137	1.0134	1.0131	1.0127	1.0124	1.0120	1.0116	1.0112	1.0107
2.9	1.0151	1.0148	1.0145	1.0142	1.0139	1.0136	1.0132	1.0128	1.0124	1.0120	1.0116	1.0111
3.0	1.0157	1.0154	1.0150	1.0147	1.0144	1.0140	1.0137	1.0133	1.0129	1.0124	1.0120	1.0116
3.1	1.0162	1.0159	1.0156	1.0152	1.0149	1.0145	1.0141	1.0137	1.0133	1.0129	1.0124	1.0120
3.2	1.0167	1.0164	1.0161	1.0158	1.0154	1.0150	1.0146	1.0142	1.0138	1.0133	1.0128	1.0124
3.3	1.0173	1.0170	1.0166	1.0163	1.0159	1.0155	1.0151	1.0147	1.0142	1.0138	1.0133	1.0128
3.4	1.0178	1.0175	1.0171	1.0168	1.0164	1.0160	1.0156	1.0151	1.0147	1.0142	1.0137	1.0132
3.5	1.0184	1.0180	1.0177	1.0173	1.0169	1.0165	1.0160	1.0156	1.0151	1.0146	1.0141	1.0136
3.6	1.0189	1.0186	1.0182	1.0178	1.0174	1.0170	1.0165	1.0161	1.0156	1.0151	1.0146	1.0140
3.7	1.0195	1.0191	1.0187	1.0183	1.0179	1.0175	1.0170	1.0165	1.0160	1.0155	1.0150	1.0144
3.8	1.0200	1.0196	1.0192	1.0188	1.0184	1.0180	1.0175	1.0170	1.0165	1.0160	1.0154	1.0148
3.9	1.0206	1.0202	1.0198	1.0194	1.0189	1.0185	1.0180	1.0175	1.0170	1.0164	1.0159	1.0153
4.0	1.0211	1.0207	1.0203	1.0199	1.0194	1.0190	1.0185	1.0180	1.0174	1.0169	1.0163	1.0157

TABLE A.26

Y_m Expansion Factors—Flange Taps

Static Pressure Mean of Upstream and Downstream

$\frac{h_w}{p_{fm}}$ Ratio	$\beta = \frac{d}{D}$ Ratio													
	0.1	0.2	0.3	0.4	0.45	0.50	0.52	0.54	0.56	0.58	0.60	0.61	0.62	0.63
0.0	1.000	1.0000	1.0000	1.0000	1.0000	1.0000	1.0000	1.0000	1.0000	1.0000	1.0000	1.0000	1.0000	1.0000
0.1	0.9998	0.9998	0.9998	0.9997	0.9997	0.9997	0.9997	0.9997	0.9997	0.9996	0.9996	0.9996	0.9996	0.9996
0.2	0.9995	0.9995	0.9995	0.9995	0.9994	0.9994	0.9994	0.9994	0.9993	0.9993	0.9993	0.9993	0.9992	0.9992
0.3	0.9993	0.9993	0.9993	0.9992	0.9992	0.9991	0.9991	0.9990	0.9990	0.9990	0.9989	0.9989	0.9989	0.9988
0.4	0.9991	0.9990	0.9990	0.9990	0.9989	0.9988	0.9988	0.9987	0.9987	0.9986	0.9986	0.9985	0.9985	0.9984
0.5	0.9988	0.9988	0.9988	0.9987	0.9986	0.9985	0.9985	0.9984	0.9984	0.9983	0.9982	0.9982	0.9981	0.9981
0.6	0.9986	0.9986	0.9986	0.9984	0.9984	0.9982	0.9982	0.9981	0.9980	0.9979	0.9978	0.9978	0.9977	0.9977
0.7	0.9984	0.9984	0.9983	0.9982	0.9981	0.9980	0.9979	0.9978	0.9977	0.9976	0.9975	0.9974	0.9974	0.9973
0.8	0.9982	0.9981	0.9981	0.9980	0.9978	0.9977	0.9976	0.9975	0.9974	0.9973	0.9971	0.9971	0.9970	0.9969
0.9	0.9979	0.9979	0.9978	0.9977	0.9976	0.9974	0.9973	0.9972	0.9971	0.9969	0.9968	0.9967	0.9966	0.9966
1.0	0.9977	0.9977	0.9976	0.9974	0.9973	0.9971	0.9970	0.9969	0.9968	0.9966	0.9964	0.9964	0.9963	0.9962
1.1	0.9975	0.9975	0.9974	0.9972	0.9970	0.9968	0.9967	0.9966	0.9964	0.9963	0.9961	0.9960	0.9959	0.9958
1.2	0.9972	0.9972	0.9972	0.9970	0.9968	0.9965	0.9964	0.9963	0.9961	0.9959	0.9958	0.9956	0.9955	0.9954
1.3	0.9970	0.9970	0.9969	0.9967	0.9965	0.9962	0.9961	0.9960	0.9958	0.9956	0.9954	0.9953	0.9952	0.9951
1.4	0.9968	0.9968	0.9967	0.9965	0.9963	0.9960	0.9958	0.9957	0.9955	0.9953	0.9951	0.9950	0.9948	0.9947
1.5	0.9966	0.9966	0.9965	0.9962	0.9960	0.9957	0.9955	0.9954	0.9952	0.9950	0.9947	0.9946	0.9945	0.9943
1.6	0.9964	0.9964	0.9962	0.9960	0.9957	0.9954	0.9952	0.9951	0.9949	0.9946	0.9944	0.9942	0.9941	0.9940
1.7	0.9962	0.9961	0.9960	0.9957	0.9955	0.9951	0.9950	0.9948	0.9946	0.9943	0.9940	0.9939	0.9938	0.9936
1.8	0.9959	0.9959	0.9958	0.9955	0.9952	0.9949	0.9947	0.9945	0.9942	0.9940	0.9937	0.9936	0.9934	0.9932
1.9	0.9957	0.9957	0.9956	0.9953	0.9950	0.9946	0.9944	0.9942	0.9939	0.9937	0.9934	0.9932	0.9930	0.9929
2.0	0.9955	0.9955	0.9954	0.9950	0.9947	0.9943	0.9941	0.9939	0.9936	0.9934	0.9930	0.9929	0.9927	0.9925
2.1	0.9953	0.9953	0.9951	0.9948	0.9945	0.9940	0.9938	0.9936	0.9933	0.9930	0.9927	0.9925	0.9923	0.9922
2.2	0.9951	0.9951	0.9949	0.9946	0.9942	0.9938	0.9936	0.9933	0.9930	0.9927	0.9924	0.9922	0.9920	0.9918
2.3	0.9949	0.9948	0.9947	0.9943	0.9940	0.9935	0.9933	0.9930	0.9927	0.9924	0.9920	0.9918	0.9916	0.9914
2.4	0.9947	0.9946	0.9945	0.9941	0.9937	0.9932	0.9930	0.9927	0.9924	0.9921	0.9917	0.9915	0.9913	0.9911
2.5	0.9945	0.9944	0.9943	0.9939	0.9935	0.9930	0.9927	0.9924	0.9921	0.9918	0.9914	0.9912	0.9910	0.9907
2.6	0.9943	0.9942	0.9941	0.9936	0.9932	0.9927	0.9924	0.9922	0.9918	0.9915	0.9911	0.9908	0.9906	0.9904
2.7	0.9940	0.9940	0.9938	0.9934	0.9930	0.9924	0.9922	0.9919	0.9915	0.9912	0.9907	0.9905	0.9903	0.9900
2.8	0.9938	0.9938	0.9936	0.9932	0.9928	0.9922	0.9919	0.9916	0.9912	0.9908	0.9904	0.9902	0.9899	0.9897
2.9	0.9936	0.9936	0.9934	0.9929	0.9925	0.9919	0.9916	0.9913	0.9910	0.9905	0.9901	0.9898	0.9896	0.9893
3.0	0.9934	0.9934	0.9932	0.9927	0.9923	0.9917	0.9914	0.9910	0.9906	0.9902	0.9898	0.9895	0.9892	0.9890
3.1	0.9932	0.9932	0.9930	0.9925	0.9920	0.9914	0.9911	0.9908	0.9904	0.9899	0.9894	0.9892	0.9889	0.9886
3.2	0.9930	0.9930	0.9928	0.9923	0.9918	0.9912	0.9908	0.9905	0.9901	0.9896	0.9891	0.9889	0.9886	0.9883
3.3	0.9928	0.9928	0.9926	0.9920	0.9916	0.9909	0.9906	0.9902	0.9898	0.9893	0.9888	0.9885	0.9882	0.9879
3.4	0.9926	0.9926	0.9924	0.9918	0.9913	0.9906	0.9903	0.9899	0.9895	0.9890	0.9885	0.9882	0.9879	0.9876
3.5	0.9924	0.9924	0.9922	0.9916	0.9911	0.9904	0.9900	0.9896	0.9892	0.9887	0.9882	0.9879	0.9876	0.9872
3.6	0.9922	0.9922	0.9920	0.9914	0.9909	0.9901	0.9898	0.9894	0.9889	0.9884	0.9879	0.9876	0.9872	0.9869
3.7	0.9921	0.9920	0.9918	0.9912	0.9906	0.9899	0.9895	0.9891	0.9886	0.9881	0.9876	0.9872	0.9869	0.9866
3.8	0.9919	0.9918	0.9916	0.9910	0.9904	0.9896	0.9893	0.9888	0.9884	0.9878	0.9872	0.9869	0.9866	0.9862
3.9	0.9917	0.9916	0.9914	0.9907	0.9902	0.9894	0.9890	0.9886	0.9881	0.9875	0.9869	0.9866	0.9863	0.9859
4.0	0.9915	0.9914	0.9912	0.9905	0.9899	0.9891	0.9887	0.9883	0.9878	0.9872	0.9866	0.9863	0.9859	0.9856

(Continued)

TABLE A.26 (Continued)

$\frac{h_w}{p_{fm}}$ Ratio	0.64	0.65	0.66	0.67	0.68	0.69	0.70	0.71	0.72	0.73	0.74	0.75
0.0	1.0000	1.0000	1.0000	1.0000	1.0000	1.0000	1.0000	1.000	1.0000	1.0000	1.0000	1.0000
0.1	0.9996	0.9996	0.9996	0.9996	0.9996	0.9995	0.9995	0.9995	0.9995	0.9995	0.9995	0.9994
0.2	0.9992	0.9992	0.9992	0.9991	0.9991	0.9991	0.9991	0.9990	0.9990	0.9990	0.9989	0.9989
0.3	0.9988	0.9988	0.9987	0.9987	0.9987	0.9986	0.0986	0.9986	0.9985	0.9985	0.9984	0.9984
0.4	0.9984	0.9984	0.9983	0.9983	0.9982	0.9982	0.9981	0.9981	0.9980	0.9980	0.9979	0.9978
0.5	0.9980	0.9980	0.9979	0.9979	0.9978	0.9977	0.9977	0.9976	0.9975	0.9975	0.9974	0.9973
0.6	0.9976	0.9976	0.9975	0.9974	0.9974	0.9973	0.9972	0.9971	0.9970	0.9970	0.9969	0.9968
0.7	0.9972	0.9972	0.9971	0.9970	0.9969	0.9968	0.9968	0.9966	0.9966	0.9964	0.9964	0.9962
0.8	0.9968	0.9968	0.9967	0.9966	0.9965	0.9964	0.9963	0.9962	0.9961	0.9960	0.9958	0.9957
0.9	0.9965	0.9964	0.9963	0.9962	0.9961	0.9960	0.9958	0.9957	0.9956	0.9955	0.9953	0.9952
1.0	0.9961	0.9960	0.9959	0.9958	0.9956	0.9955	0.9954	0.9952	0.9951	0.9950	0.9948	0.9946
1.1	0.9957	0.9956	0.9955	0.9953	0.9952	0.9951	0.9949	0.9948	0.9946	0.9945	0.9943	0.9941
1.2	0.9953	0.9952	0.9951	0.9949	0.9948	0.9946	0.9945	0.9943	0.9942	0.9940	0.9938	0.9936
1.3	0.9949	0.9948	0.9947	0.9945	0.9944	0.9942	0.9940	0.9939	0.9937	0.9935	0.9933	0.9931
1.4	0.9946	0.9944	0.9943	0.9941	0.9939	0.9938	0.9936	0.9934	0.9932	0.9930	0.9928	0.9926
1.5	0.9942	0.9940	0.9939	0.9937	0.9935	0.9933	0.9931	0.9929	0.9927	0.9925	0.9923	0.9920
1.6	0.9938	0.9936	0.9935	0.9933	0.9931	0.9929	0.9927	0.9925	0.9922	0.9920	0.9918	0.9915
1.7	0.9934	0.9932	0.9931	0.9929	0.9927	0.9925	0.9922	0.9920	0.9918	0.9915	0.9913	0.9910
1.8	0.9930	0.9929	0.9927	0.9925	0.9923	0.9920	0.9918	0.9916	0.9913	0.9910	0.9908	0.9905
1.9	0.9927	0.9925	0.9923	0.9921	0.9918	0.9916	0.9914	0.9911	0.9908	0.9906	0.9903	0.9900
2.0	0.9923	0.9921	0.9919	0.9917	0.9914	0.9912	0.9909	0.9907	0.9904	0.9901	0.9898	0.9895
2.1	0.9919	0.9917	0.9915	0.9913	0.9910	0.9908	0.9905	0.9902	0.9899	0.9896	0.9893	0.9890
2.2	0.9916	0.9914	0.9911	0.9909	0.9906	0.9903	0.9901	0.9898	0.9895	0.9891	0.9888	0.9885
2.3	0.9912	0.9910	0.9907	0.9905	0.9902	0.9899	0.9896	0.9893	0.9890	0.9887	0.9883	0.9880
2.4	0.9908	0.9906	0.9903	0.9901	0.9898	0.9895	0.9892	0.9889	0.9885	0.9882	0.9878	0.9874
2.5	0.9905	0.9902	0.9900	0.9897	0.9894	0.9891	0.9888	0.9884	0.9881	0.9877	0.9873	0.9870
2.6	0.9901	0.9898	0.9896	0.9893	0.9890	0.9887	0.9883	0.9880	0.9876	0.9872	0.9868	0.9864
2.7	0.9898	0.9895	0.9892	0.9889	0.9886	0.9882	0.9879	0.9875	0.9872	0.9868	0.9864	0.9860
2.8	0.9894	0.9891	0.9888	0.9885	0.9882	0.9878	0.9875	0.9871	0.9867	0.9863	0.9859	0.9854
2.9	0.9890	0.9887	0.9884	0.9881	0.9878	0.9874	0.9870	0.9867	0.9863	0.9858	0.9854	0.9850
3.0	0.9887	0.9884	0.9881	0.9877	0.9874	0.9870	0.9866	0.9862	0.9858	0.9854	0.9849	0.9845
3.1	0.9883	0.9880	0.9877	0.9873	0.9870	0.9866	0.9862	0.9858	0.9854	0.9849	0.9844	0.9840
3.2	0.9880	0.9876	0.9873	0.9870	0.9866	0.9862	0.9858	0.9854	0.9849	0.9845	0.9840	0.9835
3.3	0.9876	0.9873	0.9869	0.9866	0.9862	0.9858	0.9854	0.9849	0.9845	0.9840	0.9835	0.9830
3.4	0.9873	0.9869	0.9866	0.9862	0.9858	0.9854	0.9850	0.9845	0.9840	0.9835	0.9830	0.9825
3.5	0.9869	0.9866	0.9862	0.9858	0.9854	0.9850	0.9845	0.9841	0.9836	0.9831	0.9826	0.9820
3.6	0.9866	0.9862	0.9858	0.9854	0.9850	0.9846	0.9841	0.9836	0.9831	0.9826	0.9821	0.9815
3.7	0.9862	0.9858	0.9855	0.9850	0.9846	0.9842	0.9837	0.9832	0.9827	0.9822	0.9816	0.9810
3.8	0.9859	0.9855	0.9851	0.9847	0.9842	0.9838	0.9833	0.9828	0.9823	0.9817	0.9812	0.9806
3.9	0.9855	0.9851	0.9847	0.9843	0.9838	0.9834	0.9829	0.9824	0.9818	0.9813	0.9807	0.9801
4.0	0.9852	0.9848	0.9844	0.9839	0.9835	0.9830	0.9825	0.9819	0.9814	0.9808	0.9802	0.9796

A.4.2 Tables Applying to Orifice Flow Constants for Pipe Tap Installations

Table A.27 Basic Orifice Factors
Table A.28 "*b*" Values for Reynolds Number Factor Determination
Table A.29 Expansion Factors, Static Pressure Upstream
Table A.30 Expansion Factors, Static Pressure Downstream

TABLE A.27

F_b Basic Orifice Factors—Pipe Taps

Basic temperature = 60°F Flowing temperature = 60°F $\sqrt{h_w pf} = \infty$

Base pressure = 14.73 psia Specific gravity = 1.0 $h_w/pf = 0$

Pipe Sizes—Nominal and Published Inside Diameters, Inches

Orifice Diameter, in.	2			3				4	
	1.689	1.939	2.067	2.300	2.626	2.900	3.068	3.152	3.430
0.250	12.850	12.813	12.800	12.782	12.765	12.753	12.748	12.745	12.737
0.375	29.359	29.097	29.005	28.882	28.771	28.710	28.682	28.669	28.634
0.500	53.703	52.816	52.401	52.019	51.591	51.353	51.243	51.196	51.064
0.625	87.212	84.919	84.083	82.922	81.795	81.142	80.835	80.703	80.332
0.750	132.23	126.86	124.99	122.45	120.06	118.67	118.00	117.70	116.86
0.875	192.74	181.02	177.08	171.92	167.23	164.58	163.31	162.76	161.17
1.000	275.45	251.10	243.27	233.30	224.56	219.76	217.52	216.55	213.79
1.125	391.93	342.98	327.98	309.43	293.79	285.48	281.66	280.02	275.42
1.250		465.99	437.99	404.52	377.36	363.41	357.12	354.45	347.03
1.375			583.96	524.68	478.68	455.82	445.74	441.48	429.83
1.500				679.10	602.45	565.79	549.94	543.31	525.40
1.625					755.34	697.43	672.95	662.81	635.76
1.750					946.99	856.37	819.05	803.77	763.51
1.875						1050.4	993.98	971.19	911.98
2.000						1290.7	1205.6	1171.8	1085.5
2.125							1465.1	1415.0	1289.7
2.250									1532.0
2.375									1822.8

Orifice Diameter, in.	4		6				8		
	3.826	4.026	4.897	5.189	5.761	6.065	7.625	7.981	8.071
0.250	12.727	12.722							
0.375	26.598	28.584							
0.500	50.936	50.886	50.739	50.705	50.652	50.628			
0.625	79.974	79.835	79.436	79.349	79.217	79.162			
0.750	116.05	115.73	114.81	114.61	114.32	114.20			
0.875	159.57	158.94	157.11	156.71	156.13	155.89	155.10	154.99	154.96
1.000	211.03	209.91	206.62	205.91	204.84	204.41	203.00	202.80	202.75
1.125	270.90	269.10	263.71	262.51	260.71	259.98	257.62	257.28	257.20
1.250	339.87	337.05	328.73	326.85	324.02	322.86	319.10	318.56	318.44
1.375	418.79	414.51	402.06	399.30	395.08	393.33	387.62	386.81	386.62
1.500	508.76	502.38	484.20	480.23	474.20	471.69	463.39	462.19	461.92
1.625	611.11	601.80	575.73	570.14	561.73	558.24	546.61	544.92	544.53
1.750	727.54	714.16	677.38	669.63	658.08	653.33	637.51	635.19	634.65
1.875	860.17	841.19	789.99	779.40	763.77	757.39	736.34	733.23	732.52
2.000	1011.7	985.04	914.57	900.28	879.38	870.93	843.34	839.29	838.35
2.125	1185.3	1148.4	1052.3	1033.2	1005.6	994.52	958.78	953.58	952.38
2.250	1385.4	1334.4	1204.7	1179.4	1143.2	1128.8	1083.0	1076.4	1074.9
2.375	1617.2	1547.3	1373.4	1340.2	1293.1	1274.6	1216.3	1208.0	1206.1
2.500	1887.6	1792.3	1560.5	1517.2	1456.4	1432.7	1359.2	1348.8	1346.5
2.625	2206.0	2075.9	1768.3	1712.3	1634.3	1604.3	1512.0	1499.2	1496.3
2.750		2407.0	1999.8	1927.6	1828.3	1790.3	1675.4	1659.7	1656.1
2.875			2258.5	2165.9	2039.9	1992.2	1849.9	1830.6	1826.3
3.000			2548.6	2430.2	2271.2	2211.6	2036.0	2012.7	2007.3

(Continued)

TABLE A.27 (Continued)

Orifice Diameter, in.	4		6				8		
	3.826	4.026	4.897	5.189	5.761	6.065	7.625	7.981	8.071
3.125			2875.2	2724.4	2524.3	2450.1	2234.7	2206.4	2199.9
3.250			3244.8	3052.8	2801.8	2709.9	2446.5	2412.4	2404.7
3.375			3665.6	3420.9	3106.9	2993.3	2672.5	2631.6	2622.3
3.500				3835.7	3443.0	3303.0	2913.7	2864.7	2853.7
3.625				4305.7	3914.4	3642.3	3171.1	3112.7	3099.6
3.750					4226.3	4014.8	3446.0	3376.6	3361.0
3.875					4684.9	4425.1	3739.9	3657.6	3639.2
4.000					5197.7	4878.4	4054.2	3957.0	3935.2
4.250							4751.4	4616.6	4586.6
4.500							5554.7	5369.0	5327.9
4.750							6485.3	6231.1	6175.2
5.000							7571.4	7224.3	7148.7
5.250							8850.3	8376.3	8274.0
5.500								9723.8	9585.1

Orifice Diameter, in.	10			12			18		
	9.564	10.020	10.136	11.376	11.938	12.090	14.688	15.000	15.250
1.000	202.16								
1.125	256.22	256.01	255.96						
1.250	316.90	316.56	316.49	315.84	315.57	315.51			
1.375	384.29	383.79	383.68	382.66	382.30	382.22			
1.500	458.52	457.79	457.63	456.16	455.64	455.52	453.92	453.78	
1.625	539.72	538.69	538.45	536.38	535.66	535.48	533.27	533.07	532.93
1.750	628.03	626.61	626.29	623.44	622.45	622.20	619.18	618.92	618.73
1.875	723.61	721.70	721.27	717.43	716.10	715.78	711.73	711.39	711.13
2.000	826.63	824.12	823.54	818.48	816.73	816.30	810.99	810.53	810.19
2.125	937.28	934.02	933.27	926.72	924.44	923.88	917.01	916.43	915.99
2.250	1,055.7	1,051.6	1,050.6	1,042.3	1,039.4	1,038.7	1,029.9	1,092.6	1,028.6
2.375	1,182.2	1,177.0	1,175.8	1,165.3	1,161.6	1,160.7	1,149.7	1,148.8	1,148.1
2.500	1,316.9	1,310.5	1,309.0	1,295.9	1,291.4	1,290.2	1.276.5	1.275.4	1,274.5
2.625	1,460.0	1,452.1	1,450.3	1,434.3	1,428.7	1,427.4	1,410.5	1,409.1	1,408.0
2.750	1,611.8	1,602.3	1,600.1	1,580.7	1,573.9	1,572.2	1,551.7	1,549.9	1,548.6
2.875	1,772.5	1,761.0	1,758.4	1,735.1	1,726.9	1,724.9	1,700.1	1,698.1	1,696.5
3.000	1,942.5	1,928.8	1,925.6	1,897.8	1,888.1	1,885.7	1,856.1	1,853.6	1,851.7
3.125	2,122.1	2,105.7	2,102.0	2,069.0	2,057.5	2,054.7	2,019.5	2,016.6	2,014.3
3.250	2,311.6	2,292.2	2,287.8	2,248.9	2,235.4	2,232.1	2,190.7	2,187.2	2,184.5
3.375	2,511.5	2,488.6	2,483.4	2,437.7	2,421.8	2,418.0	2,369.6	2,365.5	2,362.4
3.500	2,722.3	2,695.3	2,689.1	2,635.6	2,617.2	2,612.6	2,556.5	2,551.7	2,548.1
3.625	2,944.3	2,912.7	2,905.5	2,843.0	2,821.6	2,816.3	2,751.4	2,745.9	2,741.7
3.750	3,178.1	3,141.2	3,132.7	3,060.2	3,035.3	3,029.3	2,954.5	2,948.1	2,943.3
3.875	3,424.3	3,381.3	3,371.5	3,287.4	3,258.7	3,251.7	3,165.9	3,158.6	3,153.1
4.000	3,683.5	3,633.5	3,622.1	3,524.9	3,492.0	3,483.9	3,385.8	3,377.5	3,371.2
4.250	4,243.8	4,176.8	4,161.6	4,032.8	3,989.5	3,979.0	3,851.6	3,840.9	3,832.8
4.500	4,865.1	4,776.2	4,756.1	4,587.1	4,530.8	4,517.2	4,353.4	4,339.8	4,329.6

(Continued)

TABLE A.27 (Continued)

Orifice Diameter, in.	10			12			18		
	9.564	10.020	10.136	11.376	11.938	12.090	14.688	15.000	15.250
4.750	5,554.9	5,437.9	5,411.5	5,191.5	5,119.0	5,101.5	4,892.9	4,875.8	4,862.9
5.000	6,322.2	6,169.2	6,134.9	5,850.6	5,757.8	5,735.4	5,471.9	5,450.5	5,434.3
5.250	7,177.7	6,978.9	6,934.4	6,569.4	6,451.5	6,423.2	6,092.5	6,065.9	6,045.9
5.500	8,134.1	7,877.2	7,820.0	7,354.1	7,205.1	7,169.5	6,757.0	6,724.1	6,699.4
5.750	9,207.0	8,876.3	8,803.1	8,211.4	8,024.2	7,979.6	7,468.0	7,427.6	7,397.4
6.000	10,415	9,991.2	9,897.8	9,149.5	8,915.4	8,859.8	8,228.5	8,179.2	8,142.3
6.250	11,783	11,240	11,121	10,178	9,886.1	9,817.2	9,041.6	8,981.7	8,937.0
6.500	13,340	12,644	12,492	11,307	10,945	10,860	9,911.2	9,838.7	9,764.7
6.750		14,230	14,038	12,550	12,103	11,998	10,841	10,754	10,689
7.000		16,035	15,790	13,923	13,371	13,242	11,837	11,732	11,654
7.250				15,442	14,762	14,604	12,902	12,777	12,684
7.500				17,131	16,294	16,101	14,044	13,894	13,783
7.750				19,017	17,986	17,750	15,268	15,090	14,959
8.000					19,861	19,572	16,583	16,371	16,216
8.250					21,947	21,593	17,996	17,746	17,561
8.500							19,517	19,221	19,003
8.750							21,156	20,807	20,551
9.000							22,926	22,515	22,214
9.250							24,841	24,356	24,003
9.500							26,917	26,346	25,932
9.750							29,172	28,501	28,014
10.000							31,629	30,839	30,268
10.250							34,315	33,383	32,713
10.500								36,160	35,372

Orifice Diameter, in.	20			24			30		
	18.814	19.000	19.250	22.626	23.000	23.250	28.628	29.000	29.250
2.000	806.71	806.57	806.40						
2.125	911.51	911.35	911.13						
2.250	1,022.9	1,022.7	1,022.4						
2.375	1,141.0	1,140.7	1,140.4	1,136.8	1,136.5	1,136.3			
2.500	1,265.7	1,265.4	1,265.0	1,260.6	1,260.2	1,259.9			
2.625	1,397.2	1,396.8	1,396.3	1,390.9	1,390.5	1,390.2			
2.750	1,535.5	1,535.0	1,534.4	1,527.9	1,527.3	1,527.0			
2.875	1,680.7	1,680.1	1,679.3	1,671.5	1,670.9	1,670.4	1,663.8		
3.000	1,832.7	1,832.1	1,831.2	1,821.9	1,821.1	1,820.6	1,812.7	1,812.3	1,812.0
3.125	1,991.8	1,991.0	1,990.0	1,978.9	1,978.0	1,977.4	1,968.1	1,967.7	1,967.4
3.250	2,158.0	2,157.0	2,155.8	2,142.8	2,141.7	2,141.0	2,130.2	2,129.6	2,129.3
3.375	2,331.3	2,330.2	2,328.7	2,313.5	2,312.3	2,311.5	2,298.8	2,298.2	2,297.7
3.500	2,511.9	2,510.6	2,508.8	2,491.2	2,489.7	2,488.8	2,474.1	2,473.3	2,472.9
3.625	2,699.7	2,698.2	2,696.2	2,675.8	2,674.0	2,673.0	2,656.0	2,655.2	2.654.6
3.750	2,895.0	2,893.2	2,890.9	2,867.4	2,865.4	2,864.1	2,844.6	2,843.7	2,843.0
3,875	3,097.7	3,095.7	3,093.0	3,066.0	3,063.8	3,062.3	3,040.0	3,038.9	3,038.2
4.000	3,308.0	3,305.7	3,302.7	3,271.8	3,269.2	3,267.6	3,242.2	3,240.9	3,240.1
4.250	3,751.6	3,748.7	3,744.8	3,705.0	3,701.7	3,699.6	3,666.9	3.665.3	3,664.3
4.500	4,226.8	4,223.0	4,218.1	4,167.6	4,163.4	4,160.7	4,119.3	4,117.3	4,116.0
4.750	4,734.1	4,729.4	4,723.3	4,660.0	4,654.8	4,651.4	4,599.6	4,597.1	4,595.4
5.000	5,274.6	5,268.7	5,261.1	5,183.0	5,176.4	5,172.3	5,108.2	5,105.0	5,103.0
5.250	5,849.0	5,841.9	5,832.6	5,737.1	5,729.1	5,723.9	5,645.4	5,641.4	5,639.1
5.500	6,458.6	6,449.9	6,438.7	6,322.9	6,313.2	6,307.0	6,211.8	6,207.2	6,204.2
5.750	7,104.4	7,094.0	7,080.4	6,941.3	6,929.7	6,922.2	6,807.7	6,802.1	6,798.5
6.000	7,787.9	7,775.4	7,759.1	7,592.8	7,579.0	7,570.1	7,433.6	7,426.9	7,422.6

(Continued)

TABLE A.27 (Continued)

Orifice Diameter, in.	20			24			30		
	18.814	19.000	19.250	22.626	23.000	23.250	28.628	29.000	29.250
6.250	8,510.4	8,495.4	8,476.0	8,278.3	8,262.0	8,251.5	8,089.9	8,082.0	8,076.9
6.500	9,273.4	9,255.6	9,232.5	8,998.8	8,979.5	8,967.1	8,777.2	8,768.0	8,761.9
6.750	10,079	10,058	10,030	9,755.0	9,732.4	9,717.9	9,496.0	9,485.2	9,478.1
7.000	10,928	10,903	10,871	10,548	10,522	10,505	10,247	10,234	10,226
7.250	11,823	11,794	11,756	11,379	11,348	11,329	11,030	11,016	11,006
7.500	12,767	12,733	12,689	12,249	12,214	12,191	11,847	11,830	11,819
7.750	13,762	13,722	13,670	13,160	13,119	13,093	12,697	12,678	12,665
8.000	14,810	14,763	14,703	14,113	14,065	14,035	13,582	13,560	13,546
8.250	15,914	15,860	15,791	15,109	15,054	15,020	14,501	14,477	14,461
8.500	17,078	17,015	16,935	16,150	16,087	16,048	15,457	15,429	15,411
8.750	18,305	18,232	18,129	17,237	17,166	17,121	16,450	16,418	16,397
9.000	19,598	19,515	19,408	18,373	18,292	18,241	17,480	17,444	17,421
9.250	20,963	20,866	20,743	19,560	19,468	19,409	18,548	18,508	18,482
9.500	22,402	22,292	22,151	20,800	20,695	20,628	19,656	19,611	19,582
9.750	23,923	23,796	23,634	22,094	21,976	21,900	20,805	20,754	20,721
10.000	25,529	25,384	25,198	23,447	23,312	23,227	21,995	21,938	21,901
10.250	27,227	27,061	26,849	24,859	24,708	24,612	23,228	23,165	23,124
10.500	29,023	28,834	28,592	26,335	26,164	26,056	24,505	24,434	24,358
10.750	30,925	30,709	30,434	27,878	27,685	27,563	25,827	25,749	25,698
11.000	32,940	32,694	32,381	29,490	29,273	29,136	27,196	27,109	27,052
11.250	35,078	34,798	34,443	31,175	30,932	30,779	28,613	28,156	28,453
11.500	37,348	37,030	36,626	32,938	32,666	32,494	30,080	29,972	29,903
11.750	39,761	39,400	38,941	34,783	34,478	34,285	31,598	31,479	31,402
12.000	42,330	41,920	41,399	36,714	36,373	36,158	33,169	33,038	32,953

Orifice Diameter, in.	20			24			30		
	18.814	19.000	19.250	22.626	23.000	23.250	28.628	29.000	29.250
12.500	47,991	47,461	46,790	40,855	40,429	40,161	36,478	36,318	36,215
13.000	54,463	53,778	52,914	45,406	44,877	44,544	40,024	39,829	39,704
13.500				50,420	49,763	49,352	43,823	43,589	43,437
14.000				55,959	55,147	54,638	47,898	47,615	47,433
14.500				62,099	61,094	60,468	52,271	51,932	51,714
15.000				68,929	67,687	66,915	56,967	56,562	56,301
15.500				76,562	75,025	74,074	62,017	61,533	61,223
16.000					83,231	82,055	67,453	66,878	66,509
16.500							73,314	72,630	72,193
17.000							79,641	78,831	78,313
17.500							86,485	85,525	84,913
18.000							93,900	92,765	92,042
18.500							101,950	100,610	99,758
19.000							110,720	109,130	108,130
19.500							120,300	118,420	117,230
20.000							130,780	128,560	127,150

TABLE A.28

"*b*" Values for Reynolds Number Factor F_r Determination—Pipe Taps

$$F_r = 1 + \frac{b}{\sqrt{h_w pf}}$$

Pipe Sizes—Nominal and Published Inside Diameters, Inches

Orifice Diameter, in.	2			3				4	
	1.689	1.939	2.067	2.300	2.626	2.900	3.068	3.152	3.438
0.250	0.1105	0.1091	0.1087	0.1081	0.1078	0.1078	0.1080	0.1081	0.1084
0.375	0.0890	0.0878	0.0877	0.0879	0.0888	0.0898	0.0905	0.0908	0.0918
0.050	0.0758	0.0734	0.0729	0.0728	0.0737	0.0750	0.0758	0.0763	0.0778
0.625	0.0693	0.0647	0.0635	0.0624	0.0624	0.0634	0.0642	0.0646	0.0662
0.750	0.0675	0.0608	0.0586	0.0559	0.0546	0.0548	0.0552	0.0555	0.0568
0.875	0.0684	0.0602	0.0570	0.0528	0.0497	0.0488	0.0488	0.0489	0.0496
1.000	0.0702	0.0614	0.0576	0.0522	0.0473	0.0452	0.0445	0.0443	0.0443
1.125	0.0708	0.0635	0.0595	0.0532	0.0469	0.0435	0.0422	0.0417	0.0407
1.250		0.0650	0.0616	0.0552	0.0478	0.0434	0.0414	0.0406	0.0387
1.375			0.0629	0.0574	0.0496	0.0443	0.0418	0.0408	0.0379
1.500				0.0590	0.0518	0.0460	0.0431	0.0418	0.0382
1.625					0.0539	0.0482	0.0450	0.0435	0.0392
1.750					0.0553	0.0504	0.0471	0.0456	0.0408
1.875						0.0521	0.0492	0.0477	0.0427
2.000						0.0532	0.0508	0.0495	0.0448
2.125							0.0519	0.0509	0.0467
2.250									0.0483
2.375									0.0494

Orifice Diameter, in.	4		6				8		
	3.826	4.026	4.897	5.189	5.761	6.065	7.625	7.981	8.071
0.250	0.1087	0.1091							
0.375	0.0932	0.0939							
0.500	0.0799	0.0810	0.0850	0.0862	0.0883	0.0895			
0.625	0.0685	0.0697	0.0747	0.0762	0.0789	0.0802			
0.750	0.0590	0.0602	0.0655	0.0672	0.0703	0.0718			
0.875	0.0513	0.0524	0.0575	0.0592	0.0625	0.0642	0.0716	0.0730	0.0733
1.000	0.0453	0.0461	0.0506	0.0523	0.0556	0.0573	0.0652	0.0668	0.0662
1.125	0.0408	0.0412	0.0448	0.0464	0.0495	0.0512	0.0592	0.0609	0.0613
1.250	0.0376	0.0377	0.0401	0.0413	0.0442	0.0458	0.0538	0.0555	0.0560
1.375	0.0358	0.0353	0.0363	0.0373	0.0397	0.0412	0.0489	0.0506	0.0510
1.500	0.0350	0.0340	0.0334	0.0340	0.0360	0.0372	0.0445	0.0462	0.0466

TABLE A.28 (Continued)

Orifice Diameter, in.	4		6				8		
	3.826	4.026	4.897	5.189	5.761	6.065	7.625	7.981	8.071
1.625	0.0351	0.0336	0.0313	0.0315	0.0329	0.0339	0.0404	0.0421	0.0425
1.750	0.0358	0.0340	0.0300	0.0298	0.0304	0.0311	0.0369	0.0384	0.0388
1.875	0.0371	0.0349	0.0293	0.0287	0.0285	0.0290	0.0338	0.0352	0.0355
2.000	0.0388	0.0363	0.0292	0.0281	0.0273	0.0273	0.0311	0.0323	0.0327
2.125	0.0407	0.0360	0.0297	0.0281	0.0265	0.0262	0.0288	0.0298	0.0301
2.250	0.0427	0.0398	0.0305	0.0285	0.0261	0.0258	0.0268	0.0277	0.0280
2.375	0.0445	0.0417	0.0316	0.0293	0.0262	0.0253	0.0252	0.0259	0.0261
2.500	0.0460	0.0435	0.0330	0.0304	0.0267	0.0254	0.0239	0.0244	0.0246
2.625	0.0472	0.0450	0.0345	0.0317	0.0274	0.0258	0.0230	0.0232	0.0233
2.750		0.0462	0.0362	0.0331	0.0264	0.0265	0.0224	0.0224	0.0224
2.875			0.0379	0.0347	0.0295	0.0274	0.0220	0.0218	0.0218
3.000			0.0395	0.0364	0.0308	0.0285	0.0219	0.0214	0.0213
3.125			0.0410	0.0380	0.0323	0.0297	0.0220	0.0213	0.0211
3.250			0.0422	0.0394	0.0338	0.0311	0.0223	0.0214	0.0212
3.375			0.0432	0.0408	0.0353	0.0325	0.0228	0.0216	0.0214
3.500				0.0419	0.0367	0.0339	0.0235	0.0221	0.0218
3.625				0.0428	0.0381	0.0354	0.0243	0.0227	0.0224
3.750					0.0393	0.0367	0.0252	0.0234	0.0230
3.875					0.0404	0.0380	0.0262	0.0243	0.0238
4.000					0.0413	0.0391	0.0273	0.0252	0.0246
4.250							0.0296	0.0273	0.0268
4.500							0.0321	0.0296	0.0290
4.750							0.0344	0.0320	0.0314
5.000							0.0364	0.0342	0.0336
5.250							0.0381	0.0361	0.0356
5.500								0.0377	0.0372

Orifice Diameter, in.	10			12			16		
	9.564	10.020	10.136	11.376	11.938	12.090	14.688	15.000	15.250
1.000	0.0728								
1.125	0.0674	0.0690	0.0694						
1.250	0.0624	0.0641	0.0646	0.0687	0.0704	0.0708			
1.375	0.0576	0.0594	0.0599	0.0643	0.0661	0.0666			
1.500	0.0532	0.0550	0.0555	0.0601	0.0620	0.0625	0.0697	0.0705	
1.625	0.0490	0.0509	0.0514	0.0561	0.0580	0.0585	0.0662	0.0670	0.0676

(Continued)

TABLE A.28 (Continued)

Orifice Diameter, in.	10			12			16		
	9.564	10.020	10.136	11.376	11.938	12.090	14.688	15.000	15.250
1.750	0.0452	0.0471	0.0476	0.0523	0.0543	0.0548	0.0628	0.0636	0.0642
1.875	0.0417	0.0436	0.0440	0.0488	0.0508	0.0513	0.0594	0.0603	0.0610
2.000	0.0385	0.0403	0.0407	0.0454	0.0475	0.0480	0.0563	0.0572	0.0578
2.125	0.0355	0.0372	0.0377	0.0423	0.0443	0.0449	0.0532	0.0541	0.0548
2.250	0.0329	0.0345	0.0349	0.0394	0.0414	0.0419	0.0503	0.0512	0.0519
2.375	0.0305	0.0320	0.0324	0.0367	0.0387	0.0392	0.0475	0.0484	0.0492
2.500	0.0283	0.0298	0.0301	0.0342	0.0361	0.0366	0.0449	0.0458	0.0466
2.625	0.0265	0.0277	0.0281	0.0319	0.0337	0.0342	0.0424	0.0433	0.0440
2.750	0.0248	0.0260	0.0262	0.0298	0.0316	0.0320	0.0400	0.0409	0.0417
2.875	0.0234	0.0244	0.0246	0.0279	0.0295	0.0300	0.0378	0.0387	0.0394
3.000	0.0222	0.0230	0.0232	0.0262	0.0277	0.0281	0.0356	0.0365	0.0372
3.125	0.0212	0.0218	0.0220	0.0244	0.0260	0.0264	0.0336	0.0345	0.0352
3.250	0.0204	0.0209	0.0210	0.0232	0.0245	0.0249	0.0317	0.0326	0.0332
3.375	0.0199	0.0201	0.0202	0.0220	0.0232	0.0235	0.0300	0.0308	0.0314
3.500	0.0195	0.0195	0.0196	0.0210	0.0220	0.0222	0.0263	0.0291	0.0297
3.625	0.0193	0.0191	0.0191	0.0200	0.0209	0.0212	0.0268	0.0275	0.0281
3.750	0.0192	0.0188	0.0188	0.0193	0.0200	0.0202	0.0254	0.0261	0.0267
3.875	0.0193	0.0187	0.0186	0.0187	0.0192	0.0194	0.0240	0.0247	0.0253
4.000	0.0195	0.0187	0.0186	0.0182	0.0185	0.0187	0.0228	0.0235	0.0240
4.250	0.0203	0.0192	0.0189	0.0176	0.0176	0.0177	0.0207	0.0213	0.0217
4.500	0.0215	0.0200	0.0197	0.0175	0.0172	0.0171	0.0190	0.0194	0.0198
4.750	0.0230	0.0212	0.0208	0.0178	0.0171	0.0170	0.0176	0.0180	0.0182
5.000	0.0248	0.0228	0.0223	0.0185	0.0174	0.0173	0.0166	0.0168	0.0170
5.250	0.0267	0.0244	0.0239	0.0194	0.0181	0.0178	0.0160	0.0161	0.0162
5.500	0.0287	0.0263	0.0257	0.0207	0.0190	0.0186	0.0156	0.0156	0.0156
5.750	0.0307	0.0282	0.0276	0.0221	0.0202	0.0197	0.0155	0.0154	0.0153
6.000	0.0326	0.0302	0.0295	0.0231	0.0215	0.0210	0.0157	0.0154	0.0153
6.250	0.0343	0.0320	0.0316	0.0253	0.0230	0.0224	0.0161	0.0157	0.0154
6.500	0.0358	0.0336	0.0331	0.0270	0.0246	0.0239	0.0167	0.0162	0.0159
6.750		0.0351	0.0346	0.0288	0.0262	0.0256	0.0174	0.0169	0.0164
7.000		0.0363	0.0359	0.0304	0.0279	0.0272	0.0184	0.0177	0.0172
7.250				0.0320	0.0295	0.0288	0.0195	0.0187	0.0181
7.500				0.0334	0.0310	0.0304	0.0206	0.0198	0.0191
7.750				0.0347	0.0325	0.0318	0.0219	0.0209	0.0202
8.000					0.0338	0.0332	0.0232	0.0222	0.0214
8.250					0.0349	0.0344	0.0246	0.0235	0.0227
8.500							0.0259	0.0248	0.0240

TABLE A.28 (Continued)

Orifice Diameter, in.	10			12			16		
	9.564	10.020	10.136	11.376	11.938	12.090	14.688	15.000	15.250
8.750							0.0273	0.0262	0.0253
9.000							0.0286	0.0276	0.0267
9.250							0.0299	0.0288	0.0280
9.500							0.0311	0.0300	0.0292
9.750							0.0322	0.0312	0.0304
10.000							0.0332	0.0323	0.0315
10.250							0.0341	0.0333	0.0326
10.500								0.0341	0.0335

Orifice Diameter, in.	20			24			30		
	18.814	19.000	19.250	22.626	23.000	23.250	28.628	29.000	29.250
2.000	0.0663	0.0667	0.0672						
2.125	0.0635	0.0639	0.0644						
2.250	0.0609	0.0613	0.0618						
2.375	0.0583	0.0588	0.0593	0.0658	0.0665	0.0669			
2.500	0.0558	0.0562	0.0568	0.0635	0.0642	0.0646			
2.625	0.0534	0.0539	0.0544	0.0613	0.0620	0.0624			
2.750	0.0510	0.0515	0.0520	0.0591	0.0598	0.0603			
2.875	0.0488	0.0492	0.0498	0.0570	0.0577	0.0582	0.0667		
3.000	0.0466	0.0470	0.0476	0.0549	0.0556	0.0561	0.0649	0.0654	0.0657
3.125	0.0445	0.0449	0.0455	0.0529	0.0536	0.0541	0.0630	0.0636	0.0639
3.250	0.0425	0.0429	0.0435	0.0509	0.0516	0.0521	0.0613	0.0616	0.0622
3.375	0.0406	0.0410	0.0416	0.0490	0.0497	0.0502	0.0595	0.0601	0.0604
3.500	0.0387	0.0391	0.0397	0.0471	0.0479	0.0484	0.0578	0.0584	0.0587
3.625	0.0369	0.0373	0.0379	0.0454	0.0461	0.0466	0.0561	0.0567	0.0571
3.750	0.0352	0.0356	0.0362	0.0436	0.0444	0.0449	0.0545	0.0550	0.0554
3.875	0.0336	0.0340	0.0346	0.0419	0.0427	0.0432	0.0528	0.0534	0.0538
4.000	0.0320	0.0324	0.0330	0.0403	0.0411	0.0416	0.0513	0.0518	0.0522
4.250	0.0291	0.0295	0.0301	0.0372	0.0380	0.0385	0.0482	0.0488	0.0492
4.500	0.0265	0.0269	0.0274	0.0343	0.0351	0.0356	0.0453	0.0459	0.0463
4.750	0.0242	0.0246	0.0250	0.0316	0.0324	0.0328	0.0425	0.0431	0.0435
5.000	0.0221	0.0225	0.0229	0.0292	0.0299	0.0303	0.0399	0.0405	0.0409
5.250	0.0203	0.0206	0.0210	0.0269	0.0276	0.0280	0.0374	0.0380	0.0384
5.500	0.0188	0.0190	0.0194	0.0248	0.0255	0.0259	0.0350	0.0356	0.0360

(Continued)

TABLE A.28 (Continued)

Orifice Diameter, in.	20			24			30		
	18.814	19.000	19.250	22.626	23.000	23.250	28.628	29.000	29.250
5.750	0.0175	0.0177	0.0180	0.0230	0.0236	0.0240	0.0328	0.0334	0.0338
6.000	0.0164	0.0165	0.0168	0.0212	0.0218	0.0222	0.0307	0.0313	0.0317
6.250	0.0155	0.0156	0.0158	0.0197	0.0202	0.0206	0.0287	0.0293	0.0297
6.500	0.0148	0.0149	0.0150	0.0184	0.0189	0.0192	0.0269	0.0274	0.0278
6.750	0.0143	0.0144	0.0145	0.0172	0.0176	0.0179	0.0252	0.0257	0.0260
7.000	0.0141	0.0141	0.0141	0.0162	0.0166	0.0168	0.0236	0.0241	0.0244
7.250	0.0140	0.0140	0.0139	0.0153	0.0156	0.0158	0.0221	0.0226	0.0229
7.500	0.0140	0.0140	0.0139	0.0146	0.0148	0.0150	0.0207	0.0212	0.0215
7.750	0.0142	0.0141	0.0140	0.0140	0.0142	0.0144	0.0195	0.0199	0.0202
8.000	0.0146	0.0144	0.0142	0.0136	0.0138	0.0138	0.0183	0.0187	0.0190
8.250	0.0151	0.0148	0.0146	0.0133	0.0134	0.0132	0.0173	0.0177	0.0179
8.500	0.0156	0.0154	0.0151	0.0132	0.0132	0.0130	0.0164	0.0167	0.0169
8.750	0.0163	0.0160	0.0157	0.0131	0.0130	0.0130	0.0155	0.0158	0.0161
9.000	0.0171	0.0168	0.0163	0.0131	0.0130	0.0130	0.0148	0.0151	0.0153
9.250	0.0180	0.0176	0.0171	0.0133	0.0131	0.0130	0.0142	0.0144	0.0146
9.500	0.0189	0.0185	0.0180	0.0136	0.0133	0.0132	0.0136	0.0138	0.0140
9.750	0.0198	0.0194	0.0189	0.0139	0.0136	0.0134	0.0132	0.0133	0.0134
10.000	0.0209	0.0204	0.0198	0.0143	0.0140	0.0135	0.0128	0.0129	0.0130
10.250	0.0219	0.0214	0.0208	0.0148	0.0144	0.0142	0.0125	0.0126	0.0127
10.500	0.0230	0.0225	0.0219	0.0154	0.0150	0.0147	0.0123	0.0124	0.0124
10.750	0.0241	0.0236	0.0229	0.0160	0.0155	0.0152	0.0122	0.0122	0.0122
11.000	0.0252	0.0247	0.0240	0.0168	0.0162	0.0158	0.0121	0.0121	0.0121
11.250	0.0263	0.0261	0.0251	0.0175	0.0169	0.0165	0.0122	0.0121	0.0121
11.500	0.0273	0.0268	0.0262	0.0183	0.0176	0.0172	0.0122	0.0121	0.0122
11.750	0.0284	0.0278	0.0272	0.0191	0.0184	0.0180	0.0124	0.0123	0.0122
12.000	0.0293	0.0288	0.0282	0.0200	0.0192	0.0190	0.0126	0.0124	0.0123
12.500	0.0312	0.0307	0.0301	0.0218	0.0210	0.0204	0.0132	0.0130	0.0128
13.000	0.0327	0.0323	0.0318	0.0236	0.0228	0.0222	0.0140	0.0137	0.0135
13.500				0.0254	0.0246	0.0240	0.0150	0.0146	0.0143
14.000				0.0272	0.0264	0.0258	0.0161	0.0156	0.0153
14.500				0.0289	0.0280	0.0275	0.0173	0.0168	0.0165
15.000				0.0304	0.0296	0.0291	0.0166	0.0181	0.0177

TABLE A.28 (Continued)

Orifice Diameter, in.	20			24			30		
	18.814	19.000	19.250	22.626	23.000	23.250	28.628	29.000	29.250
15.500				0.0310	0.0311	0.0306	0.0200	0.0194	0.0190
16.000					0.0323	0.0318	0.0215	0.0209	0.0204
16.500							0.0230	0.0223	0.0219
17.000							0.0244	0.0238	0.0233
17.500							0.0259	0.0252	0.0248
18.000							0.0272	0.0266	0.0261
18.500							0.0286	0.0279	0.0275
19.000							0.0298	0.0292	0.0288
19.500							0.0309	0.0303	0.0299
20.000							0.0318	0.0313	0.0310

TABLE A.29
Y_1 Expansion Factors—Pipe Taps
Static Pressure Taken from Upstream Taps

$\frac{h_w}{p_{f1}}$ Ratio	$\beta = \frac{d}{D}$ Ratio 0.1	0.2	0.3	0.4	0.45	0.50	0.52	0.54	0.56	0.58	0.60	0.61	0.62	0.63	0.64	0.65	0.66	0.67	0.68	0.69	0.70
0.0	1.0000	1.0000	1.0000	1.0000	1.0000	1.0000	1.0000	1.0000	1.0000	1.0000	1.0000	1.0000	1.0000	1.0000	1.0000	1.0000	1.0000	1.0000	1.0000	1.0000	1.0000
0.1	0.9990	0.9989	0.9988	0.9985	0.9984	0.9982	0.9981	0.9980	0.9979	0.9978	0.9977	0.9976	0.9976	0.9975	0.9974	0.9973	0.9972	0.9971	0.9970	0.9969	0.9968
0.2	0.9981	0.9979	0.9976	0.9971	0.9968	0.9964	0.9962	0.9961	0.9959	0.9957	0.9954	0.9953	0.9951	0.9950	0.9948	0.9947	0.9945	0.9943	0.9941	0.9938	0.9935
0.3	0.9971	0.9968	0.9964	0.9956	0.9952	0.9946	0.9944	0.9941	0.9938	0.9935	0.9931	0.9929	0.9927	0.9925	0.9923	0.9920	0.9917	0.9914	0.9911	0.9907	0.9903
0.4	0.9962	0.9958	0.9951	0.9942	0.9936	0.9928	0.9925	0.9921	0.9917	0.9913	0.9908	0.9906	0.9903	0.9900	0.9897	0.9893	0.9890	0.9886	0.9881	0.9876	0.9871
0.5	0.9952	0.9947	0.9939	0.9927	0.9919	0.9910	0.9906	0.9902	0.9897	0.9891	0.9885	0.9882	0.9879	0.9875	0.9871	0.9867	0.9862	0.9857	0.9851	0.9845	0.9839
0.6	0.9943	0.9937	0.9927	0.9913	0.9903	0.9892	0.9887	0.9882	0.9876	0.9870	0.9862	0.9859	0.9854	0.9850	0.9845	0.9840	0.9834	0.9828	0.9822	0.9814	0.9806
0.7	0.9933	0.9926	0.9915	0.9898	0.9887	0.9874	0.9869	0.9862	0.9856	0.9848	0.9840	0.9835	0.9830	0.9825	0.9819	0.9813	0.9807	0.9800	0.9792	0.9784	0.9774
0.8	0.9923	0.9916	0.9903	0.9883	0.9871	0.9857	0.9850	0.9843	0.9835	0.9826	0.9817	0.9811	0.9806	0.9800	0.9794	0.9787	0.9779	0.9771	0.9762	0.9753	0.9742
0.9	0.9914	0.9905	0.9891	0.9869	0.9855	0.9839	0.9831	0.9823	0.9814	0.9805	0.9794	0.9788	0.9782	0.9775	0.9768	0.9760	0.9752	0.9742	0.9733	0.9722	0.9710
1.0	0.9904	0.9895	0.9878	0.9854	0.9839	0.9821	0.9812	0.9803	0.9794	0.9783	0.9771	0.9764	0.9757	0.9750	0.9742	0.9733	0.9724	0.9714	0.9703	0.9691	0.9677
1.1	0.9895	0.9884	0.9866	0.9840	0.9823	0.9803	0.9794	0.9784	0.9773	0.9761	0.9748	0.9741	0.9733	0.9725	0.9716	0.9707	0.9696	0.9685	0.9673	0.9660	0.9645
1.2	0.9885	0.9874	0.9854	0.9825	0.9807	0.9785	0.9775	0.9764	0.9752	0.9739	0.9725	0.9717	0.9709	0.9700	0.9690	0.9680	0.9669	0.9757	0.9643	0.9629	0.9613
1.3	0.9876	0.9863	0.9842	0.9811	0.9791	0.9767	0.9756	0.9744	0.9732	0.9718	0.9702	0.9694	0.9685	0.9675	0.9664	0.9653	0.9641	0.9628	0.9614	0.9598	0.9581
1.4	0.9866	0.9853	0.9830	0.9796	0.9775	0.9749	0.9737	0.9725	0.9711	0.9696	0.9679	0.9670	0.9660	0.9650	0.9639	0.9627	0.9614	0.9599	0.9584	0.9567	0.9548
1.5	0.9857	0.9842	0.9818	0.9782	0.9758	0.9731	0.9719	0.9705	0.9690	0.9674	0.9656	0.9646	0.9636	0.9625	0.9613	0.9600	0.9586	0.9571	0.9554	0.9536	0.9516
1.6	0.9847	0.9832	0.9805	0.9767	0.9742	0.9713	0.9700	0.9685	0.9670	0.9652	0.9633	0.9623	0.9612	0.9600	0.9587	0.9573	0.9558	0.9542	0.9525	0.9505	0.9484
1.7	0.9837	0.9821	0.9793	0.9752	0.9726	0.9695	0.9681	0.9666	0.9649	0.9631	0.9610	0.9599	0.9587	0.9575	0.9561	0.9547	0.9531	0.9514	0.9495	0.9474	0.9452
1.8	0.9828	0.9811	0.9781	0.9738	0.9710	0.9677	0.9662	0.9646	0.9628	0.9609	0.9587	0.9576	0.9563	0.9550	0.9535	0.9520	0.9503	0.9485	0.9465	0.9443	0.9419
1.9	0.9818	0.9800	0.9769	0.9723	0.9694	0.9659	0.9643	0.9626	0.9608	0.9587	0.9565	0.9552	0.9539	0.9525	0.9510	0.9493	0.9476	0.9456	0.9435	0.9412	0.9387
2.0	0.9809	0.9790	0.9757	0.9709	0.9678	0.9641	0.9625	0.9607	0.9587	0.9566	0.9542	0.9529	0.9515	0.9500	0.9484	0.9467	0.9448	0.9428	0.9406	0.9381	0.9355
2.1	0.9799	0.9779	0.9745	0.9694	0.9662	0.9623	0.9606	0.9587	0.9566	0.9544	0.9519	0.9505	0.9490	0.9475	0.9458	0.9440	0.9420	0.9399	0.9376	0.9351	0.9323
2.2	0.9790	0.9768	0.9732	0.9680	0.9646	0.9605	0.9587	0.9567	0.9546	0.9522	0.9496	0.9481	0.9466	0.9450	0.9432	0.9413	0.9393	0.9371	0.9346	0.9320	0.9290
2.3	0.9780	0.9758	0.9720	0.9665	0.9630	0.9587	0.9568	0.9548	0.9525	0.9500	0.9473	0.9458	0.9442	0.9425	0.9406	0.9387	0.9365	0.9342	0.9317	0.9289	0.9258
2.4	0.9770	0.9747	0.9708	0.9650	0.9613	0.9570	0.9550	0.9528	0.9505	0.9479	0.9450	0.9434	0.9418	0.9400	0.9381	0.9360	0.9338	0.9313	0.9287	0.9258	0.9226
2.5	0.9761	0.9737	0.9696	0.9636	0.9597	0.9552	0.9531	0.9508	0.9484	0.9457	0.9427	0.9411	0.9393	0.9375	0.9355	0.9333	0.9310	0.9285	0.9257	0.9227	0.9194

$\frac{h_w}{p_{f1}}$ Ratio	$\beta = \frac{d}{D}$ Ratio 0.1	0.2	0.3	0.4	0.45	0.50	0.52	0.54	0.56	0.58	0.60	0.61	0.62	0.63	0.64	0.65	0.66	0.67	0.68	0.69	0.70
2.6	0.9751	0.9726	0.9684	0.9621	0.9581	0.9534	0.9512	0.9489	0.9463	0.9435	0.9404	0.9387	0.9369	0.9350	0.9329	0.9307	0.9282	0.9256	0.9227	0.9196	0.9161
2.7	0.9742	0.9716	0.9672	0.9607	0.9565	0.9516	0.9493	0.9469	0.9443	0.9414	0.9381	0.9364	0.9345	0.9325	0.9303	0.9280	0.9255	0.9227	0.9198	0.9165	0.9129
2.8	0.9732	0.9705	0.9659	0.9592	0.9549	0.9498	0.9475	0.9449	0.9422	0.9392	0.9358	0.9340	0.9321	0.9300	0.9277	0.9253	0.9227	0.9199	0.9168	0.9134	0.9097
2.9	0.9723	0.9695	0.9647	0.9578	0.9533	0.9480	0.9456	0.9430	0.9401	0.9370	0.9335	0.9316	0.9296	0.9275	0.9252	0.9227	0.9200	0.9170	0.9138	0.9103	0.9064
3.0	0.9713	0.9684	0.9635	0.9563	0.9517	0.9462	0.9437	0.9410	0.9381	0.9348	0.9312	0.9293	0.9272	0.9250	0.9226	0.9200	0.9172	0.9142	0.9108	0.9072	0.9032
3.1	0.9704	0.9674	0.9623	0.9549	0.9501	0.9444	0.9418	0.9390	0.9360	0.9327	0.9290	0.9269	0.9248	0.9225	0.9200	0.9173	0.9144	0.9113	0.9079	0.9041	0.9000
3.2	0.9694	0.9663	0.9611	0.9534	0.9485	0.9426	0.9400	0.9371	0.9339	0.9305	0.9267	0.9246	0.9223	0.9200	0.9174	0.9147	0.9117	0.9084	0.9049	0.9010	0.8968
3.3	0.9684	0.9653	0.9599	0.9519	0.9469	0.9408	0.9381	0.9351	0.9319	0.9283	0.9244	0.9222	0.9199	0.9175	0.9148	0.9120	0.9089	0.9056	0.9019	0.8979	0.8935
3.4	0.9675	0.9642	0.9587	0.9505	0.9452	0.9390	0.9362	0.9331	0.9298	0.9261	0.9221	0.9199	0.9175	0.9150	0.9122	0.9093	0.9062	0.9027	0.8990	0.8948	0.8903
3.5	0.9665	0.9632	0.9574	0.9490	0.9436	0.9372	0.9343	0.9312	0.9277	0.9240	0.9198	0.9175	0.9151	0.9125	0.9097	0.9067	0.9034	0.8999	0.8960	0.8918	0.8871
3.6	0.9656	0.9621	0.9562	0.9476	0.9420	0.9354	0.9324	0.9292	0.9257	0.9218	0.9175	0.9151	0.9126	0.9100	0.9071	0.9040	0.9006	0.8970	0.8930	0.8887	0.8839
3.7	0.9646	0.9611	0.9550	0.9461	0.9404	0.9336	0.9306	0.9272	0.9236	0.9196	0.9152	0.9128	0.9102	0.9075	0.9045	0.9013	0.8979	0.8941	0.8900	0.8856	0.8806
3.8	0.9637	0.9600	0.9538	0.9447	0.9388	0.9318	0.9287	0.9253	0.9216	0.9175	0.9129	0.9104	0.9078	0.9050	0.9019	0.8987	0.8951	0.8913	0.8871	0.8825	0.8774
3.9	0.9627	0.9590	0.9526	0.9432	0.9372	0.9301	0.9268	0.9233	0.9195	0.9153	0.9106	0.9081	0.9054	0.9025	0.8993	0.8960	0.8924	0.8884	0.8841	0.8794	0.8742
4.0	0.9617	0.9579	0.9514	0.9417	0.9356	0.9283	0.9249	0.9213	0.9174	0.9131	0.9083	0.9057	0.9029	0.9000	0.8968	0.8933	0.8896	0.8856	0.8811	0.8763	0.8710

TABLE A.30

Y_2 Expansion Factors—Pipe Taps

Static Pressure, Taken from Downstream Taps

$\frac{h_w}{p_{f2}}$ Ratio	$\beta = \frac{d}{D}$ Ratio																				
	0.1	0.2	0.3	0.4	0.45	0.50	0.52	0.54	0.56	0.58	0.60	0.61	0.62	0.63	0.64	0.65	0.66	0.67	0.68	0.69	0.70
0.0	1.0000	1.0000	1.0000	1.0000	1.0000	1.0000	1.0000	1.0000	1.0000	1.0000	1.0000	1.0000	1.0000	1.0000	1.0000	1.0000	1.0000	1.0000	1.0000	1.0000	1.0000
0.1	1.0008	1.0008	1.0006	1.0003	1.0002	1.0000	0.9999	0.9998	0.9997	0.9996	0.9995	0.9994	0.9994	0.9993	0.9992	0.9991	0.9990	0.9989	0.9988	0.9987	0.9986
0.2	1.0017	1.0015	1.0012	1.0007	1.0004	1.0000	0.9999	0.9997	0.9995	0.9993	0.9990	0.9989	0.9988	0.9986	0.9985	0.9983	0.9981	0.9979	0.9977	0.9974	0.9972
0.3	1.0025	1.0023	1.0018	1.0010	1.0006	1.0000	0.9998	0.9995	0.9992	0.9989	0.9986	0.9984	0.9982	0.9979	0.9977	0.9974	0.9972	0.9969	0.9965	0.9962	0.9958
0.4	1.0034	1.0030	1.0024	1.0014	1.0008	1.0001	0.9997	0.9994	0.9990	0.9986	0.9981	0.9978	0.9976	0.9972	0.9969	0.9966	0.9962	0.9958	0.9954	0.9949	0.9944
0.5	1.0042	1.0038	1.0030	1.0018	1.0010	1.0001	0.9997	0.9992	0.9988	0.9982	0.9976	0.9973	0.9970	0.9966	0.9962	0.9958	0.9953	0.9948	0.9942	0.9936	0.9930
0.6	1.0051	1.0045	1.0036	1.0021	1.0012	1.0001	0.9996	0.9991	0.9985	0.9979	0.9972	0.9968	0.9964	0.9959	0.9954	0.9949	0.9944	0.9938	0.9931	0.9924	0.9916
0.7	1.0059	1.0053	1.0041	1.0025	1.0014	1.0002	0.9996	0.9990	0.9983	0.9975	0.9967	0.9962	0.9958	0.9953	0.9947	0.9941	0.9935	0.9928	0.9920	0.9912	0.9902
0.8	1.0068	1.0060	1.0047	1.0028	1.0016	1.0002	0.9995	0.9988	0.9980	0.9972	0.9962	0.9957	0.9952	0.9946	0.9940	0.9933	0.9926	0.9918	0.9909	0.9899	0.9889
0.9	1.0076	1.0068	1.0053	1.0032	1.0018	1.0002	0.9995	0.9987	0.9978	0.9969	0.9958	0.9952	0.9946	0.9940	0.9932	0.9925	0.9917	0.9908	0.9898	0.9887	0.9875
1.0	1.0085	1.0075	1.0059	1.0036	1.0021	1.0003	0.9994	0.9986	0.9976	0.9965	0.9954	0.9947	0.9940	0.9933	0.9925	0.9917	0.9908	0.9898	0.9887	0.9875	0.9862
1.1	1.0093	1.0083	1.0065	1.0039	1.0023	1.0003	0.9994	0.9984	0.9974	0.9962	0.9949	0.9942	0.9935	0.9927	0.9918	0.9909	0.9899	0.9888	0.9876	0.9863	0.9848
1.2	1.0102	1.0091	1.0071	1.0043	1.0025	1.0004	0.9994	0.9983	0.9972	0.9959	0.9945	0.9937	0.9929	0.9920	0.9911	0.9901	0.9890	0.9878	0.9865	0.9851	0.9835
1.3	1.0110	1.0098	1.0077	1.0047	1.0027	1.0004	0.9994	0.9982	0.9970	0.9956	0.9941	0.9932	0.9924	0.9914	0.9904	0.9893	0.9881	0.9868	0.9854	0.9839	0.9822
1.4	1.0119	1.0106	1.0083	1.0051	1.0030	1.0004	0.9993	0.9981	0.9968	0.9953	0.9936	0.9928	0.9918	0.9908	0.9897	0.9885	0.9872	0.9859	0.9844	0.9827	0.9809
1.5	1.0127	1.0113	1.0089	1.0054	1.0032	1.0005	0.9993	0.9980	0.9966	0.9950	0.9932	0.9923	0.9912	0.9902	0.9890	0.9877	0.9864	0.9849	0.9833	0.9815	0.9796
1.6	1.0136	1.0121	1.0096	1.0058	1.0034	1.0006	0.9993	0.9979	0.9964	0.9947	0.9928	0.9918	0.9907	0.9896	0.9883	0.9870	0.9855	9.9840	0.9822	0.9804	0.9783
1.7	1.0144	1.0128	1.0102	1.0062	1.0036	1.0006	0.9992	0.9978	0.9962	0.9944	0.9924	0.9913	0.9902	0.9889	0.9876	0.9862	0.9847	0.9830	0.9812	0.9792	0.9770
1.8	1.0153	1.0136	1.0108	1.0066	1.0039	1.0007	0.9992	0.9977	0.9960	0.9941	0.9920	0.9908	0.9896	0.9883	0.9870	0.9854	0.9838	0.9821	0.9801	0.9780	0.9757
1.9	1.0161	1.0144	1.0114	1.0070	1.0041	1.0008	0.9992	0.9976	0.9958	0.9938	0.9916	0.9904	0.9891	0.9877	0.9863	0.9847	0.9830	0.9811	0.9791	0.9769	0.9744
2.0	1.0170	1.0151	1.0120	1.0073	1.0044	1.0008	0.9992	0.9975	0.9956	0.9935	0.9912	0.9899	0.9886	0.9872	0.9856	0.9840	0.9822	0.9802	0.9781	0.9757	0.9732
2.1	1.0178	1.0159	1.0126	1.0077	1.0046	1.0009	0.9992	0.9974	0.9954	0.9932	0.9908	0.9895	0.9881	0.9866	0.9849	0.9832	0.9813	0.9793	0.9770	0.9746	0.9719
2.2	1.0187	1.0167	1.0132	1.0081	1.0048	1.0010	0.9992	0.9973	0.9952	0.9929	0.9904	0.9890	0.9876	0.9860	0.9843	0.9825	0.9805	0.9784	0.9760	0.9734	0.9706
2.3	1.0195	1.0174	1.0138	1.0085	1.0051	1.0010	0.9992	0.9972	0.9950	0.9927	0.9900	0.9886	0.9870	0.9854	0.9836	0.9817	0.9797	0.9774	0.9750	0.9723	0.9694
2.4	1.0204	1.0182	1.0144	1.0089	1.0053	1.0011	0.9992	0.9971	0.9949	0.9924	0.9896	0.9881	0.9865	0.9848	0.9830	0.9810	0.9789	0.9765	0.9740	0.9712	0.9681
2.5	1.0212	1.0189	1.0150	1.0093	1.0056	1.0012	0.9992	0.9971	0.9947	0.9921	0.9893	0.9877	0.9860	0.9842	0.9823	0.9803	0.9780	0.9756	0.9730	0.9701	0.9669

h_w / p_{f2} Ratio	$\beta = \frac{d}{D}$ Ratio																				
	0.1	0.2	0.3	0.4	0.45	0.50	0.52	0.54	0.56	0.58	0.60	0.61	0.62	0.63	0.64	0.65	0.66	0.67	0.68	0.69	0.70
2.6	1.0221	1.0197	1.0156	1.0097	1.0058	1.0013	0.9992	0.9970	0.9945	0.9919	0.9889	0.9873	0.9855	0.9837	0.9817	0.9796	0.9772	0.9747	0.9720	0.9690	0.9657
2.7	1.0229	1.0205	1.0162	1.0101	1.0061	1.0014	0.9992	0.9969	0.9944	0.9916	0.9885	0.9868	0.9850	0.9831	0.9811	0.9788	0.9764	0.9738	0.9710	0.9679	0.9644
2.8	1.0238	1.0212	1.0169	1.0104	1.0063	1.0014	0.9992	0.9968	0.9942	0.9914	0.9882	0.9864	0.9846	0.9826	0.9804	0.9781	0.9757	0.9730	0.9700	0.9668	0.9632
2.9	1.0246	1.0220	1.0175	1.0108	1.0066	1.0015	0.9992	0.9968	0.9941	0.9911	0.9878	0.9860	0.9841	0.9820	0.9798	0.9774	0.9749	0.9721	0.9690	0.9657	0.9620
3.0	1.0255	1.0228	1.0181	1.0112	1.0068	1.0016	0.9993	0.9967	0.9939	0.9908	0.9874	0.9856	0.9836	0.9815	0.9792	0.9767	0.9741	0.9712	0.9681	0.9646	0.9608
3.1	1.0264	1.0235	1.0187	1.0116	1.0071	1.0017	0.9993	0.9966	0.9938	0.9906	0.9871	0.9852	0.9831	0.9809	0.9786	0.9760	0.9733	0.9703	0.9671	0.9635	0.9596
3.2	1.0272	1.0243	1.0193	1.0120	1.0074	1.0018	0.9993	0.9966	0.9936	0.9904	0.9867	0.9848	0.9826	0.9804	0.9780	0.9754	0.9725	0.9695	0.9661	0.9625	0.9584
3.3	1.0280	1.0250	1.0199	1.0124	1.0076	1.0019	0.9993	0.9965	0.9935	0.9901	0.9864	0.9843	0.9822	0.9798	0.9774	0.9747	0.9718	0.9686	0.9652	0.9614	0.9572
3.4	1.0289	1.0258	1.0206	1.0128	1.0079	1.0020	0.9994	0.9965	0.9933	0.9899	0.9860	0.9839	0.9817	0.9793	0.9768	0.9740	0.9710	0.9678	0.9642	0.9603	0.9561
3.5	1.0298	1.0266	1.0212	1.0133	1.0082	1.0021	0.9994	0.9964	0.9932	0.9896	0.9857	0.9835	0.9812	0.9788	0.9762	0.9733	0.9702	0.9669	0.9633	0.9593	0.9549
3.6	1.0306	1.0273	1.0218	1.0137	1.0084	1.0022	0.9994	0.9964	0.9931	0.9894	0.9854	0.9832	0.9808	0.9783	0.9756	0.9727	0.9695	0.9661	0.9623	0.9582	0.9537
3.7	1.0314	1.0281	1.0224	1.0141	1.0087	1.0024	0.9994	0.9963	0.9929	0.9892	0.9850	0.9828	0.9803	0.9778	0.9750	0.9720	0.9688	0.9652	0.9614	0.9572	0.9526
3.8	1.0323	1.0289	1.0230	1.0145	1.0090	1.0025	0.9995	0.9963	0.9928	0.9890	0.9847	0.9824	0.9799	0.9772	0.9744	0.9713	0.9680	0.9644	0.9605	0.9562	0.9514
3.9	1.0332	1.0296	1.0237	1.0149	1.0093	1.0026	0.9995	0.9963	0.9927	0.9888	0.9844	0.9820	0.9794	0.9767	0.9738	0.9707	0.9673	0.9636	0.9596	0.9551	0.9503
4.0	1.0340	1.0304	1.0243	1.0153	1.0095	1.0027	0.9996	0.9962	0.9926	0.9885	0.9840	0.9816	0.9790	0.9762	0.9732	0.9700	0.9665	0.9628	0.9586	0.9541	0.9491

A.4.3 Tables Applying to Orifice Flow Constants for Both Flange Tap and Pipe Tap Installations

TABLE A.31
F_{pb} Factors to Change from a Pressure Base of 14.73 psia to Other Pressure Bases

Pressure Base, psia	F_{pb}
14.4	1.0229
14.525	1.0141
14.65	1.0055
14.70	1.0020
14.73	1.0000
14.775	0.9970
14.90	0.9886
15.025	0.9804
15.15	0.9723
15.225	0.9675
15.275	0.9643
15.325	0.9612
15.40	0.9565
15.525	0.9488
15.65	0.9412
15.775	0.9338
15.90	0.9264
16.025	0.9192
16.15	0.9121
16.275	0.9051
16.40	0.8982
16.70	0.8820

TABLE A.32
F_{tb} Factors to Change from a Temperature Base of 60°F to Other Temperature Bases

Temperature, °F	F_{tb}	Temperature, °F	F_{tb}
40	0.9615	65	1.0096
41	0.9625	66	1.0115
42	0.9654	67	1.0135
43	0.9673	68	1.0154
44	0.9692	69	1.0173
45	0.9712	70	1.0192
46	0.9731	71	1.0212
47	0.9750	72	1.0231
48	0.9769	73	1.0250
49	0.9788	74	1.0269
50	0.9808	75	1.0288
51	0.9827	76	1.0308
52	0.9846	77	1.0327
53	0.9865	78	1.0346
54	0.9885	79	1.0365
55	0.9904	80	1.0385
56	0.9923	81	1.0404
57	0.9942	82	1.0423
58	0.9962	83	1.0442
59	0.9981	84	1.0462
60	1.0000	85	1.0481
61	1.0019	86	1.0500
62	1.0038	87	1.0519
63	1.0058	88	1.0538
64	1.0077	89	1.0558
		90	1.0577

TABLE A.33

F_{tf} Factors to Change from Flowing Temperature of 60°F to Actual Flowing Temperature

Temp., °F	F_{tf}	Temp., °F	F_{tf}	Temp., °F	F_{tf}
1	1.0621	51	1.0088	101	0.9628
2	1.0609	52	1.0078	102	0.9619
3	1.0598	53	1.0068	103	0.9610
4	1.0586	54	1.0058	104	0.9602
5	1.0575	55	1.0048	105	0.9594
6	1.0564	56	1.0039	106	0.9585
7	1.0552	57	1.0029	107	0.9577
8	1.0541	58	1.0019	108	0.9568
9	1.0530	59	1.0010	109	0.9560
10	1.0518	60	1.0000	110	0.9551
11	1.0507	61	0.9990	111	0.9543
12	1.0496	62	0.9981	112	0.9535
13	1.0485	63	0.9971	113	0.9526
14	1.0474	64	0.9962	114	0.9518
15	1.0463	65	0.9952	115	0.9510
16	1.0452	66	0.9943	116	0.9501
17	1.0441	67	0.9933	117	0.9493
18	1.0430	68	0.9924	118	0.9485
19	1.0419	69	0.9915	119	0.9477
20	1.0408	70	0.9905	120	0.9469
21	1.0398	71	0.9896	121	0.9460
22	1.0387	72	0.9887	122	0.9452
23	1.0376	73	0.9877	123	0.9444
24	1.0365	74	0.9868	124	0.9436
25	1.0355	75	0.9859	125	0.9428
26	1.0344	76	0.9850	126	0.9420
27	1.0333	77	0.9840	127	0.9412
28	1.0323	78	0.9831	128	0.9404
29	1.0312	79	0.9822	129	0.9396
30	1.0302	80	0.9813	130	0.9388
31	1.0291	81	0.9804	131	0.9380
32	1.0281	82	0.9795	132	0.9372
33	1.0270	83	0.9786	133	0.9364
34	1.0260	84	0.9777	134	0.9356
35	1.0249	85	0.9768	135	0.9349

TABLE A.33 (Continued)

Temp., °F	F_{tf}	Temp., °F	F_{tf}	Temp., °F	F_{tf}
36	1.0239	86	0.9759	136	0.9341
37	1.0229	87	0.9750	137	0.9333
38	1.0218	88	0.9741	138	0.9325
39	1.0208	89	0.9732	139	0.9317
40	1.0198	90	0.9723	140	0.9309
41	1.0188	91	0.9715	141	0.9302
42	1.0178	92	0.9706	142	0.9294
43	1.0168	93	0.9697	143	0.9286
44	1.0158	94	0.9688	144	0.9279
45	1.0147	95	0.9680	145	0.9271
46	1.0137	96	0.9671	146	0.9263
47	1.0127	97	0.9662	147	0.9256
48	1.0117	98	0.9653	148	0.9248
49	1.0108	99	0.9645	149	0.9240
50	1.0098	100	0.9636	150	0.9233

TABLE A.34

F_g Factors to Adjust for Specific Gravity

Specific Gravity, G	0.000	0.001	0.002	0.003	0.004	0.005	0.006	0.007	0.008	0.009
0.550	1.3484	1.3472	1.3460	1.3447	1.3435	1.3423	1.3411	1.3399	1.3387	1.3375
0.560	1.3363	1.3351	1.3339	1.3327	1.3316	1.3304	1.3292	1.3280	1.3269	1.3257
0.570	1.3245	1.3234	1.3222	1.3211	1.3199	1.3188	1.3176	1.3165	1.3153	1.3142
0.580	1.3131	1.3119	1.3108	1.3097	1.3086	1.3074	1.3063	1.3052	1.3041	1.3030
0.590	1.3019	1.3008	1.2997	1.2986	1.2975	1.2964	1.2953	1.2942	1.2932	1.2921
0.600	1.2910	1.2899	1.2888	1.2878	1.2867	1.2856	1.2846	1.2835	1.2825	1.2814
0.610	1.2804	1.2793	1.2783	1.2772	1.2762	1.2752	1.2741	1.2731	1.2720	1.2710
0.620	1.2700	1.2690	1.2680	1.2669	1.2659	1.2649	1.2639	1.2629	1.2619	1.2609
0.630	1.2599	1.2589	1.2579	1.2569	1.2559	1.2549	1.2539	1.2529	1.2520	1.2510
0.640	1.2500	1.2490	1.2480	1.2471	1.2461	1.2451	1.2442	1.2432	1.2423	1.2413
0.650	1.2403	1.2394	1.2384	1.2375	1.2365	1.2356	1.2347	1.2337	1.2328	1.2318
0.660	1.2309	1.2300	1.2290	1.2281	1.2272	1.2263	1.2254	1.2244	1.2235	1.2266
0.670	1.2217	1.2208	1.2199	1.2190	1.2181	1.2172	1.2163	1.2154	1.2145	1.2136
0.680	1.2127	1.2118	1.2109	1.2100	1.2091	1.2082	1.2074	1.2065	1.2056	1.2047
0.690	1.2039	1.2030	1.2021	1.2012	1.2004	1.1995	1.1986	1.1978	1.1969	1.1961
0.700	1.1952	1.1944	1.1935	1.1927	1.1918	1.1910	1.1901	1.1893	1.1884	1.1876
0.710	1.1868	1.1859	1.1851	1.1843	1.1834	1.1826	1.1818	1.1810	1.1802	1.1793
0.720	1.1785	1.1777	1.1769	1.1761	1.1752	1.1744	1.1736	1.1728	1.1720	1.1712
0.730	1.1704	1.1696	1.1688	1.1680	1.1672	1.1664	1.1656	1.1648	1.1640	1.1633
0.740	1.1625	1.1617	1.1609	1.1601	1.1593	1.1586	1.1578	1.1570	1.1562	1.1555
0.750	1.1547	1.1539	1.1532	1.1524	1.1516	1.1509	1.1501	1.1493	1.1486	1.1478
0.760	1.1471	1.1463	1.1456	1.1448	1.1441	1.1433	1.1426	1.1418	1.1411	1.1403
0.770	1.1396	1.1389	1.1381	1.1374	1.1366	1.1359	1.1352	1.1345	1.1337	1.1330
0.780	1.1323	1.1316	1.1308	1.1301	1.1294	1.1287	1.1279	1.1272	1.1265	1.1258
0.790	1.1251	1.1244	1.1237	1.1230	1.1222	1.1215	1.1208	1.1201	1.1194	1.1187
0.800	1.1180	1.1173	1.1166	1.1159	1.1152	1.1146	1.1139	1.1132	1.1125	1.1118
0.810	1.1111	1.1104	1.1097	1.1090	1.1084	1.1077	1.1070	1.1063	1.1057	1.1050
0.820	1.1043	1.1036	1.1030	1.1023	1.1016	1.1010	1.1003	1.0996	1.0990	1.0983
0.830	1.0976	1.0970	1.0963	1.0957	1.0950	1.0944	1.0937	1.0930	1.0924	1.0917
0.840	1.0911	1.0904	1.0898	1.0891	1.0885	1.0878	1.0872	1.0866	1.0859	1.0853
0.850	1.0846	1.0840	1.0834	1.0827	1.0821	1.0815	1.0808	1.0802	1.0796	1.0790
0.860	1.0783	1.0777	1.0771	1.0764	1.0758	1.0752	1.0746	1.0740	1.0733	1.0727
0.870	1.0721	1.0715	1.0709	1.0703	1.0696	1.0690	1.0684	1.0678	1.0672	1.0666
0.880	1.0660	1.0654	1.0648	1.0642	1.0636	1.0630	1.0624	1.0618	1.0612	1.0606
0.890	1.0600	1.0594	1.0588	1.0582	1.0576	1.0570	1.0564	1.0558	1.0553	1.0547
0.900	1.0541	1.0535	1.0529	1.0523	1.0518	1.0512	1.0506	1.0500	1.0494	1.0489
0.910	1.0483	1.0477	1.0471	1.0466	1.0460	1.0454	1.0448	1.0443	1.0437	1.0431
0.920	1.0426	1.0420	1.0414	1.0409	1.0403	1.0398	1.0392	1.0386	1.0381	1.0375
0.930	1.0370	1.0364	1.0358	1.0353	1.0347	1.0342	1.0336	1.0331	1.0325	1.0320
0.940	1.0314	1.0309	1.0303	1.0298	1.0292	1.0287	1.0281	1.0276	1.0270	1.0265
0.950	1.0260	1.0254	1.0249	1.0244	1.0238	1.0233	1.0228	1.0222	1.0217	1.0212
0.960	1.0206	1.0201	1.0196	1.0190	1.0185	1.0180	1.0174	1.0169	1.0164	1.0159
0.970	1.0153	1.0148	1.0143	1.0138	1.0132	1.0127	1.0122	1.0117	1.0112	1.0107
0.980	1.0102	1.0096	1.0091	1.0086	1.0081	1.0076	1.0071	1.0066	1.0060	1.0055
0.990	1.0050	1.0045	1.0040	1.0035	1.0030	1.0025	1.0020	1.0015	1.0010	1.0005
1.000	1.0000									

TABLE A.35(a)

F_{pv} Supercompressibility Factors

Base Data—0.6 Specific Gravity Hydrocarbon Gas

p_f	Temperature, °F							
psig	−40	−35	−30	−25	−20	−15	−10	−5
0	1.0000	1.0000	1.0000	1.0000	1.0000	1.0000	1.0000	1.0000
20	1.0031	1.0030	1.0029	1.0028	1.0027	1.0026	1.0025	1.0024
40	1.0062	1.0060	1.0059	1.0057	1.0055	1.0053	1.0051	1.0049
60	1.0093	1.0091	1.0089	1.0086	1.0083	1.0080	1.0077	1.0074
80	1.0125	1.0122	1.0119	1.0115	1.0111	1.0107	1.0103	1.0099
100	1.0158	1.0154	1.0148	1.0145	1.0139	1.0134	1.0129	1.0125
120	1.0192	1.0186	1.0178	1.0175	1.0168	1.0162	1.0156	1.0151
140	1.0227	1.0218	1.0209	1.0205	1.0198	1.0190	1.0183	1.0177
160	1.0262	1.0251	1.0241	1.0235	1.0228	1.0218	1.0210	1.0202
180	1.0297	1.0285	1.0274	1.0265	1.0258	1.0246	1.0237	1.0228
200	1.0333	1.0319	1.0307	1.0296	1.0288	1.0275	1.0265	1.0255
220	1.0369	1.0335	1.0340	1.0328	1.0317	1.0304	1.0291	1.0282
240	1.0406	1.0388	1.0373	1.0360	1.0347	1.0334	1.0321	1.0309
260	1.0444	1.0424	1.0407	1.0392	1.0377	1.0364	1.0350	1.0337
280	1.0482	1.0461	1.0442	1.0425	1.0408	1.0394	1.0379	1.0365
300	1.0522	1.0499	1.0478	1.0459	1.0441	1.0425	1.0409	1.0393
320	1.0562	1.0537	1.0514	1.0494	1.0474	1.0456	1.0439	1.0422
340	1.0602	1.0575	1.0551	1.0529	1.0507	1.0488	1.0469	1.0451
360	1.0642	1.0614	1.0589	1.0564	1.0541	1.0520	1.0500	1.0480
380	1.0684	1.0654	1.0627	1.0601	1.0576	1.0553	1.0531	1.0510
400	1.0727	1.0695	1.0666	1.0638	1.0611	1.0586	1.0563	1.0540
420	1.0771	1.0737	1.0706	1.0675	1.0646	1.0620	1.0595	1.0571
440	1.0816	1.0779	1.0746	1.0713	1.0682	1.0654	1.0627	1.0601
460	1.0862	1.0822	1.0787	1.0752	1.0719	1.0688	1.0660	1.0632
480	1.0909	1.0866	1.0828	1.0791	1.0756	1.0723	1.0693	1.0664
500	1.0956	1.0910	1.0869	1.0830	1.0793	1.0759	1.0727	1.0696
520	1.1004	1.0956	1.0911	1.0869	1.0830	1.0794	1.0761	1.0728
540	1.1055	1.1002	1.0955	1.0910	1.0868	1.0830	1.0795	1.0760
560	1.1106	1.1051	1.1000	1.0952	1.0908	1.0868	1.0830	1.0793
580	1.1159	1.1100	1.1045	1.0995	1.0948	1.0906	1.0865	1.0826
600	1.1213	1.1149	1.1091	1.1038	1.0989	1.0944	1.0901	1.0860
620	1.1267	1.1200	1.1138	1.1082	1.1030	1.0982	1.0937	1.0894
640	1.1323	1.1252	1.1186	1.1127	1.1072	1.1021	1.0973	1.0928
660	1.1379	1.1305	1.1236	1.1172	1.1114	1.1060	1.1010	1.0963
680	1.1439	1.1359	1.1286	1.1218	1.1156	1.1099	1.1047	1.0998

TABLE A.35(a) (Continued)

p_f	Temperature, °F							
psig	−40	−35	−30	−25	−20	−15	−10	−5
700	1.1499	1.1413	1.1336	1.1265	1.1199	1.1138	1.1083	1.1033
720	1.1562	1.1469	1.1388	1.1313	1.1245	1.1181	1.1123	1.1069
740	1.1626	1.1528	1.1442	1.1363	1.1291	1.1225	1.1162	1.1106
760	1.1692	1.1587	1.1496	1.1413	1.1337	1.1267	1.1202	1.1143
780	1.1759	1.1647	1.1551	1.1464	1.1384	1.1311	1.1242	1.1180
800	1.1826	1.1708	1.1607	1.1516	1.1432	1.1355	1.1283	1.1217
820	1.1894	1.1769	1.1663	1.1568	1.1480	1.1399	1.1324	1.1255
840	1.1967	1.1835	1.1723	1.1622	1.1528	1.1443	1.1365	1.1293
860	1.2041	1.1901	1.1783	1.1676	1.1577	1.1488	1.1407	1.1332
880	1.2116	1.1968	1.1843	1.1731	1.1627	1.1533	1.1449	1.1373
900	1.2191	1.2035	1.1903	1.1786	1.1677	1.1579	1.1491	1.1410
920	1.2269	1.2103	1.1965	1.1842	1.1728	1.1625	1.1534	1.1450
940	1.2347	1.2173	1.2028	1.1899	1.1780	1.1674	1.1577	1.1490
960	1.2427	1.2245	1.2093	1.1956	1.1832	1.1721	1.1620	1.1530
980	1.2509	1.2318	1.2157	1.2014	1.1884	1.1768	1.1663	1.1570
1000	1.2591	1.2391	1.2221	1.2072	1.1936	1.1815	1.1706	1.1610
1020	1.2673	1.2464	1.2286	1.2131	1.1990	1.1864	1.1751	1.1650
1040	1.2756	1.2537	1.2351	1.2190	1.2044	1.1913	1.1796	1.1690
1060	1.2839	1.2611	1.2418	1.2250	1.2098	1.1962	1.1841	1.1731
1080	1.2922	1.2685	1.2485	1.2310	1.2152	1.2011	1.1886	1.1772
1100	1.3008	1.2759	1.2552	1.2370	1.2206	1.2060	1.1933	1.1813
1120	1.3091	1.2834	1.2619	1.2431	1.2260	1.2109	1.1978	1.1854
1140	1.3176	1.2909	1.2686	1.2492	1.2315	1.2159	1.2023	1.1896
1160	1.3259	1.2985	1.2753	1.2552	1.2370	1.2209	1.2068	1.1939
1180	1.3337	1.3056	1.2820	1.2612	1.2425	1.2258	1.2111	1.1979
1200	1.3412	1.3127	1.2883	1.2669	1.2477	1.2305	1.2154	1.2018
1220	1.3486	1.3196	1.2946	1.2726	1.2529	1.2352	1.2197	1.2058
1240	1.3559	1.3264	1.3009	1.2783	1.2580	1.2399	1.2240	1.2098
1260	1.3628	1.3329	1.3071	1.2839	1.2631	1.2446	1.2283	1.2138
1280	1.3692	1.3390	1.3128	1.2894	1.2682	1.2493	1.2326	1.2176
1300	1.3754	1.3448	1.3184	1.2947	1.2732	1.2540	1.2369	1.2214
1320	1.3812	1.3505	1.3240	1.3000	1.2782	1.2586	1.2411	1.2252
1340	1.3867	1.3561	1.3294	1.3053	1.2832	1.2631	1.2451	1.2289
1360	1.3917	1.3611	1.3344	1.3101	1.2878	1.2675	1.2491	1.2326
1380	1.3961	1.3655	1.3388	1.3145	1.2920	1.2715	1.2530	1.2362
1400	1.4002	1.3699	1.3432	1.3186	1.2960	1.2754	1.2568	1.2398
1420	1.4037	1.3738	1.3473	1.3228	1.3000	1.2792	1.2604	1.2432

TABLE A.35(a) (Continued)

p_f	Temperature, °F							
psig	−40	−35	−30	−25	−20	−15	−10	−5
1440	1.4069	1.3774	1.3508	1.3264	1.3038	1.2830	1.2640	1.2466
1460	1.4096	1.3805	1.3540	1.3298	1.3072	1.2864	1.2673	1.2498
1480	1.4118	1.3833	1.3571	1.3331	1.3105	1.2894	1.2703	1.2530
1500	1.4137	1.3857	1.3597	1.3357	1.3132	1.2924	1.2735	1.2558
1520	1.4152	1.3878	1.3621	1.3384	1.3161	1.2954	1.2763	1.2586
1540	1.4164	1.3896	1.3643	1.3408	1.3186	1.2979	1.2788	1.2612
1560	1.4172	1.3910	1.3661	1.3428	1.3207	1.3004	1.2813	1.2638
1580	1.4177	1.3922	1.3677	1.3445	1.3228	1.3027	1.2838	1.2661
1600	1.4179	1.3930	1.3690	1.3462	1.3247	1.3047	1.2860	1.2683
1620	1.4179	1.3936	1.3700	1.3476	1.3263	1.3064	1.2878	1.2702
1640	1.4176	1.3938	1.3708	1.3488	1.3278	1.3079	1.2895	1.2720
1660	1.4170	1.3939	1.3713	1.3497	1.3289	1.3094	1.2912	1.2738
1680	1.4162	1.3936	1.3716	1.3504	1.3300	1.3108	1.2928	1.2755
1700	1.4151	1.3932	1.3718	1.3510	1.3309	1.3119	1.2940	1.2769
1720	1.4139	1.3926	1.3715	1.3513	1.3317	1.3130	1.2951	1.2782
1740	1.4126	1.3919	1.3712	1.3514	1.3321	1.3137	1.2961	1.2793
1760	1.4111	1.3909	1.3707	1.3513	1.3321	1.3143	1.2970	1.2804
1780	1.4094	1.3897	1.3701	1.3511	1.3322	1.3148	1.2977	1.2812
1800	1.4075	1.3884	1.3693	1.3507	1.3323	1.3151	1.2983	1.2819
1820	1.4056	1.3870	1.3684	1.3502	1.3324	1.3153	1.2988	1.2826
1840	1.4035	1.3855	1.3673	1.3496	1.3321	1.3153	1.2990	1.2831
1860	1.4012	1.3837	1.3661	1.3488	1.3317	1.3152	1.2991	1.2835
1880	1.3989	1.3818	1.3647	1.3478	1.3312	1.3150	1.2992	1.2838
1900	1.3965	1.3799	1.3632	1.3468	1.3305	1.3146	1.2990	1.2839
1920	1.3940	1.3779	1.3617	1.3457	1.3298	1.3142	1.2989	1.2840
1940	1.3914	1.3758	1.3601	1.3444	1.3289	1.3136	1.2986	1.2841
1960	1.3888	1.3737	1.3584	1.3431	1.3279	1.3129	1.2982	1.2839
1980	1.3861	1.3714	1.3566	1.3416	1.3267	1.3120	1.2977	1.2836
2000	1.3834	1.3691	1.3547	1.3400	1.3254	1.3110	1.2971	1.2833
2020	1.3806	1.3667	1.3527	1.3384	1.3241	1.3100	1.2963	1.2828
2040	1.3778	1.3642	1.3506	1.3368	1.3228	1.3089	1.2955	1.2823
2060	1.3749	1.3617	1.3484	1.3351	1.3212	1.3078	1.2947	1.2817
2080	1.3720	1.3591	1.3462	1.3332	1.3196	1.3065	1.2937	1.2809
2100	1.3690	1.3565	1.3439	1.3312	1.3180	1.3052	1.2926	1.2801
2120	1.3660	1.3539	1.3416	1.3292	1.3164	1.3039	1.2915	1.2793
2140	1.3630	1.3513	1.3392	1.3271	1.3147	1.3025	1.2903	1.2784

(Continued)

TABLE A.35(a) (Continued)

p_f	Temperature, °F							
psig	−40	−35	−30	−25	−20	−15	−10	−5
2160	1.3600	1.3486	1.3367	1.3250	1.3129	1.3010	1.2891	1.2774
2180	1.3569	1.3459	1.3343	1.3228	1.3110	1.2994	1.2878	1.2764
2200	1.3538	1.3431	1.3318	1.3206	1.3091	1.2978	1.2864	1.2753
2220	1.3507	1.3402	1.3295	1.3184	1.3071	1.2961	1.2850	1.2741
2240	1.3476	1.3373	1.3268	1.3162	1.3051	1.2943	1.2835	1.2729
2260	1.3444	1.3344	1.3243	1.3139	1.3031	1.2925	1.2820	1.2716
2280	1.3412	1.3315	1.3217	1.3116	1.3011	1.2907	1.2804	1.2702
2300	1.3380	1.3286	1.3191	1.3092	1.2990	1.2889	1.2788	1.2688
2320	1.3349	1.3257	1.3164	1.3068	1.2969	1.2870	1.2772	1.2674
2340	1.3317	1.3228	1.3137	1.3044	1.2947	1.2851	1.2755	1.2659
2360	1.3285	1.3199	1.3110	1.3019	1.2925	1.2831	1.2737	1.2643
2380	1.3254	1.3170	1.3083	1.2994	1.2903	1.2811	1.2719	1.2627
2400	1.3223	1.3141	1.3056	1.2969	1.2880	1.2790	1.2700	1.2611
2420	1.3191	1.3112	1.3029	1.2944	1.2857	1.2769	1.2682	1.2594
2440	1.3159	1.3082	1.3002	1.2919	1.2734	1.2748	1.2663	1.2577
2460	1.3128	1.3052	1.2975	1.2894	1.2811	1.2727	1.2644	1.2560
2480	1.3096	1.3022	1.2948	1.2869	1.2788	1.2706	1.2624	1.2542
2500	1.3064	1.2992	1.2921	1.2843	1.2764	1.2684	1.2604	1.2524
2520	1.3033	1.2963	1.2893	1.2817	1.2741	1.2663	1.2585	1.2506
2540	1.3001	1.2934	1.2864	1.2792	1.2717	1.2642	1.2566	1.2488
2560	1.2970	1.2904	1.2835	1.2766	1.2693	1.2620	1.2546	1.2470
2580	1.2939	1.2875	1.2807	1.2740	1.2669	1.2597	1.2525	1.2451
2600	1.2909	1.2846	1.2780	1.2714	1.2645	1.2575	1.2505	1.2433
2620	1.2878	1.2817	1.2753	1.2687	1.2620	1.2553	1.2484	1.2414
2640	1.2847	1.2787	1.2725	1.2661	1.2596	1.2530	1.2462	1.2394
2660	1.2816	1.2758	1.2697	1.2635	1.2572	1.2507	1.2441	1.2375
2680	1.2785	1.2729	1.2670	1.2609	1.2547	1.2484	1.2420	1.2356
2700	1.2754	1.2700	1.2643	1.2584	1.2523	1.2461	1.2399	1.2336
2720	1.2723	1.2670	1.2614	1.2557	1.2498	1.2438	1.2377	1.2315
2740	1.2693	1.2641	1.2587	1.2531	1.2473	1.2414	1.2355	1.2295
2760	1.2663	1.2612	1.2559	1.2505	1.2448	1.2391	1.2334	1.2275
2780	1.2633	1.2584	1.2532	1.2479	1.2424	1.2368	1.2312	1.2255
2800	1.2603	1.2555	1.2504	1.2454	1.2400	1.2345	1.2290	1.2234
2820	1.2573	1.2526	1.2476	1.2427	1.2374	1.2322	1.2268	1.2213
2840	1.2543	1.2497	1.2448	1.2401	1.2349	1.2298	1.2246	1.2193
2860	1.2513	1.2469	1.2421	1.2375	1.2324	1.2274	1.2224	1.2172
2880	1.2483	1.2441	1.2394	1.2349	1.2300	1.2251	1.2202	1.2152

TABLE A.35(a) (Continued)

p_f	Temperature, °F							
psig	−40	−35	−30	−25	−20	−15	−10	−5
2900	1.2454	1.2413	1.2368	1.2324	1.2276	1.2228	1.2180	1.2131
2920	1.2424	1.2384	1.2341	1.2298	1.2252	1.2205	1.2158	1.2110
2940	1.2395	1.2356	1.2314	1.2272	1.2227	1.2181	1.2135	1.2089
2960	1.2366	1.2328	1.2287	1.2246	1.2202	1.2157	1.2112	1.2067
2980	1.2338	1.2301	1.2261	1.2221	1.2178	1.2134	1.2091	1.2047
3000	1.2309	1.2273	1.2234	1.2195	1.2153	1.2111	1.2069	1.2027

Note: Factors for intermediate values of pressure and temperature should be interpolated.

TABLE A.35(a) (Continued)
F_{pv} Supercompressibility Factors
Base Data—0.6 Specific Gravity Hydrocarbon Gas

p_f	Temperature, °F											
psig	0	5	10	15	20	25	30	35	40	45	50	55
0	1.0000	1.0000	1.0000	1.0000	1.0000	1.0000	1.0000	1.0000	1.0000	1.0000	1.0000	1.0000
20	1.0023	1.0022	1.0022	1.0021	1.0020	1.0020	1.0019	1.0018	1.0018	1.0017	1.0016	1.0016
40	1.0048	1.0047	1.0045	1.0044	1.0042	1.0041	1.0040	1.0038	1.0037	1.0036	1.0034	1.0033
60	1.0071	1.0069	1.0067	1.0065	1.0063	1.0061	1.0059	1.0057	1.0054	1.0053	1.0051	1.0049
80	1.0096	1.0093	1.0090	1.0087	1.0084	1.0081	1.0078	1.0076	1.0073	1.0070	1.0068	1.0066
100	1.0121	1.0117	1.0113	1.0109	1.0105	1.0102	1.0098	1.0095	1.0091	1.0088	1.0085	1.0083
120	1.0146	1.0141	1.0136	1.0131	1.0127	1.0122	1.0118	1.0114	1.0110	1.0106	1.0103	1.0100
140	1.0170	1.0164	1.0158	1.0152	1.0148	1.0142	1.0138	1.0132	1.0128	1.0124	1.0120	1.0116
160	1.0195	1.0188	1.0182	1.0176	1.0169	1.0163	1.0158	1.0152	1.0147	1.0142	1.0138	1.0133
180	1.0220	1.0213	1.0206	1.0198	1.0191	1.0184	1.0178	1.0171	1.0166	1.0160	1.0155	1.0150
200	1.0245	1.0237	1.0229	1.0220	1.0213	1.0206	1.0198	1.0192	1.0185	1.0179	1.0173	1.0167
220	1.0272	1.0263	1.0254	1.0244	1.0235	1.0227	1.0219	1.0211	1.0204	1.0197	1.0191	1.0184
240	1.0298	1.0288	1.0277	1.0267	1.0257	1.0248	1.0239	1.0231	1.0223	1.0215	1.0208	1.0201
260	1.0324	1.0313	1.0302	1.0291	1.0280	1.0270	1.0260	1.0250	1.0242	1.0234	1.0226	1.0219
280	1.0351	1.0339	1.0327	1.0315	1.0303	1.0292	1.0281	1.0271	1.0261	1.0252	1.0244	1.0236
300	1.0379	1.0365	1.0352	1.0339	1.0326	1.0314	1.0303	1.0291	1.0281	1.0271	1.0262	1.0253
320	1.0406	1.0391	1.0377	1.0363	1.0349	1.0336	1.0324	1.0312	1.0300	1.0290	1.0280	1.0270
340	1.0434	1.0417	1.0401	1.0386	1.0372	1.0358	1.0344	1.0332	1.0320	1.0308	1.0298	1.0287
360	1.0462	1.0444	1.0427	1.0411	1.0395	1.0380	1.0366	1.0353	1.0340	1.0328	1.0316	1.0305
380	1.0491	1.0471	1.0453	1.0436	1.0420	1.0404	1.0388	1.0374	1.0361	1.0347	1.0334	1.0322
400	1.0519	1.0498	1.0479	1.0461	1.0444	1.0427	1.0410	1.0395	1.0381	1.0366	1.0352	1.0340
420	1.0548	1.0526	1.0506	1.0486	1.0468	1.0450	1.0433	1.0417	1.0401	1.0386	1.0371	1.0358
440	1.0577	1.0553	1.0531	1.0511	1.0492	1.0472	1.0453	1.0437	1.0421	1.0405	1.0389	1.0375
460	1.0606	1.0581	1.0558	1.0536	1.0516	1.0496	1.0476	1.0458	1.0441	1.0425	1.0408	1.0393
480	1.0636	1.0609	1.0585	1.0562	1.0540	1.0519	1.0498	1.0479	1.0461	1.0444	1.0427	1.0411
500	1.0667	1.0639	1.0613	1.0588	1.0565	1.0543	1.0521	1.0501	1.0482	1.0464	1.0446	1.0429
520	1.0697	1.0667	1.0639	1.0613	1.0588	1.0565	1.0543	1.0522	1.0503	1.0484	1.0465	1.0447
540	1.0727	1.0696	1.0667	1.0640	1.0613	1.0588	1.0564	1.0543	1.0523	1.0503	1.0483	1.0465
560	1.0759	1.0726	1.0695	1.0666	1.0639	1.0612	1.0587	1.0565	1.0544	1.0523	1.0502	1.0483
580	1.0790	1.0757	1.0724	1.0693	1.0665	1.0637	1.0611	1.0587	1.0565	1.0543	1.0521	1.0501
600	1.0822	1.0787	1.0753	1.0721	1.0691	1.0661	1.0634	1.0609	1.0586	1.0562	1.0540	1.0519
620	1.0853	1.0816	1.0781	1.0747	1.0716	1.0685	1.0656	1.0631	1.0607	1.0582	1.0559	1.0538
640	1.0886	1.0848	1.0811	1.0775	1.0742	1.0710	1.0680	1.0653	1.0628	1.0602	1.0578	1.0556
660	1.0919	1.0879	1.0840	1.0802	1.0767	1.0735	1.0704	1.0675	1.0649	1.0623	1.0598	1.0574
680	1.0953	1.0910	1.0869	1.0830	1.0793	1.0760	1.0728	1.0698	1.0670	1.0643	1.0617	1.0593
700	1.0986	1.0941	1.0898	1.0857	1.0819	1.0784	1.0751	1.0720	1.0691	1.0663	1.0636	1.0611
720	1.1020	1.0973	1.0928	1.0885	1.0847	1.0810	1.0775	1.0742	1.0712	1.0684	1.0656	1.0630
740	1.1054	1.1005	1.0958	1.0914	1.0873	1.0835	1.0799	1.0766	1.0734	1.0704	1.0675	1.0648
760	1.1089	1.1038	1.0989	1.0943	1.0900	1.0860	1.0822	1.0788	1.0756	1.0725	1.0694	1.0667
780	1.1124	1.1070	1.1019	1.0972	1.0927	1.0885	1.0846	1.0810	1.0777	1.0745	1.0714	1.0685

TABLE A.35(a) (Continued)

p_f	Temperature, °F											
psig	0	5	10	15	20	25	30	35	40	45	50	55
800	1.1159	1.1103	1.1050	1.1000	1.0954	1.0911	1.0870	1.0833	1.0798	1.0765	1.0733	1.0704
820	1.1193	1.1135	1.1080	1.1029	1.0981	1.0936	1.0894	1.0856	1.0819	1.0785	1.0752	1.0722
840	1.1229	1.1169	1.1112	1.1057	1.1008	1.0962	1.0919	1.0879	1.0841	1.0805	1.0771	1.0740
860	1.1265	1.1202	1.1143	1.1087	1.1037	1.0989	1.0943	1.0902	1.0863	1.0826	1.0792	1.0759
880	1.1301	1.1236	1.1175	1.1117	1.1064	1.1015	1.0968	1.0925	1.0885	1.0847	1.0811	1.0778
900	1.1337	1.1270	1.1206	1.1146	1.1091	1.1040	1.0991	1.0947	1.0906	1.0867	1.0830	1.0795
920	1.1373	1.1303	1.1237	1.1175	1.1118	1.1066	1.1016	1.0970	1.0928	1.0887	1.0849	1.0813
940	1.1410	1.1338	1.1269	1.1205	1.1146	1.1092	1.1041	1.0994	1.0950	1.0908	1.0868	1.0832
960	1.1448	1.1372	1.1301	1.1234	1.1175	1.1119	1.1065	1.1016	1.0971	1.0928	1.0887	1.0850
980	1.1485	1.1407	1.1334	1.1265	1.1203	1.1145	1.1090	1.1039	1.0992	1.0948	1.0906	1.0868
1000	1.1520	1.1440	1.1365	1.1294	1.1230	1.1170	1.1114	1.1062	1.1013	1.0968	1.0925	1.0885
1020	1.1558	1.1475	1.1397	1.1324	1.1258	1.1196	1.1138	1.1084	1.1035	1.0988	1.0945	1.0904
1040	1.1595	1.1509	1.1428	1.1353	1.1285	1.1222	1.1163	1.1107	1.1057	1.1008	1.0964	1.0922
1060	1.1633	1.1544	1.1461	1.1383	1.1313	1.1249	1.1188	1.1131	1.1078	1.1028	1.0983	1.0940
1080	1.1669	1.1578	1.1492	1.1411	1.1340	1.1273	1.1211	1.1153	1.1099	1.1048	1.1001	1.0957
1100	1.1707	1.1612	1.1524	1.1441	1.1368	1.1299	1.1235	1.1175	1.1120	1.1069	1.1020	1.0976
1120	1.1744	1.1647	1.1555	1.1471	1.1395	1.1325	1.1259	1.1198	1.1141	1.1088	1.1038	1.0993
1140	1.1781	1.1681	1.1587	1.1501	1.1423	1.1350	1.1282	1.1220	1.1163	1.1109	1.1057	1.1011
1160	1.1819	1.1716	1.1619	1.1531	1.1451	1.1377	1.1307	1.1243	1.1184	1.1128	1.1075	1.1028
1180	1.1858	1.1751	1.1651	1.1559	1.1478	1.1402	1.1331	1.1265	1.1205	1.1148	1.1094	1.1046
1200	1.1895	1.1784	1.1682	1.1588	1.1505	1.1427	1.1354	1.1287	1.1225	1.1167	1.1113	1.1063
1220	1.1932	1.1819	1.1714	1.1617	1.1532	1.1453	1.1377	1.1308	1.1245	1.1186	1.1131	1.1080
1240	1.1968	1.1852	1.1745	1.1646	1.1558	1.1477	1.1401	1.1331	1.1266	1.1206	1.1149	1.1097
1260	1.2005	1.1886	1.1776	1.1675	1.1585	1.1502	1.1425	1.1353	1.1287	1.1225	1.1167	1.1114
1280	1.2040	1.1918	1.1805	1.1703	1.1611	1.1526	1.1446	1.1374	1.1307	1.1244	1.1184	1.1130
1300	1.2075	1.1951	1.1836	1.1730	1.1637	1.1550	1.1469	1.1395	1.1327	1.1263	1.1202	1.1147
1320	1.2109	1.1983	1.1867	1.1758	1.1663	1.1574	1.1492	1.1417	1.1347	1.1281	1.1219	1.1163
1340	1.2144	1.2016	1.1897	1.1786	1.1689	1.1599	1.1514	1.1437	1.1366	1.1299	1.1237	1.1180
1360	1.2178	1.2048	1.1926	1.1814	1.1714	1.1622	1.1536	1.1458	1.1386	1.1317	1.1253	1.1195
1380	1.2210	1.2078	1.1954	1.1840	1.1739	1.1645	1.1557	1.1476	1.1404	1.1334	1.1270	1.1211
1400	1.2244	1.2108	1.1983	1.1866	1.1763	1.1667	1.1577	1.1496	1.1422	1.1352	1.1287	1.1226
1420	1.2276	1.2137	1.2010	1.1892	1.1786	1.1689	1.1598	1.1516	1.1441	1.1369	1.1303	1.1241
1440	1.2307	1.2166	1.2037	1.1918	1.1810	1.1712	1.1602	1.1536	1.1459	1.1386	1.1318	1.1256
1460	1.2336	1.2193	1.2062	1.1942	1.1833	1.1732	1.1639	1.1554	1.1476	1.1402	1.1333	1.1270
1480	1.2365	1.2220	1.2088	1.1966	1.1856	1.1754	1.1658	1.1572	1.1493	1.1418	1.1349	1.1285
1500	1.2394	1.2247	1.2112	1.1989	1.1877	1.1774	1.1678	1.1591	1.1510	1.1434	1.1364	1.1299
1520	1.2421	1.2273	1.2137	1.2012	1.1900	1.1795	1.1697	1.1608	1.1526	1.1450	1.1378	1.1313
1540	1.2447	1.2298	1.2160	1.2034	1.1921	1.1815	1.1716	1.1626	1.1543	1.1466	1.1393	1.1327
1560	1.2469	1.2320	1.2182	1.2054	1.1940	1.1834	1.1733	1.1642	1.1559	1.1480	1.1407	1.1340
1580	1.2492	1.2343	1.2204	1.2075	1.1960	1.1853	1.1752	1.1660	1.1575	1.1495	1.1420	1.1352

(Continued)

TABLE A.35(a) (Continued)

p_f	Temperature, °F											
psig	0	5	10	15	20	25	30	35	40	45	50	55
1600	1.2514	1.2365	1.2225	1.2095	1.1979	1.1871	1.1769	1.1676	1.1590	1.1510	1.1435	1.1366
1620	1.2535	1.2386	1.2245	1.2114	1.1998	1.1889	1.1786	1.1692	1.1606	1.1524	1.1448	1.1378
1640	1.2555	1.2406	1.2265	1.2132	1.2015	1.1905	1.1802	1.1707	1.1620	1.1537	1.1461	1.1390
1660	1.2573	1.2423	1.2282	1.2149	1.2032	1.1921	1.1817	1.1722	1.1633	1.1550	1.1473	1.1401
1680	1.2591	1.2441	1.2299	1.2166	1.2049	1.1938	1.1832	1.1736	1.1647	1.1563	1.1485	1.1413
1700	1.2606	1.2457	1.2315	1.2182	1.2064	1.1953	1.1847	1.1751	1.1661	1.1575	1.1496	1.1424
1720	1.2620	1.2471	1.2331	1.2198	1.2079	1.1967	1.1861	1.1764	1.1674	1.1587	1.1508	1.1435
1740	1.2633	1.2485	1.2345	1.2213	1.2093	1.1980	1.1874	1.1776	1.1686	1.1600	1.1519	1.1445
1760	1.2645	1.2497	1.2357	1.2227	1.2106	1.1993	1.1887	1.1787	1.1697	1.1610	1.1529	1.1455
1780	1.2656	1.2509	1.2370	1.2239	1.2118	1.2005	1.1899	1.1799	1.1708	1.1621	1.1539	1.1464
1800	1.2665	1.2519	1.2381	1.2251	1.2130	1.2017	1.1910	1.1810	1.1718	1.1631	1.1549	1.1473
1820	1.2674	1.2529	1.2392	1.2262	1.2141	1.2028	1.1921	1.1821	1.1728	1.1640	1.1558	1.1482
1840	1.2680	1.2538	1.2401	1.2272	1.2151	1.2038	1.1930	1.1830	1.1738	1.1649	1.1566	1.1490
1860	1.2685	1.2545	1.2410	1.2281	1.2160	1.2047	1.1939	1.1839	1.1747	1.1658	1.1575	1.1498
1880	1.2690	1.2551	1.2417	1.2289	1.2169	1.2056	1.1948	1.1848	1.1755	1.1667	1.1583	1.1506
1900	1.2694	1.2556	1.2424	1.2296	1.2177	1.2064	1.1956	1.1856	1.1763	1.1675	1.1591	1.1514
1920	1.2697	1.2561	1.2429	1.2303	1.2184	1.2072	1.1964	1.1864	1.1770	1.1682	1.1598	1.1521
1940	1.2699	1.2564	1.2434	1.2309	1.2191	1.2078	1.1971	1.1871	1.1777	1.1689	1.1605	1.1528
1960	1.2700	1.2566	1.2438	1.2314	1.2197	1.2085	1.1978	1.1877	1.1784	1.1696	1.1612	1.1534
1980	1.2700	1.2568	1.2441	1.2318	1.2203	1.2092	1.1985	1.1884	1.1790	1.1701	1.1617	1.1540
2000	1.2699	1.2569	1.2443	1.2321	1.2207	1.2097	1.1990	1.1890	1.1796	1.1707	1.1623	1.1545
2020	1.2698	1.2570	1.2445	1.2324	1.2210	1.2101	1.1994	1.1894	1.1801	1.1712	1.1628	1.1550
2040	1.2695	1.2569	1.2446	1.2326	1.2213	1.2104	1.1998	1.1898	1.1805	1.1716	1.1632	1.1554
2060	1.2691	1.2568	1.2446	1.2327	1.2215	1.2107	1.2002	1.1902	1.1809	1.1721	1.1637	1.1559
2080	1.2686	1.2565	1.2445	1.2328	1.2216	1.2109	1.2005	1.1906	1.1813	1.1725	1.1640	1.1563
2100	1.2680	1.2561	1.2443	1.2328	1.2217	1.2110	1.2008	1.1909	1.1816	1.1728	1.1643	1.1566
2120	1.2674	1.2556	1.2440	1.2327	1.2217	1.2111	1.2009	1.1912	1.1819	1.1730	1.1646	1.1569
2140	1.2666	1.2551	1.2437	1.2325	1.2117	1.2212	1.2010	1.1914	1.1821	1.1733	1.1649	1.1572
2160	1.2658	1.2545	1.2433	1.2322	1.2216	1.2112	1.2011	1.1915	1.1823	1.1735	1.1651	1.1574
2180	1.2650	1.2538	1.2428	1.2319	1.2214	1.2111	1.2011	1.1916	1.1824	1.1736	1.1653	1.1576
2200	1.2640	1.2531	1.2423	1.2315	1.2212	1.2110	1.2011	1.1916	1.1825	1.1737	1.1654	1.1577
2220	1.2631	1.2523	1.2417	1.2311	1.2209	1.2108	1.2010	1.1916	1.1825	1.1738	1.1655	1.1578
2240	1.2621	1.2514	1.2410	1.2307	1.2206	1.2106	1.2009	1.1916	1.1825	1.1738	1.1655	1.1579
2260	1.2610	1.2505	1.2402	1.2302	1.2202	1.2103	1.2007	1.1915	1.1825	1.1738	1.1656	1.1579
2280	1.2600	1.2495	1.2394	1.2296	1.2197	1.2100	1.2004	1.1913	1.1824	1.1738	1.1656	1.1579
2300	1.2588	1.2485	1.2386	1.2289	1.2192	1.2096	1.2001	1.1911	1.1823	1.1737	1.1656	1.1579
2320	1.2576	1.2475	1.2378	1.2282	1.2186	1.2092	1.1998	1.1909	1.1821	1.1736	1.1655	1.1579
2340	1.2563	1.2465	1.2369	1.2275	1.2180	1.2087	1.1995	1.1906	1.1819	1.1734	1.1654	1.1578
2360	1.2549	1.2454	1.2360	1.2267	1.2173	1.2081	1.1991	1.1903	1.1816	1.1732	1.1652	1.1576
2380	1.2535	1.2442	1.2350	1.2258	1.2166	1.2076	1.1987	1.1899	1.1813	1.1730	1.1650	1.1574

TABLE A.35(a) (Continued)

p_f	Temperature, °F											
psig	0	5	10	15	20	25	30	35	40	45	50	55
2400	1.2521	1.2430	1.2339	1.2249	1.2158	1.2070	1.1983	1.1895	1.1810	1.1727	1.1648	1.1572
2420	1.2507	1.2418	1.2329	1.2240	1.2150	1.2063	1.1977	1.1891	1.1806	1.1724	1.1646	1.1570
2440	1.2491	1.2405	1.2318	1.2231	1.2142	1.2056	1.1971	1.1886	1.1802	1.1720	1.1643	1.1567
2460	1.2475	1.2391	1.2306	1.2221	1.2134	1.2049	1.1965	1.1881	1.1798	1.1716	1.1639	1.1565
2480	1.2459	1.2377	1.2294	1.2210	1.2125	1.2041	1.1958	1.1875	1.1793	1.1712	1.1636	1.1562
2500	1.2443	1.2363	1.2282	1.2199	1.2115	1.2033	1.1951	1.1869	1.1787	1.1708	1.1631	1.1558
2520	1.2427	1.2349	1.2269	1.2188	1.2106	1.2025	1.1944	1.1863	1.1782	1.1703	1.1626	1.1555
2540	1.2411	1.2335	1.2256	1.2176	1.2096	1.2016	1.1936	1.1856	1.1776	1.1698	1.1621	1.1551
2560	1.2395	1.2320	1.2242	1.2164	1.2086	1.2007	1.1928	1.1849	1.1770	1.1693	1.1617	1.1546
2580	1.2378	1.2304	1.2228	1.2152	1.2075	1.1997	1.1919	1.1842	1.1764	1.1687	1.1612	1.1542
2600	1.2361	1.2287	1.2214	1.2140	1.2064	1.1987	1.1910	1.1834	1.1757	1.1681	1.1606	1.1537
2620	1.2343	1.2271	1.2200	1.2127	1.2052	1.1977	1.1901	1.1825	1.1749	1.1675	1.1601	1.1532
2640	1.2325	1.2256	1.2185	1.2113	1.2040	1.1966	1.1892	1.1817	1.1742	1.1668	1.1596	1.1527
2660	1.2308	1.2240	1.2170	1.2100	1.2028	1.1955	1.1882	1.1808	1.1734	1.1661	1.1590	1.1522
2680	1.2290	1.2223	1.2155	1.2086	1.2015	1.1944	1.1872	1.1798	1.1725	1.1653	1.1584	1.1516
2700	1.2272	1.2206	1.2139	1.2072	1.2003	1.1933	1.1862	1.1789	1.1717	1.1646	1.1577	1.1510
2720	1.2253	1.2189	1.2124	1.2058	1.1990	1.1922	1.1852	1.1780	1.1709	1.1638	1.1569	1.1503
2740	1.2234	1.2172	1.2108	1.2044	1.1977	1.1910	1.1841	1.1771	1.1700	1.1630	1.1562	1.1497
2760	1.2216	1.2155	1.2092	1.2029	1.1964	1.1898	1.1830	1.1761	1.1691	1.1622	1.1554	1.1490
2780	1.2197	1.2138	1.2077	1.2014	1.1950	1.1885	1.1818	1.1750	1.1682	1.1613	1.1547	1.1483
2800	1.2178	1.2120	1.2060	1.1999	1.1936	1.1872	1.1806	1.1740	1.1672	1.1605	1.1539	1.1476
2820	1.2159	1.2102	1.2044	1.1983	1.1922	1.1859	1.1794	1.1729	1.1662	1.1596	1.1531	1.1468
2840	1.2140	1.2084	1.2027	1.1968	1.1908	1.1846	1.1782	1.1718	1.1652	1.1586	1.1522	1.1461
2860	1.2120	1.2066	1.2010	1.1952	1.1893	1.1832	1.1770	1.1707	1.1642	1.1577	1.1513	1.1453
2880	1.2100	1.2048	1.1993	1.1937	1.1878	1.1818	1,1757	1.1696	1.1632	1.1568	1.1504	1.1445
2900	1.2081	1.2029	1.1976	1.1921	1.1863	1.1804	1.1744	1.1685	1.1621	1.1558	1.1495	1.1437
2920	1.2062	1.2011	1.1959	1.1905	1.1848	1.1790	1.1731	1.1673	1.1610	1.1547	1.1486	1.1428
2940	1.2042	1.1992	1.1942	1.1888	1.1832	1.1776	1.1718	1.1660	1.1599	1.1537	1.1477	1.1420
2960	1.2023	1.1974	1.1924	1.1872	1.1817	1.1761	1.1705	1.1648	1.1587	1.1527	1.1468	1.1411
2980	1.2004	1.1956	1.1907	1.1856	1.1802	1.1747	1.1692	1.1635	1.1576	1.1516	1.1458	1.1402
3000	1.1984	1.1937	1.1889	1.1839	1.1786	1.1733	1.1678	1.1622	1.1564	1.1505	1.1448	1.1393

Note: Factors for intermediate values of pressure and temperature should be interpolated.

TABLE A.35(a) (Continued)
F_{pv} Supercompressibility Factors
Base Data—0.6 Specific Gravity Hydrocarbon Gas

p_f	Temperature °F												
psig	60	65	70	75	80	85	90	95	100	105	110	115	120
0	1.0000	1.0000	1.0000	1.0000	1.0000	1.0000	1.0000	1.0000	1.000	1.000	1.0000	1.000	1.0000
20	1.0016	1.0015	1.0014	1.0014	1.0014	1.0013	1.0013	1.0012	1.0012	1.0012	1.0011	1.0011	1.0010
40	1.0032	1.0031	1.0030	1.0029	1.0028	1.0027	1.0027	1.0026	1.0025	1.0024	1.0023	1.0022	1.0022
60	1.0047	1.0046	1.0045	1.0043	1.0042	1.0040	1.0039	1.0038	1.0037	1.0036	1.0035	1.0033	1.0032
80	1.0064	1.0062	1.0061	1.0058	1.0056	1.0054	1.0052	1.0051	1.0049	1.0047	1.0046	1.0044	1.0043
100	1.0080	1.0078	1.0075	1.0073	1.0071	1.0068	1.0066	1.0064	1.0061	1.0059	1.0058	1.0056	1.0055
120	1.0097	1.0094	1.0091	1.0088	1.0085	1.0082	1.0079	1.0076	1.0073	1.0071	1.0069	1.0067	1.0065
140	1.0112	1.0109	1.0105	1.0102	1.0099	1.0095	1.0092	1.0088	1.0085	1.0083	1,0080	1.0078	1.0076
160	1.0129	1.0125	1.0121	1.0117	1.0112	1.0108	1.0105	1.0101	1.0098	1.0095	1.0092	1.0089	1.0087
180	1.0145	1.0140	1.0136	1.0131	1.0126	1.0122	1.0118	1.0114	1.0111	1.0107	1.0103	1.0100	1.0098
200	1.0162	1.0156	1.0151	1.0146	1.0140	1.0135	1.0131	1.0127	1.0123	1.0119	1.0115	1.0111	1.0108
220	1.0178	1.0172	1.0166	1.0160	1.0154	1.0149	1.0145	1.0140	1.0136	1.0131	1.0126	1.0122	1.0119
240	1.0194	1.0188	1.0181	1.0175	1.0168	1.0163	1.0158	1.0153	1.0148	1.0143	1.0138	1.0133	1.0129
260	1.0211	1.0204	1.0197	1.0190	1.0183	1.0177	1.0171	1.0165	1.0160	1.0155	1.0150	1.0144	1.0139
280	1.0228	1.0220	1.0212	1.0205	1.0197	1.0191	1.0185	1.0178	1.0173	1.0167	1.0162	1.0155	1.0150
300	1.0244	1.0236	1.0228	1.0220	1.0212	1.0205	1.0199	1.0192	1.0185	1.0179	1.0173	1.0167	1.0162
320	1.0261	1.0252	1.0243	1.0235	1.0227	1.0219	1.0212	1.0205	1.0198	1.0191	1.0185	1.0178	1.0173
340	1.0277	1.0267	1.0258	1.0249	1.0241	1.0233	1.0225	1.0217	1.0209	1.0203	1.0196	1.0189	1.0183
360	1.0294	1.0284	1.0273	1.0264	1.0256	1.0247	1.0238	1.0230	1.0222	1.0215	1.0207	1.0200	1.0194
380	1.0311	1.0300	1.0289	1.0279	1.0270	1.0261	1.0252	1.0243	1.0234	1.0227	1.0219	1.0211	1.0204
400	1.0328	1.0317	1.0305	1.0294	1.0285	1.0275	1.0265	1.0256	1.0246	1.0238	1.0230	1.0223	1.0215
420	1.0345	1.0333	1.0321	1.0309	1.0299	1.0289	1.0279	1.0269	1.0259	1.0250	1.0242	1.0234	1.0226
440	1.0361	1.0349	1.0336	1.0324	1.0313	1.0302	1.0292	1.0281	1.0272	1.0262	1.0253	1.0244	1.0236

460	1.0378	1.0365	1.0351	1.0339	1.0327	1.0315	1.0305	1.0294	1.0285	1.0275	1.0265	1.0255	1.0247
480	1.0395	1.0381	1.0367	1.0254	1.0341	1.0329	1.0318	1.0307	1.0297	1.0287	1.0276	1.0267	1.0258
500	1.0413	1.0398	1.0384	1.0370	1.0356	1.0344	1.0332	1.0320	1.0309	1.0298	1.0288	1.0278	1.0269
520	1.0430	1.0414	1.0399	1.0385	1.0371	1.0357	1.0345	1.0333	1.0321	1.0310	1.0299	1.0289	1.0279
540	1.0447	1.0431	1.0415	1.0400	1.0385	1.0371	1.0358	1.0346	1.0334	1.0322	1.0310	1.0300	1.0289
560	1.0465	1.0448	1.0432	1.0416	1.0400	1.0385	1.0372	1.0359	1.0346	1.0334	1.0322	1.0311	1.0300
580	1.0482	1.0464	1.0447	1.0431	1.0415	1.0399	1.0385	1.0372	1.0358	1.0346	1.0333	1.0322	1.0310
600	1.0499	1.0481	1.0463	1.0446	1.0430	1.0414	1.0399	1.0384	1.0370	1.0358	1.0345	1.0333	1.0321
620	1.0517	1.0497	1.0479	1.0461	1.0445	1.0428	1.0412	1.0397	1.0383	1.0369	1.0356	1.0344	1.0331
640	1.0534	1.0514	1.0495	1.0476	1.0460	1.0442	1.0246	1.0410	1.0396	1.0381	1.0368	1.0355	1.0341
660	1.0552	1.0530	1.0511	1.0492	1.0474	1.0456	1.0439	1.0423	1.0408	1.0393	1.0379	1.0366	1.0352
680	1.0570	1.0547	1.0527	1.0507	1.0488	1.0470	1.0453	1.0436	1.0420	1.0405	1.0390	1.0377	1.0363
700	1.0587	1.0563	1.0543	1.0522	1.0502	1.0483	1.0466	1.0449	1.0432	1.0416	1.0401	1.0387	1.0373
720	1.0605	1.0580	1.0559	1.0537	1.0517	1.0497	1.0479	1.0461	1.0444	1.0428	1.0412	1.0398	1.0383
740	1.0622	1.0597	1.0575	1.0553	1.0531	1.0510	1.0492	1.0474	1.0456	1.0440	1.0424	1.0409	1.0393
760	1.0640	1.0614	1.0591	1.0568	1.0546	1.0524	1.0505	1.0487	1.0468	1.0451	1.0435	1.0419	1.0403
780	1.0658	1.0631	1.0607	1.0583	1.0560	1.0538	1.0519	1.0500	1.0480	1.0463	1.0446	1.0430	1.0414
800	1.0676	1.0648	1.0623	1.0598	1.0575	1.0552	1.0532	1.0513	1.0492	1.0474	1.0456	1.0440	1.0424
820	1.0693	1.0665	1.0639	1.0613	1.0589	1.0566	1.0545	1.0524	1.0504	1.0485	1.0467	1.0450	1.0434
840	1.0711	1.0681	1.0654	1.0628	1.0603	1.0580	1.0558	1.0536	1.0517	1.0497	1.0478	1.0460	1.0443
860	1.0728	1.0697	1.0670	1.0643	1.0617	1.0593	1.0571	1.0549	1.0529	1.0500	1.0489	1.0471	1.0453
880	1.0745	1.0714	1.0686	1.0658	1.0631	1.0607	1.0584	1.0562	1.0540	1.0519	1.0500	1.0481	1.0463
900	1.0762	1.0730	1.0701	1.0673	1.0646	1.0620	1.0597	1.0574	1.0552	1.0530	1.0510	1.0491	1.0473
920	1.0779	1.0746	1.0716	1.0688	1.0660	1.0634	1.0610	1.0586	1.0563	1.0541	1.0520	1.0501	1.0482
940	1.0797	1.0763	1.0733	1.0703	1.0675	1.0649	1.0623	1.0599	1.0575	1.0553	1.0531	1.0511	1.0492
960	1.0814	1.0779	1.0748	1.0718	1.0689	1.0662	1.0636	1.0610	1.0586	1.0563	1.0541	1.0521	1.0501
980	1.0831	1.0795	1.0763	1.0732	1.0703	1.0675	1.0648	1.0622	1.0597	1.0574	1.0552	1.0530	1.0510

(Continued)

TABLE A.35(a) (Continued)

p_f	Temperature °F												
psig	60	65	70	75	80	85	90	95	100	105	110	115	120
1000	1.0847	1.0811	1.0778	1.0746	1.0717	1.0687	1.0660	1.0634	1.0608	1.0585	1.0562	1.0539	1.0519
1020	1.0865	1.0827	1.0794	1.0761	1.0730	1.0701	1.0673	1.0646	1.0619	1.0595	1.0572	1.0549	1.0529
1040	1.0882	1.0843	1.0809	1.0775	1.0744	1.0714	1.0685	1.0658	1.0631	1.0606	1.0582	1.0559	1.0538
1060	1.0900	1.0860	1.0825	1.0790	1.0758	1.0727	1.0697	1.0670	1.0642	1.0617	1.0592	1.0569	1.0547
1080	1.0916	1.0875	1.0839	1.0804	1.0771	1.0740	1.0709	1.0681	1.0654	1.0628	1.0602	1.0578	1.0556
1100	1.0933	1.0891	1.0854	1.0819	1.0785	1.0753	1.0722	1.0692	1.0665	1.0638	1.0612	1.0588	1.0565
1120	1.0950	1.0908	1.0870	1.0834	1.0800	1.0766	1.0734	1.0703	1.0676	1.0649	1.0623	1.0598	1.0574
1140	1.0966	1.0924	1.0885	1.0848	1.0814	1.0779	1.0746	1.0716	1.0687	1.0659	1.0633	1.0607	1.0583
1160	1.0983	1.0939	1.0899	1.0862	1.0826	1.0791	1.0758	1.0727	1.0698	1.0669	1.0643	1.0616	1.0592
1180	1.1000	1.0955	1.0914	1.0875	1.0839	1.0804	1.0771	1.0738	1.0708	1.0679	1.0652	1.0625	1.0601
1200	1.1016	1.0970	1.0928	1.0889	1.0851	1.0816	1.0782	1.0750	1.0718	1.0689	1.0661	1.0634	1.0610
1220	1.1032	1.0985	1.0942	1.0902	1.0864	1.0828	1.0794	1.0760	1.0729	1.0699	1.0671	1.0643	1.0618
1240	1.1048	1.1001	1.0957	1.0916	1.0876	1.0840	1.0805	1.0771	1.0739	1.0709	1.0681	1.0652	1.0626
1260	1.1064	1.1015	1.0971	1.0929	1.0889	1.0852	1.0816	1.0781	1.0748	1.0719	1.0690	1.0661	1.0635
1280	1.1079	1.1030	1.0985	1.0942	1.0901	1.0863	1.0827	1.0791	1.0758	1.0728	1.0699	1.0670	1.0643
1300	1.1094	1.1044	1.0999	1.0955	1.0913	1.0875	1.0838	1.0802	1.0768	1.0737	1.0707	1.0678	1.0651
1320	1.1110	1.1059	1.1012	1.0968	1.0925	1.0886	1.0849	1.0812	1.0778	1.0746	1.0716	1.0686	1.0659
1340	1.1125	1.1073	1.1025	1.0980	1.0937	1.0897	1.0859	1.0822	1.0788	1.0755	1.0725	1.0695	1.0667
1360	1.1140	1.1087	1.1039	1.0993	1.0949	1.0909	1.0870	1.0833	1.0797	1.0764	1.0733	1.0703	1.0675
1380	1.1154	1.1100	1.1052	1.1005	1.0961	1.0920	1.0881	1.0843	1.0806	1.0773	1.0741	1.0711	1.0682
1400	1.1168	1.1114	1.1065	1.1017	1.0973	1.0931	1.0891	1.0853	1.0816	1.0782	1.0750	1.0719	1.0690
1420	1.1183	1.1128	1.1078	1.1030	1.0985	1.0941	1.0902	1.0863	1.0825	1.0791	1.0759	1.0727	1.0697
1440	1.1197	1.1141	1.1090	1.1042	1.0995	1.0952	1.0912	1.0873	1.0834	1.0800	1.0767	1.0735	1.0705
1460	1.1210	1.1154	1.1103	1.1053	1.1006	1.0962	1.0921	1.0882	1.0843	1.0808	1.0775	1.0742	1.0712
1480	1.1225	1.1167	1.1115	1.1064	1.1016	1.0973	1.0931	1.0891	1.0852	1.0816	1.0783	1.0750	1.0719

1500	1.1238	1.1179	1.1126	1.1075	1.1027	1.0983	1.0941	1.0900	1.0861	1.0825	1.0791	1.0758	1.0727
1520	1.1251	1.1191	1.1138	1.1087	1.1038	1.0993	1.0950	1.0909	1.0870	1.0833	1.0799	1.0766	1.0734
1540	1.1263	1.1204	1.1150	1.1098	1.1049	1.1003	1.0960	1.0918	1.0879	1.0842	1.0807	1.0773	1.0741
1560	1.1276	1.1215	1.1161	1.1108	1.1059	1.1012	1.0969	1.0927	1.0887	1.0850	1.0815	1.0780	1.0748
1580	1.1288	1.1227	1.1172	1.1119	1.1068	1.1022	1.0978	1.0935	1.0896	1.0858	1.0823	1.0788	1.0755
1600	1.1301	1.1238	1.1183	1.1129	1.1078	1.1031	1.0987	1.0944	1.0904	1.0866	1.0830	1.0795	1.0762
1620	1.1312	1.1249	1.1193	1.1139	1.1088	1.1041	1.0995	1.0952	1.0912	1.0873	1.0837	1.0802	1.0768
1640	1.1323	1.1260	1.1203	1.1149	1.1097	1.1049	1.1004	1.0960	1.0920	1.0881	1.0844	1.0809	1.0775
1660	1.1334	1.1270	1.1213	1.1158	1.1106	1.1058	1.1012	1.0968	1.0927	1.0888	1.0851	1.0815	1.0781
1680	1.1345	1.1281	1.1223	1.1167	1.1115	1.1066	1.1020	1.0976	1.0934	1.0895	1.0858	1.0822	1.0787
1700	1.1335	1.1290	1.1232	1.1176	1.1124	1.1074	1.1028	1.0984	1.0942	1.0903	1.0865	1.0828	1.0793
1720	1.1366	1.1300	1.1241	1.1185	1.1132	1.1082	1.1036	1.0992	1.0950	1.0910	1.0872	1.0835	1.0799
1740	1.1376	1.1309	1.1250	1.1193	1.1139	1.1089	1.1044	1.0999	1.0957	1.0917	1.0878	1.0841	1.0805
1760	1.1385	1.1318	1.1258	1.1201	1.1147	1.1097	1.1051	1.1006	1.0964	1.0923	1.0884	1.0847	1.0811
1780	1.1393	1.1326	1.1266	1.1209	1.1154	1.1104	1.1058	1.1012	1.0970	1.0929	1.0890	1.0853	1.0816
1800	1.1402	1.1334	1.1273	1.1216	1.1161	1.1111	1.1064	1.1019	1.0976	1.0935	1.0896	1.0858	1.0821
1820	1.1410	1.1342	1.1281	1.1223	1.1168	1.1118	1.1071	1.1025	1.0982	1.0941	1.0902	1.0863	1.0826
1840	1.1418	1.1349	1.1288	1.1230	1.1175	1.1124	1.1077	1.1031	1.0988	1.0947	1.0907	1.0868	1.0831
1860	1.1426	1.1357	1.1295	1.1237	1.1181	1.1130	1.1083	1.1037	1.0094	1.0952	1.0911	1.0873	1.0836
1880	1.1433	1.1364	1.1302	1.1243	1.1187	1.1137	1.1089	1.1043	1.0999	1.0957	1.0916	1.0877	1.0840
1900	1.1440	1.1371	1.1309	1.1249	1.1193	1.1142	1.1094	1.1048	1.1004	1.0962	1.0920	1.0881	1.0844
1920	1.1447	1.1378	1.1315	1.1255	1.1199	1.1148	1.1099	1.1053	1.1009	1.0967	1.0925	1.0886	1.0848
1940	1.1454	1.1384	1.1321	1.1261	1.1204	1.1153	1.1104	1.1058	1.1014	1.0971	1.0929	1.0890	1.0852
1960	1.1460	1.1389	1.1326	1.1266	1.1209	1.1158	1.1109	1.1063	1.1019	1.0976	1.0934	1.0894	1.0856
1980	1.1465	1.1394	1.1331	1.1271	1.1214	1.1163	1.1114	1.1068	1.1023	1.0980	1.0938	1.0898	1.0860
2000	1.1470	1.1399	1.1336	1.1276	1.1219	1.1168	1.1119	1.1073	1.1027	1.0984	1.0942	1.0902	1.0864
2020	1.1475	1.1403	1.1340	1.1280	1.1223	1.1172	1.1123	1.1077	1.1031	1.0988	1.0946	1.0906	1.0867
2040	1.1480	1.1408	1.1344	1.1284	1.1227	1.1176	1.1127	1.1081	1.1034	1.0992	1.0950	1.0909	1.0870
2060	1.1484	1.1412	1.1349	1.1288	1.1231	1.1180	1.1131	1.1085	1.1038	1.0995	1.0953	1.0912	1.0873
2080	1.1488	1.1416	1.1353	1.1291	1.1234	1.1184	1.1134	1.1088	1.1042	1.0999	1.0956	1.0915	1.0876

(Continued)

TABLE A.35(a) (Continued)

p_f	Temperature °F												
psig	60	65	70	75	80	85	90	95	100	105	110	115	120
2100	1.1491	1.1419	1.1355	1.1294	1.1237	1.1186	1.1137	1.1091	1.1045	1.1002	1.0959	1.0917	1.0878
2120	1.1494	1.1422	1.1358	1.1297	1.1240	1.1189	1.1140	1.1094	1.1048	1.1004	1.0961	1.0919	1.0880
2140	1.1497	1.1425	1.1361	1.1300	1.1243	1.1192	1.1143	1.1096	1.1050	1.1006	1.0963	1.0921	1.0882
2160	1.1499	1.1427	1.1363	1.1302	1.1245	1.1194	1.1145	1.1098	1.1052	1.1008	1.0965	1.0923	1.0884
2180	1.1501	1.1429	1.1365	1.1304	1.1248	1.1196	1.1147	1.1100	1.1054	1.1010	1.0967	1.0925	1.0886
2200	1.1503	1.1431	1.1367	1.1306	1.1250	1.1198	1.1149	1.1102	1.1056	1.1012	1.0969	1.0927	1.0888
2220	1.1504	1.1432	1.1368	1.1308	1.1252	1.1200	1.1151	1.1104	1.1058	1.1014	1.0971	1.0928	1.0890
2240	1.1505	1.1433	1.1369	1.1310	1.1254	1.1201	1.1152	1.1105	1.1059	1.1015	1.0972	1.0930	1.0891
2260	1.1505	1.1434	1.1370	1.1311	1.1256	1.1203	1.1153	1.1107	1.1060	1.1016	1.0973	1.0931	1.0892
2280	1.1505	1.1434	1.1371	1.1312	1.1257	1.1204	1.1154	1.1108	1.1061	1.1017	1.0974	1.0932	1.0893
2300	1.1505	1.1434	1.1371	1.1312	1.1258	1.1205	1.1155	1.1109	1.1062	1.1018	1.0975	1.0933	1.0894
2320	1.1504	1.1434	1.1371	1.1312	1.1258	1.1205	1.1156	1.1110	1.1063	1.1019	1.0976	1.0934	1.0895
2340	1.1503	1.1433	1.1371	1.1312	1.1258	1.1205	1.1156	1.1110	1.1063	1.1020	1.0977	1.0935	1.0896
2360	1.1502	1.1432	1.1370	1.1312	1.1258	1.1205	1.1156	1.1110	1.1063	1.1020	1.0978	1.0936	1.0897
2380	1.1501	1.1431	1.1369	1.1312	1.1257	1.1205	1.1156	1.1110	1.1063	1.1020	1.0978	1.0937	1.0897
2400	1.1499	1.1429	1.1368	1.1311	1.1256	1.1205	1.1156	1.1110	1.1063	1.1020	1.0978	1.0937	1.0897
2420	1.1497	1.1428	1.1367	1.1310	1.1256	1.1205	1.1156	1.1110	1.1063	1.1020	1.0978	1.0937	1.0897
2440	1.1495	1.1426	1.1366	1.1309	1.1255	1.1204	1.1155	1.1109	1.1063	1.1020	1.0978	1.0937	1.0897
2460	1.1493	1.1424	1.1365	1.1308	1.1254	1.1203	1.1154	1.1108	1.1062	1.1019	1.0977	1.0936	1.0896
2480	1.1491	1.1422	1.1363	1.1306	1.1253	1.1201	1.1153	1.1107	1.1061	1.1018	1.0976	1.0935	1.0895
2500	1.1488	1.1420	1.1361	1.1304	1.1251	1.1200	1.1152	1.1106	1.1060	1.1017	1.0975	1.0934	1.0894
2520	1.1485	1.1417	1.1358	1.1302	1.1249	1.1198	1.1151	1.1105	1.1059	1.1016	1.0974	1.0933	1.0893
2540	1.1482	1.1414	1.1356	1.1300	1.1247	1.1196	1.1149	1.1103	1.1057	1.1014	1.0973	1.0932	1.0892
2560	1.1478	1.1411	1.1352	1.1297	1.1244	1.1194	1.1147	1.1101	1.1055	1.1013	1.0972	1.0931	1.0891
2580	1.1474	1.1408	1.1349	1.1294	1.1242	1.1191	1.1145	1.1099	1.1053	1.1011	1.0970	1.0930	1.0890

2600	1.1470	1.1404	1.1345	1.1290	1.1239	1.1189	1.1142	1.1097	1.1051	1.1010	1.0968	1.0929	1.0889
2620	1.1466	1.1400	1.1341	1.1287	1.1236	1.1186	1.1139	1.1094	1.1049	1.1008	1.0967	1.0927	1.0887
2640	1.1461	1.1396	1.1337	1.1283	1.1232	1.1183	1.1136	1.1091	1.1047	1.1006	1.0965	1.0925	1.0885
2660	1.1456	1.1392	1.1333	1.1279	1.1229	1.1180	1.1133	1.1088	1.1045	1.1004	1.0963	1.0923	1.0883
2680	1.1450	1.1387	1.1329	1.1275	1.1225	1.1177	1.1130	1.1085	1.1042	1.1001	1.0960	1.0921	1.0881
2700	1.1445	1.1382	1.1325	1.1270	1.1221	1.1173	1.1127	1.1082	1.1039	1.0998	1.0958	1.0918	1.0879
2720	1.1440	1.1377	1.1320	1.1266	1.1217	1.1170	1.1123	1.1079	1.1036	1.0995	1.0955	1.0917	1.0876
2740	1.1434	1.1371	1.1315	1.1262	1.1213	1.1166	1.1120	1.1076	1.1033	1.0992	1.0952	1.0914	1.0874
2760	1.1428	1.1366	1.1310	1.1257	1.1208	1.1162	1.1116	1.1072	1.1030	1.0989	1.0949	1.0911	1.0872
2780	1.1421	1.1360	1.1305	1.1252	1.1204	1.1157	1.1112	1.1069	1.1027	1.0986	1.0946	1.0908	1.0869
2800	1.1414	1.1354	1.1299	1.1247	1.1199	1.1153	1.1108	1.1065	1.1024	1.0983	1.0943	1.0904	1.0866
2820	1.1408	1.1349	1.1294	1.1242	1.1194	1.1148	1.1104	1.1061	1.1020	1.0979	1.0940	1.0901	1.0863
2840	1.1401	1.1343	1.1288	1.1237	1.1188	1.1144	1.1099	1.1057	1.1016	1.0975	1.0936	1.0898	1.0860
2860	1.1394	1.1336	1.1282	1.1231	1.1183	1.1139	1.1094	1.1052	1.1012	1.0972	1.0933	1.0895	1.0857
2880	1.1387	1.1330	1.1276	1.1225	1.1177	1.1134	1.1090	1.1047	1.1008	1.0968	1.0929	1.0892	1.0854
2900	1.1379	1.1324	1.1270	1.1219	1.1172	1.1128	1.1085	1.1042	1.1003	1.0964	1.0925	1.0888	1.0851
2920	1.1371	1.1316	1.1263	1.1213	1.1166	1.1123	1.1079	1.1037	1.0998	1.0959	1.0921	1.0884	1.0847
2940	1.1364	1.1309	1.1256	1.1207	1.1160	1.1117	1.1074	1.1032	1.0993	1.0955	1.0917	1.0880	1.0844
2960	1.1355	1.1302	1.1249	1.1201	1.1155	1.1111	1.1069	1.1027	1.0988	1.0950	1.0913	1.0876	1.0840
2980	1.1347	1.1294	1.1242	1.1194	1.1149	1.1105	1.1063	1.1022	1.0983	1.0945	1.0908	1.0871	1.0835
3000	1.1339	1.1288	1.1235	1.1187	1.1142	1.1099	1.1058	1.1017	1.0978	1.0941	1.0904	1.0867	1.0831

Note: Factors for intermediate values of pressure and temperature should be interpolated.

TABLE A.35(a) (Continued)
F_{pv} Supercompressibility Factors
Base Data—0.6 Specific Gravity Hydrocarbon Gas

p_f	Temperature, °F												
psig	125	130	135	140	145	150	155	160	165	170	175	180	185
0	1.0000	1.0000	1.0000	1.0000	1.0000	1.0000	1.0000	1.0000	1.0000	1.0000	1.0000	1.0000	1.0000
20	1.0010	1.0010	1.0010	1.0010	1.0009	1.0009	1.0009	1.0008	1.0008	1.0008	1.0008	1.0007	1.0007
40	1.0022	1.0020	1.0020	1.0020	1.0019	1.0018	1.0018	1.0017	1.0016	1.0016	1.0016	1.0015	1.0014
60	1.0032	1.0030	1.0030	1.0029	1.0028	1.0027	1.0027	1.0026	1.0024	1.0023	1.0023	1.0022	1.0021
80	1.0042	1.0040	1.0039	1.0039	1.0038	1.0036	1.0035	1.0034	1.0032	1.0031	1.0030	1.0029	1.0028
100	1.0053	1.0051	1.0049	1.0048	1.0047	1.0045	1.0044	1.0042	1.0040	1.0039	1.0038	1.0037	1.0035
120	1.0063	1.0061	1.0059	1.0057	1.0056	1.0054	1.0052	1.0050	1.0048	1.0047	1.0045	1.0044	1.0042
140	1.0074	1.0071	1.0068	1.0066	1.0065	1.0063	1.0060	1.0058	1.0056	1.0055	1.0053	1.0051	1.0049
160	1.0084	1.0081	1.0078	1.0076	1.0074	1.0072	1.0069	1.0067	1.0064	1.0063	1.0061	1.0058	1.0056
180	1.0094	1.0091	1.0088	1.0085	1.0083	1.0081	1.0078	1.0075	1.0072	1.0070	1.0068	1.0065	1.0063
200	1.0104	1.0101	1.0097	1.0094	1.0092	1.0089	1.0086	1.0083	1.0080	1.0078	1.0075	1.0073	1.0070
220	1.0115	1.0111	1.0107	1.0104	1.0101	1.0098	1.0095	1.0092	1.0088	1.0086	1.0083	1.0080	1.0077
240	1.0125	1.0121	1.0117	1.0114	1.0110	1.0107	1.0103	1.0100	1.0096	1.0094	1.0090	1.0087	1.0084
260	1.0135	1.0132	1.0128	1.0123	1.0119	1.0116	1.0112	1.0109	1.0104	1.0102	1.0098	1.0095	1.0091
280	1.0146	1.0142	1.0137	1.0132	1.0128	1.0125	1.0121	1.0117	1.0112	1.0109	1.0105	1.0102	1.0098
300	1.0157	1.0152	1.0146	1.0141	1.0137	1.0134	1.0130	1.0125	1.0121	1.0116	1.0112	1.0109	1.0105
320	1.0167	1.0161	1.0156	1.0151	1.0146	1.0142	1.0138	1.0133	1.0129	1.0124	1.0119	1.0116	1.0112
340	1.0177	1.0171	1.0165	1.0160	1.0155	1.0151	1.0146	1.0141	1.0137	1.0132	1.0127	1.0122	1.0118
360	1.0187	1.0181	1.0175	1.0169	1.0164	1.0159	1.0154	1.0149	1.0144	1.0139	1.0134	1.0129	1.0125
380	1.0197	1.0191	1.0185	1.0179	1.0173	1.0168	1.0163	1.0157	1.0152	1.0146	1.0141	1.0136	1.0131
400	1.0208	1.0201	1.0195	1.0189	1.0182	1.0177	1.0171	1.0165	1.0160	1.0154	1.0149	1.0143	1.0138
420	1.0218	1.0211	1.0204	1.0198	1.0191	1.0185	1.0179	1.0173	1.0167	1.0161	1.0156	1.0150	1.0144
440	1.0228	1.0220	1.0213	1.0207	1.0200	1.0193	1.0187	1.0181	1.0175	1.0168	1.0162	1.0156	1.0151

460	1.0238	1.0229	1.0222	1.0216	1.0219	1.0202	1.0196	1.0189	1.0182	1.0175	1.0169	1.0163	1.0157
480	1.0248	1.0239	1.0232	1.0225	1.0218	1.0211	1.0204	1.0197	1.0190	1.0183	1.0176	1.0169	1.0163
500	1.0259	1.0249	1.0242	1.0234	1.0227	1.0220	1.0212	1.0205	1.0197	1.0190	1.0183	1.0176	1.0170
520	1.0269	1.0258	1.0251	1.0243	1.0235	1.0228	1.0220	1.0212	1.0204	1.0197	1.0190	1.0183	1.0177
540	1.0279	1.0268	1.0260	1.0252	1.0243	1.0236	1.0228	1.0220	1.0212	1.0204	1.0197	1.0190	1.0183
560	1.0289	1.0278	1.0269	1.0261	1.0252	1.0245	1.0236	1.0227	1.0219	1.0211	1.0204	1.0196	1.0189
580	1.0299	1.0288	1.0279	1.0270	1.0261	1.0253	1.0244	1.0235	1.0226	1.0218	1.0210	1.0202	1.0195
600	1.0309	1.0298	1.0288	1.0279	1.0270	1.0261	1.0251	1.0242	1.0233	1.0225	1.0217	1.0209	1.0201
620	1.0319	1.0308	1.0298	1.0288	1.0278	1.0269	1.0259	1.0250	1.0241	1.0232	1.0223	1.0215	1.0207
640	1.0329	1.0317	1.0307	1.0296	1.0287	1.0277	1.0267	1.0257	1.0248	1.0239	1.0230	1.0222	1.0214
660	1.0340	1.0327	1.0316	1.0305	1.0295	1.0285	1.0275	1.0265	1.0255	1.0246	1.0237	1.0228	1.0220
680	1.0350	1.0337	1.0325	1.0314	1.0304	1.0293	1.0282	1.0272	1.0262	1.0253	1.0244	1.0235	1.0226
700	1.0359	1.0346	1.0334	1.0323	1.0312	1.0301	1.0290	1.0279	1.0269	1.0259	1.0250	1.0241	1.0231
720	1.0369	1.0355	1.0343	1.0331	1.0320	1.0309	1.0298	1.0287	1.0276	1.0266	1.0257	1.0247	1.0237
740	1.0379	1.0365	1.0352	1.0340	1.0328	1.0316	1.0305	1.0294	1.0283	1.0273	1.0263	1.0253	1.0243
760	1.0388	1.0374	1.0361	1.0349	1.0336	1.0324	1.0313	1.0301	1.0290	1.0280	1.0269	1.0259	1.0249
780	1.0398	1.0384	1.0371	1.0358	1.0344	1.0332	1.0320	1.0308	1.0297	1.0286	1.0275	1.0265	1.0255
800	1.0408	1.0393	1.0380	1.0366	1.0353	1.0340	1.0327	1.0315	1.0303	1.0292	1.0281	1.0271	1.0260
820	1.0418	1.0402	1.0388	1.0374	1.0360	1.0347	1.0334	1.0322	1.0310	1.0299	1.0287	1.0277	1.0266
840	1.0427	1.0412	1.0396	1.0382	1.0368	1.0355	1.0342	1.0329	1.0317	1.0306	1.0294	1.0283	1.0272
860	1.0437	1.0421	1.0405	1.0391	1.0376	1.0362	1.0349	1.0336	1.0324	1.0312	1.0300	1.0288	1.0277
880	1.0446	1.0430	1.0414	1.0399	1.0384	1.0370	1.0356	1.0343	1.0330	1.0318	1.0306	1.0294	1.0282
900	1.0455	1.0439	1.0423	1.0407	1.0392	1.0377	1.0363	1.0350	1.0336	1.0324	1.0311	1.0299	1.0287
920	1.0464	1.0448	1.0431	1.0415	1.0400	1.0385	1.0371	1.0357	1.0343	1.0330	1.0317	1.0305	1.0293
940	1.0474	1.0457	1.0440	1.0423	1.0408	1.0393	1.0378	1.0363	1.0350	1.0336	1.0323	1.0310	1.0298
960	1.0483	1.0465	1.0448	1.0431	1.0415	1.0400	1.0385	1.0370	1.0356	1.0342	1.0329	1.0315	1.0303
980	1.0492	1.0473	1.0456	1.0439	1.0422	1.0407	1.0391	1.0376	1.0363	1.0348	1.0334	1.0321	1.0308

(Continued)

TABLE A.35(a) (Continued)

p_f	Temperature, °F												
psig	125	130	135	140	145	150	155	160	165	170	175	180	185
1000	1.0501	1.0481	1.0463	1.0446	1.0429	1.0413	1.0398	1.0383	1.0369	1.0354	1.0340	1.0326	1.0313
1020	1.0509	1.0489	1.0471	1.0454	1.0437	1.0420	1.0404	1.0389	1.0375	1.0360	1.0345	1.0331	1.0318
1040	1.0518	1.0498	1.0480	1.0461	1.0444	1.0428	1.0412	1.0396	1.0381	1.0365	1.0350	1.0336	1.0323
1060	1.0527	1.0506	1.0487	1.0469	1.0452	1.0435	1.0419	1.0403	1.0387	1.0371	1.0356	1.0341	1.0328
1080	1.0535	1.0514	1.0495	1.0476	1.0459	1.0442	1.0425	1.0409	1.0393	1.0377	1.0361	1.0346	1.0333
1100	1.0544	1.0522	1.0503	1.0484	1.0465	1.0448	1.0431	1.0415	1.0399	1.0383	1.0367	1.0352	1.0338
1120	1.0552	1.0530	1.0510	1.0491	1.0472	1.0454	1.0437	1.0420	1.0404	1.0388	1.0372	1.0357	1.0343
1140	1.0561	1.0538	1.0517	1.0498	1.0478	1.0460	1.0443	1.0426	1.0410	1.0394	1.0378	1.0362	1.0348
1160	1.0569	1.0546	1.0525	1.0505	1.0485	1.0467	1.0450	1.0432	1.0415	1.0399	1.0382	1.0367	1.0352
1180	1.0577	1.0554	1.0533	1.0512	1.0492	1.0473	1.0456	1.0438	1.0421	1.0404	1.0388	1.0372	1.0357
1200	1.0585	1.0562	1.0540	1.0519	1.0499	1.0479	1.0461	1.0443	1.0426	1.0410	1.0393	1.0377	1.0361
1220	1.0593	1.0569	1.0547	1.0526	1.0506	1.0486	1.0467	1.0449	1.0432	1.0415	1.0398	1.0381	1.0365
1240	1.0601	1.0577	1.0554	1.0532	1.0512	1.0492	1.0473	1.0455	1.0437	1.0420	1.0403	1.0386	1.0370
1260	1.0610	1.0585	1.0561	1.0539	1.0518	1.0498	1.0479	1.0461	1.0443	1.0425	1.0407	1.0391	1.0375
1280	1.0618	1.0592	1.0568	1.0546	1.0524	1.0504	1.0485	1.0466	1.0448	1.0430	1.0412	1.0395	1.0379
1300	1.0626	1.0600	1.0575	1.0552	1.0530	1.0510	1.0490	1.0471	1.0453	1.0435	1.0417	1.0399	1.0383
1320	1.0633	1.0607	1.0583	1.0559	1.0536	1.0515	1.0496	1.0477	1.0457	1.0439	1.0421	1.0404	1.0386
1340	1.0640	1.0614	1.0590	1.0565	1.0542	1.0521	1.0501	1.0482	1.0462	1.0444	1.0426	1.0408	1.0390
1360	1.0647	1.0621	1.0597	1.0572	1.0548	1.0527	1.0506	1.0487	1.0467	1.0449	1.0431	1.0412	1.0394
1380	1.0654	1.0628	1.0603	1.0578	1.0554	1.0532	1.0511	1.0492	1.0472	1.0453	1.0435	1.0416	1.0398
1400	1.0661	1.0635	1.0610	1.0585	1.0560	1.0537	1.0516	1.0497	1.0477	1.0458	1.0439	1.0420	1.0402
1420	1.0669	1.0641	1.0616	1.0591	1.0566	1.0543	1.0522	1.0501	1.0481	1.0462	1.0443	1.0424	1.0406
1440	1.0676	1.0648	1.0622	1.0596	1.0571	1.0548	1.0527	1.0506	1.0486	1.0466	1.0447	1.0428	1.0410
1460	1.0683	1.0655	1.0629	1.0603	1.0577	1.0554	1.0532	1.0510	1.0490	1.0470	1.0451	1.0432	1.0414
1480	1.0690	1.0662	1.0635	1.0609	1.0583	1.0560	1.0538	1.0515	1.0494	1.0474	1.0454	1.0435	1.0417

1500	1.0697	1.0668	1.0641	1.0614	1.0588	1.0565	1.0542	1.0520	1.0499	1.0478	1.0458	1.0439	1.0421
1520	1.0704	1.0675	1.0647	1.0620	1.0594	1.0570	1.0547	1.0525	1.0503	1.0482	1.0462	1.0443	1.0425
1540	1.0711	1.0681	1.0653	1.0626	1.0600	1.0575	1.0552	1.0530	1.0507	1.0486	1.0466	1.0446	1.0428
1560	1.0717	1.0687	1.0658	1.0631	1.0605	1.0579	1.0556	1.0534	1.0511	1.0490	1.0470	1.0450	1.0432
1580	1.0724	1.0693	1.0664	1.0637	1.0610	1.0584	1.0561	1.0538	1.0516	1.0494	1.0473	1.0453	1.0435
1600	1.0730	1.0699	1.0670	1.0642	1.0615	1.0589	1.0566	1.0543	1.0520	1.0498	1.0477	1.0457	1.0439
1620	1.0736	1.0705	1.0675	1.0647	1.0620	1.0593	1.0570	1.0547	1.0524	1.0502	1.0481	1.0460	1.0442
1640	1.0743	1.0711	1.0681	1.0652	1.0625	1.0597	1.0574	1.0551	1.0528	1.0505	1.0484	1.0464	1.0445
1660	1.0748	1.0716	1.0686	1.0657	1.0630	1.0602	1.0578	1.0554	1.0531	1.0509	1.0488	1.0467	1.0447
1680	1.0754	1.0721	1.0691	1.0662	1.0634	1.0606	1.0582	1.0558	1.0535	1.0512	1.0491	1.0470	1.0450
1700	1.0759	1.0726	1.0696	1.0667	1.0639	1.0611	1.0586	1.0562	1.0539	1.0516	1.0494	1.0473	1.0453
1720	1.0765	1.0732	1.0701	1.0672	1.0643	1.0615	1.0591	1.0567	1.0542	1.0519	1.0497	1.0476	1.0456
1740	1.0770	1.0737	1.0707	1.0677	1.0648	1.0620	1.0595	1.0571	1.0546	1.0522	1.0500	1.0479	1.0459
1760	1.0776	1.0742	1.0711	1.0681	1.0652	1.0624	1.0598	1.0574	1.0549	1.0525	1.0503	1.0482	1.0462
1780	1.0781	1.0747	1.0716	1.0686	1.0656	1.0628	1.0602	1.0577	1.0553	1.0528	1.0505	1.0484	1.0464
1800	1.0786	1.0752	1.0720	1.0690	1.0659	1.0631	1.0605	1.0580	1.0556	1.0531	1.0508	1.0487	1.0466
1820	1.0791	1.0757	1.0725	1.0694	1.0663	1.0635	1.0608	1.0583	1.0559	1.0534	1.0511	1.0490	1.0469
1840	1.0796	1.0761	1.0729	1.0698	1.0667	1.0639	1.0612	1.0586	1.0562	1.0537	1.0514	1.0492	1.0471
1860	1.0800	1.0765	1.0733	1.0702	1.0671	1.0643	1.0615	1.0589	1.0564	1.0539	1.0516	1.0495	1.0473
1880	1.0805	1.0769	1.0737	1.0706	1.0675	1.0647	1.0618	1.0592	1.0567	1.0542	1.0519	1.0497	1.0475
1900	1.0809	1.0773	1.0741	1.0709	1.0678	1.0650	1.0621	1.0595	1.0569	1.0544	1.0521	1.0499	1.0477
1920	1.0813	1.0777	1.0745	1.0713	1.0682	1.0653	1.0625	1.0598	1.0572	1.0546	1.0523	1.0501	1.0479
1940	1.0817	1.0781	1.0749	1.0716	1.0685	1.0656	1.0628	1.0600	1.0574	1.0548	1.0524	1.0503	1.0480
1960	1.0820	1.0784	1.0752	1.0719	1.0688	1.0659	1.0630	1.0602	1.0576	1.0550	1.0526	1.0505	1.0482
1980	1.0823	1.0787	1.0754	1.0721	1.0690	1.0661	1.0632	1.0605	1.0578	1.0552	1.0528	1.0506	1.0483
2000	1.0826	1.0790	1.0757	1.0724	1.0693	1.0664	1.0635	1.0607	1.0580	1.0554	1.0530	1.0508	1.0485
2020	1.0830	1.0793	1.0760	1.0727	1.0696	1.0667	1.0638	1.0609	1.0582	1.0556	1.0532	1.0509	1.0487
2040	1.0833	1.0796	1.0762	1.0729	1.0698	1.0669	1.0640	1.0611	1.0584	1.0558	1.0534	1.0511	1.0488
2060	1.0836	1.0799	1.0765	1.0732	1.0701	1.0671	1.0642	1.0613	1.0586	1.0560	1.0536	1.0512	1.0489
2080	1.0839	1.0801	1.0767	1.0734	1.0703	1.0673	1.0644	1.0615	1.0588	1.0562	1.0537	1.0514	1.0490

(Continued)

TABLE A.35(a) (Continued)

p_f	Temperature, °F												
psig	125	130	135	140	145	150	155	160	165	170	175	180	185
2100	1.0841	1.0804	1.0770	1.0736	1.0705	1.0674	1.0645	1.0617	1.0589	1.0563	1.0538	1.0515	1.0491
2120	1.0843	1.0807	1.0772	1.0738	1.0707	1.0676	1.0647	1.0619	1.0590	1.0565	1.0540	1.0516	1.0492
2140	1.0845	1.0809	1.0774	1.0740	1.0709	1.0677	1.0648	1.0620	1.0591	1.0566	1.0541	1.0517	1.0493
2160	1.0847	1.0811	1.0776	1.0742	1.0711	1.0679	1.0650	1.0621	1.0592	1.0567	1.0542	1.0518	1.0493
2180	1.0849	1.0813	1.0778	1.0744	1.0712	1.0680	1.0651	1.0622	1.0593	1.0568	1.0543	1.0519	1.0494
2200	1.0851	1.0814	1.0780	1.0746	1.0713	1.0681	1.0652	1.0623	1.0594	1.0569	1.0544	1.0520	1.0495
2220	1.0853	1.0816	1.0781	1.0747	1.0714	1.0682	1.0653	1.0624	1.0595	1.0569	1.0544	1.0521	1.0495
2240	1.0854	1.0817	1.0782	1.0748	1.0715	1.0683	1.0654	1.0625	1.0596	1.0570	1.0545	1.0521	1.0495
2260	1.0855	1.0818	1.0783	1.0749	1.0716	1.0684	1.0654	1.0625	1.0596	1.0570	1.0545	1.0521	1.0496
2280	1.0856	1.0819	1.0784	1.0750	1.0717	1.0685	1.0655	1.0626	1.0597	1.0571	1.0546	1.0521	1.0496
2300	1.0856	1.0819	1.0784	1.0750	1.0718	1.0685	1.0655	1.0626	1.0597	1.0571	1.0546	1.0521	1.0496
2320	1.0857	1.0820	1.0785	1.0751	1.0719	1.0686	1.0655	1.0626	1.0597	1.0571	1.0546	1.0522	1.0496
2340	1.0857	1.0821	1.0785	1.0752	1.0719	1.0687	1.0656	1.0627	1.0598	1.0572	1.0547	1.0522	1.0497
2360	1.0858	1.0821	1.0786	1.0752	1.0719	1.0687	1.0656	1.0627	1.0598	1.0572	1.0547	1.0522	1.0497
2380	1.0858	1.0821	1.0786	1.0752	1.0719	1.0687	1.0656	1.0627	1.0598	1.0572	1.0547	1.0522	1.0497
2400	1.0859	1.0822	1.0787	1.0752	1.0719	1.0687	1.0657	1.0628	1.0599	1.0572	1.0546	1.0521	1.0496
2420	1.0859	1.0822	1.0787	1.0752	1.0719	1.0687	1.0657	1.0628	1.0599	1.0572	1.0546	1.0521	1.0496
2440	1.0859	1.0822	1.0787	1.0752	1.0719	1.0687	1.0657	1.0628	1.0599	1.0572	1.0546	1.0521	1.0496
2460	1.0858	1.0822	1.0786	1.0751	1.0718	1.0687	1.0657	1.0627	1.0598	1.0571	1.0545	1.0521	1.0495
2480	1.0858	1.0822	1.0786	1.0751	1.0718	1.0687	1.0657	1.0627	1.0598	1.0571	1.0545	1.0520	1.0495
2500	1.0857	1.0821	1.0785	1.0750	1.0717	1.0686	1.0657	1.0627	1.0598	1.0571	1.0545	1.0510	1.0494
2520	1.0856	1.0820	10.784	1.0749	1.0716	1.0685	1.0656	1.0627	1.0598	1.0571	1.0544	1.0519	1.0493
2540	1.0855	1.0819	1.0783	1.0748	1.0716	1.0684	1.0655	1.0626	1.0597	1.0570	1.0543	1.0518	1.0493
2560	1.0854	1.0818	1.0782	1.0747	1.0715	1.0683	1.0654	1.0625	1.0596	1.0569	1.0542	1.0517	1.0492
2580	1.0853	1.0816	1.0781	1.0746	1.0714	1.0682	1.0653	1.0624	1.0595	1.0567	1.0541	1.0516	1.0491

p_f	Temperature, °F												
psig	125	130	135	140	145	150	155	160	165	170	175	180	185
2600	1.0852	1.0814	1.0779	1.0745	1.0713	1.0681	1.0651	1.0622	1.0594	1.0566	1.0540	1.0515	1.0490
2620	1.0850	1.0813	1.0778	1.0744	1.0712	1.0680	1.0650	1.0621	1.0593	1.0565	1.0539	1.0514	1.0489
2640	1.0848	1.0811	1.0776	1.0742	1.0710	1.0679	1.0648	1.0619	1.0591	1.0563	1.0537	1.0513	1.0488
2660	1.0846	1.0809	1.0774	1.0741	1.0709	1.0678	1.0647	1.0618	1.0590	1.0562	1.0536	1.0512	1.0487
2680	1.0844	1.0807	1.0772	1.0739	1.0707	1.0676	1.0645	1.0616	1.0588	1.0560	1.0534	1.0511	1.0485
2700	1.0842	1.0805	1.0770	1.0737	1.0705	1.0675	1.0644	1.0615	1.0587	1.0559	1.0533	1.0509	1.0484
2720	1.0840	1.0803	1.0768	1.0734	1.0703	1.0673	1.0643	1.0614	1.0586	1.0558	1.0532	1.0507	1.0482
2740	1.0838	1.0801	1.0766	1.0732	1.0701	1.0671	1.0641	1.0612	1.0584	1.0556	1.0530	1.0505	1.0480
2760	1.0835	1.0799	1.0764	1.0730	1.0699	1.0669	1.0639	1.0610	1.0582	1.0555	1.0528	1.0503	1.0478
2780	1.0833	1.0796	1.0762	1.0728	1.0697	1.0667	1.0638	1.0608	1.0580	1.0553	1.0527	1.0501	1.0476
2800	1.0830	1.0793	1.0759	1.0726	1.0695	1.0664	1.0635	1.0606	1.0577	1.0551	1.0525	1.0499	1.0474
2820	1.0827	1.0791	1.0757	1.0724	1.0693	1.0662	1.0633	1.0604	1.0575	1.0549	1.0523	1.0497	1.0472
2840	1.0824	1.0788	1.0754	1.0721	1.0690	1.0659	1.0631	1.0602	1.0573	1.0547	1.0521	1.0495	1.0470
2860	1.0822	1.0786	1.0752	1.0719	1.0688	1.0656	1.0629	1.0600	1.0571	1.0544	1.0518	1.0493	1.0468
2880	1.0819	1.0783	1.0749	1.0716	1.0685	1.0653	1.0626	1.0598	1.0569	1.0542	1.0516	1.0491	1.0466
2900	1.0816	1.0780	1.0746	1.0713	1.0682	1.0650	1.0623	1.0595	1.0566	1.0539	1.0513	1.0488	1.0463
2920	1.0812	1.0777	1.0743	1.0710	1.0679	1.0648	1.0620	1.0592	1.0564	1.0537	1.0511	1.0486	1.0461
2940	1.0808	1.0773	1.0740	1.0707	1.0676	1.0646	1.0617	1.0589	1.0561	1.0534	1.0509	1.0484	1.0458
2960	1.0805	1.0770	1.0737	1.0704	1.0673	1.0643	1.0614	1.0586	1.0558	1.0531	1.0506	1.0481	1.0456
2980	1.0801	1.0767	1.0734	1.0701	1.0670	1.0640	1.0611	1.0583	1.0556	1.0529	1.0504	1.0479	1.0454
3000	1.0797	1.0764	1.0731	1.0698	1.0667	1.0637	1.0608	1.0580	1.0554	1.0527	1.0501	1.0476	1.0451

Note: Factors for intermediate values of pressure and temperature should be interpolated.

TABLE A.35(b)

Supercompressibility Pressure Adjustments, ΔP

Based on carbon dioxide and nitrogen contents and specific gravity pressure adjustment index = $f_{pg} = G - 13.84X_c + 5.420X_n$

Pressure Adjustment Index, f_{pg}	Pressure, psig															
	0	200	400	600	800	1000	1200	1400	1600	1800	2000	2200	2400	2600	2800	3000
−0.7	0	−11.32	−22.65	−33.97	−45.30	−56.62	−67.94	−79.27	−90.59	−101.92	−113.24	−124.56	−135.89	−147.21	−158.54	−169.86
−0.6	0	−10.50	−21.00	−31.49	−41.99	−52.49	−62.99	−73.49	−83.98	−94.48	−104.98	−115.48	−125.98	−136.47	−146.97	−157.47
−0.5	0	−9.67	−19.33	−29.00	−38.66	−48.33	−58.00	−67.66	−77.33	−86.99	−96.66	−106.33	−115.99	−125.66	−135.32	−144.99
−0.4	0	−8.83	−17.65	−26.48	−35.30	−44.13	−52.96	−61.78	−70.61	−79.43	−88.26	−97.09	−105.91	−114.74	−123.56	−132.39
−0.3	0	−7.98	−15.96	−23.93	−31.91	−39.89	−47.87	−55.85	−63.82	−71.80	−79.78	−87.76	−95.74	−103.71	−111.69	−119.67
−0.2	0	−7.12	−14.25	−21.37	−28.50	−35.62	−42.74	−49.87	−56.99	−64.12	−71.24	−78.36	−85.49	−92.61	−99.74	−106.86
−0.1	0	−6.26	−12.52	−18.78	−25.04	−31.30	−37.56	−43.82	−50.08	−56.34	−62.60	−68.86	−75.12	−81.38	−87.64	−93.90
0	0	−5.39	−10.78	−16.17	−21.56	−26.95	−32.34	−37.73	−43.12	−48.51	−53.90	−59.29	−64.68	−70.07	−75.46	−80.85
+0.1	0	−4.51	−9.02	−13.54	−18.05	−22.56	−27.07	−31.58	−36.10	−40.61	−45.12	−49.63	−54.14	−58.66	−63.17	−67.68
+0.2	0	−3.63	−7.25	−10.88	−14.50	−18.13	−21.76	−25.38	−29.01	−32.63	−36.26	−39.89	−43.51	−47.14	−50.76	−54.39
+0.3	0	−2.73	−5.46	−8.20	−10.93	−13.66	−16.39	−19.12	−21.86	−24.59	−27.32	−30.05	−32.78	−35.52	−38.25	−40.98
+0.4	0	−1.83	−3.66	−5.49	−7.32	−9.15	−10.98	−12.81	−14.64	−16.47	−18.30	−20.13	−21.96	−23.79	−25.62	−27.45
+0.5	0	−0.92	−1.84	−2.76	−3.68	−4.60	−5.52	−6.43	−7.35	−8.27	−9.19	−10.11	−11.03	−11.95	−12.87	−13.79
+0.6	0	0	0	0	0	0	0	0	0	0	0	0	0	0	0	0
+0.7	0	0.93	1.86	2.78	3.71	4.64	5.57	6.49	7.42	8.35	9.28	10.20	11.13	12.06	12.99	13.91
+0.8	0	1.86	3.73	5.59	7.46	9.32	11.18	13.05	14.91	16.78	18.64	20.50	22.37	24.23	26.10	27.96
+0.9	0	2.81	5.62	8.42	11.23	14.04	16.85	19.66	22.46	25.27	28.08	30.89	33.70	36.50	39.31	42.12
+1.0	0	3.76	7.52	11.29	15.05	18.81	22.57	26.33	30.10	33.86	37.62	41.38	45.14	48.91	52.67	56.43
+1.1	0	4.73	9.45	14.18	18.90	23.63	28.36	33.08	37.81	42.53	47.26	51.99	56.71	61.44	66.16	70.89
+1.2	0	5.70	11.40	17.09	22.79	28.49	34.19	39.89	45.58	51.28	56.98	62.68	68.38	74.07	79.77	85.47
+1.3	0	6.68	13.36	20.04	26.72	33.40	40.08	46.76	53.44	60.12	66.80	73.48	80.16	86.84	93.52	100.20
+1.4	0	7.67	15.34	23.01	30.68	38.35	46.02	53.69	61.36	69.03	76.70	84.37	92.04	99.71	107.38	115.05
+1.5	0	8.67	17.34	26.01	34.68	43.35	52.02	60.69	69.36	78.03	86.70	95.37	104.04	112.71	121.38	130.05
+1.6	0	9.68	19.36	29.04	38.72	48.40	58.08	67.76	77.44	87.12	96.80	106.48	116.16	125.84	135.52	145.20
+1.7	0	10.70	21.40	32.10	42.80	53.50	64.20	74.90	85.60	96.30	107.00	117.70	128.40	139.10	149.80	160.50
+1.8	0	11.73	23.46	35.19	46.92	58.65	70.38	82.11	93.84	105.57	117.30	129.03	140.76	152.49	164.22	175.95
+1.9	0	12.77	25.54	38.31	51.08	63.85	76.62	89.39	102.16	114.93	127.70	140.47	153.24	166.01	178.78	191.55
+2.0	0	13.82	27.64	41.46	55.28	69.10	82.92	96.74	110.56	124.38	138.20	152.02	165.84	179.66	193.48	207.30

Note: Factors for intermediate values of pressure adjustment index and pressure should be interpolated.

TABLE A.35(c)

Supercompressibility Pressure Adjustments, ΔT

Based on carbon dioxide and nitrogen contents and specific gravity pressure adjustment index = $f_{tg} = G - 0.472X_c - 0.793X_n$

Temperature Adjustment Index, f_{tg}	Temperature, °F										
	0	20	40	60	80	100	120	140	160	180	200
0.45	75.16	78.43	81.70	84.97	88.24	91.50	94.77	98.04	101.31	104.58	107.84
0.46	69.41	72.43	75.45	78.47	81.49	84.50	87.52	90.54	93.56	96.58	99.59
0.47	63.76	66.53	69.30	72.07	74.84	77.62	80.39	83.16	85.93	88.70	91.48
0.48	58.24	60.77	63.30	65.83	68.36	70.90	73.43	75.96	78.49	81.02	83.56
0.49	52.81	55.10	57.40	59.70	61.99	64.29	66.58	68.88	71.18	73.47	75.77
0.50	47.52	49.58	51.65	53.72	55.78	57.85	59.91	61.98	64.05	66.11	68.18
0.51	42.33	44.17	46.01	47.85	49.69	51.53	53.37	55.21	57.05	58.89	60.73
0.52	37.25	38.87	40.48	42.10	43.72	45.34	46.96	48.58	50.20	51.82	53.44
0.53	32.26	33.67	35.07	36.47	37.88	39.28	40.68	42.08	43.49	44.89	46.29
0.54	27.38	28.57	29.76	30.95	32.14	33.33	34.52	35.71	36.90	38.09	39.28
0.55	22.60	23.58	24.56	25.54	26.52	27.51	28.49	29.47	30.45	31.44	32.42
0.56	17.90	18.68	19.46	20.23	21.01	21.79	22.57	23.35	24.12	24.90	25.68
0.57	13.29	13.87	14.45	15.03	15.61	16.18	16.76	17.34	17.92	18.50	19.07
0.58	8.78	9.16	9.54	9.92	10.30	10.68	11.07	11.45	11.83	12.21	12.59
0.59	4.35	4.54	4.73	4.92	5.10	5.29	5.48	5.67	5.86	6.05	6.24
0.60	0	0	0	0	0	0	0	0	0	0	0
0.61	− 4.27	− 4.45	− 4.64	− 4.82	− 5.01	− 5.19	− 5.38	− 5.57	− 5.75	− 5.94	− 6.12
0.62	− 8.45	− 8.82	− 9.19	− 9.56	− 9.93	−10.29	−10.66	−11.03	−11.40	−11.76	−12.13
0.63	−12.57	−13.11	−13.66	−14.21	−14.75	−15.30	−15.85	−16.39	−16.94	−17.48	−18.03
0.64	−16.61	−17.33	−18.05	−18.77	−19.49	−20.22	−20.94	−21.66	−22.38	−23.10	−23.83
0.65	−20.57	−21.47	−22.36	−23.25	−24.15	−25.04	−25.94	−26.83	−27.73	−28.62	−29.52
0.66	−24.47	−25.53	−26.60	−27.66	−28.72	−29.79	−30.85	−31.91	−32.98	−34.04	−35.11
0.67	−28.29	−29.52	−30.76	−31.99	−33.22	−34.45	−35.68	−36.91	−38.14	−39.37	−40.60
0.68	−32.06	−33.45	−34.84	−36.24	−37.63	−39.03	−40.42	−41.81	−43.21	−44.60	−46.00
0.69	−35.75	−37.30	−38.86	−40.41	−41.97	−43.52	−45.08	−46.63	−48.19	−49.74	−51.30
0.70	−39.38	−41.10	−42.81	−44.52	−46.23	−47.95	−49.66	−51.37	−53.08	−54.80	−56.51
0.71	−42.95	−44.82	−46.69	−48.56	−50.42	−52.29	−54.16	−56.03	−57.90	−59.76	−61.63
0.72	−46.46	−48.48	−50.50	−52.52	−54.54	−56.56	−58.58	−60.60	−62.62	−64.64	−66.66
0.73	−49.91	−52.08	−54.25	−56.42	−58.59	−60.76	−62.93	−65.10	−67.27	−69.44	−71.61
0.74	−53.31	−55.63	−57.95	−60.27	−62.59	−64.90	−67.22	−69.54	−71.86	−74.18	−76.49
0.75	−56.67	−59.14	−61.60	−64.06	−66.53	−68.99	−71.46	−73.92	−76.38	−78.85	−81.31

Note: Factors forintermediate values of temperature adjustment index and temperture should be interpolated.

TABLE A.35(d)

Supercompressibility Pressure Adjustments, ΔP

Based on carbon dioxide content, heating value and specific gravity pressure adjustment index = $f_{ph} = G - 0.5688\ H_w/1000 - 3.690X_c$

Pressure Adjustment Index, f_{pb}	Pressure, psig															
	0	200	400	600	800	1000	1200	1400	1600	1800	2000	2200	2400	2600	2800	3000
−0.22	0	−11.25	−22.51	−33.76	−45.02	−56.27	−67.52	−78.78	−90.03	−101.29	−112.54	−123.79	−135.05	−146.30	−157.56	−168.81
−0.20	0	−10.27	−20.54	−30.80	−41.07	−51.34	−61.61	−71.88	−82.14	− 92.41	−102.68	−112.95	−123.22	−133.48	−143.75	−154.02
−0.18	0	− 9.27	−18.54	−27.82	−37.09	−46.36	−55.63	−64.90	−74.18	− 83.45	− 92.72	−101.99	−111.26	−120.54	−129.81	−139.08
−0.16	0	− 8.27	−16.53	−24.80	−33.06	−41.33	−49.60	−57.86	−66.13	− 74.39	− 82.66	− 90.93	− 99.19	−107.46	−115.72	−123.99
−0.14	0	− 7.25	−14.50	−21.74	−28.99	−36.24	−43.49	−50.74	−57.98	− 65.23	− 72.48	− 79.73	− 86.98	− 94.22	−101.47	−108.72
−0.12	0	− 6.22	−12.44	−18.66	−24.88	−31.10	−37.32	−43.54	−49.76	− 55.98	− 62.20	− 68.42	−74.64	− 80.86	− 87.08	− 93.30
−0.10	0	− 5.18	−10.36	−15.55	−20.73	−25.91	−31.09	−36.27	−41.46	− 46.64	− 51.82	− 57.00	− 62.18	− 67.37	− 72.55	− 77.73
−0.08	0	− 4.13	− 8.26	−12.40	−16.53	−20.66	−24.79	−28.92	−33.06	− 37.19	− 41.32	− 45.45	− 49.58	− 53.72	− 57.85	− 61.98
−0.06	0	− 3.07	− 6.14	− 9.21	−12.28	−15.35	−18.42	−21.49	−24.56	− 27.63	− 30.70	− 33.77	− 36.84	− 39.91	− 42.98	− 46.05
−0.04	0	− 2.00	− 3.99	− 5.99	− 7.99	− 9.98	−11.98	−13.98	−15.97	− 17.97	− 19.97	− 21.96	− 23.96	− 25.96	− 27.96	− 29.95
−0.02	0	− 0.91	− 1.82	− 2.74	− 3.65	− 4.56	− 5.47	− 6.38	− 7.29	− 8.21	− 9.12	− 10.03	− 10.94	− 11.85	− 12.76	− 13.68
0.00	0	0.18	0.37	0.56	0.74	0.92	1.11	1.30	1.48	1.67	1.85	2.04	2.22	2.41	2.59	2.78
+0.02	0	1.29	2.59	3.88	5.18	6.47	7.77	9.06	10.35	11.65	12.95	14.24	15.53	16.82	18.12	19.41
+0.04	0	2.42	4.83	7.25	9.66	12.08	14.50	16.91	19.33	21.74	24.16	26.58	28.99	31.41	33.82	36.24
+0.06	0	3.55	7.10	10.65	14.20	17.75	21.30	24.85	28.40	31.95	35.50	39.05	42.60	46.15	49.70	53.25
+0.08	0	4.70	9.39	14.09	18.78	23.48	28.18	32.87	37.57	42.26	46.96	51.66	56.35	61.05	65.74	70.44
+0.10	0	5.86	11.71	17.57	23.42	29.28	35.14	40.99	46.85	52.70	58.56	64.42	70.27	76.13	81.98	87.84
+0.12	0	7.03	14.06	21.08	28.11	35.14	42.17	49.20	56.22	63.25	70.28	77.31	84.34	91.36	98.39	105.52
+0.14	0	8.22	16.43	24.65	32.86	41.08	49.30	57.51	65.73	73.94	82.16	90.38	98.59	106.81	115.02	123.24
+0.16	0	9.42	18.83	28.25	37.66	47.08	56.50	65.91	75.33	84.74	94.16	103.58	112.99	122.41	131.82	141.24
+0.18	0	10.63	21.26	31.89	42.52	53.15	63.78	74.41	85.04	95.67	106.30	116.93	127.56	138.19	148.82	159.45
+0.20	0	11.86	23.72	35.57	47.43	59.29	71.15	83.01	94.86	106.72	118.58	130.44	142.30	154.15	166.01	177.87
+0.22	0	13.10	26.20	39.30	52.40	65.50	78.60	91.70	104.80	117.90	131.00	144.10	157.20	170.30	183.40	196.50

Note: Factors for intermediate values of pressure adjustment index and pressure should be interpolated.

TABLE A.35(e)

Supercompressibility Temperature Adjustments, ΔT

Based on carbon dioxide content, heating value and specific gravity temperature adjustment index = $f_{tb} = G + 1.814\ H_w/1000 + 2.641 X_c$

Temperature Adjustment Index, f_{tb}	Temperature, °F										
	0	20	40	60	80	100	120	140	160	180	200
2.10	56.03	58.46	60.90	63.34	65.77	68.21	70.64	73.08	75.52	77.95	80.39
2.12	53.13	55.44	57.75	60.06	62.37	64.68	66.99	69.30	71.61	73.92	76.23
2.14	50.19	52.37	54.55	56.73	58.91	61.10	63.28	65.46	67.64	69.82	72.01
2.16	47.29	49.34	51.40	53.46	55.51	57.57	59.62	61.68	63.74	65.79	67.85
2.18	44.46	46.40	48.33	50.26	52.20	54.13	56.06	58.00	59.93	61.86	63.80
2.20	41.64	43.45	45.26	47.08	48.89	50.70	52.51	54.32	56.13	57.94	59.75
2.22	38.86	40.54	42.24	43.92	45.61	47.30	48.99	50.68	52.37	54.06	55.75
2.24	36.10	37.66	39.24	40.80	42.37	43.94	45.51	47.08	48.65	50.22	51.79
2.26	33.37	34.82	36.28	37.73	39.18	40.63	42.08	43.53	44.98	46.43	47.88
2.28	30.67	32.01	33.34	34.67	36.01	37.34	38.67	40.01	41.34	42.68	44.01
2.30	28.01	29.23	30.44	31.66	32.88	34.10	35.32	36.53	37.75	38.97	40.19
2.32	25.37	26.47	27.58	28.68	29.78	30.88	31.99	33.09	34.19	35.30	36.40
2.34	22.76	23.75	24.74	25.73	26.72	27.71	28.70	29.69	30.68	31.67	31.66
2.36	20.18	21.05	21.93	22.81	23.68	24.56	25.44	26.32	27.19	28.07	28.95
2.38	17.62	18.39	19.16	19.92	20.69	21.45	22.22	22.99	23.75	24.52	25.28
2.40	15.09	15.75	16.40	17.06	17.72	18.37	19.03	19.69	20.34	21.00	21.65
2.42	12.59	13.14	13.69	14.24	14.78	15.33	15.88	16.43	16.98	17.52	18.07
2.44	10.12	10.56	11.00	11.44	11.88	12.32	12.76	13.20	13.64	14.08	14.52
2.46	7.67	8.00	8.34	8.67	9.00	9.34	9.67	10.00	10.34	10.67	11.00
2.48	5.24	5.47	5.70	5.93	6.16	6.38	6.61	6.84	7.07	7.30	7.52
2.50	2.85	2.97	3.10	3.22	3.34	3.46	3.59	3.71	3.84	3.96	4.08
2.52	0.47	0.49	0.51	0.53	0.55	0.57	0.60	0.62	0.64	0.66	0.68
2.54	−1.88	−1.96	−2.04	−2.12	−2.20	−2.29	−2.37	−2.45	−2.53	−2.61	−2.69
2.56	−4.20	−4.39	−4.57	−4.75	−4.94	−4.12	−5.30	−5.48	−5.67	−5.85	−6.03
2.58	−6.50	−6.79	−7.07	−7.35	−7.64	−7.92	−8.20	−8.48	−8.77	−9.05	−9.33
2.60	−8.79	−9.17	−9.55	−9.93	−10.31	−10.70	−11.08	−11.46	−11.84	−12.22	−12.61
2.62	−11.04	−11.52	−12.00	−12.48	−12.96	−13.44	−13.92	−14.41	−14.89	−15.37	−15.85
2.64	−13.28	−13.85	−14.43	−15.01	−15.58	−16.16	−16.74	−17.32	−17.89	−18.47	−19.05
2.66	−15.49	−16.16	−16.84	−17.51	−18.18	−18.86	−19.53	−20.20	−20.88	−21.55	−22.22
2.68	−17.68	−18.45	−19.22	−19.98	−20.75	−21.52	−22.29	−23.06	−23.83	−24.60	−25.36
2.70	−19.85	−20.71	−21.58	−22.44	−23.30	−24.16	−25.03	−25.89	−26.75	−27.62	−28.48
2.72	−22.00	−22.95	−23.91	−24.87	−25.82	−26.78	−27.74	−28.69	−29.65	−30.60	−31.56
2.74	−24.12	−25.17	−26.22	−27.27	−28.32	−29.37	−30.42	−31.46	−32.51	−33.56	−34.61
2.76	−26.23	−27.37	−28.51	−29.65	−30.79	−31.93	−33.07	−34.21	−35.35	−36.49	−37.63
2.78	−28.31	−29.54	−30.78	−32.01	−33.24	−34.47	−35.70	−36.93	−38.16	−39.39	−40.62
2.80	−30.38	−31.70	−33.02	−34.34	−35.66	−36.98	−38.30	−39.62	−40.94	−42.26	−43.59
2.82	−32.42	−33.84	−35.24	−36.65	−38.06	−39.47	−40.88	−42.29	−43.70	−45.11	−46.52
2.84	−34.45	−35.95	−37.45	−38.95	−40.45	−41.94	−43.44	−44.94	−46.44	−47.94	−49.43
2.86	−36.46	−38.04	−39.63	−41.22	−42.80	−44.38	−45.97	−47.56	−49.14	−50.73	−52.31
2.88	−38.45	−40.12	−41.80	−43.47	−45.14	−46.81	−48.48	−40.15	−51.82	−53.50	−55.17

(Continued)

TABLE A.35(e) (Continued

Temperature Adjustment Index, f_{tb}	Temperature, °F										
	0	20	40	60	80	100	120	140	160	180	200
2.90	−40.42	−42.18	−43.94	−45.69	−47.45	−49.21	−50.96	−52.72	−54.48	−56.24	−57.99
2.92	−42.37	−44.21	−46.06	−47.90	−49.74	−51.58	−53.42	−55.27	−57.11	−58.95	−60.79
2.94	−44.31	−46.23	−48.16	−50.09	−52.01	−53.94	−55.86	−57.79	−59.72	−61.64	−63.57
2.96	−46.23	−48.24	−50.25	−52.26	−54.27	−56.28	−58.29	−60.30	−62.31	−64.32	−66.33
2.98	−48.12	−50.21	−52.30	−54.39	−56.48	−58.58	−60.67	−62.76	−64.85	−66.94	−69.04
3.00	−50.00	−52.18	−54.35	−56.52	−58.70	−60.87	−63.05	−65.22	−67.39	−69.57	−71.74
3.02	−51.89	−54.14	−56.40	−58.66	−60.91	−63.17	−65.42	−67.68	−69.94	−72.19	−74.45
3.04	−53.73	−56.06	−58.40	−60.74	−63.07	−65.40	−67.74	−70.08	−72.42	−74.75	−77.09
306	−55.57	−57.98	−60.40	−62.82	−65.23	−67.65	−70.06	−72.48	−74.90	−77.31	−79.73
3.08	−57.36	−59.86	−62.35	−64.84	−67.34	−69.83	−72.33	−74.82	−77.31	−79.81	−82.30
3.10	−59.16	−61.73	−64.30	−66.87	−69.44	−72.02	−74.59	−77.16	−79.73	−82.30	−84.88
3.12	−60.95	−63.60	−66.25	−68.90	−71.55	−74.20	−76.85	−79.50	−82.15	−84.80	−87.45
3.14	−62.70	−65.42	−68.15	−70.88	−73.60	−76.33	−79.05	−81.78	−84.51	−87.23	−89.96
3.16	−64.45	−67.25	−70.05	−72.85	−75.65	−78.46	−81.26	−84.06	−86.86	−89.66	−92.47
3.18	−66.19	−69.07	−71.95	−74.83	−77.71	−80.58	−83.46	−86.34	−89.22	−92.10	−94.97
3.20	−67.90	−70.85	−73.80	−76.75	−79.70	−82.66	−85.61	−88.56	−91.51	−94.46	−97.42

Note: Factors for intermediate values of temperature adjustment index and temperature should be interpolated.

TABLE A.36
F_m Manometer Factors

Specific Gravity, G	Flowing Pressure, psig						
	0	500	1000	1500	2000	2500	3000
			Ambient Temperature = 0°F				
0.55	1.0000	0.9989	0.9976	0.9960	0.9943	0.9930	0.9921
0.60	1.0000	0.9988	0.9972	0.9952	0.9932	0.9919	0.9910
0.65	1.0000	0.9987	0.9967	0.9941	0.9920	0.9908	0.9900
0.70	1.0000	0.9985	0.9961	0.9927	0.9907	0.9896	0.9890
0.75	1.0000						
			Ambient Temperature = 40°F				
0.55	1.0000	0.9990	0.9979	0.9967	0.9954	0.9942	0.9932
0.60	1.0000	0.9989	0.9976	0.9962	0.9946	0.9933	0.9923
0.65	1.0000	0.9988	0.9973	0.9955	0.9937	0.9923	0.9913
0.70	1.0000	0.9987	0.9970	0.9947	0.9926	0.9912	0.9903
0.75	1.0000	0.9986	0.9965	0.9937	0.9915	0.9902	0.9893
			Ambient Temperature = 80°F				
0.55	1.0000	0.9991	0.9981	0.9971	0.9960	0.9950	0.9941
0.60	1.0000	0.9990	0.9979	0.9967	0.9955	0.9943	0.9933
0.65	1.0000	0.9989	0.9977	0.9963	0.9948	0.9935	0.9925
0.70	1.0000	0.9988	0.9974	0.9958	0.9940	0.9926	0.9915
0.75	1.0000	0.9987	0.9971	0.9951	0.9931	0.9916	0.9906
			Ambient Temperature = 120°F				
0.55	1.0000	0.9992	0.9983	0.9974	0.9965	0.9956	0.9948
0.60	1.0000	0.9991	0.9981	0.9971	0.9960	0.9950	0.9941
0.65	1.0000	0.9990	0.9979	0.9967	0.9955	0.9944	0.9934
0.70	1.0000	0.9989	0.9977	0.9963	0.9950	0.9937	0.9926
0.75	1.0000	0.9988	0.9975	0.9959	0.9943	0.9929	0.9918

Note: Factors for itermediate values of pressure, temperature, and specific gravity should be interpolated.
Note: This table is for use with mercury manometer type recording gauges that have gas in contact with the mercury surface.

TABLE A.37

F_l—Gauge Location Factors Gravitation Correction Factors for Manometer Factor Adjustment

Based on Elevation and Latitude, Applicable to Unadjusted Factors in Preceding Table

	Gauge elevation above sea level—lineal feet					
Degrees latitude	Sea level	2,000′	4,000′	6,000′	8,000′	10,000′
0 (Equator)	0.9987	0.9986	0.9985	0.9984	0.9983	0.9982
5	0.9987	0.9986	0.9985	0.9984	0.9983	0.9982
10	0.9988	0.9987	0.9986	0.9985	0.9984	0.9983
15	0.9989	0.9988	0.9987	0.9986	0.9985	0.9984
20	0.9990	0.9989	0.9988	0.9987	0.9986	0.9985
25	0.9991	0.9990	0.9989	0.9988	0.9987	0.9986
30	0.9993	0.9992	0.9991	0.9990	0.9989	0.9988
35	0.9995	0.9994	0.9993	0.9992	0.9991	0.9990
40	0.9998	0.9997	0.9996	0.9995	0.9994	0.9993
45	1.0000	0.9999	0.9998	0.9997	0.9996	0.9995
50	1.0002	1.0001	1.0000	0.9999	0.9998	0.9997
55	1.0004	1.0003	1.0002	1.0001	1.0000	0.9999
60	1.0007	1.0006	1.0005	1.0004	1.0003	1.0002
65	1.0008	1.0007	1.0006	1.0005	1.0004	1.0003
70	1.0010	1.0009	1.0008	1.0007	1.0006	1.0005
75	1.0011	1.0010	1.0009	1.0008	1.0007	1.0006
80	1.0012	1.0011	1.0010	1.0009	1.0008	1.0007
85	1.0013	1.0012	1.0011	1.0010	1.0009	1.0008
90 (Pole)	1.0013	1.0012	1.0011	1.0010	1.0009	1.0008

Note: While F_1 values are strictly manometer factors, to account for guages being operated under gravitational forces that depart from standard location; it is suggested that it be combined with other flow constants. In which instance. F_1 becomes a location factor constant and F_m, the manometer factor agreeable with standard gravity remains a variable factor, subject to change with specific gravity, ambient temperature, and *static pressure*.

A.5 EXTENDENDER SUKKER CORNELL INTEGRAL FOR BOTTOM-HOLE PRESSURE CALCULATION FROM MESSER, RAGHAVAN, AND RAMEY

TABLE A.38(a)
Extended Sukkar–Cornell Integral for Bottom-hole Pressure Calculation

	Reduced Temperature for $B = 0.00$														
	1.1	1.2	1.3	1.4	1.5	1.6	1.7	1.8	1.9	2.0	2.2	2.4	2.6	2.8	3.0
0.20	0.0000	0.0000	0.0000	0.0000	0.0000	0.0000	0.0000	0.0000	0.0000	0.0000	0.0000	0.0000	0.0000	0.0000	0.0000
0.50	0.8387	0.8582	0.8719	0.8824	0.8897	0.8966	0.9017	0.9079	0.9082	0.9108	0.9147	0.9177	0.9194	0.9206	0.9218
1.00	1.3774	1.4440	1.4836	1.5129	1.5334	1.5514	1.5654	1.5781	1.5823	1.5889	1.5966	1.6059	1.5111	1.6148	1.6184
1.50	1.6048	1.7373	1.8078	1.8565	1.8911	1.9192	1.9422	1.9609	1.9693	1.9798	1.9951	2.0063	2.0151	2.0211	2.0274
2.00	1.7149	1.9116	2.0157	2.0642	2.1331	2.1709	2.2023	2.2273	2.2397	2.2536	2.2744	2.2893	2.3013	2.3100	2.3104
2.50	1.7995	2.0298	2.1631	2.2507	2.3138	2.3607	2.3996	2.4307	2.4469	2.4641	2.4900	2.5081	2.5234	2.5347	2.5452
3.00	1.8750	2.1255	2.2778	2.3813	2.4570	2.5125	2.5583	2.5947	2.6148	2.6354	2.6654	2.6863	2.7050	2.7189	2.7134
3.50	1.9473	2.2101	2.3746	2.4898	2.5762	2.6190	2.6909	2.7325	2.7561	2.7798	2.8138	2.8382	2.8589	2.8752	2.8896
4.00	2.0178	2.2882	2.4603	2.5945	2.6793	2.7480	2.8052	2.8515	2.8784	2.9050	2.9426	2.9699	2.9928	3.0114	3.0274
4.50	2.0689	2.3622	2.5390	2.6698	2.7715	2.8449	2.9065	2.9569	2.9867	3.0158	3.0571	3.0871	3.1119	3.1322	3.1496
5.00	2.1547	2.4330	2.6128	2.7484	2.8558	2.9330	2.9982	3.0523	3.0645	3.1158	3.1605	3.1930	3.2195	3.2413	3.2597
5.50	2.2214	2.5013	2.6833	2.8222	2.9341	3.0146	3.0828	3.1400	3.1742	3.2074	3.2557	3.2899	3.3178	3.3408	3.3600
6.00	2.2872	2.5677	2.7512	2.8926	3.0079	3.0911	3.1616	3.2215	3.2575	3.2924	3.3428	3.3795	3.4085	3.4325	3.4524
6.50	2.3522	2.6329	2.8171	2.9603	3.0781	3.1635	3.2360	3.2980	3.3355	3.3720	3.4245	3.4629	3.4931	3.5176	3.5381
7.00	2.4165	2.6971	2.8814	3.0258	3.1452	3.2324	3.3065	3.3704	3.4092	3.4470	3.6012	3.5411	3.5722	3.5973	3.6181
7.50	2.4802	2.7602	2.5442	3.0893	3.2100	3.2985	3.3740	3.4393	3.4792	3.5180	3.5738	3.6148	3.6467	3.6723	3.6934
8.00	2.5432	2.8223	3.0058	3.1612	3.2727	3.3623	3.4367	3.5052	3.5460	3.5657	3.6486	3.6847	3.7173	3.7432	3.7646
8.50	2.6057	2.8336	3.0864	3.2118	3.3338	3.4239	3.5012	3.5665	3.6101	3.6504	3.7144	3.7512	3.7844	3.8108	3.8323
9.00	2.6676	2.9441	3.1260	3.2713	3.3934	3.4838	3.5617	3.6297	3.6718	3.7126	3.7775	3.8148	3.8484	3.8750	3.8969
9.50	2.7289	3.0039	3.1847	3.3296	3.4516	3.5422	3.6204	3.6889	3.7315	3.7727	3.6382	3.8760	3.9099	3.9367	3.9588
10.00	2.7896	3.0630	3.2427	3.3870	3.5087	3.5993	3.6776	3.7465	3.7894	3.8308	3.8969	3.9350	3.9690	3.9961	4.0182
10.50	2.8499	3.1215	3.2999	3.4436	3.5647	3.6552	3.7336	3.8026	3.8456	3.8872	3.9538	3.9921	4.0262	2.0583	4.0755
11.00	2.9096	3.1794	3.3565	3.4993	3.6198	3.7100	3.7883	3.8573	3.9004	3.9421	4.0090	4.0473	4.0814	4.1086	4.1309
11.50	2.9690	3.2369	3.4126	3.5543	3.6741	3.7640	3.8420	3.9108	3.9540	3.9958	4.0627	4.1010	4.1351	4.1622	4.1845
12.00	3.0280	3.2940	3.4681	3.6086	3.7277	3.8171	3.8948	3.9634	4.0065	4.0432	4.1150	4.1532	4.1872	4.2143	4.2366

(Continued)

TABLE A.38(a) (Continued)

P_r	Reduced Temperature for $B = 0.00$														
	1.1	1.2	1.3	1.4	1.5	1.6	1.7	1.8	1.9	2.0	2.2	2.4	2.6	2.8	3.0
12.50	3.0867	3.3506	3.5231	3.6523	3.7806	3.8694	3.9467	4.0150	4.0579	4.0994	4.1660	4.2041	4.2380	4.2650	4.2872
13.00	3.1452	3.4068	3.5777	3.7154	3.8328	3.9211	3.9977	4.0557	4.1084	4.1495	4.2158	4.2567	4.2875	4.3144	4.3365
13.50	2.2033	3.4627	3.6319	3.7880	3.8644	3.9721	4.0480	4.1155	4.1580	4.1989	4.2645	4.3021	4.3357	4.3625	4.3846
14.00	3.2612	3.5183	3.6857	3.8200	3.9354	4.0224	4.0977	4.1547	4.2067	4.2472	4.3122	4.3494	4.3829	4.4095	4.4316
14.50	3.3189	3.5735	3.7391	3.8716	3.9859	4.0722	4.1480	4.2131	4.2546	4.2947	4.3589	4.3957	4.4289	4.4555	4.4775
15.00	3.3763	3.6285	3.7922	3.9228	4.0359	4.1215	4.1950	4.2609	4.3018	4.3414	4.4047	4.4410	4.4741	4.5005	4.6224
15.50	3.4335	3.6832	3.8450	3.9736	4.0866	4.1702	4.2428	4.8080	4.3483	4.3874	4.4497	4.4855	4.5183	4.5446	4.5663
16.00	3.4906	3.7376	3.8974	4.0240	4.1346	4.2185	4.2900	4.3546	4.3942	4.4327	4.4939	4.5291	4.5617	4.5878	4.6094
16.50	3.5474	3.7919	4.9497	4.0740	4.1833	4.2663	4.3368	4.4007	4.4395	4.4773	4.5374	4.5720	4.6042	4.6302	4.6518
17.00	3.6041	3.8459	4.0016	4.1237	4.2316	4.3138	4.3830	4.4462	4.4843	4.5213	4.5802	4.6141	4.6461	4.6719	4.6933
17.50	3.6606	3.8996	4.0533	4.1731	4.2795	4.3608	4.4289	4.4913	4.5285	4.5648	4.6223	4.6555	4.5872	4.7129	4.7341
18.00	3.7170	3.9532	4.1048	4.2221	4.3271	4.4075	4.4743	4.5359	4.5722	4.6077	4.6638	4.6963	4.7276	4.7532	4.7743
18.50	3.7732	4.0066	4.1560	4.2709	4.3744	4.4538	4.5193	4.5801	4.6154	4.6501	4.7048	4.7365	4.7675	4.7928	4.8138
19.00	3.8293	4.0599	4.2071	4.3195	4.4214	4.6998	4.5640	4.6239	4.6582	4.6921	4.7451	5.7761	4.8067	4.8319	4.8527
19.50	3.8853	4.1129	4.2579	4.3678	4.4681	4.5455	4.6053	4.6574	4.7006	4.7335	4.7850	4.8151	4.8454	4.8704	4.8911
20.00	3.9411	4.1658	4.3086	4.4158	4.5145	4.5909	4.6522	4.7104	4.7425	4.7746	4.8244	4.8536	4.8835	4.9083	4.9288
20.50	3.9969	4.2186	4.3590	4.4636	4.5606	4.6360	4.6959	4.7531	4.7841	4.8152	4.8633	4.8916	4.9211	4.9457	4.9661
21.00	4.0525	4.2712	4.4094	4.5112	4.6065	4.6808	4.7392	4.7955	4.8253	4.8554	4.9017	4.9291	4.9582	4.9827	5.0029
21.50	4.1080	4.3237	4.4595	4.5586	4.6522	4.7254	4.7822	4.8376	4.8662	4.8953	4.9397	4.9662	4.9969	5.0192	5.0392
22.00	4.1634	4.3760	4.5095	4.6058	4.6976	4.7697	4.8250	4.8794	4.9068	4.9348	4.9774	5.0027	5.0311	5.0552	5.0751
22.50	4.2187	4.4282	4.5594	4.6528	4.7428	4.8138	4.8675	4.9209	4.9470	4.9739	5.0146	5.0391	5.0670	5.0908	5.1105
23.00	4.2739	4.4803	4.6091	4.6996	4.7879	4.8577	4.9098	4.9621	4.9869	5.0128	5.0514	5.0750	5.1024	5.1260	5.1455
23.50	4.3291	4.5323	4.6587	4.7463	4.8327	4.9014	4.9518	5.0031	5.0265	5.0513	5.0879	5.1104	5.1374	5.1808	5.1802
24.00	4.3841	4.5842	4.7081	4.7928	4.8773	4.9449	4.9935	5.0438	5.0659	5.0895	5.1241	5.1455	5.1720	5.1953	5.2144
24.50	4.4391	4.6360	4.7575	4.8391	4.9217	4.9882	5.0351	5.0843	5.1050	5.1275	5.1599	5.1803	5.2083	5.2294	5.2483
25.00	4.4940	4.6877	4.8067	4.8853	4.9660	5.0312	5.0764	5.1245	5.1438	5.1651	5.1955	5.2147	5.2403	5.2631	5.2819
25.50	4.5488	4.7392	4.8558	4.9314	5.0101	5.0741	5.1176	5.1646	5.1824	5.2025	5.2307	4.2488	5.2739	5.2965	5.3151
26.00	4.6036	4.7907	4.9048	4.9772	5.0541	5.1169	5.1585	5.2044	5.2208	5.2397	5.2656	5.2826	5.3073	5.3296	5.3480
26.50	4.6583	4.8421	4.9536	5.0230	5.0979	5.1594	5.1993	5.2440	5.2589	5.2766	5.3003	5.3162	5.3403	5.3624	5.3806

27.00	4.7129	4.8934	5.0024	5.0686	5.1415	5.2019	5.2398	5.2634	5.2968	5.3132	5.3347	5.3494	5.3780	5.3950	5.4129
27.50	4.7675	4.9447	5.0511	5.1142	5.1850	5.2441	5.2802	5.3227	5.3345	5.3497	5.3688	5.3823	5.4054	5.4272	5.4460
28.00	4.8220	4.9958	5.0997	5.1595	5.2284	5.2862	5.3204	5.3617	5.3720	5.3859	5.4027	4.4150	5.4376	5.4591	5.4767
28.50	4.8764	5.0469	5.1482	5.2048	5.2716	5.3282	5.3605	5.4006	5.4096	5.4219	5.4363	5.4475	5.4695	5.4903	5.5082
29.00	4.9308	5.0979	5.1966	5.2500	5.3147	5.3700	5.4004	5.4393	5.4465	5.4577	5.4697	5.4796	5.5012	5.5223	5.5394
29.50	4.9851	5.1488	5.2450	5.2950	5.3577	5.4117	5.4401	5.4779	5.4834	5.4933	5.5029	5.5116	5.5326	5.5935	5.5704
30.00	5.0394	5.1997	5.2932	5.3400	5.4005	5.4532	5.4797	5.5163	5.5202	5.5287	5.5369	5.5433	5.5638	5.5844	5.6011

TABLE A.38(b)
Extended Sukkar–Cornell Integral for Bottom-hole Pressure Calculation

P_r	Reduced Temperature for $B = 5.0$														
	1.1	1.2	1.3	1.4	1.5	1.6	1.7	1.8	1.9	2.0	2.2	2.4	2.6	2.8	3.0
0.20	0.0000	0.0000	0.0000	0.0000	0.0000	0.0000	0.0000	0.0000	0.0000	0.0000	0.0000	0.0000	0.0000	0.0000	0.0000
0.50	0.0226	0.0220	0.0216	0.0214	0.0212	0.0210	0.0209	0.0207	0.0207	0.0205	0.0205	0.0206	0.0204	0.0204	0.0204
1.00	0.1036	0.0983	0.0954	0.0934	0.0921	0.0909	0.0901	0.0894	0.0890	0.0886	0.0881	0.0877	0.0874	0.0871	0.0869
1.50	0.2121	0.2052	0.1995	0.1954	0.1924	0.1901	0.1882	0.1868	0.1859	0.1850	0.1938	0.1829	0.1822	0.1816	0.1811
2.00	0.3002	0.3125	0.3102	0.366	0.3034	0.3007	0.2983	0.2965	0.2954	0.2943	0.2926	0.2914	0.2904	0.2896	0.2889
2.50	0.3741	0.4046	0.4126	0.4133	0.4124	0.4107	0.4090	0.4076	0.4066	0.4056	0.4048	0.4030	0.4020	0.4012	0.4005
3.00	0.4419	0.4854	0.5032	0.5105	0.5137	0.5144	0.5143	0.5140	0.5138	0.5134	0.5125	0.5118	0.5112	0.5108	0.5103
3.50	0.5074	0.5594	0.5847	0.5983	0.6065	0.6101	0.6123	0.6138	0.6147	0.6152	0.6154	0.6155	0.6155	0.6157	0.6156
4.00	0.5715	0.6291	0.6594	0.6785	0.6915	0.6982	0.7029	0.7064	0.7087	0.7104	0.7121	0.7133	0.7140	0.7149	0.7154
4.50	0.6346	0.6957	0.7294	0.7530	0.7702	0.7797	0.7868	0.7927	0.7964	0.7994	0.8027	0.8051	0.8068	0.8084	0.8094
5.00	0.6966	0.7601	0.7960	0.8229	0.8440	0.8560	0.8653	0.8734	0.8785	0.8827	0.8879	0.8916	0.8941	0.8965	0.8980
5.50	0.7579	0.8225	0.8601	0.8895	0.9138	0.9280	0.9393	0.9493	0.9558	0.9611	0.9682	0.9732	0.9765	0.9795	0.9315
6.00	0.8185	0.8836	0.9222	0.9536	0.9803	0.9965	1.0095	1.0213	1.0289	1.0354	1.0441	1.0504	1.0544	1.0580	1.0604
6.50	0.8784	0.9437	0.9829	1.0156	1.0442	1.0620	1.0764	1.0896	1.0984	1.1060	1.1162	1.1236	1.1284	1.1324	1.1351
7.00	0.9378	1.0030	1.0423	1.0758	1.1058	1.1249	1.1406	1.1552	1.1649	1.1734	1.1848	1.1932	1.1987	1.2031	1.2060
7.50	0.9967	1.0614	1.1005	1.1346	1.1656	1.1857	1.2024	1.2182	1.2286	1.2379	1.2504	1.2597	1.2657	1.2704	1.2737
8.00	1.0551	1.1191	1.1578	1.1921	1.2237	1.2447	1.2621	1.2788	1.2900	1.2999	1.3167	1.3234	1.3299	1.3349	1.3383
8.50	1.1131	1.1761	1.2142	1.2486	1.2805	1.3020	1.3201	1.3374	1.3492	1.3596	1.3773	1.3845	1.3914	1.3967	1.4003
9.00	1.1706	1.2325	1.2698	1.3041	1.3361	1.3579	1.3764	1.3943	1.4066	1.4173	1.4357	1.4434	1.4506	1.4561	1.4599

(Continued)

TABLE A.38(b) (Continued)

P_r	Reduced Temperature for B = 5.0														
	1.1	1.2	1.3	1.4	1.5	1.6	1.7	1.8	1.9	2.0	2.2	2.4	2.6	2.8	3.0
9.50	1.2275	1.2883	1.3248	1.3687	1.3907	1.4125	1.4313	1.4497	1.4623	1.4733	1.4922	1.5003	1.5077	1.5135	1.5174
10.00	2.2841	1.3435	1.3791	1.4126	1.4443	1.4661	1.4851	1.5037	1.5165	1.5278	1.5472	1.5555	1.5630	1.5689	1.5729
10.50	1.3403	1.3983	1.4328	1.4658	1.4970	1.5187	1.5377	1.5564	1.5694	1.5808	1.6006	1.6090	1.5167	1.6226	1.6267
11.00	1.3961	1.4526	1.4860	1.5182	1.5490	1.5705	1.5894	1.6081	1.6211	1.6326	1.6526	1.6611	1.6687	1.6747	1.6789
11.50	1.4515	1.5065	1.5387	1.5701	1.6002	1.6214	1.6401	1.6587	1.6718	1.6833	1.7034	1.7118	1.7195	1.7254	1.7296
12.00	1.5067	1.5601	1.5910	1.6214	1.6509	1.6717	1.6901	1.7085	1.7215	1.7330	1.7530	1.7613	1.7689	1.7749	1.7790
12.50	1.5616	1.6133	1.6429	1.6721	1.7010	1.7213	1.7393	1.7575	1.7704	1.7817	1.8015	1.8097	1.8172	1.8231	1.8271
13.00	1.6163	1.6662	1.6944	1.7224	1.7505	1.7704	1.7879	1.8067	1.8184	1.8295	1.8489	1.8569	1.8644	1.8701	1.8742
13.50	1.6708	1.7188	1.7456	1.7722	1.7995	1.8188	1.8358	1.8532	1.8656	1.8765	1.8954	1.9032	1.9105	1.9161	1.9201
14.00	1.7250	1.7711	1.7965	1.8216	1.8480	1.8667	1.8830	1.9001	1.9121	1.9227	1.9410	1.9485	1.9556	1.9612	1.9651
14.50	1.7791	1.8212	1.8470	1.8706	1.8960	1.9142	1.9298	1.9463	1.9580	1.9681	1.9858	1.9927	1.9998	2.0053	2.0091
15.00	1.8330	1.8750	1.8973	1.9192	1.9436	1.9612	1.9760	1.9920	2.0032	2.0128	2.0298	2.0364	2.0432	2.0485	2.0523
15.50	1.8867	1.9266	1.9472	1.9675	1.9909	2.0077	2.0217	2.0372	2.0478	2.0570	2.0730	2.0792	2.0857	2.0910	2.0946
16.00	1.9402	1.9780	1.9970	2.0154	2.0377	2.0538	2.0669	2.0818	2.0918	2.1005	2.1155	2.1212	2.1275	2.1326	2.1362
16.50	1.9936	2.0292	2.0465	2.0631	2.0842	2.0996	2.1117	2.1260	2.1353	2.1434	2.1574	2.1626	2.1686	2.1736	2.1770
17.00	2.0469	2.0802	2.0958	2.1104	2.1303	2.1450	2.1561	2.1697	2.1783	2.1858	2.1987	2.2032	2.7090	2.2138	2.2172
17.50	2.1000	2.1311	2.1449	2.1575	2.1762	2.1900	2.2000	2.2131	2.2209	2.2276	2.2394	2.2433	2.2488	2.2535	2.2567
18.00	2.1530	2.1817	2.1937	2.2043	2.2217	2.2347	2.2437	2.2560	2.2630	2.2690	2.2795	2.2828	2.2880	2.2925	2.2956
18.50	2.2059	2.2323	2.2424	2.2509	2.2670	2.2791	2.2869	2.2965	2.3046	2.3100	2.3191	2.3217	2.3266	2.3309	2.3339
19.00	2.2587	2.2826	2.2909	2.2973	2.3120	2.3233	2.3299	2.3407	2.3459	2.3505	2.3582	2.3600	2.3646	2.3688	2.3717
19.50	2.3113	2.3329	2.3393	2.3434	2.3567	2.3671	2.3725	2.3825	2.3868	2.3906	2.3969	2.3979	2.4022	2.4062	2.4089
20.00	2.3639	2.3830	2.3875	2.3893	2.4012	2.4107	2.4148	2.4241	2.4273	2.4303	2.4350	2.4353	2.4392	2.4431	2.4456
20.50	2.4164	2.4329	2.4355	2.4350	2.4455	2.4541	2.4568	2.4653	2.4675	2.4696	2.4728	2.4723	2.4758	2.4795	2.4819
21.00	2.4688	2.4828	2.4834	2.4306	2.4895	2.4972	2.4986	2.5062	2.5074	2.5086	2.5101	2.5088	2.5119	2.5155	2.5177
21.50	2.5210	2.5325	2.5311	2.5259	2.5333	2.5400	2.5401	2.5468	2.5470	2.5472	2.5471	2.5449	2.5477	2.5510	2.5531
22.00	2.5733	2.5822	2.5788	2.5711	2.5770	2.5827	2.5814	2.5872	2.5862	2.5855	2.5837	2.5806	2.5830	2.5861	2.5881
22.50	2.6254	2.6317	2.6263	2.6161	2.6204	2.6252	2.6224	2.6273	2.6252	2.6235	2.6199	2.6159	2.6179	2.6209	2.6226
23.00	2.6774	2.6811	2.6736	2.6610	2.6637	2.6674	2.6632	2.6672	2.6639	2.6612	2.6558	2.6508	2.6524	2.6552	2.6568
23.50	2.7294	2.7304	2.7209	2.7057	2.7068	2.7095	2.7038	2.7068	2.7023	2.6986	2.6913	2.6854	2.5866	2.6892	2.6906
24.00	2.7813	2.7796	2.7680	2.7503	2.7497	2.7514	2.7441	2.7462	2.7405	2.7357	2.7266	2.7197	2.7204	2.7229	2.7241

24.50	2.8332	2.8288	2.8151	2.7047	2.7924	2.7931	2.7843	2.7854	2.7784	2.7726	2.7615	2.7536	2.7540	2.7562	2.7573
25.00	2.8849	2.8778	2.8620	2.8390	2.8351	2.8346	2.8243	2.8244	2.8161	2.8092	2.7961	2.7872	2.7872	2.7892	2.7901
25.50	2.9367	2.9268	2.9088	2.8832	2.8775	2.8760	2.8640	2.8532	2.8536	2.8456	2.8305	2.8206	2.8200	2.8192	2.8226
26.00	2.9883	2.9757	2.9556	2.9272	2.9198	2.9172	2.9037	2.9018	2.8908	2.8818	2.8646	2.8536	2.8526	2.8543	2.8548
26.50	3.0399	3.0245	3.0022	2.9711	2.9620	2.9583	2.9431	2.9402	2.9279	2.9177	2.8985	2.8864	2.8850	2.8864	2.6867
27.00	3.0915	3.0733	3.0488	3.0149	3.0040	2.9993	2.9824	2.9785	2.9648	2.9534	2.9320	2.9189	2.9170	2.9182	2.9184
27.50	3.1429	3.1220	3.0953	3.0586	3.0459	3.0400	3.0215	3.0165	3.0014	2.9889	2.9654	2.9512	2.9488	2.9498	2.9497
28.00	3.1944	3.1706	3.1417	3.1022	3.0877	3.0807	3.0604	3.0544	3.0379	3.0242	2.9985	2.9832	2.9803	2.9811	2.9809
28.50	3.2458	3.2191	3.1880	3.1457	3.1294	3.1212	3.0992	3.0922	3.0742	3.0593	3.0314	3.0149	3.0116	3.0122	3.0117
29.00	3.2971	3.2676	3.2343	3.1891	3.1710	3.1616	3.1379	3.1297	3.1103	3.0942	3.0641	3.0465	3.0426	3.0430	3.0424
29.50	3.3484	3.3160	3.2804	3.2324	3.2124	3.2019	3.1764	3.1672	3.1463	3.1289	3.0966	3.0778	3.0735	3.0736	3.0728
30.00	3.3997	3.3644	3.3265	3.2756	3.2537	3.2421	3.2148	3.2045	3.1821	3.1635	3.1288	3.1089	3.1040	3.1040	3.1029

TABLE A.38(c)
Extended Sukkar–Cornell Integral for Bottom-hole Pressure Calculation

P_r	Reduced Temperature for $B = 10.0$														
	1.1	1.2	1.3	1.4	1.5	1.6	1.7	1.8	1.9	2.0	2.2	2.4	2.6	2.8	3.0
0.20	0.0000	0.0000	0.0000	0.0000	0.0000	0.0000	0.0000	0.0000	0.0000	0.0000	0.0000	0.0000	0.0000	0.0000	0.0000
0.50	0.0115	0.0112	0.0110	0.0108	0.0107	0.0107	0.0106	0.0105	0.0105	0.0105	0.0104	0.0104	0.0104	0.0103	0.0103
1.00	0.0561	0.0525	0.0507	0.0494	0.0486	0.0479	0.0474	0.0470	0.0468	0.0465	0.0462	0.0460	0.0458	0.0456	0.0455
1.50	0.1292	0.1187	0.1132	0.1098	0.1074	0.1056	0.1041	0.1031	0.1024	0.1018	0.1009	0.1003	0.0997	0.0994	0.0990
2.00	0.2028	0.1968	0.1891	0.1837	0.1797	0.1767	0.1743	0.1725	0.1713	0.1703	0.1687	0.1676	0.1667	0.1660	0.1653
2.50	0.2684	0.2723	0.2677	0.2624	0.2578	0.2543	0.2513	0.2490	0.2475	0.2461	0.2440	0.2426	0.2413	0.2403	0.2394
3.00	0.3300	0.3422	0.3427	0.3399	0.3364	0.3332	0.3302	0.3278	0.3263	0.3248	0.3225	0.3210	0.3195	0.3184	0.3174
3.50	0.3897	0.4080	0.4130	0.4135	0.4123	0.4102	0.4080	0.4061	0.4047	0.4035	0.4014	0.3999	0.3985	0.3974	0.3964
4.00	0.4485	0.4708	0.4793	0.4832	0.4846	0.4841	0.4830	0.4820	0.4812	0.4803	0.4787	0.4776	0.4764	0.4755	0.4746
4.50	0.5065	0.5315	0.5423	0.5492	0.5533	0.5545	0.5547	0.5549	0.5549	0.5546	0.5538	0.5532	0.5523	0.5517	0.5511
5.00	0.5638	0.5904	0.6029	0.6122	0.6189	0.6217	0.6233	0.6248	0.6256	0.6260	0.6262	0.6263	0.6258	0.6256	0.6252
5.50	0.6204	0.6480	0.6617	0.6729	0.6818	0.6861	0.6891	0.6919	0.6934	0.6946	0.6959	0.6967	0.6967	0.6968	0.6967
6.00	0.6765	0.7045	0.7190	0.7316	0.7424	0.7481	0.7522	0.7563	0.7586	0.7605	0.7629	0.7645	0.7650	0.7654	0.7655
6.50	0.7321	0.7602	0.7752	0.7808	0.8010	0.8079	0.8131	0.8182	0.8214	0.8240	0.8273	0.8297	0.8307	0.8314	0.8317

(Continued)

TABLE A.38(c) (Continued)

P_r	Reduced Temperature for $B = 10.0$														
	1.1	1.2	1.3	1.4	1.5	1.6	1.7	1.8	1.9	2.0	2.2	2.4	2.6	2.8	3.0
7.00	0.7873	0.8153	0.8304	0.6447	0.8580	0.8659	0.8720	0.8781	0.8819	0.8852	0.8895	0.8925	0.8940	0.8950	0.8955
7.50	0.8421	0.8697	0.8846	0.8994	0.9134	0.9221	0.9290	0.9360	0.9404	0.9443	0.9494	0.9531	0.9550	0.9562	0.9566
8.00	0.8965	0.9236	0.9381	0.9531	0.9676	0.9770	0.9845	0.9921	0.9971	1.0015	1.0092	1.0115	1.0138	1.0152	1.0160
8.50	0.9506	0.9769	0.9909	1.0059	1.0207	1.0305	1.0385	1.0467	1.0522	1.0569	1.0653	1.0681	1.0706	1.0723	1.0732
9.00	1.0043	1.0296	1.0431	1.0580	1.0729	1.0829	1.0912	1.0999	1.1057	1.1108	1.1197	1.1228	1.1256	1.1275	1.1286
9.50	1.0575	1.0819	1.0947	1.1094	1.1242	1.1342	1.1428	1.1518	1.1579	1.1633	1.1726	1.1760	1.1790	1.1810	1.1822
10.00	1.1104	1.1338	1.1458	1.1601	1.1747	1.1847	1.1935	1.2027	1.2090	1.2145	1.2242	1.2278	1.2309	1.2331	1.2343
10.50	1.1630	1.1852	1.1964	1.2102	1.2245	1.2344	1.2432	1.2525	1.2689	1.2645	1.2746	1.2783	1.2814	1.2836	1.2850
11.00	1.2153	1.2363	1.2466	1.2598	1.2736	1.2834	1.2920	1.3013	1.3078	1.3135	1.3238	1.3275	1.3907	1.3329	1.3343
11.50	1.2674	1.2871	1.2964	1.3089	1.3222	1.3317	1.3402	1.3494	1.3559	1.3616	1.3719	1.3756	1.3788	1.3810	1.3824
12.00	1.3192	1.3376	1.3458	1.3574	1.3702	1.3794	1.3876	1.3967	1.4032	1.4088	1.4190	1.4227	1.4258	1.4280	1.4294
12.50	1.3708	1.3877	1.3949	1.4056	1.4178	1.4266	1.4345	1.4433	1.4497	1.4552	1.4653	1.4688	1.4719	1.4740	1.4753
13.00	1.4222	1.4377	1.4437	1.4533	1.4649	1.4733	1.4807	1.4893	1.4955	1.5008	1.5106	1.5140	1.5169	1.5139	1.5202
13.50	1.4734	1.4873	1.4921	1.5006	1.5115	1.5194	1.5264	1.5346	1.5406	1.5457	1.5551	1.5582	1.5611	1.5630	1.5642
14.00	1.5244	1.5368	1.5403	1.5476	1.5577	1.5652	1.5716	1.5794	1.5851	1.5899	1.5988	1.6016	1.6043	1.6062	1.6074
14.50	1.5753	1.5860	1.5883	1.5942	1.6035	1.6104	1.6163	1.6237	1.6290	1.6335	1.6417	1.6443	1.6468	1.6486	1.6497
15.00	1.6261	1.6351	1.6360	1.6405	1.6490	1.6553	1.6605	1.6575	1.6723	1.6764	1.6840	1.6862	1.6885	1.6902	1.6912
15.50	1.6767	1.6839	1.6835	1.6865	1.6941	1.6999	1.7043	1.7108	1.7151	1.7188	1.7256	1.7274	1.7296	1.7311	1.7320
16.00	1.7271	1.7326	1.7308	1.7323	1.7389	1.7440	1.7477	1.7537	1.7575	1.7607	1.7666	1.7679	1.7699	1.7713	1.7722
16.50	1.7775	1.7811	1.7778	1.7778	1.7834	1.7878	1.7906	1.7961	1.7993	1.8020	1.8070	1.8078	1.8096	1.8109	1.8116
17.00	1.8277	1.8294	1.8247	1.8230	1.8275	1.8314	1.8333	1.8382	1.8407	1.8429	1.8469	1.8472	1.8487	1.8499	1.8505
17.50	1.8778	1.8777	1.8714	1.8680	1.8714	1.8746	1.8756	1.8799	1.8818	1.8833	1.8862	1.8859	1.8872	1.8883	1.8888
18.00	1.9278	1.9257	1.9179	1.9127	1.9151	1.9175	1.9175	1.9212	1.9224	1.9232	1.9251	1.9242	1.9252	1.9261	1.9265
18.50	1.9777	1.9737	1.9643	1.9573	1.9585	1.9602	1.9592	1.9622	1.9626	1.9628	1.9634	1.9619	1.9626	1.9634	1.9637
19.00	2.0276	2.0215	2.0105	2.0017	2.0016	2.0026	2.0005	2.0029	2.0025	2.0020	2.0013	1.9992	1.9996	2.0002	2.0004
19.50	2.0773	2.0592	2.0566	2.0458	2.0446	2.0447	2.0416	2.0433	2.0420	2.0408	2.0388	2.0359	2.0360	2.0365	2.0366
20.00	2.1269	2.1167	2.1026	2.0898	2.0873	2.0867	2.0824	2.0833	2.0812	2.0792	2.0759	2.0723	2.0721	2.0724	2.0723
20.50	2.1765	2.1642	2.1484	2.1336	2.1298	2.1284	2.1229	2.1232	2.1201	2.1173	2.1126	2.1082	2.1077	2.1079	2.1077
21.00	2.2260	2.2116	2.1941	2.1773	2.1722	2.1699	2.1632	2.1627	2.1587	2.1551	2.1489	2.1438	2.1429	2.1429	2.1425
21.50	2.2754	2.2588	2.2396	2.2207	2.2143	2.2112	2.2033	2.2020	2.1970	2.1926	2.1848	2.1789	2.1777	2.1775	2.1770

22.00	2.3248	2.3060	2.2851	2.2641	2.2563	2.2523	2.2432	2.2411	2.2350	2.2298	2.2204	2.2137	2.2121	2.2118	2.2111
22.50	2.3741	2.3531	2.3304	2.3073	2.2981	2.2932	2.2828	2.2799	2.2728	2.2667	2.2557	2.2461	2.2462	2.2457	2.2449
23.00	2.4233	2.4001	2.3757	2.3503	2.3397	2.3340	2.3222	2.3185	2.3103	2.3033	2.2906	2.2822	2.2799	2.2792	2.2783
23.50	2.4725	2.4470	2.4208	2.3932	2.3812	2.3745	2.3615	2.3569	2.3476	2.3397	2.3253	2.3160	2.3133	2.3124	2.3133
24.00	2.5216	2.4938	2.4659	2.4360	2.4226	2.4149	2.4005	2.3951	2.3847	2.3758	2.3597	2.3494	2.3463	2.3453	2.3440
24.50	2.5706	2.5406	2.5108	2.4787	2.4637	2.4952	2.4394	2.4331	2.4215	2.4117	2.3937	2.3826	2.3791	2.3779	2.3765
25.00	2.6196	2.5873	2.5557	2.5212	2.5048	2.4953	2.4761	2.4709	2.4581	2.4473	2.4275	2.4155	2.4115	2.4102	2.4086
25.50	2.6685	2.6339	2.6005	2.5637	2.5457	2.5353	2.5166	2.5085	2.4946	2.4827	2.4611	2.4481	2.4437	2.4422	2.4404
26.00	2.7174	2.6805	2.6452	2.6060	2.5865	2.5751	2.5550	2.5459	2.5308	2.5179	2.4944	2.4804	2.4756	2.4739	2.4719
26.50	2.7663	2.7269	2.6898	2.6482	2.6272	2.6148	2.5932	2.5832	2.5668	2.5529	2.5275	2.5124	2.5073	2.5053	2.5032
27.00	2.8151	2.7734	2.7343	2.6904	2.6677	2.6543	2.6312	2.6203	2.6027	2.5877	2.5603	2.5443	2.5386	2.5365	2.5342
27.50	2.8638	2.8197	2.7788	2.7324	2.7082	2.6938	2.6691	2.6573	2.6384	2.6223	2.5929	2.5758	2.5698	2.5675	2.5650
28.00	2.9125	2.8660	2.8232	2.7743	2.7485	2.7331	2.7069	2.6941	2.6739	3.6567	2.6253	2.6072	2.6007	2.5982	2.5955
28.50	2.9612	2.9123	2.8675	2.8162	2.7887	2.7723	2.7446	2.7307	2.7092	2.6909	2.6575	2.6383	2.6314	2.6286	2.6258
29.00	3.0098	2.9585	2.9118	2.8579	2.8288	2.8114	2.7821	2.7673	2.7444	2.7250	2.6895	2.6692	2.6618	2.6589	2.6558
29.50	3.0584	3.0046	2.9560	2.8996	2.8689	2.8504	2.8194	2.8036	2.7794	2.7589	2.7212	2.6999	2.6970	2.6889	2.6857
30.00	3.1069	3.0507	3.0001	2.9412	2.9088	2.8892	2.8567	2.8399	2.0143	2.7926	2.7528	2.7304	2.7221	2.7187	2.7153

TABLE A.38(d)

Extended Sukkar–Cornell Integral for Bottom-hole Pressure Calculation

P_r	Reduced Temperature for B = 15.0														
	1.1	1.2	1.3	1.4	1.5	1.6	1.7	1.8	1.9	2.0	2.2	2.4	2.6	2.8	3.0
0.20	0.0000	0.0000	0.0000	0.0000	0.0000	0.0000	0.0000	0.0000	0.0000	0.0000	0.0000	0.0000	0.0000	0.0000	0.0000
0.50	0.0077	0.0075	0.0074	0.0073	0.0072	0.0071	0.0071	0.0071	0.0070	0.0070	0.0070	0.0070	0.0069	0.0069	0.0069
1.00	0.0385	0.0359	0.0345	0.0336	0.0330	0.0325	0.0322	0.3119	0.0317	0.0316	0.0313	0.0311	0.0310	0.0309	0.0308
1.50	0.0939	0.0838	0.0793	0.0765	0.0746	0.0732	0.0721	0.0713	0.0708	0.0703	0.0696	0.0692	0.0687	0.0685	0.0682
2.00	0.1571	0.1453	0.1371	0.1319	0.1282	0.1257	0.1236	0.1220	0.1211	0.1202	0.1189	0.1180	0.1172	0.1167	0.1161
2.50	0.2162	0.2093	0.2008	0.1943	0.1892	0.1857	0.1827	0.1804	0.1790	0.1777	0.1758	0.1745	0.1733	0.1724	0.1716
3.00	0.2725	0.2710	0.2648	0.2587	0.2533	0.2493	0.2458	0.2431	0.2413	0.2397	0.2374	0.2357	0.2342	0.2331	0.2320
3.50	0.3275	0.3302	0.3267	0.3222	0.3176	0.3138	0.3102	0.3074	0.3055	0.3038	0.3012	0.2994	0.2978	0.2964	0.2952
4.00	0.3818	0.3874	0.3862	0.3837	0.3805	0.3774	0.3743	0.3717	0.3699	0.3683	0.3657	0.3679	0.3622	0.3608	0.3596

(Continued)

TABLE A.38(d) (Continued)

P_r	Reduced Temperature for $B = 15.0$													
	1.1	1.2	1.3	1.4	1.5	1.6	1.7	1.8	1.9	2.0	2.2	2.4	2.6	2.8	3.0
4.50	0.4355	0.4430	0.4435	0.4431	0.4415	0.4393	0.4369	0.4349	0.4335	0.4320	0.4298	0.4281	0.4265	0.4252	0.4240
5.00	0.4887	0.4975	0.4992	0.5004	0.5006	0.4994	0.4978	0.4966	0.4956	0.4945	0.4928	0.4914	0.4900	0.4888	0.4877
5.50	0.5413	0.5508	0.5535	0.5561	0.5579	0.5577	0.5570	0.5566	0.5561	0.5554	0.5543	0.5534	0.5522	0.5512	0.5503
6.00	0.5936	0.6034	0.6066	0.6103	0.6135	0.6143	0.6144	0.6149	0.6149	0.6147	0.6143	0.6138	0.6129	0.6121	0.6113
6.50	0.6454	0.6553	0.6590	0.6634	0.6676	0.6694	0.6703	0.6715	0.6720	0.6724	0.6726	0.6727	0.6721	0.6715	0.6708
7.00	0.6969	0.7068	0.7105	0.7155	0.7205	0.7230	0.7246	0.7256	0.7276	0.7284	0.7293	0.7299	0.7296	0.7291	0.7286
7.50	0.7482	0.7577	0.7613	0.7666	0.7722	0.7754	0.7776	0.7802	0.7817	0.7829	0.7844	0.7854	0.7855	0.7852	0.7848
8.00	0.7991	0.8082	0.8114	0.8170	0.8230	0.8266	0.8293	0.8324	0.8344	0.8360	0.8391	0.8395	0.8398	0.8397	0.8394
8.50	0.8497	0.8582	0.8611	0.8666	0.8729	0.8768	0.8799	0.8835	0.8858	0.8878	0.8914	0.8920	0.8926	0.8927	0.8925
9.00	0.9000	0.9078	0.9102	0.9157	0.9220	0.9261	0.9295	0.9334	0.9360	0.9382	0.9423	0.9432	0.9440	0.9442	0.9441
9.50	0.9500	0.9570	0.9588	0.9641	0.9704	0.9746	0.9782	0.9824	0.9852	0.9876	0.9920	0.9932	0.9941	0.9444	0.9944
10.00	0.9998	1.0059	1.0071	1.0121	1.0181	1.0223	1.0260	1.0304	1.0334	1.0359	1.0407	1.0420	1.0430	1.0434	1.0435
10.50	1.0492	1.0544	1.0549	1.0595	1.0653	1.0694	1.0731	1.0776	1.0806	1.0833	1.0883	1.0897	1.0908	1.0913	1.0914
11.00	1.0985	1.1026	1.1024	1.1065	1.1119	1.1159	1.1195	1.1239	1.1271	1.1298	1.1349	1.1364	1.1375	1.1380	1.1381
11.50	1.1475	1.1506	1.1496	1.1530	1.1580	1.1618	1.1653	1.1696	1.1728	1.1755	1.1807	1.1822	1.1832	1.1837	1.1839
12.00	1.1963	1.1983	1.1964	1.1992	1.2037	1.2072	1.2105	1.2147	1.2178	1.2205	1.2256	1.2270	1.2281	1.2285	1.2287
12.50	1.2449	1.2458	1.2430	1.2449	1.2490	1.2522	1.2551	1.2592	1.2622	1.2648	1.2698	1.2711	1.2720	1.2724	1.2725
13.00	1.2934	1.2931	1.2893	1.2903	1.2939	1.2967	1.2993	1.3031	1.3060	1.3084	1.3131	1.3143	1.3152	1.3155	1.3156
13.50	1.3417	1.3402	1.3354	1.3354	1.3384	1.3408	1.3430	1.3465	1.3492	1.3514	1.3558	1.3567	1.3575	1.3578	1.3578
14.00	1.3899	1.3870	1.3812	1.3802	1.3825	1.3845	1.3862	1.3694	1.3918	1.3938	1.3977	1.3984	1.3991	1.3993	1.3992
14.50	1.4380	1.4337	1.4268	1.4247	1.4263	1.4278	1.4290	1.4319	1.4339	1.4356	1.4390	1.4395	1.4400	1.4401	1.4400
15.00	1.4860	1.4803	1.4722	1.4689	1.4698	1.4708	1.4714	1.4739	1.4756	1.4769	1.4797	1.4798	1.4802	1.4802	1.4800
15.50	1.5338	1.5266	1.5174	1.5129	1.5130	1.5135	1.5134	1.5155	1.5168	1.5177	1.5198	1.5196	1.5197	1.5197	1.5194
16.00	1.5815	1.5728	1.5625	1.5566	1.5559	1.5558	1.5551	1.5567	1.5575	1.5580	1.5594	1.5587	1.5587	1.5585	1.5582
16.50	1.6291	1.6189	1.6073	1.6001	1.5985	1.5979	1.5964	1.5976	1.5978	1.5979	1.5984	1.5973	1.5971	1.5968	1.5964
17.00	1.6766	1.6649	1.6520	1.6434	1.6409	1.6397	1.6374	1.6381	1.6378	1.6373	1.6370	1.6354	1.6350	1.6346	1.6341
17.50	1.7241	1.7107	1.6966	1.6865	1.6830	1.6812	1.6781	1.6783	1.6773	1.6764	1.6750	1.6730	1.6723	1.6718	1.6712
18.00	1.7714	1.7564	1.7410	1.7293	1.7249	1.7225	1.7186	1.7181	1.7166	1.7150	1.7127	1.7100	1.7091	1.7085	1.7078
18.50	1.8187	1.8020	1.7853	1.7720	1.7666	1.7635	1.7587	1.7577	1.7554	1.7533	1.7499	1.7466	1.7455	1.7447	1.7439
19.00	1.8659	1.8475	1.8294	1.8146	1.8081	1.8043	1.7986	1.7970	1.7940	1.7912	1.7866	1.7828	1.7814	1.7805	1.7796

19.50	1.9130	1.8929	1.8734	1.8569	1.8493	1.8449	1.8382	1.8360	1.8322	1.8288	1.8230	1.8186	1.8169	1.8158	1.8148
20.00	1.9600	1.9382	1.9173	1.8991	1.8904	1.8853	1.8776	1.8747	1.8702	1.8661	1.8590	1.8540	1.8519	1.8508	1.8496
20.50	2.0070	1.9834	1.9611	1.9412	1.9314	1.9255	1.9168	1.9132	1.9079	1.9031	1.8947	1.8889	1.8866	1.8853	1.8840
21.00	2.0539	2.0285	2.0048	1.9831	1.9721	1.9655	1.9557	1.9515	1.9453	1.9397	1.9300	1.9236	1.9209	1.9195	1.9180
21.50	2.1007	2.0736	2.0484	2.0248	2.0127	2.0054	1.9944	1.9895	1.9824	1.9761	1.9650	1.9578	1.9549	1.9532	1.9517
22.00	2.1475	2.1185	2.0918	2.0665	2.0531	2.0450	2.0330	2.0273	2.0193	2.0122	1.9997	1.9917	1.9884	1.9867	1.9850
22.50	2.1943	2.1634	2.1352	2.1080	2.0934	2.0845	2.0713	2.0649	2.0560	2.0481	2.0341	2.0253	2.0217	2.0148	2.0179
23.00	2.2410	2.2082	2.1785	2.1494	2.1335	2.1239	2.1095	2.1024	2.0924	2.0837	2.0681	2.0586	2.0546	2.0525	2.0506
23.50	2.2876	2.2529	2.2217	2.1906	2.1735	2.1631	2.1475	2.1346	2.1286	2.1191	2.1019	2.0916	2.0872	2.0850	2.0829
24.00	2.3342	2.2976	2.2648	2.2318	2.2134	2.2021	2.1853	2.1766	2.1646	2.1542	2.1355	2.1242	2.1196	2.1171	2.1149
24.50	2.3807	2.3422	2.3079	2.2728	2.2531	2.2410	2.2229	2.2135	2.2005	2.1891	2.1687	2.1567	2.1516	2.1490	2.1466
25.00	2.4272	2.3867	2.3509	2.3138	2.2927	2.2798	2.2604	2.2502	2.2361	2.2238	2.2017	2.1888	2.1834	2.1806	2.1780
25.50	2.4736	2.4312	2.3937	2.3546	2.3322	2.3184	2.2978	2.2867	2.2715	2.2583	2.2345	2.2207	2.2149	2.2119	2.2092
26.00	2.5200	1.4756	2.4366	2.3953	2.3716	2.3569	2.3550	2.3230	2.3067	2.2927	2.2671	2.2523	2.2461	2.2430	2.2401
26.50	2.5664	2.5200	2.4793	2.4360	2.4109	2.3953	2.3720	2.3592	2.3418	2.3268	2.2994	2.2837	2.2771	2.2738	2.2707
27.00	2.6127	2.5643	2.5220	2.4766	2.4501	2.4336	2.4089	2.3953	2.3767	2.3607	2.3315	2.3149	2.3078	2.3044	2.3011
27.50	2.6590	2.6086	2.5646	2.5170	2.4891	2.4718	2.4457	2.4312	2.4115	2.3944	2.3634	2.3458	2.3384	2.3347	2.3313
28.00	2.7053	2.6520	2.6072	2.5574	2.5281	2.5098	2.4824	2.4670	2.4460	2.4280	2.3951	2.3765	2.3687	2.3648	2.3612
28.50	2.7515	2.6969	2.6497	2.5977	2.5669	2.5478	2.5189	2.5026	2.4805	2.4614	2.4266	2.4070	2.3987	2.3947	2.3909
29.00	2.7977	2.7410	2.6921	2.6380	2.6057	2.5856	2.5553	2.5382	2.5148	2.4947	2.4579	2.4373	2.4286	2.4244	2.4205
29.50	2.8438	2.7851	2.7345	2.6781	2.6444	2.6234	2.5916	2.5736	2.5489	2.5278	2.4890	2.4674	2.4583	2.4538	2.4497
30.00	2.8899	2.8291	2.7769	2.7182	2.6830	2.6610	2.6278	2.6088	2.5829	2.5607	2.5200	2.4974	2.4878	2.4831	2.4788

TABLE A.38(e)

Extended Sukkar–Cornell Integral for Bottom-hole Pressure Calculation

P_r	Reduced Temperature for B = 20.0														
	1.1	1.2	1.3	1.4	1.5	1.6	1.7	1.8	1.9	2.0	2.2	2.4	2.6	2.8	3.0
0.20	0.0000	0.0000	0.0000	0.0000	0.0000	0.0000	0.0000	0.0000	0.0000	0.0000	0.0000	0.0000	0.0000	0.0000	0.0000
0.50	0.0058	0.0056	0.0055	0.0055	0.0054	0.0054	0.0053	0.0053	0.0053	0.0053	0.0052	0.0052	0.0052	0.0052	0.0052
1.00	0.0294	0.0272	0.0262	0.0255	0.0250	0.0246	0.0243	0.0241	0.0240	0.0239	0.0237	0.0236	0.0235	0.0234	0.0233
1.50	0.0740	0.0649	0.0610	0.0587	0.0572	0.0561	0.0561	0.0545	0.0541	0.0537	0.0532	0.0528	0.0525	0.0522	0.0520

(Continued)

TABLE A.38(e) (Continued)

P_r	Reduced Temperature for B = 20.0														
	1.1	1.2	1.3	1.4	1.5	1.6	1.7	1.8	1.9	2.0	2.2	2.4	2.6	2.8	3.0
2.00	0.1295	0.1156	0.1077	0.1030	0.0998	0.0976	0.0958	0.0945	0.0937	0.0930	0.0918	0.0911	0.0905	0.0900	0.0895
2.50	0.1832	0.1712	0.1614	0.1547	0.1498	0.1465	0.1438	0.1417	0.1404	0.1393	0.1376	0.1364	0.1354	0.1346	0.1339
3.00	0.2350	0.2264	0.2172	0.2099	0.2040	0.1999	0.1964	0.1937	0.1920	0.1904	0.1882	0.1867	0.1853	0.1842	0.1832
3.50	0.2860	0.2801	0.2725	0.2657	0.2597	0.2553	0.2514	0.2484	0.2463	0.2445	0.2419	0.2401	0.2384	0.2371	0.2359
4.00	0.3365	0.3326	0.3264	0.3208	0.3154	0.3111	0.3073	0.3041	0.3020	0.3000	0.2972	0.2952	0.2934	0.2919	0.2906
4.50	0.3865	0.3841	0.3790	0.3747	0.3703	0.3664	0.3629	0.3599	0.3578	0.3559	0.3531	0.3510	0.3492	0.3476	0.3462
5.00	0.4360	0.4346	0.4305	0.4273	0.4240	0.4208	0.4177	0.4151	0.4132	0.4114	0.4088	0.4068	0.4050	0.4034	0.4021
5.50	0.4852	0.4843	0.4809	0.4787	0.4765	0.4740	0.4714	0.4594	0.4678	0.4662	0.4639	0.4622	0.4604	0.4589	0.4577
6.00	0.5341	0.5335	0.5305	0.5291	0.5279	0.5261	0.5241	0.5226	0.5213	0.5201	0.5182	0.5167	0.5151	0.5137	0.5125
6.50	0.5827	0.5821	0.5794	0.5786	0.5783	0.5771	0.5756	0.5747	0.5738	0.5729	0.5714	0.5703	0.5689	0.5676	0.5665
7.00	0.6310	0.6304	0.6277	0.6274	0.6276	0.6270	0.6261	0.6257	0.6252	0.6246	0.6236	0.6228	0.6216	0.6205	0.6194
7.50	0.6791	0.6782	0.6755	0.6754	0.6761	0.6760	0.6755	0.6756	0.6754	0.6752	0.6746	0.6741	0.6732	0.6722	0.6712
8.00	0.7269	0.7257	0.7227	0.7228	0.7238	0.7241	0.7240	0.7245	0.7247	0.7247	0.7251	0.7244	0.7237	0.7227	0.7219
8.50	0.7745	0.7728	0.7695	0.7696	0.7708	0.7714	0.7716	0.7725	0.7729	0.7732	0.7740	0.7735	0.7730	0.7722	0.7714
9.00	0.8219	0.8196	0.8159	0.8160	0.8172	0.8179	0.8184	0.8195	0.8202	0.8207	0.8218	0.8216	0.8212	0.8205	0.8198
9.50	0.8690	0.8661	0.8620	0.8618	0.8631	0.8638	0.8644	0.8658	0.8666	0.8673	0.8687	0.8687	0.8684	0.8678	0.8672
10.00	0.9159	0.9123	0.9077	0.9073	0.9083	0.9091	0.9098	0.9113	0.9123	0.9131	0.9147	0.9148	0.9146	0.9141	0.9135
10.50	0.9626	0.9582	0.9530	0.9523	0.9531	0.9538	0.9545	0.9561	0.9571	0.9580	0.9599	0.9601	0.9599	0.9595	0.9589
11.00	1.0091	1.0089	0.9981	0.9969	0.9975	0.9980	0.9987	1.0002	1.0014	1.0023	1.0043	1.0045	1.0043	1.0039	1.0034
11.50	1.0554	1.0494	1.0429	1.0412	1.0414	1.0418	1.0423	1.0438	1.0450	1.0459	1.0479	1.0481	1.0479	1.0475	1.0470
12.00	1.1016	1.0946	1.0874	1.0851	1.0849	1.0851	1.0855	1.0868	1.0879	1.0688	1.0908	1.0909	1.0908	1.0903	1.0698
12.50	1.1476	1.1397	1.1317	1.1288	1.1282	1.1280	1.1282	1.1294	1.1304	1.1312	1.1331	1.1331	1.1328	1.1323	1.1318
13.00	1.1935	1.1846	1.1758	1.1721	1.1710	1.1706	1.1704	1.1714	1.1723	1.1730	1.1746	1.1745	1.1742	1.1736	1.1731
13.50	1.2392	1.2293	1.2197	1.2151	1.2136	1.2128	1.2122	1.2130	1.2137	1.2142	1.2156	1.2153	1.2149	1.2143	1.2136
14.00	1.2849	1.2739	1.2833	1.2579	1.2558	1.2547	1.2537	1.2542	1.2547	2.2549	1.2559	1.2564	1.2549	1.2542	1.2535
14.50	1.3304	1.3183	1.3068	1.3005	1.2977	1.2962	1.2948	1.2949	1.2952	1.2952	1.2957	1.2949	1.2463	1.2935	1.2928
15.00	1.3759	1.3625	1.3501	1.3428	1.3394	1.3375	1.3355	1.3353	1.3352	1.3349	1.3349	1.3339	1.3331	1.3322	1.3315
15.50	1.4212	1.4067	1.3933	1.3849	1.3808	1.3784	1.3759	1.3754	1.3749	1.3743	1.3736	1.3723	1.3713	1.3704	1.3695
16.00	1.4665	1.4507	1.4363	1.4267	1.4220	1.4191	1.4150	1.4151	1.4142	1.4132	1.4118	1.4101	1.4090	1.4080	1.4071
16.50	1.5116	1.4945	1.4792	1.4684	1.4629	1.4595	1.4558	1.4544	1.4531	1.4517	1.4496	1.4475	1.4462	1.4451	1.4441

P_r	Reduced Temperature for B = 20.0														
	1.1	1.2	1.3	1.4	1.5	1.6	1.7	1.8	1.9	2.0	2.2	2.4	2.6	2.8	3.0
17.00	1.5567	1.5383	1.5219	1.5099	1.5036	1.4997	1.4953	1.4935	1.4916	1.4898	1.4869	1.4844	1.4829	1.4817	1.4806
17.50	1.6017	1.5820	1.5645	1.5512	1.5441	1.5397	1.5345	1.5323	1.5298	1.5275	1.5238	1.5208	1.5191	1.5178	1.5166
18.00	1.6467	1.6256	1.6069	1.5924	1.5844	1.5794	1.5735	1.5708	1.5678	1.5649	1.5603	1.5588	1.5549	1.5584	1.5522
18.50	1.6916	1.6691	1.6493	1.6334	1.6245	1.6190	1.6123	1.6090	1.6054	1.6020	1.5964	1.5924	1.5902	1.5837	1.5973
19.00	1.7364	1.7125	1.6915	1.6742	1.6644	1.6583	1.6508	1.6470	1.6427	1.6388	1.6321	1.6275	1.5252	1.6235	1.6220
19.50	1.7811	1.7558	1.7336	1.7149	1.7042	1.6975	1.6891	1.6847	1.6797	1.6752	1.6675	1.6623	1.6597	1.6579	1.6563
20.00	1.8258	1.7990	1.7757	1.7555	1.7438	1.7364	1.7271	1.7222	1.7165	1.7114	1.7025	1.6967	1.6938	1.6919	1.6902
20.50	1.8705	1.8421	1.8176	1.7959	1.7832	1.7752	1.7650	1.7595	1.7530	1.7473	1.7372	1.7308	1.7276	1.7256	1.7238
21.00	1.9150	1.8852	1.8594	1.8362	1.8225	1.8139	1.8027	1.7965	1.7893	1.7829	1.7716	1.7645	1.7611	1.7589	1.7570
21.50	1.9596	1.9282	1.9012	1.8763	1.8616	1.8523	1.8401	1.8334	1.8254	1.8183	1.8056	1.7979	1.7942	1.7918	1.7898
22.00	2.0041	1.9711	1.9429	1.9164	1.9006	1.8906	1.8774	1.8700	1.8612	1.8534	1.8394	1.8310	1.8270	1.8245	1.8223
22.50	2.0485	2.0140	1.9844	1.9563	1.9395	1.9288	1.9146	1.9065	1.8968	1.8882	1.8730	1.8638	1.8595	1.8568	1.8545
23.00	2.0929	2.0568	2.0259	1.9982	1.9782	1.9668	1.9516	1.9428	1.9322	1.9229	1.9062	1.8963	1.8916	1.8889	1.8864
23.50	2.1372	2.0995	2.0674	2.0359	2.0168	2.0047	1.9684	1.9789	1.9674	1.9573	1.9392	1.9286	1.9235	1.9206	1.9180
24.00	2.1815	2.1422	2.1087	2.0756	2.0553	2.0425	2.0250	2.0149	2.0025	1.9916	1.9719	1.9605	1.9551	1.9521	2.9493
24.50	2.2258	2.1849	2.1500	2.1151	2.0937	2.0801	2.0615	2.0507	2.0373	2.0256	2.0044	1.9922	1.9865	1.9832	1.9804
25.00	2.2700	2.2274	2.1912	2.1546	2.1319	2.1176	2.0979	2.0863	2.0719	2.0594	2.0367	2.0237	2.0176	2.0142	2.0112
25.50	2.3142	2.2700	2.2324	2.1939	2.1701	2.1550	2.1341	2.1218	2.1064	2.0980	2.0687	2.0549	2.0484	2.0449	2.0417
26.00	2.3584	2.3124	2.2735	2.2332	2.2002	2.1923	2.1702	2.1671	2.1408	2.1265	2.1005	2.0858	2.0790	2.0753	2.0720
26.50	2.4025	2.3549	2.3145	2.2724	2.2461	2.2295	2.2062	2.1923	2.1749	2.1598	2.1321	2.1166	2.1094	2.1055	2.1020
27.00	2.4466	2.3973	2.3555	2.3115	2.2840	2.2665	2.2420	2.2274	2.2089	2.1929	2.1636	2.1471	2.1395	2.1355	2.1318
27.50	2.4907	2.4396	2.3964	2.3505	2.3218	2.3035	2.2778	2.2823	2.2428	2.2258	2.1968	2.1774	2.1695	2.1652	2.1614
28.00	2.5347	2.4819	2.4373	2.3895	2.3595	2.3404	2.3134	2.2971	2.2765	2.2586	2.2258	2.2075	2.1992	2.1948	2.1908
28.50	2.5787	2.5742	2.4781	2.4284	2.3971	2.3772	2.3409	2.3118	2.3100	2.2912	2.2566	2.2375	2.2287	2.2241	2.2220
29.00	2.6227	2.5664	2.5189	2.4672	2.4146	2.4119	2.3843	2.3664	2.3435	2.3217	2.2873	2.2677	2.2540	2.2662	2.2480
29.50	2.6666	2.6085	2.5596	2.5060	2.4720	2.4504	2.4195	2.4008	2.3768	2.3560	2.3178	2.2967	2.2871	2.2822	2.2777
30.00	2.7106	2.6507	2.6003	2.5447	2.5094	2.4870	2.4547	2.4352	2.4100	2.3882	2.3481	2.3261	2.3161	2.3109	2.3063

TABLE A.38(f)
Extended Sukkar–Cornell Integral for Bottom-hole Pressure Calculation

P_r	Reduced Temperature for $B = 25.0$														
	1.1	1.2	1.3	1.4	1.5	1.6	1.7	1.8	1.9	2.0	2.2	2.4	2.6	2.8	3.0
0.20	0.0000	0.0000	0.0000	0.0000	0.0000	0.0000	0.0000	0.0000	0.0000	0.0000	0.0000	0.0000	0.0000	0.0000	0.0000
0.50	0.0047	0.0045	0.0044	0.0044	0.0043	0.0043	0.0043	0.0042	0.0042	0.0042	0.0042	0.0042	0.0042	0.0042	0.0042
1.00	0.0237	0.0219	0.0211	0.0205	0.0201	0.0198	0.0196	0.0194	0.0193	0.0192	0.0191	0.0189	0.0189	0.0198	0.0187
1.50	0.0611	0.0529	0.0496	0.0477	0.0464	0.0454	0.0446	0.0441	0.0438	0.0435	0.0430	0.0427	0.0424	0.0422	0.0420
2.00	0.1106	0.0961	0.0888	0.0846	0.0818	0.0798	0.0783	0.0771	0.0764	0.0758	0.0749	0.0742	0.0737	0.0733	0.0729
2.50	0.1598	0.1453	0.1352	0.1287	0.1241	0.1211	0.1186	0.1188	0.1156	0.1146	0.1131	0.1171	0.1111	0.1104	0.1098
3.00	0.2079	0.1952	0.1846	0.1769	0.1711	0.1670	0.1637	0.1612	0.1596	0.1581	0.1561	0.1547	0.1534	0.1524	0.1515
3.50	0.2554	0.2444	0.2346	0.2267	0.2202	0.2156	0.2117	0.2087	0.2067	0.2049	0.2024	0.2007	0.1991	0.1978	0.1967
4.00	0.3025	0.2930	0.2840	0.2766	0.2702	0.2654	0.2613	0.2579	0.2557	0.2537	0.2508	0.2488	0.2470	0.2455	0.2442
4.50	0.3492	0.3408	0.3325	0.3260	0.3200	0.3154	0.3112	0.3078	0.3055	0.3036	0.3004	0.2982	0.2962	0.2946	0.2932
5.00	0.3957	0.3879	0.3803	0.3745	0.3693	0.3650	0.3610	0.3578	0.3555	0.3536	0.3503	0.3481	0.3461	0.3444	0.3429
5.50	0.4418	0.4345	0.4274	0.4223	0.4178	0.4139	0.4103	0.4073	0.4052	0.4031	0.4002	0.3980	0.3961	0.3943	0.3929
6.00	0.4878	0.4806	0.4739	0.4694	0.4656	0.4622	0.4589	0.4563	0.4543	0.4525	0.4498	0.4477	0.4450	0.4441	0.4428
6.50	0.5335	0.5263	0.5198	0.5158	0.5126	0.5097	0.5068	0.5045	0.5028	0.5012	0.4988	0.4969	0.4951	0.4935	0.4922
7.00	0.5790	0.5718	0.5653	0.5616	0.5589	0.5564	0.5539	0.5520	0.5506	0.5492	0.5471	0.5454	0.5437	0.5422	0.5409
7.50	0.6243	0.6169	0.6104	0.6069	0.6045	0.6024	0.6003	0.5987	0.5975	0.5966	0.5946	0.5932	0.5917	0.5902	0.5690
8.00	0.6694	0.6618	0.6550	0.6516	0.6495	0.6477	0.6459	0.6447	0.6437	0.6428	0.6415	0.6401	0.6388	0.6374	0.6362
8.50	0.7143	0.7063	0.6993	0.6960	0.6940	0.6924	0.6908	0.6899	0.6892	0.6884	0.6874	0.6882	0.6850	0.6837	0.6826
9.00	0.7591	0.7506	0.7433	0.7399	0.7380	0.7365	0.7351	0.7344	0.7338	0.7333	0.7325	0.7315	0.7304	0.7292	0.7282
9.50	0.8036	0.7946	0.7870	0.7834	0.7814	0.7800	0.7788	0.7783	0.7778	0.7774	0.7769	0.7760	0.7750	0.7739	0.7830
10.00	0.8480	0.8384	0.8303	0.8266	0.8245	0.8231	0.8219	0.8215	0.8212	0.8208	0.8205	0.8198	0.8189	0.8178	0.8169
10.50	0.8922	0.8520	0.8735	0.8695	0.8671	0.8657	0.8645	0.8641	0.8639	0.8636	0.8635	0.8628	0.8619	0.8609	0.8600
11.00	0.9362	0.9254	0.9163	0.9120	0.9094	0.9078	0.9056	0.9063	0.9061	0.9058	0.9058	0.9052	0.9043	0.9033	0.9024
11.50	0.9801	0.9686	0.9590	0.9542	0.9514	0.9496	0.9483	0.9679	0.9477	0.9475	0.9475	0.9468	0.9459	0.9449	0.9440
12.00	1.0239	1.0117	1.0014	0.9961	0.9930	0.9910	0.9896	0.9891	0.9889	0.9886	0.9885	0.9879	0.9864	0.9854	0.9850
12.50	1.0676	1.0545	1.0437	1.0378	1.0343	1.0321	1.0304	1.0298	1.0295	1.0292	1.0240	1.0283	1.0273	1.0262	1.0253
13.00	1.1111	1.0973	1.0857	1.0792	1.0753	1.0729	1.0709	1.0701	1.0698	1.0693	1.0689	1.0681	1.0670	1.0659	1.0650
13.50	1.1546	1.1398	1.1276	1.1204	1.1161	1.1134	1.1111	1.1101	1.1095	1.1089	1.1083	1.1073	1.1062	1.1050	1.1040
14.00	1.1979	1.1823	1.1693	1.1614	1.1566	1.1535	1.1509	1.1496	1.1489	1.1481	1.1472	1.1459	1.1447	1.1435	1.1425

14.50	1.2412	1.2246	1.2109	1.2021	1.1968	1.1934	1.1904	1.1889	1.1879	1.1868	1.1855	1.1840	1.1827	1.1615	1.1604
15.00	1.2844	1.2668	1.2523	1.2427	1.2368	1.2331	1.2296	1.2278	1.2265	1.2252	1.2234	1.2217	1.2202	1.2189	1.2177
15.50	1.3275	1.3089	1.2936	1.2830	1.2766	1.2725	1.2685	1.2663	1.2647	1.2631	1.2608	1.2588	1.2572	1.2558	1.2546
16.00	1.3705	1.3509	1.3347	1.3232	1.3161	1.3116	1.3071	1.3046	1.3026	1.3007	1.2978	1.2954	1.2937	1.2922	1.2909
16.50	1.4135	1.3928	1.3757	1.3632	1.3555	1.3505	1.3455	1.3426	1.3402	1.3379	1.3343	1.3316	1.3298	1.3281	1.3268
17.00	1.4564	1.4346	1.4166	1.4031	1.3947	1.3892	1.3836	1.3803	1.3775	1.3748	1.3705	1.3674	1.3653	1.3637	1.3623
17.50	1.4992	1.4763	1.4574	1.4428	1.4336	1.4278	1.4215	1.4178	1.4145	1.4114	1.4062	1.4028	1.4005	1.3987	1.3973
18.00	1.5420	1.5180	1.4981	1.4823	1.4724	1.4661	1.4591	1.4550	1.4512	1.4476	1.4617	1.4377	1.4353	1.4334	1.4318
18.50	1.5847	1.5595	1.5387	1.5217	1.5111	1.5042	1.4965	1.4920	1.4876	1.4835	1.4767	1.4728	1.4697	1.4677	1.4660
19.00	1.6274	1.6010	1.5792	1.5610	1.5496	1.5422	1.5338	1.5287	1.5238	1.5192	1.5114	1.5065	1.5036	1.5015	1.4998
19.50	1.6700	1.6424	1.6196	1.6002	1.5879	1.5800	1.5708	1.5653	1.5597	1.5546	1.5458	1.5404	1.5373	1.5351	1.5332
20.00	1.7126	1.6837	1.6597	1.6392	1.6261	1.6176	1.6076	1.6016	1.5954	1.5997	1.5799	1.5739	1.5706	1.5692	1.5663
20.50	1.7551	1.7250	1.7001	1.6781	1.6641	1.6551	1.6443	1.6377	1.6308	1.6246	1.6137	1.6071	1.6035	1.6011	1.5990
21.00	1.7975	1.7662	1.7403	1.7169	1.7020	1.6924	1.6808	1.6736	1.6660	1.6592	1.6472	1.6400	1.5362	1.6336	1.6614
21.50	1.8400	1.8073	1.7803	1.7556	1.7398	1.7296	1.7171	1.7094	1.7011	1.6936	1.6804	1.6726	1.5685	1.6658	1.6635
22.00	1.8824	1.8484	1.8203	1.7942	1.7775	1.7667	1.7532	1.7450	1.7359	1.7278	1.7134	1.7049	1.7005	1.6977	1.6953
22.50	1.9247	1.8895	1.8603	1.8327	1.8150	1.8036	1.7892	1.7804	1.7705	1.7617	1.7460	1.7370	1.7322	1.7243	1.7267
23.00	1.9670	1.9304	1.9001	1.8711	1.8524	1.8404	1.8251	1.8156	1.8049	1.7955	1.7785	1.7687	1.7637	1.7606	1.7579
23.50	2.0093	1.9714	1.9399	1.9094	1.8898	1.8771	1.8608	1.8507	1.8392	1.8290	1.8107	1.8002	1.7949	1.7916	1.7889
24.00	2.0516	2.0122	1.9797	1.9477	1.9270	1.9136	1.8964	1.8856	1.8733	1.8623	1.8427	1.8315	1.8258	1.8224	1.8195
24.50	2.0938	2.0531	2.0193	1.9858	1.9641	1.9501	1.9318	1.9204	1.9072	1.8955	1.8744	1.8625	1.8565	1.8530	1.8499
25.00	2.1360	2.0938	2.0590	2.0239	2.0011	1.9864	1.9671	1.9550	1.9409	1.9285	1.9060	1.8933	1.8870	1.8833	1.8801
25.50	2.1761	2.1346	2.0985	2.0618	2.0380	2.0226	2.0023	1.9895	1.9745	1.9613	1.9373	1.9238	1.9172	1.9133	1.9100
26.00	2.2202	2.1753	2.1380	2.0998	2.0749	2.0588	2.0373	2.0239	2.0079	1.9939	1.9684	1.9542	1.9472	1.9431	1.9397
26.50	2.2623	2.2159	2.1775	2.1376	2.1116	2.0948	2.0723	2.0581	2.0412	2.0264	1.9994	1.9843	1.9769	1.9728	1.9692
27.00	2.3044	2.2566	2.2169	2.1754	2.1403	2.1307	2.1071	2.0923	2.0744	2.0587	2.0301	2.0142	2.0065	2.0022	1.9964
27.50	2.3464	2.2971	2.2562	2.2131	2.1848	2.1666	2.1418	2.1263	2.1074	2.0909	2.0607	2.0440	2.0359	2.0314	2.0275
28.00	2.3885	2.3377	2.2955	2.2507	2.2213	2.2024	2.1764	2.1601	2.1403	2.1229	2.0911	2.0735	2.0650	2.0603	2.0563
28.50	2.4305	2.3782	2.3348	2.2883	2.2578	2.2380	2.2110	2.1939	2.1730	2.1548	2.1213	2.1028	2.0940	2.0891	2.0849
29.00	2.4724	2.4186	2.3740	2.3258	2.2941	2.2736	2.2454	2.2276	2.2056	2.1865	2.1913	2.1320	2.1228	2.1178	2.1134
29.50	2.5144	2.4591	2.4132	2.3632	2.3304	2.3091	2.2797	2.2611	2.2331	2.2181	2.1812	2.1610	2.1514	2.1462	2.1417
30.00	2.5563	2.4995	2.4523	2.4006	2.3666	2.3446	2.3139	2.2946	2.2705	2.2496	2.2110	2.1898	2.1798	2.1744	2.1698

TABLE A.38(g)
Extended Sukkar–Cornell Integral for Bottom-hole Pressure Calculation

P_r	Reduced Temperature for $B = 30.0$														
	1.1	1.2	1.3	1.4	1.5	1.6	1.7	1.8	1.9	2.0	2.2	2.4	2.6	2.8	3.0
0.20	0.0000	0.0000	0.0000	0.0000	0.0000	0.0000	0.0000	0.0000	0.0000	0.0000	0.0000	0.0000	0.0000	0.0000	0.0000
0.50	0.0039	0.0038	0.0037	0.0037	0.0036	0.0036	0.0036	0.0035	0.0035	0.0035	0.0035	0.0035	0.0035	0.0035	0.0035
1.00	0.0199	0.0184	0.0176	0.0172	0.0168	0.0166	0.0164	0.0162	0.0162	0.0161	0.0169	0.0159	0.0158	0.0157	0.0157
1.50	0.0521	0.0447	0.0418	0.0401	0.0390	0.0382	0.0375	0.0371	0.0368	0.0365	0.0361	0.0358	0.0356	0.0355	0.0353
2.00	0.0967	0.0823	0.0755	0.0718	0.0692	0.0676	0.0672	0.0652	0.0646	0.0640	0.0632	0.0626	0.0621	0.0618	0.0615
2.50	0.1422	0.1264	0.1164	0.1103	0.1060	0.1033	0.1010	0.0993	0.0963	0.0974	0.0960	0.0951	0.0943	0.0937	0.0931
3.00	0.1870	0.1719	0.1608	0.1531	0.1474	0.1436	0.1404	0.1381	0.1366	0.1353	0.1334	0.1321	0.1309	0.1300	0.1292
3.50	0.2314	0.2174	0.2063	0.1980	0.1914	0.1869	0.1831	0.1801	0.1782	0.1765	0.1741	0.1725	0.1710	0.1697	0.1687
4.00	0.2756	0.2625	0.2519	0.2436	0.2367	0.2318	0.2275	0.2242	0.2219	0.2199	0.2172	0.2152	0.2135	0.2120	0.2108
4.50	0.3195	0.3071	0.2970	0.2891	0.2823	0.2773	0.2729	0.2693	0.2669	0.2647	0.2617	0.2594	0.2675	0.2559	0.2545
5.00	0.3632	0.3513	0.3416	0.3343	0.3278	0.3229	0.3186	0.3149	0.3124	0.3101	0.3069	0.3046	0.3025	0.3008	0.2993
5.50	0.4067	0.3951	0.3858	0.3789	0.3729	0.3683	0.3641	0.3605	0.3580	0.3558	0.3525	0.3501	0.3480	0.3462	0.3448
6.00	0.4500	0.4386	0.4295	0.4230	0.4175	0.4132	0.4092	0.4059	0.4035	0.4013	0.3981	0.3957	0.3937	0.3919	0.3904
6.50	0.4931	0.4817	0.4728	0.4667	0.4616	0.4576	0.4539	0.4508	0.4486	0.4465	0.4435	0.4412	0.4392	0.4374	0.4359
7.00	0.5361	0.5247	0.5158	0.5099	0.5052	0.5015	0.4981	0.4952	0.4932	0.4913	0.4884	0.4863	0.4843	0.4826	0.4812
7.50	0.5789	0.5574	0.5584	0.5527	0.5483	0.5449	0.5417	0.5391	0.5372	0.5355	0.5329	0.5309	0.5291	0.5274	0.5260
8.00	0.6216	0.6098	0.6007	0.5951	0.5909	0.5877	0.5848	0.5824	0.5808	0.5792	0.5767	0.5749	0.5732	0.5716	0.5703
8.50	0.6642	0.6521	0.6428	0.6372	0.6331	0.6301	0.6273	0.6252	0.6237	0.6223	0.6200	0.6194	0.6168	0.6152	0.6139
9.00	0.7066	0.6941	0.6846	0.6789	0.6749	0.6719	0.6693	0.6674	0.6660	0.6647	0.6627	0.6612	0.6597	0.6582	0.6570
9.50	0.7488	0.7360	0.7261	0.7204	0.7163	0.7134	0.7109	0.7091	0.7078	0.7066	0.7048	0.7034	0.7020	0.7006	0.6994
10.00	0.7909	0.7775	0.7674	0.7615	0.7573	0.7544	0.7520	0.7503	0.7491	0.7480	0.7463	0.7451	0.7436	0.7423	0.7411
10.50	0.8329	0.8181	0.8085	0.8026	0.7980	0.7951	0.7926	0.7910	0.7899	0.7888	0.7873	0.7861	0.7847	0.7833	0.7822
11.00	0.8747	0.8604	0.8494	0.8430	0.8384	0.8354	0.8329	0.8313	0.8302	0.8292	0.8277	0.8265	0.8251	0.8238	0.8227
11.50	0.9165	0.9016	0.8901	0.8833	0.8785	0.8754	0.8728	0.8711	0.8700	0.8690	0.8676	0.8664	0.8650	0.8637	0.8626
12.00	0.9581	0.9426	0.9306	0.9234	0.9183	0.9150	0.9123	0.9106	0.9095	0.9084	0.9070	0.9057	0.9043	0.9030	0.9019
12.50	0.9996	0.9835	0.9710	0.9633	0.9579	0.9544	0.9515	0.9497	0.9485	0.9474	0.9459	0.9446	0.9431	0.9417	0.9406
13.00	1.0411	1.0242	1.0112	1.0030	0.9973	0.9936	0.9904	0.9884	0.9872	0.9860	0.9842	0.9828	0.9813	0.9799	0.9787
13.50	1.0824	1.0649	1.0513	1.0425	1.0364	1.0324	1.0290	1.0268	1.0264	1.0241	1.0222	1.0206	1.0191	1.0176	1.0164
14.00	1.1237	1.1054	1.0912	1.0318	1.0753	1.0710	1.0673	1.0649	1.0634	1.0618	1.0596	1.0579	1.0563	1.0547	1.0535

14.50	1.1649	1.1459	1.1310	1.1209	1.1139	1.1094	1.1054	1.1027	1.1009	1.0992	1.0966	1.0947	1.0930	1.0914	1.0901
15.00	1.2060	1.1862	1.1707	1.1598	1.1524	1.1475	1.1431	1.1402	1.1382	1.1362	1.1332	1.1311	1.1293	1.1276	1.1263
15.50	1.2471	1.2264	1.2102	1.1986	1.1907	1.1855	1.1606	1.1774	1.1751	1.1729	1.1694	1.1670	1.1651	1.1633	1.1620
16.00	1.2881	1.2666	1.2497	1.2372	1.2287	1.2232	1.2179	1.2144	1.2117	1.2092	1.2052	1.2026	1.2005	1.1987	1.1972
16.50	1.3291	1.3067	1.2890	1.2757	1.2666	1.2607	1.2549	1.2511	1.2481	1.2453	1.2407	1.2377	1.2354	1.2335	1.2320
17.00	1.3700	1.3467	1.3282	1.3140	1.3044	1.2981	1.2917	1.2876	1.2842	1.2610	1.2757	1.2724	1.2700	1.2680	1.2665
17.50	1.4109	1.3866	1.3674	1.3522	1.3419	1.3352	1.3283	1.3238	1.3200	1.3164	1.3105	1.3067	1.3042	1.3021	1.3005
18.00	1.4517	1.4264	1.4064	1.3903	1.3794	1.3722	1.3847	1.3598	1.3555	1.3515	1.3449	1.3407	1.3380	1.3358	1.3341
18.50	1.4924	1.4662	1.4454	1.4282	1.4167	1.4091	1.4009	1.3956	1.3908	1.3864	1.3789	1.3744	1.3714	1.3692	1.3674
19.00	1.5332	1.5059	1.4843	1.4661	1.4538	1.4457	1.4370	1.4312	1.4259	1.4211	1.4127	1.4077	1.4045	1.4022	1.4003
19.50	1.5738	1.5456	1.5231	1.5038	1.4908	1.4823	1.4728	1.4666	1.4608	1.4554	1.4462	1.4407	1.4373	1.4349	1.4329
20.00	1.6145	1.5852	1.5618	1.5414	1.5277	1.5187	1.5085	1.5019	1.4954	1.4896	1.4794	1.4734	1.4698	1.4672	1.4652
20.50	1.6551	1.6247	1.6005	1.5789	1.5644	1.5549	1.5440	1.5369	1.5298	1.5235	1.5123	1.5058	1.5019	1.4993	1.4971
21.00	1.6956	1.6642	1.6391	1.6163	1.6011	1.5910	1.5794	1.5718	1.5641	1.5572	1.5449	1.5379	1.5338	1.5310	1.5288
21.50	1.7361	1.7037	1.6776	1.6537	1.6376	1.6270	1.6146	1.6065	1.5981	1.5906	1.5773	1.5697	1.5654	1.5625	1.5601
22.00	1.7766	1.7431	1.7160	1.6909	1.6740	1.6629	1.6497	1.6410	1.6320	1.6239	1.6095	1.6013	1.5967	1.5937	1.5912
22.50	1.8171	1.7824	1.7544	1.7281	1.7103	1.6987	1.6846	1.6754	1.6657	1.6570	1.6414	1.6326	1.6277	1.6246	1.6220
23.00	1.8575	1.8217	1.7928	1.7651	1.7485	1.7343	1.7194	1.7096	1.6992	1.6899	1.6731	1.6636	1.6585	1.6552	1.6525
23.50	1.8979	1.8610	1.8311	1.8021	1.7826	1.7698	1.7541	1.7437	1.7325	1.7226	1.7046	1.6945	1.6890	1.6856	1.6828
24.00	1.9383	1.9002	1.8693	1.8390	1.8186	1.8053	1.7886	1.7777	1.7657	1.7551	1.7358	1.7250	1.7193	1.7158	1.7128
24.50	1.9786	1.9393	1.9075	1.8759	1.8546	1.8406	1.8230	1.8115	1.7987	1.7874	1.7669	1.7554	1.7494	1.7457	1.7426
25.00	2.0189	1.9785	1.9456	1.9127	1.8904	1.8758	1.8573	1.8452	1.8316	1.8196	1.7977	1.7855	1.7792	1.7754	1.7722
25.50	2.0592	2.0176	1.9837	1.9493	1.9262	1.9110	1.8915	1.8788	1.8644	1.8516	1.8284	1.8155	1.8088	1.8048	1.8015
26.00	2.0995	2.0566	2.0217	1.9860	1.9618	1.9460	1.9256	1.9123	1.8970	1.8835	1.8589	1.8452	1.8382	1.8341	1.8306
26.50	2.1397	2.0957	2.0597	2.0226	1.9974	1.9810	1.9596	1.9456	1.9294	1.9152	1.8891	1.8747	1.8674	1.8631	1.8595
27.00	2.1799	2.1346	2.0976	2.0591	2.0330	2.0159	1.9934	1.9788	1.9618	1.9468	1.9192	1.9040	1.8964	1.8920	1.8882
27.50	2.2201	2.1736	2.1355	2.0955	2.0684	2.0507	2.0272	2.0119	1.9940	1.9782	1.9492	1.9332	1.9252	1.9206	1.9167
28.00	2.2603	2.2125	2.1734	2.1319	2.1038	2.0854	2.0609	2.0449	2.0261	2.0095	1.9790	1.9672	1.9538	1.9491	1.9451
28.50	2.3005	2.2514	2.2112	2.1682	2.1391	2.1200	2.0945	2.0779	2.0580	2.0407	2.0086	1.9910	1.9823	1.9774	1.9732
29.00	2.3406	2.2903	2.2490	2.2045	2.1743	2.1546	2.1280	2.1107	2.0899	2.0717	2.0380	2.0196	2.0105	2.0055	2.0012
29.50	2.3807	2.3291	2.2868	2.2407	2.2095	2.1891	2.1614	2.1434	2.1216	2.1026	2.0673	2.0481	2.0386	2.0334	2.0289
30.00	2.4208	2.3679	2.3245	2.2769	2.2446	2.2235	2.1947	2.1760	2.1533	2.1334	2.0965	2.0764	2.0666	2.0612	2.0566

TABLE A.38(h)
Extended Sukkar–Cornell Integral for Bottom-hole Pressure Calculation

P_r	Reduced Temperature for B = 35.0														
	1.1	1.2	1.3	1.4	1.5	1.6	1.7	1.8	1.9	2.0	2.2	2.4	2.6	2.8	3.0
0.20	0.0000	0.0000	0.0000	0.0000	0.0000	0.0000	0.0000	0.0000	0.0000	0.0000	0.0000	0.0000	0.0000	0.0000	0.0000
0.50	0.0033	0.0032	0.0032	0.0031	0.0031	0.0031	0.0031	0.0030	0.0030	0.0030	0.0030	0.0030	0.0030	0.0030	0.0030
1.00	0.0171	0.0158	0.0152	0.0148	0.0145	0.0143	0.0141	0.0139	0.0139	0.0139	0.0137	0.0136	0.0136	0.0135	0.0135
1.50	0.0454	0.0387	0.0361	0.0346	0.0336	0.0329	0.0323	0.0320	0.0317	0.0315	0.0311	0.0309	0.0307	0.0305	0.0304
2.00	0.0861	0.0720	0.0657	0.0623	0.0601	0.0585	0.0573	0.0564	0.0559	0.0554	0.0546	0.0542	0.0537	0.0534	0.0531
2.50	0.1283	0.1119	0.1022	0.0965	0.0925	0.0900	0.0879	0.0864	0.8055	0.0847	0.0834	0.0826	0.0819	0.0813	0.0808
3.00	0.1703	0.1538	0.1425	0.1350	0.1295	0.1259	0.1230	0.1208	0.1194	0.1182	0.1165	0.1153	0.1142	0.1134	0.1127
3.50	0.2120	0.1960	0.1844	0.1759	0.1694	0.1650	0.1613	0.1585	0.1567	0.1550	0.1528	0.1513	0.1499	0.1487	0.1478
4.00	0.2536	0.2382	0.2266	0.2179	0.2108	0.2059	0.2017	0.1984	0.1962	0.1942	0.1916	0.1897	0.1880	0.1866	0.1855
4.50	0.2950	0.2800	0.2688	0.2601	0.2529	0.2477	0.2433	0.2396	0.2372	0.2350	0.2320	0.2298	0.2279	0.2263	0.2250
5.00	0.3362	0.3216	0.3106	0.3023	0.2951	0.2899	0.2854	0.2816	0.2790	0.2766	0.2734	0.2710	0.2690	0.2672	0.2658
5.50	0.3773	0.3630	0.3522	0.3442	0.3373	0.3321	0.3276	0.3238	0.3211	0.3187	0.3153	0.3128	0.3107	0.3089	0.3074
6.00	0.4183	0.4040	0.3934	0.3857	0.3791	0.3742	0.3698	0.3660	0.3634	0.3610	0.3576	0.3550	0.3529	0.3910	0.3495
6.50	0.4591	0.4449	0.4344	0.4270	0.4207	0.4159	0.4117	0.4080	0.4055	0.4032	0.3998	0.3972	0.3951	0.3932	0.3918
7.00	0.4999	0.4856	0.4752	0.4679	0.4618	0.4573	0.4532	0.4498	0.4473	0.4451	0.4418	0.4394	0.4373	0.4354	0.4339
7.50	0.5405	0.5261	0.5156	0.5085	0.5026	0.4983	0.4944	0.4912	0.4889	0.4867	0.4836	0.4812	0.4792	0.4774	0.4759
8.00	0.5810	0.5665	0.5558	0.5487	0.5431	0.5390	0.5352	0.5822	0.5300	0.5280	0.5247	0.5227	0.5208	0.5190	0.5175
8.50	0.6214	0.6066	0.5959	0.5888	0.5832	0.5792	0.5756	0.5727	0.5707	0.5688	0.5657	0.5638	0.5619	0.5602	0.5588
9.00	0.6617	0.6466	0.6357	0.6285	0.6230	0.6191	0.6156	0.6129	0.6109	0.6091	0.6062	0.6044	0.6026	0.6009	0.5996
9.50	0.7018	0.6865	0.6753	0.6681	0.6625	0.6586	0.6552	0.6526	0.6507	0.6490	0.6462	0.6445	0.6428	0.6412	0.6398
10.00	0.7419	0.7262	0.7147	0.7073	0.7017	0.6978	0.6945	0.6919	0.6901	0.6885	0.6858	0.6842	0.6825	0.6809	0.6796
10.50	0.7818	0.7657	0.7539	0.7464	0.7406	0.7367	0.7334	0.7308	0.7291	0.7275	0.7250	0.7234	0.7217	0.7201	0.7189
11.00	0.8217	0.8051	0.7930	0.7852	0.7793	0.7753	0.7719	0.7694	0.7677	0.7661	0.7637	0.7621	0.7604	0.7589	0.7576
11.50	0.8614	0.8444	0.8319	0.8239	0.8177	0.8136	0.8102	0.8076	0.8059	0.8043	0.8019	0.8004	0.7987	0.7971	0.7958
12.00	0.9011	0.8836	0.8607	0.8623	0.8559	0.8517	0.8481	0.8655	0.8438	0.3422	0.8398	0.8381	0.8364	0.8349	0.8336
12.50	0.9407	0.9227	0.9094	0.9006	0.8939	0.8895	0.8858	0.8831	0.8813	0.8797	0.8771	0.8755	0.8737	0.8721	0.8708
13.00	0.9803	0.9617	0.9479	0.9386	0.9317	0.9271	0.9232	0.9204	0.9185	0.9168	0.9141	0.9124	0.9106	0.9089	0.9076
13.50	1.0197	1.0006	0.9863	0.9765	0.9693	0.9645	0.9604	0.9574	0.9554	0.9535	0.9507	0.9483	0.9470	0.9453	0.9439
14.00	1.0591	1.0394	1.0246	1.0143	1.0067	1.0017	0.9973	0.9941	0.9920	0.9900	0.9869	0.9848	0.9829	0.9812	0.9798

14.50	1.0985	1.0781	1.0627	1.0519	1.0439	1.0386	1.0340	1.0305	1.0282	1.0261	1.0226	1.0205	1.0184	1.0167	1.0153
15.00	1.1377	1.1167	1.1008	1.0893	1.0809	1.0754	1.0704	1.0667	1.0642	1.0618	1.0580	1.0557	1.0536	1.0517	1.0503
15.50	1.1770	1.1552	1.1388	1.1266	1.1178	1.1120	1.1066	1.1027	1.0999	1.0973	1.0931	1.0905	1.0883	1.0864	1.0849
16.00	1.2162	1.1937	1.1767	1.1638	1.1549	1.1484	1.1426	1.1384	1.1354	1.1325	1.1278	1.1247	1.1226	1.1206	1.1191
16.50	1.2553	1.2321	1.2144	1.2008	1.1911	1.1846	1.1784	1.1739	1.1705	1.1674	1.1622	1.1590	1.1566	1.1545	1.1529
17.00	1.2944	1.2705	1.2521	1.2378	1.2275	1.2207	1.2140	1.2092	1.2055	1.2020	1.1962	1.1928	1.1901	1.1880	1.1864
17.50	1.3334	1.3087	1.2898	1.2746	1.2638	1.2566	1.2494	1.2443	1.2402	1.2364	1.2300	1.2262	1.2234	1.2212	1.2195
18.00	1.3725	1.3470	1.3273	1.3113	1.2999	1.2923	1.2846	1.2792	1.2747	1.2705	1.2634	1.2592	1.2563	1.2540	1.2522
18.50	1.4114	1.3851	1.3648	1.3479	1.3359	1.3280	1.3197	1.3139	1.3089	1.3044	1.2966	1.2970	1.2889	1.2865	1.2847
19.00	1.4504	1.4232	1.4022	1.3844	1.3718	1.3634	1.3546	1.3484	1.3430	1.3380	1.3294	1.3245	1.3212	1.3187	1.3168
19.50	1.4893	1.4613	1.4395	1.4208	1.4075	1.3988	1.3893	1.3828	1.3769	1.3714	1.3620	1.3566	1.3531	1.3506	1.3485
20.00	1.5281	1.4993	1.4768	1.4571	1.4432	1.4340	1.4239	1.4170	1.4105	1.4046	1.3944	1.3885	1.3848	1.3822	1.3800
20.50	1.5670	1.5373	1.5140	1.4933	1.4788	1.4691	1.4584	1.4510	1.4440	1.4376	1.4265	1.4201	1.4162	1.4135	1.4112
21.00	1.6058	1.5752	1.5511	1.5294	1.5142	1.5041	1.4927	1.4849	1.4773	1.4704	1.4583	1.4515	1.4473	1.4445	1.4422
21.50	1.6446	1.6130	1.5882	1.5655	1.5495	1.5390	1.5269	1.5186	1.5104	1.5030	1.4900	1.4826	1.4782	1.4752	1.4720
22.00	1.6833	1.6509	1.6252	1.6014	1.5848	1.5738	1.5609	1.5522	1.5434	1.5355	1.5214	1.5134	1.5088	1.5057	1.5032
22.50	1.7220	1.6887	1.6622	1.6373	1.6199	1.6084	1.5948	1.5856	1.5762	1.5677	1.5525	1.5440	1.5391	1.5360	1.5333
23.00	1.7607	1.7264	1.6991	1.6732	1.6550	1.6430	1.6286	1.6189	1.6088	1.5998	1.5835	1.5744	1.5693	1.5660	1.5632
23.50	1.7994	1.7641	1.7360	1.7089	1.6900	1.6775	1.6623	1.6521	1.6413	1.6317	1.6143	1.6046	1.5992	1.5957	1.5929
24.00	1.8381	1.8018	1.7729	1.7446	1.7249	1.7118	1.6959	1.6851	1.6736	1.6634	1.6448	1.6345	1.6288	1.6253	1.6223
24.50	1.8767	1.8394	1.8097	1.7802	1.7597	1.7461	1.7294	1.7180	1.7058	1.6950	1.6752	1.6642	1.6583	1.6546	1.6515
25.00	1.9153	1.8771	1.8464	1.8158	1.7944	1.7803	1.7627	1.7508	1.7379	1.7264	1.7054	1.6937	1.6875	1.6837	1.6805
25.50	1.9539	1.9146	1.8831	1.8513	1.8291	1.8144	1.7960	1.7835	1.7698	1.7577	1.7354	1.7231	1.7165	1.7126	1.7093
26.00	1.9924	1.9522	1.9198	1.8867	1.8637	1.8484	1.8291	1.8161	1.8016	1.7888	1.7652	1.7522	1.7454	1.7413	1.7378
26.50	2.0310	1.9897	1.9564	1.9221	1.8982	1.8824	1.8622	1.8486	1.8333	1.8198	1.7949	1.7812	1.7740	1.7698	1.7662
27.00	2.0695	2.0272	1.9930	1.9574	1.9326	1.9163	1.8951	1.8810	1.8649	1.8506	1.8244	1.8100	1.8025	1.7981	1.7944
27.50	2.1080	2.0647	2.0295	1.9927	1.9670	1.9501	1.9280	1.9133	1.8963	1.8814	1.8537	1.8386	1.8306	1.8262	1.8224
28.00	2.1465	2.1021	2.0661	2.0279	2.0014	1.9838	1.9608	1.9454	1.9277	1.9119	1.8829	1.8670	1.8589	1.8542	1.8502
28.50	2.1850	2.1395	2.1025	2.0631	2.0356	2.0175	1.9935	1.9775	1.9589	1.9424	1.9119	1.8953	1.8868	1.8820	1.8779
29.00	2.2234	2.1769	2.1390	2.0983	2.0698	2.0511	2.0261	2.0095	1.9900	1.9726	1.9408	1.9234	1.9146	1.9096	1.9053
29.50	2.2619	2.2142	2.1754	2.1333	2.1040	2.0846	2.0587	2.0414	2.0210	2.0030	1.9696	1.9513	1.9422	1.9370	1.9327
30.00	2.3003	2.2516	2.2118	2.1684	2.1381	2.1180	2.0912	2.0732	2.0519	2.0331	1.9982	1.9791	1.9696	1.9643	1.9598

TABLE A.38(i)
Extended Sukkar–Cornell Integral for Bottom-hole Pressure Calculation

P_r	Reduced Temperature for B = 40.0														
	1.1	1.2	1.3	1.4	1.5	1.6	1.7	1.8	1.9	2.0	2.2	2.4	2.6	2.8	3.0
0.20	0.0000	0.0000	0.0000	0.0000	0.0000	0.0000	0.0000	0.0000	0.0000	0.0000	0.0000	0.0000	0.0000	0.0000	0.0000
0.50	0.0029	0.0028	0.0028	0.0027	0.0027	0.0027	0.0027	0.0027	0.0027	0.0026	0.0026	0.0026	0.0026	0.0026	0.0026
1.00	0.0150	0.0139	0.0133	0.0129	0.0127	0.0125	0.0123	0.0122	0.0122	0.0121	0.0120	0.0119	0.0119	0.0118	0.0118
1.50	0.0403	0.0341	0.0318	0.0305	0.0296	0.0290	0.0284	0.0281	0.0279	0.0276	0.0273	0.0271	0.0270	0.0268	0.0267
2.00	0.0776	0.0640	0.0582	0.0551	0.0530	0.0517	0.0505	0.0497	0.0493	0.0488	0.0482	0.0477	0.0473	0.0471	0.0468
2.50	0.1170	0.1005	0.0912	0.0858	0.0821	0.0798	0.0779	0.0765	0.0756	0.0749	0.0738	0.0730	0.0724	0.0718	0.0714
3.00	0.1565	0.1393	0.1281	0.1208	0.1156	0.1122	0.1095	0.2074	0.1061	0.1050	0.1034	0.1023	0.1013	0.1005	0.0999
3.50	0.1958	0.1787	0.1668	0.1584	0.1520	0.1477	0.1442	0.1416	0.1398	0.1383	0.1362	0.1348	0.1335	0.1324	0.1315
4.00	0.2351	0.2182	0.2062	0.1973	0.1901	0.1853	0.1812	0.1780	0.1758	0.1740	0.1714	0.1696	0.1681	0.1667	0.1656
4.50	0.2743	0.2576	0.2457	0.2367	0.2292	0.2240	0.2195	0.2159	0.2135	0.2113	0.2084	0.2063	0.2045	0.2029	0.2017
5.00	0.3133	0.2969	0.2851	0.2762	0.2686	0.2633	0.2586	0.2548	0.2521	0.2498	0.2465	0.2442	0.2422	0.2405	0.2391
5.50	0.3523	0.3360	0.3244	0.3156	0.3081	0.3028	0.2980	0.2941	0.2913	0.2889	0.2854	0.2829	0.2808	0.2790	0.2775
6.00	0.3912	0.3750	0.3634	0.3549	0.3476	0.3423	0.3376	0.3336	0.3308	0.3283	0.3247	0.3221	0.3199	0.3181	0.3166
6.50	0.4300	0.4138	0.4023	0.3939	0.3868	0.3816	0.3770	0.3731	0.3703	0.3678	0.3642	0.3616	0.3594	0.3575	0.3560
7.00	0.4687	0.4525	0.4410	0.4328	0.4258	0.4208	0.4163	0.4124	0.4097	0.4073	0.4037	0.4011	0.3989	0.3970	0.3955
7.50	0.0573	0.4910	0.4795	0.4714	0.4646	0.4597	0.4553	0.4516	0.4490	0.4466	0.4431	0.4405	0.4383	0.4365	0.4350
8.00	0.5458	0.5294	0.5179	0.5097	0.5031	0.4983	0.4941	0.4905	0.4879	0.4856	0.4819	0.4797	0.4776	0.4758	0.4743
8.50	0.5843	0.5677	0.5560	0.5479	0.5413	0.5367	0.5325	0.5290	0.5266	0.5244	0.5208	0.5187	0.5166	0.5148	0.5133
9.00	0.6227	0.6059	0.5940	0.5859	0.5793	0.5747	0.5707	0.5673	0.5650	0.5628	0.5593	0.5573	0.5553	0.5535	0.5521
9.50	0.6609	0.6439	0.6319	0.6237	0.6171	0.6125	0.6085	0.6052	0.6030	0.6009	0.5975	0.5955	0.5936	0.5918	0.5904
10.00	0.6991	0.6818	0.6696	0.6612	0.6546	0.6500	0.6461	0.6429	0.6407	0.6386	0.6353	0.6334	0.6315	0.6298	0.6284
10.50	0.7372	0.7196	0.7071	0.6987	0.6919	0.6873	0.6833	0.6802	0.6780	0.6760	0.6728	0.6710	0.6690	0.6673	0.6660
11.00	0.7753	0.7573	0.7446	0.7359	0.7290	0.7243	9.7203	0.7172	0.7150	0.7130	0.7099	0.7081	0.7052	0.7045	0.7031
11.50	0.8132	0.7949	0.7819	0.7729	0.7659	0.7611	0.7571	0.7539	0.7517	0.7498	0.7466	0.7448	0.7429	0.7412	0.7398
12.00	0.8511	0.8324	0.8190	0.8098	0.8026	0.7977	0.7936	0.7903	0.7882	0.7862	0.7830	0.7812	0.7792	0.7775	0.7762
12.50	0.8890	0.8696	0.8561	0.8466	0.8391	0.8341	0.8299	0.8265	0.8243	0.8223	0.8190	0.8171	0.8152	0.8134	0.8121
13.00	0.9268	0.9072	0.8931	0.8832	0.8755	0.8703	0.8659	0.8624	0.8602	0.8580	0.8547	0.8527	0.8507	0.8490	0.8476
13.50	0.9645	0.9445	0.9299	0.9196	0.9117	0.9063	0.9017	0.8981	0.8957	0.8935	0.8900	0.8879	0.8859	0.8841	0.8827
14.00	1.0022	0.9816	0.9667	0.9559	0.9477	0.9421	0.9373	0.9335	0.9310	0.9287	0.9250	0.9228	0.9207	0.9188	0.9174

14.50	1.0398	0.0188	1.0034	0.9921	0.9835	0.9778	0.9727	0.9588	0.9661	0.9636	0.9596	0.9572	0.9551	0.9532	0.9517
15.00	1.0774	1.0558	1.0400	1.0282	1.0193	1.0133	1.0079	1.0037	1.0009	0.9982	0.9939	0.9914	0.9891	0.9872	0.9856
15.50	1.1149	1.0928	1.0765	1.0641	1.0548	1.0486	1.0429	1.0385	1.0355	1.0328	1.0279	1.0251	1.0228	1.0208	1.0192
16.00	1.1525	1.1297	1.1129	1.1000	1.0903	1.0837	1.0777	1.0731	1.0698	1.0667	1.0516	1.0586	1.0561	1.0541	1.0525
16.50	1.1899	1.1666	1.1492	1.1357	1.1255	1.1187	1.1123	1.1075	1.1039	1.1005	1.0949	1.0917	1.0891	1.0870	1.0853
17.00	1.2274	1.2034	1.1855	1.1713	1.1607	1.1536	1.1468	1.1417	1.1378	1.1341	1.1280	1.1245	1.1218	1.1196	1.1179
17.50	1.2648	1.2402	1.2217	1.2068	1.1958	1.1884	1.1811	1.1757	1.1714	1.1675	1.1608	1.1570	1.1541	1.1519	1.1501
18.00	1.3021	1.2769	1.2579	1.2422	1.2307	1.2230	1.2152	1.2095	1.2049	1.2006	1.1934	1.1892	1.1862	1.1839	1.1820
18.50	1.3395	1.3136	1.2940	1.2776	1.2655	1.2574	1.2492	1.2432	1.2382	1.2336	1.2256	1.2211	1.2180	1.2155	1.2136
19.00	1.3768	1.3502	1.3300	1.3128	1.3002	1.2918	1.2831	1.2767	1.2713	1.2663	1.2577	1.2528	1.2494	1.2469	1.2450
19.50	1.4140	1.3868	1.3659	1.3480	1.3349	1.3261	1.3168	1.3101	1.3042	1.2988	1.2894	1.2842	1.2806	1.2780	1.2760
20.00	1.4513	1.4233	1.4019	1.3831	1.3694	1.3602	1.3504	1.3433	1.3369	1.3311	1.3210	1.3153	1.3116	1.3089	1.3068
20.50	1.4885	1.4598	1.4377	1.4181	1.4038	1.3942	1.3838	1.3763	1.3695	1.3633	1.3523	1.3462	1.3422	1.3395	1.3373
21.00	1.5257	1.4963	1.4735	1.4530	1.4381	1.4281	1.4171	1.4093	1.4019	1.3952	1.3834	1.3768	1.3727	1.3698	1.3675
21.50	1.5629	1.5327	1.5093	1.4879	1.4723	1.4620	1.4503	1.4421	1.4341	1.4270	1.4143	1.4072	1.4028	1.3999	1.3975
22.00	1.6001	1.5691	1.5450	1.5227	1.5065	1.4957	1.4834	1.4747	1.4662	1.4586	1.4449	1.4373	1.4328	1.4297	1.4272
22.50	1.6372	1.6054	1.5807	1.5574	1.5406	1.5293	1.5164	1.5072	1.4982	1.4900	1.4754	1.4673	1.4625	1.4593	1.4567
23.00	1.6743	1.6417	1.6163	1.5920	1.5746	1.5629	1.5492	1.5396	1.5300	1.5213	1.5057	1.4970	1.4920	1.4887	1.4860
23.50	1.7114	1.6780	1.6519	1.6266	1.6085	1.5963	1.5820	1.5719	1.5617	1.5525	1.5358	1.5265	1.5213	1.5178	1.5151
24.00	1.7485	1.7143	1.6874	1.6612	1.6423	1.6297	1.6146	1.6041	1.5932	1.5834	1.5657	1.5559	1.5503	1.5468	1.5439
24.50	1.7855	1.7505	1.7229	1.6957	1.6761	1.6630	1.6472	1.6362	1.6246	1.6143	1.5954	1.5850	1.5792	1.5755	1.5725
25.00	1.8226	1.7867	1.7584	1.7301	1.7098	1.6962	1.6797	1.6582	1.6559	1.6450	1.6249	1.6139	1.5078	1.6041	1.6010
25.50	1.8596	1.8229	1.7938	1.7645	1.7434	1.7293	1.7120	1.7000	1.6871	1.6755	1.6543	1.6427	1.6363	1.6324	1.6292
26.00	1.8966	1.8591	1.8292	1.7988	1.7770	1.7624	1.7443	1.7318	1.7181	1.7059	1.6836	1.6713	1.6646	1.6606	1.6572
26.50	1.9336	1.8952	1.8645	1.8331	1.8105	1.7954	1.7765	1.7634	1.7491	1.7362	1.7126	1.6997	1.6927	1.6886	1.6851
27.00	1.9705	1.9313	1.8999	1.8673	1.8439	1.8283	1.8086	1.7950	1.7799	1.7664	1.7415	1.7279	1.7207	1.7164	1.7128
27.50	2.0075	1.9674	1.9352	1.9015	1.8773	1.8612	1.8406	1.8265	1.8106	1.7965	1.7703	1.7560	1.7484	1.7440	1.7403
28.00	2.0444	2.0034	1.9704	1.9356	1.9107	1.8940	1.8726	1.8579	1.8412	1.8264	1.7989	1.7839	1.7760	1.7715	1.7676
28.50	2.0813	2.0194	2.0057	1.9697	1.9439	1.9267	1.9044	1.8892	1.8717	1.8562	1.8274	1.8116	1.8035	1.7988	1.7946
29.00	2.1182	2.0755	2.0409	2.0038	1.9771	1.9594	1.9362	1.9204	1.9021	1.8859	1.8557	1.8393	1.8308	1.8259	1.8216
29.50	2.1551	2.1114	2.0761	2.0378	2.0103	1.9920	1.9680	1.9516	1.9325	1.9155	1.8840	1.8667	1.8579	1.8529	1.8487
30.00	2.1920	2.1474	2.1112	2.0717	2.0434	2.0246	1.9996	1.9826	1.9627	1.9450	1.9120	1.8940	1.8849	1.8797	1.8754

TABLE A.38(j)
Extended Sukkar–Cornell Integral for Bottom-hole Pressure Calculation

P_r	Reduced Temperature for $B = 45.0$														
	1.1	1.2	1.3	1.4	1.5	1.6	1.7	1.8	1.9	2.0	2.2	2.4	2.6	2.8	3.0
0.20	0.0000	0.0000	0.0000	0.0000	0.0000	0.0000	0.0000	0.0000	0.0000	0.0000	0.0000	0.0000	0.0000	0.0000	0.0000
0.50	0.0026	0.0025	0.0025	0.0024	0.0024	0.0024	0.0024	0.0024	0.0024	0.0024	0.0023	0.0023	0.0023	0.0023	0.0023
1.00	0.0134	0.0124	0.0119	0.0115	0.0113	0.0111	0.0110	0.0109	0.0108	0.0108	0.0107	0.0106	0.0106	0.0105	0.0105
1.50	0.0362	0.0305	0.0284	0.0272	0.0264	0.0258	0.0254	0.0250	0.0248	0.0247	0.0244	0.0242	0.0240	0.0239	0.0238
2.00	0.0707	0.0576	0.0522	0.0494	0.0475	0.0462	0.0452	0.0445	0.0440	0.0436	0.0430	0.0426	0.0423	0.0420	0.0418
2.50	0.1016	0.0912	0.0823	0.0772	0.0738	0.0716	0.0699	0.0586	0.0678	0.0671	0.0661	0.0654	0.0648	0.0644	0.0640
3.00	0.1449	0.1273	0.1163	0.1093	0.1043	0.1012	0.0986	0.0967	0.0955	0.0944	0.0930	0.0919	0.0910	0.0903	0.0897
3.50	0.1821	0.1643	0.1523	0.1441	0.1378	0.1338	0.1304	0.1279	0.1263	0.1248	0.1229	0.1215	0.1203	0.1193	0.1185
4.00	0.2193	0.2015	0.1892	0.1803	0.1732	0.1685	0.1645	0.1614	0.1594	0.1576	0.1552	0.1534	0.1520	0.1507	0.1496
4.50	0.2565	0.2388	0.2264	0.2172	0.2096	0.2045	0.2001	0.1966	0.1942	0.1921	0.1893	0.1872	0.1855	0.1840	0.1828
5.00	0.2936	0.2760	0.2637	0.2544	0.2466	0.2412	0.2366	0.2327	0.2301	0.2278	0.2246	0.2223	0.2204	0.2187	0.2174
5.50	0.3306	0.3131	0.3009	0.2917	0.2838	0.2783	0.2735	0.2695	0.2667	0.2643	0.2608	0.2583	0.2562	0.2544	0.2530
6.00	0.3676	0.3501	0.3380	0.3289	0.3211	0.3158	0.3107	0.3066	0.3038	0.3012	0.2976	0.2949	0.2928	0.2909	0.2895
6.50	0.4045	0.3871	0.3750	0.3660	0.3583	0.3528	0.3480	0.3439	0.3410	0.3384	0.3347	0.3319	0.3297	0.3278	0.3264
7.00	0.4414	0.4239	0.4118	0.4029	0.3954	0.3900	0.3852	0.3811	0.3782	0.3757	0.3719	0.3692	0.3669	0.3650	0.3635
7.50	0.4782	0.4607	0.4486	0.4397	0.4323	0.4270	0.4223	0.4182	0.4154	0.4129	0.4092	0.4064	0.4042	0.4023	0.4008
8.00	0.5150	0.4973	0.4852	0.4763	0.4690	0.4638	0.4592	0.4552	0.4525	0.4500	0.4459	0.4436	0.4414	0.4395	0.4360
8.50	0.5517	0.5339	0.5216	0.5128	0.5055	0.5004	0.4959	0.4920	0.4893	0.4869	0.4828	0.4806	0.4785	0.4766	0.4751
9.00	0.5883	0.5704	0.5580	0.5492	0.5419	0.5368	0.5323	0.5286	0.5259	0.5235	0.5196	0.5174	0.5153	0.5135	0.5120
9.50	0.6248	0.6067	0.5942	0.5853	0.5780	0.5730	0.5686	0.5649	0.5623	0.5599	0.5561	0.5540	0.5519	0.5501	0.5486
10.00	0.6613	0.6430	0.6304	0.6214	0.6140	0.6090	0.6046	0.6009	0.5984	0.5961	0.5923	0.5903	0.5882	0.5864	0.5650
10.50	0.6978	0.6792	0.6664	0.6573	0.6498	0.6447	0.6606	0.6367	0.6342	0.6320	0.6283	0.6262	0.6242	0.6224	0.6210
11.00	0.7342	0.7153	0.7023	0.6930	0.6854	0.6803	0.6759	0.6723	0.6698	0.6676	0.6639	0.6619	0.6598	0.6580	0.6566
11.50	0.7705	0.7514	0.7381	0.7286	0.7209	0.7157	0.7113	0.7076	0.7051	0.7029	0.6993	0.6977	0.6952	0.6934	0.6920
12.00	0.8068	0.7874	0.7738	0.7641	0.7562	0.7509	0.7464	0.7427	0.7402	0.7380	0.7343	0.7323	0.7302	0.7284	0.7270
12.50	0.8430	0.8233	0.8094	0.7994	0.7914	0.7860	0.7814	0.7776	0.7751	0.7728	0.7690	0.7670	0.7649	0.7630	0.7616
13.00	0.8792	0.8591	0.8449	0.8347	0.8264	0.8209	0.8161	0.8122	0.8097	0.8073	0.8035	0.8013	0.7992	0.7974	0.7959
13.50	0.9153	0.8949	0.8804	0.8698	0.8613	0.8556	0.8507	0.8467	0.8440	0.8416	0.8376	0.8354	0.8332	0.8313	0.8299
14.00	0.9514	0.9306	0.9157	0.9048	0.8961	0.8902	0.8851	0.8809	0.8782	0.8756	0.8715	0.8691	0.8669	0.8650	0.8635

14.50	0.9875	0.9663	0.9510	0.9396	0.9307	0.9246	0.9193	0.9150	0.9121	0.9094	0.9050	0.9025	0.9002	0.8983	0.8968
15.00	1.0235	1.0019	0.9863	0.9744	0.9652	0.9589	0.9533	0.9489	0.9458	0.9429	0.9382	0.9355	0.9332	0.9312	0.9297
15.50	1.0595	1.0374	1.0214	1.0091	0.9995	0.9931	0.9872	0.9825	0.9793	0.9762	0.9712	0.9684	0.9660	0.9639	0.9623
16.00	1.0955	1.0729	1.0565	1.0437	1.0338	1.0271	1.0209	1.0160	1.0125	1.0093	1.0039	1.0009	0.9984	0.9963	0.9946
16.50	1.1315	1.1084	1.0915	1.0782	1.0679	1.0609	1.0544	1.0494	1.0456	1.0422	1.0364	1.0331	1.0305	1.0283	1.0266
17.00	1.1674	1.1438	1.1265	1.1126	1.1019	1.0947	1.0878	1.0825	1.0785	1.0748	1.0685	1.0650	1.0623	1.0600	1.0583
17.50	1.2032	1.1791	1.1614	1.1469	1.1358	1.1283	1.1211	1.1155	1.1112	1.1072	1.1005	1.0967	1.0938	1.0915	1.0697
18.00	1.2391	1.2145	1.1962	1.1811	1.1696	1.1619	1.1542	1.1484	1.1437	1.1394	1.1321	1.1281	1.1250	1.1227	1.1208
18.50	1.2749	1.2497	1.2310	1.2153	1.2033	1.1953	1.1872	1.1811	1.1761	1.1715	1.1636	1.1592	1.1560	1.1536	1.1517
19.00	1.3107	1.2850	1.2658	1.2494	1.2370	1.2286	1.2200	1.2136	1.2082	1.2033	1.1948	1.1901	1.1867	1.1842	1.1823
19.50	1.3465	1.3202	1.3005	1.2834	1.2705	1.2618	1.2528	1.2460	1.2403	1.2350	1.2258	1.2207	1.2172	1.2146	1.2126
20.00	1.3823	1.3554	1.3351	1.3173	1.3039	1.2949	1.2854	1.2783	1.2721	1.2665	1.2566	1.2511	1.2474	1.2447	1.2426
20.50	1.4180	1.3905	1.3697	1.3512	1.3373	1.3279	1.3179	1.3105	1.3038	1.2978	1.2871	1.2812	1.2774	1.2746	1.2724
21.00	1.4538	1.4256	1.4043	1.3850	1.3706	1.3608	1.3503	1.3425	1.3354	1.3290	1.3175	1.3112	1.3071	1.3043	1.3020
21.50	1.4895	1.4607	1.4388	1.4187	1.4038	1.3937	1.3825	1.3744	1.3668	1.3599	1.3477	1.3409	1.3367	1.3337	1.3314
22.00	1.5251	1.4958	1.4733	1.4524	1.4369	1.4264	1.4147	1.4062	1.3981	1.3908	1.3776	1.3704	1.3660	1.3629	1.3605
22.50	1.5608	1.5308	1.5077	1.4860	1.4699	1.4591	1.4468	1.4379	1.4292	1.4215	1.4074	1.3997	1.3951	1.3919	1.3894
23.00	1.5965	1.5658	1.5421	1.5196	1.5029	1.4916	1.4788	1.4694	1.4603	1.4520	1.4371	1.4288	1.4239	1.4207	1.4181
23.50	1.6321	1.6008	1.5765	1.5531	1.5358	1.5242	1.5106	1.5009	1.4912	1.4824	1.4665	1.4577	1.4526	1.4493	1.4466
24.00	1.6677	1.6357	1.6108	1.5866	1.5687	1.5566	1.5424	1.5323	1.5219	1.5127	1.4958	1.4865	1.4811	1.4776	1.4748
24.50	1.7033	1.6706	1.6451	1.6200	1.6015	1.5890	1.5741	1.5635	1.5526	1.5428	1.5249	1.5150	1.5094	1.5058	1.5029
25.00	1.7389	1.7055	1.6794	1.6534	1.6342	1.6212	1.6057	1.5947	1.5831	1.5728	1.5538	1.5436	1.5375	1.5338	1.5308
25.50	1.7745	1.7404	1.7136	1.6867	1.6668	1.6535	1.6373	1.6257	1.6136	1.6027	1.5826	1.5716	1.5655	1.5617	1.5585
26.00	1.8100	1.7752	1.7478	1.7200	1.6995	1.6856	1.6687	1.6567	1.6439	1.6324	1.6112	1.5996	1.5933	1.5893	1.5861
26.50	1.8456	1.8101	1.7820	1.7532	1.7320	1.7177	1.7001	1.6876	1.6741	1.6621	1.6397	1.6275	1.6209	1.6168	1.6134
27.00	1.8811	1.8449	1.8162	1.7864	1.7645	1.7498	1.7314	1.7184	1.7042	1.6916	1.6681	1.6552	1.6483	1.6441	1.6406
27.50	1.9166	1.8797	1.8503	1.8195	1.7969	1.7817	1.7626	1.7491	1.7343	1.7210	1.6963	1.6828	1.6756	1.6712	1.6677
28.00	1.9521	1.9144	1.8844	1.8526	1.8293	1.8136	1.7937	1.7798	1.7642	1.7503	1.7244	1.7102	1.7027	1.6982	1.6945
28.50	1.9876	1.9492	1.9184	1.8857	1.8617	1.8455	1.8248	1.8103	1.7940	1.7795	1.7523	1.7375	1.7297	1.7251	1.7212
29.00	2.0231	1.9839	1.9525	1.9187	1.8940	1.8773	1.8558	1.8408	1.8238	1.8086	1.7801	1.7646	1.7565	1.7518	1.7478
29.50	2.0586	2.0186	1.9865	1.9517	1.9262	1.9091	1.8868	1.8712	1.8534	1.8376	1.8078	1.7916	1.7832	1.7783	1.7742
30.00	2.0941	2.0533	2.0205	1.9847	1.9584	1.9408	1.9176	1.9016	1.8830	1.8664	1.8354	1.8184	1.8097	1.8047	1.8005

TABLE A.38(k)
Extended Sukkar–Cornell Integral for Bottom-hole Pressure Calculation

P_r	Reduced Temperature for $B = 50.0$														
	1.1	1.2	1.3	1.4	1.5	1.6	1.7	1.8	1.9	2.0	2.2	2.4	2.6	2.8	3.0
0.20	0.0000	0.0000	0.0000	0.0000	0.0000	0.0000	0.0000	0.0000	0.0000	0.0000	0.0000	0.0000	0.0000	0.0000	0.0000
0.50	0.0023	0.0023	0.0022	0.0022	0.0022	0.0022	0.0021	0.0021	0.0021	0.0021	0.0021	0.0021	0.0021	0.0021	0.0021
1.00	0.0121	0.0111	0.0107	0.0104	0.0102	0.0100	0.0099	0.0098	0.0098	0.0097	0.0096	0.0096	0.0095	0.0095	0.0095
1.50	0.0328	0.0276	0.0257	0.0246	0.0238	0.0233	0.0229	0.0226	0.0224	0.0222	0.0220	0.0218	0.0217	0.0216	0.0215
2.00	0.0649	0.0524	0.0474	0.0447	0.0430	0.0418	0.0409	0.0402	0.0398	0.0395	0.0389	0.0385	0.0382	0.0380	0.0378
2.50	0.0997	0.0835	0.0750	0.0702	0.0670	0.0650	0.0634	0.0622	0.0615	0.0608	0.0599	0.0593	0.0587	0.0583	0.0579
3.00	0.1350	0.1173	0.1066	0.0998	0.0951	0.0921	0.0897	0.0879	0.0868	0.0858	0.0844	0.0835	0.0827	0.0820	0.0814
3.50	0.1703	0.1521	0.1402	0.1322	0.1261	0.1222	0.1191	0.1167	0.1151	0.1138	0.1119	0.1100	0.1095	0.1085	0.1078
4.00	0.2057	0.1873	0.1749	0.1660	0.1591	0.1545	0.1507	0.1477	0.1457	0.1440	0.1417	0.1401	0.1387	0.1375	0.1365
4.50	0.2410	0.2226	0.2101	0.2008	0.1933	0.1882	0.1839	0.1804	0.1781	0.1761	0.1734	0.1714	0.1697	0.1683	0.1671
5.00	0.2763	0.2579	0.2454	0.2359	0.2281	0.2227	0.2181	0.2143	0.2117	0.2094	0.2063	0.2040	0.2022	0.2006	0.1993
5.50	0.3116	0.2933	0.2807	0.2712	0.2632	0.2577	0.2529	0.2488	0.2461	0.2436	0.2402	0.2377	0.2356	0.2339	0.2326
6.00	0.3469	0.3285	0.3161	0.3066	0.2985	0.2929	0.2880	0.2838	0.2809	0.2784	0.2747	0.2721	0.2700	0.2681	0.2667
6.50	0.3821	0.3638	0.3513	0.3419	0.3339	0.3282	0.3233	0.3190	0.3161	0.3135	0.3097	0.3069	0.3048	0.3029	0.3014
7.00	0.4173	0.3990	0.3865	0.3772	0.3692	0.3636	0.3587	0.3544	0.3514	0.3488	0.3450	0.3421	0.3399	0.3380	0.3365
7.50	0.4525	0.4341	0.4216	0.4123	0.4044	0.3989	0.3940	0.3897	0.3868	0.3841	0.3803	0.3774	0.3752	0.3733	0.3718
8.00	0.4876	0.4692	0.4567	0.4474	0.4395	0.4340	0.4292	0.4250	0.4221	0.4194	0.4151	0.4128	0.4105	0.4086	0.4071
8.50	0.5227	0.5042	0.4916	0.4823	0.4745	0.4690	0.4643	0.4601	0.4573	0.4547	0.4504	0.4481	0.4458	0.4439	0.4424
9.00	0.5577	0.5391	0.5264	0.5171	0.5093	0.5039	0.4992	0.4951	0.4923	0.4897	0.4855	0.4832	0.4810	0.4791	0.4777
9.50	0.5927	0.5739	0.5612	0.5518	0.5440	0.5386	0.5340	0.5299	0.5271	0.5246	0.5204	0.5182	0.5160	0.5142	0.5127
10.00	0.6277	0.6087	0.5959	0.5864	0.5786	0.5732	0.5685	0.5645	0.5618	0.5593	0.5552	0.5530	0.5508	0.5490	0.5475
10.50	0.6626	0.6435	0.6304	0.6209	0.6130	0.6076	0.6029	0.5990	0.5962	0.5938	0.5897	0.5875	0.5854	0.5835	0.5821
11.00	0.6974	0.6781	0.6649	0.6553	0.6473	0.6418	0.6372	0.6332	0.6305	0.6280	0.6240	0.6219	0.6197	0.6179	0.6184
11.50	0.7323	0.7127	0.6994	0.6896	0.6815	0.6759	0.6712	0.6672	0.6645	0.6621	0.6581	0.6558	0.6537	0.6519	0.6505
12.00	0.7670	0.7473	0.7337	0.7237	0.7155	0.7099	0.7051	0.7011	0.6984	0.6959	0.6919	0.6897	0.6875	0.6857	0.6842
12.50	0.8018	0.7818	0.7680	0.7578	0.7494	0.7437	0.7388	0.7347	0.7320	0.7295	0.7254	0.7232	0.7210	0.7192	0.7177
13.00	0.8365	0.8163	0.8022	0.7917	0.7832	0.7774	0.7724	0.7582	0.7654	0.7629	0.7587	0.7565	0.7542	0.7523	0.7509
13.50	0.8712	0.8507	0.8363	0.8256	0.8169	0.8109	0.8058	0.8015	0.7987	0.7960	0.7917	0.7894	0.7672	0.7852	0.7838
14.00	0.9059	0.8850	0.8704	0.8594	0.8504	0.8443	0.8391	0.8347	0.8317	0.8290	0.8245	0.8221	0.8198	0.8178	0.8183

14.50	0.9405	0.9193	0.9044	0.8930	0.8839	0.8776	0.8722	0.8576	0.8645	0.8617	0.8570	0.8545	0.8521	0.8502	0.8486
15.00	0.9751	0.9536	0.9384	0.9266	0.9172	0.9108	0.9051	0.9004	0.8972	0.8942	0.8893	0.8866	0.8842	0.8822	0.8806
15.50	1.0097	0.9878	0.9722	0.9601	0.9504	0.9438	0.9379	0.9331	0.9297	0.9265	0.9213	0.9185	0.9160	0.9139	0.9123
16.00	1.0442	1.0220	1.0061	0.9935	0.9836	0.9768	0.9706	0.9656	0.9620	0.9586	0.9531	0.9501	0.9475	0.9454	0.9438
16.50	1.0788	1.0561	1.0399	1.0269	1.0166	1.0096	1.0031	0.9979	0.9941	0.9906	0.9847	0.9814	0.9788	0.9766	0.9749
17.00	1.1133	1.0902	1.0736	1.0601	1.0495	1.0423	1.0355	1.0301	1.0260	1.0223	1.0160	1.0125	1.0097	1.0075	1.0058
17.50	1.1477	1.1243	1.1073	1.0933	1.0824	1.0749	1.0678	1.0621	1.0578	1.0538	1.0471	1.0434	1.0405	1.0382	1.0364
18.00	1.1822	1.1583	1.1409	1.1266	1.1151	1.1074	1.0999	1.0940	1.0894	1.0852	1.0779	1.0740	1.0709	1.0686	1.0668
18.50	1.2167	1.1923	1.1745	1.1595	1.1478	1.1398	1.1320	1.1258	1.1209	1.1164	1.1086	1.1043	1.1012	1.0988	1.0969
19.00	1.2511	1.2263	1.2081	1.1925	1.1804	1.1721	1.1639	1.1575	1.1522	1.1474	1.1390	1.1345	1.1312	1.1287	1.1268
19.50	1.2855	1.2602	1.2416	1.2254	1.2129	1.2044	1.1957	1.1890	1.1834	1.1783	1.1693	1.1644	1.1609	1.1584	1.1564
20.00	1.3199	1.2942	1.2751	1.2583	1.2453	1.2365	1.2274	1.2204	1.2144	1.2090	1.1993	1.1941	1.1905	1.1878	1.1858
20.50	1.3542	1.3280	1.3085	1.2911	1.2777	1.2686	1.2590	1.2517	1.2453	1.2395	1.2292	1.2236	1.2198	1.2171	1.2149
21.00	1.3886	1.3619	1.3419	1.3238	1.3100	1.3005	1.2905	1.2829	1.2761	1.2699	1.2589	1.2528	1.2489	1.2461	1.2439
21.50	1.4229	1.3957	1.3753	1.3565	1.3422	1.3324	1.3219	1.3140	1.3067	1.3001	1.2884	1.2819	1.2778	1.2749	1.2726
22.00	1.4573	1.4295	1.4086	1.3892	1.3743	1.3643	1.3532	1.3449	1.3372	1.3302	1.3177	1.3108	1.3065	1.3035	1.3011
22.50	1.4916	1.4633	1.4419	1.4218	1.4064	1.3960	1.3844	1.3758	1.3676	1.3602	1.3468	1.3395	1.3350	1.3319	1.3295
23.00	1.5259	1.4971	1.4752	1.4543	1.4385	1.4277	1.4155	1.4066	1.3979	1.3900	1.3758	1.3680	1.3633	1.3601	1.3576
23.50	1.5602	1.5308	1.5084	1.4868	1.4704	1.4593	1.4456	1.4372	1.4280	1.4197	1.4046	1.3964	1.3914	1.3881	1.3855
24.00	1.5944	1.5646	1.5416	1.5193	1.5024	1.4908	1.4775	1.4678	1.4581	1.4493	1.4333	1.4245	1.4193	1.4160	1.4133
24.50	1.6287	1.5983	1.5748	1.5517	1.5342	1.5223	1.5084	1.4983	1.4880	1.4788	1.4618	1.4525	1.4471	1.4436	1.4408
25.00	1.6629	1.6319	1.6079	1.5841	1.5660	1.5537	1.5392	1.5287	1.5178	1.5081	1.4902	1.4803	1.4747	1.4711	1.4682
25.50	1.6972	1.6656	1.6410	1.6164	1.5978	1.5851	1.5700	1.5590	1.5476	1.5373	1.5184	1.5080	1.5021	1.4984	1.4954
26.00	1.7314	1.6992	1.6741	1.6487	1.6295	1.6164	1.6006	1.5892	1.5772	1.5664	1.5465	1.5355	1.5294	1.5256	1.5225
26.50	1.7656	1.7329	1.7072	1.6809	1.6611	1.6476	1.6312	1.6194	1.6068	1.5954	1.5744	1.5629	1.5565	1.5526	1.5494
27.00	1.7998	1.7665	1.7403	1.7131	1.6927	1.6788	1.6617	1.6494	1.6362	1.6243	1.6022	1.5901	1.5835	1.5794	1.5761
27.50	1.8340	1.8001	1.7733	1.7453	1.7243	1.7100	1.6922	1.6794	1.6656	1.6531	1.6299	1.6172	1.6103	1.6061	1.6027
28.00	1.8882	1.8337	1.8063	1.7775	1.7558	1.7410	1.7226	1.7094	1.6948	1.6818	1.6574	1.6441	1.6389	1.6328	1.6291
28.50	1.9024	1.8672	1.8393	1.8096	1.7872	1.7721	1.7529	1.7392	1.7240	1.7104	1.6849	1.6709	1.6634	1.6590	1.6553
29.00	1.9366	1.9008	1.8722	1.8616	1.8187	1.8030	1.7831	1.7690	1.7531	1.7389	1.7122	1.6976	1.6898	1.6853	1.6815
29.50	1.9707	1.9341	1.9052	1.8787	1.8500	1.8340	1.8133	1.7987	1.7821	1.7673	1.7394	1.7241	1.7160	1.7114	1.7075
30.00	2.0049	1.9678	1.9381	1.9057	1.8814	1.8649	1.8435	1.8284	1.8111	1.7956	1.7664	1.7505	1.7421	1.7373	1.7333

TABLE A.38(l)
Extended Sukkar–Cornell Integral for Bottom-hole Pressure Calculation

P_r	Reduced Temperature for B = 60.0														
	1.1	1.2	1.3	1.4	1.5	1.6	1.7	1.8	1.9	2.0	2.2	2.4	2.6	2.8	3.0
0.20	0.0000	0.0000	0.0000	0.0000	0.0000	0.0000	0.0000	0.0000	0.0000	0.0000	0.0000	0.0000	0.0000	0.0000	0.0000
0.50	0.0019	0.0019	0.0019	0.0018	0.0018	0.0018	0.0018	0.0018	0.0018	0.0018	0.0018	0.0018	0.0017	0.0017	0.0017
1.00	0.0101	0.0093	0.0089	0.0087	0.0085	0.0084	0.0083	0.0082	0.0081	0.0081	0.0080	0.0080	0.0080	0.0079	0.0079
1.50	0.0277	0.0232	0.0215	0.0206	0.0200	0.0195	0.0192	0.0189	0.0188	0.0186	0.0184	0.0183	0.0181	0.0181	0.0180
2.00	0.0559	0.0443	0.0399	0.0376	0.0361	0.0351	0.0343	0.0338	0.0334	0.0331	0.0326	0.0323	0.0321	0.0319	0.0317
2.50	0.0870	0.0715	0.0637	0.0594	0.0566	0.0549	0.0535	0.0524	0.0518	0.0512	0.0504	0.0499	0.0494	0.0490	0.0487
3.00	0.1189	0.1014	0.0913	0.0851	0.0808	0.0781	0.0760	0.0745	0.0734	0.0726	0.0714	0.0705	0.0698	0.0692	0.0687
3.50	0.1509	0.1325	0.1211	0.1135	0.1079	0.1043	0.1014	0.0993	0.0979	0.0966	0.0950	0.0989	0.0928	0.0920	0.0913
4.00	0.1831	0.1642	0.1521	0.1435	0.1369	0.1326	0.1291	0.1263	0.1245	0.1229	0.1209	0.1194	0.1181	0.1170	0.1161
4.50	0.2153	0.1962	0.1837	0.1745	0.1672	0.1624	0.1583	0.1551	0.1529	0.1510	0.1485	0.1466	0.1451	0.1438	0.1428
5.00	0.2475	0.2283	0.2157	0.2062	0.1984	0.1931	0.1887	0.1850	0.1826	0.1804	0.1775	0.1783	0.1736	0.1721	0.1709
5.50	0.2798	0.2606	0.2479	0.2382	0.2301	0.2245	0.2198	0.2158	0.2132	0.2108	0.2075	0.2051	0.2032	0.2016	0.2003
6.00	0.3120	0.2928	0.2801	0.2703	0.2620	0.2563	0.2515	0.2472	0.2444	0.2419	0.2383	0.2357	0.2337	0.2320	0.2306
6.50	0.3443	0.3251	0.3124	0.3026	0.2942	0.2884	0.2834	0.2791	0.2761	0.2735	0.2697	0.2670	0.2648	0.2630	0.2616
7.00	0.3766	0.3574	0.3446	0.3348	0.3264	0.3206	0.3156	0.3111	0.3081	0.3054	0.3015	0.2986	0.2964	0.2946	0.2932
7.50	0.4088	0.3896	0.3769	0.3671	0.3587	0.3529	0.3478	0.3433	0.3403	0.3375	0.3336	0.3306	0.3284	0.3265	0.3251
8.00	0.4411	0.4219	0.4091	0.3994	0.3910	0.3851	0.3801	0.3756	0.3725	0.3697	0.3651	0.3628	0.3605	0.3586	0.3572
8.50	0.4734	0.4541	0.4413	0.4316	0.4232	0.4174	0.4123	0.4079	0.4048	0.4020	0.3976	0.3951	0.3928	0.3909	0.3894
9.00	0.5056	0.4863	0.4735	0.4637	0.4554	0.4496	0.4445	0.4401	0.5370	0.4343	0.4297	0.4273	0.4251	0.4231	0.4217
9.50	0.5378	0.5185	0.5056	0.4958	0.4875	0.4817	0.4767	0.4722	0.4692	0.4665	0.4619	0.4596	0.4573	0.4554	0.4539
10.00	0.5701	0.5507	0.5377	0.5279	0.5195	0.5137	0.5087	0.5043	0.5013	0.4985	0.4940	0.4917	0.4894	0.4875	0.4861
10.50	0.6023	0.5828	0.5698	0.5599	0.5515	0.5457	0.5407	0.5363	0.5333	0.5305	0.5260	0.5237	0.5215	0.5196	0.5181
11.00	0.6344	0.6149	0.6018	0.5718	0.5833	0.5775	0.5725	0.5581	0.5651	0.5624	0.5579	0.5556	0.5534	0.5515	0.5500
11.50	0.6666	0.6469	0.6337	0.6237	0.6151	0.6093	0.6042	0.5998	0.5988	0.5942	0.5896	0.5873	0.5851	0.5832	0.5818
12.00	0.6987	0.6790	0.6656	0.6555	0.6469	0.6409	0.6359	0.6314	0.6284	0.6257	0.6212	0.6189	0.6166	0.6148	0.6133
12.50	0.7309	0.7110	0.6975	0.6872	0.6785	0.6725	0.6674	0.6629	0.6599	0.6571	0.6526	0.6503	0.6480	0.6461	0.6446
13.00	0.7630	0.7429	0.7293	0.7189	0.7101	0.7040	0.6986	0.6943	0.6912	0.6884	0.6838	0.6815	0.6792	0.6773	0.6758
13.50	0.7951	0.7749	0.7611	0.7505	0.7415	0.7354	0.7301	0.7255	0.7224	0.7196	0.7149	0.7125	0.7101	0.7032	0.7067
14.00	0.8272	0.8068	0.7929	0.7820	0.7730	0.7667	0.7613	0.7566	0.7534	0.7505	0.7457	0.7482	0.7409	0.7389	0.7374

14.50	0.8592	0.8387	0.8246	0.8135	0.8043	0.7979	0.7924	0.7876	0.7843	0.7813	0.7764	0.7738	0.7714	0.7694	0.7679
15.00	0.8913	0.8705	0.8562	0.8449	0.8355	0.8291	0.8233	0.8184	0.8151	0.8120	0.8069	0.8042	0.8017	0.7997	0.7982
15.50	0.9233	0.9024	0.8879	0.8763	0.8667	0.8601	0.8542	0.8492	0.8457	0.8425	0.8371	0.8343	0.8318	0.8298	0.8282
16.00	0.9554	0.9342	0.9195	0.9076	0.8978	0.8911	0.8850	0.8798	0.8762	0.8728	0.8672	0.8643	0.8617	0.8596	0.8580
16.50	0.9874	0.9660	0.9510	0.9389	0.9288	0.9219	0.9156	0.9103	0.9065	0.9030	0.8971	0.8940	0.8914	0.8892	0.8876
17.00	1.0194	0.9977	0.9826	0.9701	0.9598	0.9527	0.9462	0.9408	0.9368	0.9331	0.9269	0.9236	0.9208	0.9186	0.9170
17.50	1.0514	1.0295	1.0141	1.0012	0.9907	0.9835	0.9767	0.9711	0.9668	0.9630	0.9564	0.9529	0.9501	0.9478	0.9461
18.00	1.0834	1.0612	1.0455	1.0323	1.0215	1.0141	1.0070	1.0013	0.9968	0.9928	0.9858	0.9820	0.9791	0.9768	0.9751
18.50	1.1153	1.0929	1.0769	1.0634	1.0523	1.0447	1.0373	1.0313	1.0267	1.0224	1.0150	1.0110	1.0080	1.0056	1.0038
19.00	1.1473	1.1246	1.1083	1.0944	1.0830	1.0752	1.0675	1.0613	1.0564	1.0519	1.0440	1.0398	1.0366	1.0342	1.0324
19.50	1.1792	1.1562	1.1397	1.1253	1.1137	1.1056	1.0976	1.0912	1.0860	1.0812	1.0728	1.0683	1.0651	1.0626	1.0607
20.00	1.2112	1.1879	1.1711	1.1562	1.1443	1.1360	1.1277	1.1210	1.1155	1.1104	1.1015	1.0967	1.0933	1.0908	1.0889
20.50	1.2431	1.2195	1.2024	1.1871	1.1748	1.1663	1.1576	1.1507	1.1449	1.1395	1.1301	1.1250	1.1214	1.1188	1.1168
21.00	1.2750	1.2511	1.2337	1.2179	1.2053	1.1965	1.1875	1.1803	1.1741	1.1685	1.1584	1.1530	1.1493	1.1466	1.1446
21.50	1.3069	1.2827	1.2650	1.2487	1.2357	1.2267	1.2173	1.2099	1.2033	1.1974	1.1867	1.1800	1.1770	1.1743	1.1721
22.00	1.3388	1.3143	1.2962	1.2795	1.2661	1.2568	1.2470	1.2393	1.2324	1.2261	1.2147	1.2086	1.2046	1.2018	1.1995
22.50	1.3707	1.3458	1.3274	1.3102	1.2964	1.2869	1.2766	1.2667	1.2614	1.2547	1.2427	1.2361	1.2319	1.2291	1.2268
23.00	1.4026	1.3774	1.3586	1.3409	1.3267	1.3169	1.3062	1.2979	1.2902	1.2832	1.2705	1.2635	1.2592	1.2562	1.2538
23.50	1.4344	1.4089	1.3898	1.3715	1.3569	1.3469	1.3357	1.3271	1.3190	1.3116	1.2981	1.2908	1.2862	1.2832	1.2807
24.00	1.4663	1.4404	1.4210	1.4021	1.3871	1.3768	1.3652	1.3563	1.3477	1.3399	1.3256	1.3179	1.3131	1.3100	1.3074
24.50	1.4982	1.4719	1.4521	1.4327	1.4173	1.4066	1.3945	1.3853	1.3763	1.3681	1.3530	1.3448	1.3399	1.3366	1.3340
25.00	1.5300	1.0534	1.4832	1.4632	1.4474	1.4364	1.4238	1.4143	1.4048	1.3962	1.3803	1.3716	1.3664	1.3631	1.3604
25.50	1.5619	1.5349	1.5143	1.4937	1.4774	1.4662	1.4531	1.4432	1.4332	1.4242	1.4074	1.3983	1.3929	1.3895	1.3867
26.00	1.5937	1.5664	1.5454	1.5242	1.5075	1.4959	1.4823	1.4721	1.4616	1.4521	1.4344	1.4248	1.4192	1.4157	1.4128
26.50	1.6255	1.5978	1.5765	1.5547	1.5374	1.5255	1.5114	1.5008	1.4898	1.4799	1.4613	1.4512	1.4454	1.4417	1.4388
27.00	1.6574	1.6292	1.6075	1.5851	1.5674	1.5552	1.5405	1.5295	1.5180	1.5076	1.4881	1.4775	1.4714	1.4677	1.4646
27.50	1.6892	1.6607	1.6385	1.6155	1.5973	1.5847	1.5695	1.5582	1.5461	1.5353	1.5148	1.5036	1.4973	1.4935	1.4903
28.00	1.7210	1.6921	1.6695	1.6459	1.6272	1.6143	1.5985	1.5868	1.5742	1.5628	1.5413	1.5296	1.5231	1.5191	1.5159
28.50	1.7528	1.7235	1.7005	1.6762	1.6570	1.6438	1.6274	1.6153	1.6021	1.5903	1.5678	1.5555	1.5407	1.5447	1.5413
29.00	1.7846	1.7549	1.7315	1.7065	1.6868	1.6732	1.6563	1.6438	1.6300	1.6176	1.5941	1.5813	1.5742	1.5701	1.5666
29.50	1.8164	1.7863	1.7625	1.7168	1.7166	1.7026	1.6851	1.6722	1.6579	1.6449	1.6204	1.6070	1.5997	1.5954	1.5918
30.00	1.8482	1.8177	1.7934	1.7671	1.7463	1.7320	1.7139	1.7005	1.6856	1.6722	1.6465	1.6325	1.6249	1.6205	1.6168

TABLE A.38(m)
Extended Sukkar–Cornell Integral for Bottom-hole Pressure Calculation

P_r	Reduced Temperature for $B = 70.0$														
	1.1	1.2	1.3	1.4	1.5	1.6	1.7	1.8	1.9	2.0	2.2	2.4	2.6	2.8	3.0
0.20	0.0000	0.0000	0.0000	0.0000	0.0000	0.0000	0.0000	0.0000	0.0000	0.0000	0.0000	0.0000	0.0000	0.0000	0.0000
0.50	0.0017	0.0016	0.0016	0.0016	0.0016	0.0015	0.0015	0.0015	0.0015	0.0015	0.0015	0.0015	0.0015	0.0015	0.0015
1.00	0.0087	0.0080	0.0077	0.0074	0.0073	0.0072	0.0071	0.0070	0.0070	0.0070	0.0069	0.0069	0.0068	0.0068	0.0068
1.50	0.0240	0.0199	0.0185	0.0177	0.0172	0.0168	0.0165	0.0163	0.0161	0.0160	0.0158	0.0157	0.0156	0.0155	0.0154
2.00	0.0491	0.0385	0.0345	0.0325	0.0312	0.0303	0.0296	0.0291	0.0288	0.0285	0.0281	0.0278	0.0276	0.0274	0.0273
2.50	0.0772	0.0625	0.0554	0.0515	0.0490	0.0475	0.0462	0.0453	0.0448	0.0443	0.0435	0.0431	0.0426	0.0423	0.0420
3.00	0.1063	0.0894	0.0799	0.0742	0.0703	0.0679	0.0660	0.0646	0.0637	0.0629	0.0618	0.0611	0.0604	0.0599	0.0595
3.50	0.1356	0.1175	0.1066	0.0994	0.0943	0.0910	0.0884	0.0864	0.0851	0.0840	0.0825	0.0815	0.0806	0.0798	0.0792
4.00	0.1651	0.1464	0.1346	0.1264	0.1202	0.1162	0.1129	0.1104	0.1087	0.1073	0.1054	0.1040	0.1029	0.1018	0.1010
4.50	0.1947	0.1756	0.1634	0.1545	0.1475	0.1429	0.1391	0.1360	0.1340	0.1322	0.1299	0.1282	0.1268	0.1256	0.1246
5.00	0.2243	0.2050	0.1926	0.1833	0.1756	0.1706	0.1664	0.1629	0.1606	0.1585	0.1558	0.1538	0.1522	0.1508	0.1497
5.50	0.2540	0.2347	0.2221	0.2125	0.2045	0.1991	0.1946	0.1907	0.1881	0.1859	0.1827	0.1805	0.1787	0.1772	0.1760
6.00	0.2838	0.2644	0.2517	0.2420	0.2337	0.2281	0.2233	0.2192	0.2164	0.2140	0.2106	0.2081	0.2061	0.2045	0.2032
6.50	0.3135	0.2941	0.2815	0.2716	0.2632	0.2574	0.2525	0.2482	0.2453	0.2427	0.2390	0.2363	0.2343	0.2326	0.2313
7.00	0.3433	0.3239	0.3113	0.3014	0.2929	0.2870	0.2820	0.2775	0.2745	0.2718	0.2680	0.2652	0.2630	0.2613	0.2599
7.50	0.3732	0.3538	0.3411	0.3312	0.3226	0.3167	0.3116	0.3071	0.3040	0.3013	0.2973	0.2944	0.2922	0.2904	0.2890
8.00	0.4030	0.3836	0.3710	0.3611	0.3525	0.3465	0.3414	0.3368	0.3337	0.3309	0.3262	0.3239	0.3217	0.3198	0.3184
8.50	0.4328	0.4135	0.4009	0.3909	0.3824	0.3764	0.3713	0.3667	0.3635	0.3607	0.3560	0.3536	0.3514	0.3495	0.3481
9.00	0.4627	0.4434	0.4307	0.4208	0.4122	0.4063	0.4011	0.3965	0.3934	0.3905	0.3858	0.3834	0.3812	0.3793	0.3779
9.50	0.4926	0.4733	0.4606	0.4507	0.4421	0.4362	0.4310	0.4264	0.4233	0.4204	0.4157	0.4133	0.4110	0.4092	0.4077
10.00	0.5225	0.5031	0.4905	0.4805	0.4720	0.4660	0.4609	0.4563	0.4531	0.4503	0.4456	0.4432	0.4409	0.4390	0.4376
10.50	0.5523	0.5330	0.5203	0.5104	0.5018	0.4958	0.4907	0.4861	0.4830	0.4801	0.4754	0.4730	0.4708	0.4689	0.4675
11.00	0.5822	0.5629	0.5502	0.5402	0.5316	0.5256	0.5204	0.5159	0.5127	0.5099	0.5052	0.5028	0.5005	0.4987	0.4972
11.50	0.6121	0.5927	0.5800	0.5700	0.5613	0.5553	0.5502	0.5456	0.5424	0.5396	0.5349	0.5325	0.5303	0.5284	0.5270
12.00	0.6420	0.6226	0.6098	0.5997	0.5910	0.5850	0.5798	0.5752	0.5721	0.5692	0.5645	0.5621	0.5599	0.5580	0.5566
12.50	0.6718	0.6524	0.6396	0.6294	0.6207	0.6146	0.6094	0.6047	0.6016	0.5987	0.5940	0.5916	0.5893	0.5875	0.5860
13.00	0.7017	0.6822	0.6693	0.6591	0.6503	0.6442	0.6389	0.6342	0.6311	0.6282	0.6234	0.6210	0.6187	0.6168	0.6154
13.50	0.7316	0.7121	0.6991	0.6887	0.6798	0.6737	0.6683	0.6636	0.6604	0.6575	0.6527	0.6502	0.6479	0.6460	0.6445
14.00	0.7615	0.7419	0.7288	0.7183	0.7093	0.7031	0.6977	0.6929	0.6897	0.6867	0.6818	0.6793	0.6770	0.6750	0.6736

14.50	0.7913	0.7717	0.7585	0.7479	0.7388	0.7325	0.7270	0.7222	0.7189	0.7158	0.7108	0.7062	0.7059	0.7039	0.7024
15.00	0.8212	0.8014	0.7881	0.7774	0.7682	0.7619	0.7562	0.7513	0.7479	0.7448	0.7397	0.7370	0.7346	0.7326	0.7311
15.50	0.8510	0.8312	0.8178	0.8069	0.7976	0.7911	0.7854	0.7804	0.7769	0.7737	0.7684	0.7656	0.7632	0.7612	0.7597
16.00	0.8809	0.8609	0.8474	0.8363	0.8269	0.8203	0.8145	0.8094	0.8058	0.8025	0.7969	0.7941	0.7916	0.7898	0.7880
16.50	0.9107	0.8907	0.8770	0.8658	0.8662	0.8495	0.8435	0.8363	0.8345	0.8311	0.8254	0.8224	0.8198	0.8178	0.8162
17.00	0.9406	0.9204	0.9066	0.8951	0.8854	0.8786	0.8724	0.8671	0.8632	0.8597	0.8537	0.8505	0.8479	0.8458	0.8442
17.50	0.9704	0.9501	0.9362	0.9245	0.9146	0.9076	0.9013	0.8958	0.8918	0.8881	0.8818	0.8765	0.8758	0.8737	0.8721
18.00	1.0002	0.9798	0.9657	0.9538	0.9437	0.9366	0.9300	0.9245	0.9203	0.9164	0.9098	0.9064	0.9036	0.9014	0.8997
18.50	1.0300	1.0095	0.9953	0.9831	0.9728	0.9656	0.9588	0.9530	0.9486	0.9446	0.9377	0.9340	0.9311	0.9289	0.9272
19.00	1.0599	1.0392	1.0248	1.0123	1.0018	0.9945	0.9874	0.9815	0.9769	0.9727	0.9654	0.9615	0.9586	0.9563	0.9545
19.50	1.0897	1.0689	1.0543	1.0415	1.0308	1.0233	1.0160	1.0099	1.0051	1.0007	0.9930	0.9889	0.9858	0.9835	0.9817
20.00	1.1195	1.0985	1.0837	1.0707	1.0597	1.0521	1.0445	1.0383	1.0332	1.0286	1.0204	1.0181	1.0129	1.0105	1.0087
20.50	1.1493	1.1282	1.1132	1.0999	1.0886	1.0808	1.0730	1.0665	1.0612	1.0564	1.0478	1.0432	1.0398	1.0374	1.0355
21.00	1.1791	1.1578	1.1426	1.1290	1.1175	1.1095	1.1014	1.0947	1.0892	1.0841	1.0749	1.0701	1.0666	1.0641	1.0622
21.50	1.2089	1.1874	1.1721	1.1581	1.1463	1.1381	1.1297	1.1229	1.1170	1.1116	1.1020	1.0968	1.0933	1.0907	1.0887
22.00	1.2387	1.2170	1.2015	1.1871	1.1751	1.1667	1.1580	1.1509	1.1448	1.1391	1.1289	1.1235	1.1198	1.1171	1.1151
22.50	1.2685	1.2466	1.2309	1.2162	1.2039	1.1953	1.1862	1.1789	1.1724	1.1665	1.1558	1.1500	1.1461	1.1434	1.1413
23.00	1.2982	1.2762	1.2602	1.2452	1.2326	1.2238	1.2144	1.2069	1.2000	1.1938	1.1825	1.1763	1.1723	1.1695	1.1674
23.50	1.3280	1.3058	1.2896	1.2742	1.2613	1.2522	1.2425	1.2347	1.2276	1.2210	1.2090	1.2026	1.1984	1.1955	1.1933
24.00	1.3578	1.3354	1.3190	1.3031	1.2899	1.2807	1.2706	1.2625	1.2550	1.2482	1.2355	1.2287	1.2243	1.2214	1.2191
24.50	1.3876	1.3650	1.3483	1.3321	1.3185	1.3090	1.2986	1.2903	1.2824	1.2752	1.2619	1.2546	1.2501	1.2471	1.2447
25.00	1.4173	1.3946	1.3776	1.3610	1.3471	1.3374	1.3265	1.3180	1.3097	1.3022	1.2881	1.2805	1.2758	1.2727	1.2702
25.50	1.4471	1.4241	1.4069	1.3899	1.3757	1.3657	1.3544	1.3456	1.3369	1.3290	1.3142	1.3062	1.3013	1.2981	1.2956
26.00	1.4769	1.4537	1.4362	1.4187	1.4042	1.3940	1.3823	1.3732	1.3641	1.3658	1.3403	1.3318	1.3267	1.3235	1.3209
26.50	1.5066	1.4832	1.4655	1.4476	1.4327	1.4222	1.4101	1.4007	1.3912	1.3625	1.3662	1.3573	1.3520	1.3487	1.3460
27.00	1.5364	1.5127	1.4948	1.4764	1.4611	1.4504	1.4379	1.4282	1.4182	1.4092	1.3920	1.3827	1.3772	1.3738	1.3710
27.50	1.5661	1.5423	1.5240	1.5052	1.4895	1.4786	1.4656	1.4556	1.4452	1.4357	1.4178	1.4079	1.4023	1.3987	1.3759
28.00	1.5959	1.5718	1.5533	1.5340	1.5179	1.5067	1.4933	1.4829	1.4721	1.4622	1.4434	1.4331	1.4272	1.4236	1.4206
28.50	1.6256	1.6013	1.5825	1.5627	1.5463	1.5348	1.5209	1.5102	1.4989	1.4886	1.4690	1.4581	1.4520	1.4483	1.4452
29.00	1.6554	1.6308	1.6117	1.5915	1.5747	1.5629	1.5485	1.5375	1.5257	1.5150	1.4944	1.4831	1.4768	1.4729	1.4698
29.50	1.6851	1.6603	1.6410	1.6202	1.6030	1.5909	1.5761	1.5547	1.5524	1.5412	1.5198	1.5079	1.5014	1.4974	1.4942
30.00	1.7148	1.6898	1.6702	1.6489	1.6313	1.6189	1.6036	1.5919	1.5791	1.5675	1.5450	1.5327	1.5259	1.5218	1.5165

INDEX